5 Springer Series in Solid-State Sciences

Edited by M. Cardona, P. Fulde, and H.-J. Queisser

Springer Series in Solid-State Sciences

Editors: M. Cardona P. Fulde H.-J. Queisser

F. Rosenberger

Fundamentals of Crystal Growth I

Macroscopic Equilibrium
and Transport Concepts

With 271 Figures

Springer-Verlag Berlin Heidelberg New York 1979

Professor Dr. Franz E. Rosenberger

Department of Physics and Department of Materials Science and Engineering,
University of Utah, Salt Lake City, Utah 84112, USA

Editors:

Professor Dr. Manuel Cardona
Professor Dr. Peter Fulde
Professor Dr. Hans-Joachim Queisser

Max-Planck-Institut für Festkörperforschung
Büsnauer Strasse 171, D-7000 Stuttgart 80, Fed. Rep. of Germany

ISBN 3-540-09023-1 Springer-Verlag Berlin Heidelberg New York
ISBN 0-387-09023-1 Springer-Verlag New York Heidelberg Berlin

Library of Congress Cataloging in Publication Data. Rosenberger, Franz E. 1933—. Fundamentals of crystal growth. (Springer series in solid-state sciences; v. 5). Includes bibliographies and index. Contents: 1. Macroscopic equilibrium and transport concepts. 1. Crystals—Growth. I. Title. II. Series. QD921.R617 548'.5 79-13697

Offset printing: Zechnersche Buchdruckerei, Speyer. Bookbinding: J. Schäffer OHG, Grünstadt.
2153/3020-543210

Preface

Crystal growth concepts have been fundamental to many areas of science for quite some time. Disciplines as diverse as mineralogy, meteorology, biology, astrophysics and chemical engineering have contributed to and benefited from crystal growth concepts. It was, however, not before the onset of the materials science era and the mushrooming development of the electronic and optical materials industries that crystal growth began to develop into a discipline in its own right. In concerted efforts, solid-state and crystal-growth research have led to many new materials and devices. Their applications have had a great impact on modern life. Today crystal growth forms one of the main pillars of modern technology.

Crystal growth research demands an unusually broad background, cutting across the boundaries of many tranditional disciplines. It requires detailed understanding of areas in solid-state physics and chemistry, thermodynamics and statistical mechanics, fluid dynamics and numerical methods, analytical and preparative chemistry, surface physics and catalysis, crystallography and electrochemistry. Crystal-growth techniques involve an immense arsenal of laboratory methods ranging from ultra-high vacuum to ultra-high pressures and from cryogenic to plasma temperatures.

This truly interdisciplinary nature of crystal growth presents a major hurdle for education and training. The relevant research literature is scattered over a large number of specialized journals and books. For a broad and fundamental study of crystal growth these sources require a prohibitive amount of supplementary reading. This makes it very difficult for both students *and* established workers to gain knowledge in areas of crystal growth that are not along the lines of their traditional training. Consequently, overspecialization is widespread in crystal growth and a perturbing gap has developed between theoretical efforts and daily laboratory practice. The amount of duplication of effort in closely related areas, due to a lack of emphasis on the common fundamentals, is sometimes amazing.

This text, based upon the author's extensive teaching experience in crystal growth, attempts to present for the first time the entire field from a unifying viewpoint and in a form that appeals to students as well as to seasoned workers who desire professional broadening.

The attempt of a single author to present such a wide area in a representative manner is, of course, a daring enterprise. This book, as any other, must reflect the prejudices and research interests of the writer. Although a multi-author scheme could certainly offer a far more extensive range of expertise, it is hoped that the advantages of a unifying and pedagogically proven treatise counterbalance, to some extent, the shortcomings of one author. I shall be grateful for any comments from readers.

In the presentation of the material I have drawn heavily upon the scientific literature. I feel a debt of gratitude to the authors of the works referred to. Numerous students and coworkers have contributed materially to the final text with suggestions on structure and contents, conceptual criticism, solutions to problems and proofreading. Among them I am particularly indebted to Glenn H. Westphal, Matthew C. DeLong, Jerry M. Olson, David M. Greenwell, Jerry Youngblood, and Eddy N. Dalal. Also, I wish to thank Mrs. Florence Anderson and in particular Mrs. Elaine Wagner for the skillful and patient typing of the drafts and manuscript.

Support during parts of the writing by the National Science Foundation under Grant number DMR 75-03175, by the National Aeronautics and Space Administration under Grant number NSG-1534 and by the Crystal Products Department, Untion Carbide Corporation, are also gratefully acknowledged.

My wife, Renate, deserves a very special acknowledgment for her continuous encouragement – in spite of having to live as a "book widow" during the past several hundred weekends.

Salt Lake City, Utah
January 1979

Franz E. Rosenberger

Contents

Appendices

1. Introduction

The intrinsic properties of a solid, i.e., the properties that result from its specific structure, can be largely modified by crystallographic and chemical defects. The formation of these defects is governed by the heat and mass transfer conditions which prevail on and near a crystal-nutrient interface during crystallization. Hence, both the growth of highly perfect crystals and the preparation of samples having predetermined defect-induced (extrinsic) properties require a thorough understanding of the reaction and transport mechanisms that govern crystallization from vapors, solutions and melts. Crystal growth, as a science, is therefore mostly concerned with the chemistry and physics of heat and mass transport in these fluid-solid phase transitions. Solid-solid transitions are, at this time, not widely employed for high quality single-crystal production.

Transport concepts are largely built upon equilibrium considerations, i.e., on thermodynamic and phase equilibrium concepts. Hence to supply a "workable" foundation for the succeeding discussions, this text begins in Chapter 2 with a concise treatment of thermodynamics which emphasizes applications to materials preparation. After working through this chapter, the reader should feel at ease with often (particularly among physicists) unfamiliar entities such as chemical potentials, fugacities, activities, etc. Special sections on thermochemical calculations (and their pitfalls) and compilations of thermochemical data conclude the second chapter.

Crystal growth can be called, in a wide sense, the science and technology of controlling phase transitions that lead to (single crystalline) solids. Hence a thorough knowledge of the thermodynamics of phase equilibria and skill in deducing from phase diagrams their consequences for crystal growth are among the keys to efficient work in that area. Chapter 3 supplies the necessary background. First, in the discussion of liquid-solid equilibria, the treatment follows conventional lines. Then, taking into account the ever-increasing importance of vapor-condensed phase equilibria for materials preparation, vapor-liquid-solid phase diagrams are particularly emphasized. To facilitate the presentation of this complex matter, a unique sequence of presentations in composition-temperature-pressure space has been developed.

This general presentation of possible phase diagrams is followed, in Chapter 4, by a broad discussion of the phase equilibria of numerous materials systems of current interest. Here emphasis is put upon the information that can be deduced from a specific phase diagram concerning both the materials transport problems to be expected and the choice of methods and conditions advantageous to crystal growth.

It has long been recognized that the growth of a solid, i.e., the transfer of "building blocks" from their position in a nutrient to more or less permanent sites on or in a crystal lattice consists of a complex sequence of steps involving fluid dynamics and interfacial kinetics. However, the thermal and fluid dynamic configurations encountered in most crystal growth systems generate problems which are among the most difficult to solve in mathematical physics. Thus, traditionally, crystal growth theories have concentrated on interfacial morphology and kinetics while relatively little work has been done concerning transport in the nutrient. Only during the last 10 years or so has the extent to which the chemical and crystallographic qualities of solids are determined by the mass and heat transfer conditions existing during their preparation become obvious. In particular non-steady convection is now recognized as being largely responsible for inhomogeneities in solids. Such inhomogeneities currently hamper device performance as well as basic research efforts. This insight and the availability of high speed computers have stimulated a wave of fluid dynamics work in crystal growth configurations. These efforts culminated in some rather promising results obtained recently under low-gravity conditions aboard various space missions. In response to this trend, Chapter 5 presents a detailed introduction to mass and heat transfer concepts. Rather than starting out with a discussion of specialized geometries, the presentation begins in each section with the most generally valid description of the respective topic. This allows for a full appraisal of the (often all-too necessary!) simplifications typically tacitly made in the literature. Extensive use of dimensionless groups in the text allows for emphasis on similarities and differences in the fluid dynamic behavior across the boundaries imposed by the traditional separation into areas such as "solution growth" and "vapor growth." Numerous compilations of data on relevant fluid properties supplement the text.

Chapter 6 is concerned with segregation, i.e. with the fact that the chemical composition of crystals differs in general, even in equilibrium, from that of the coexisting fluid. Segregation is of particular practical importance in materials purification and device manufacturing. A thorough discussion of the thermodynamics of segregation is followed by a detailed presentation of its fluid dynamic aspects. The extensive illustrative material contains also results most recently obtained in low-gravity environments. Chapter 6 concludes this text on macroscopic equilibrium and transport concepts.

Kinetic and morphological crystal growth concepts will be treated in a forthcoming volume [1.1]. This volume will contain a concise discussion of the rate theory, which is basic to some of the later material. A treatment of spontaneous and stimulated nucleation phenomena follows, stressing experimental developments which demonstrate the limited validity of some of the current theoretical models of nucleation. A broad discussion of interface morphology and its connections to growth mechanisms will be presented. Following an introduction to the thermodynamics of interfaces, atomistic models of surfaces and the concept of atomistic interface roughness will be treated. Another chapter will deal with the kinetics of the various crystal-growth mechanisms. The presentation, beginning with the still valid classical models, leads up to the revealing results of recent computer simulation work. The important practical question of shape stability of a growing interface and its dependence on growth parameters will conclude the second volume [1.1].

A third volume [1.2] will be concerned with the techniques of crystal growth from melts, vapors, and solutions. Though this grouping may suggest a rather traditional approach, the presentation will be oriented toward the fundamental questions: How can the partly idealized concepts developed in the earlier parts be most advantageously used for guidance in the choice, design, and conduction of crystal-growth processes? What are the practical limitations imposed by the chemo-physical properties of actual materials and equipment components? Special emphasis will be put on the origins of defect formation that prevail in specific classes of techniques. This facilitates the choice of a technique according to a selected group of properties desired in the end product. No efforts will be made to encyclopedically cover all materials grown to date. However a rather complete compilation of information sources on the preparation and growth of specific crystals will be given at the end of [1.2].

This three-part text has been developed by the author, over the past 7 years or so, as the backbone of a broadly oriented Graduate Curriculum in

Crystal Growth within the interdisciplinary Chemical Physics and Materials Science programs at the University of Utah. The material is covered within a one-year course series. Drafts of the text have been used by numerous graduate and advanced undergraduate students with physics, materials science, chemistry, and enginnering background. This direct interaction with students of such diverse training has contributed considerably to the current form of the presentation.

Each of the three volumes is largely self-contained. The study of any volume does not necessarily require the other volumes. Depending on the specific background and knowledge of the reader, certain volumes or chapters can be omitted. However, since a relatively novel approach is taken in most chapters, the reader may find it beneficial to work through the material in the sequence suggested by the overall arrangement.

The choice of topics and illustrative material has been based on the following guidelines:

1) A textbook on a developing field must, to some extent, form a current research monograph. Yet, just as important, it must also contain sufficiently diverse background material to allow a traditionally trained reader to "catch up" on basics without extensive additional reading.
2) In order to reduce the communication gap between the various specialized areas of crystal growth, emphasis should be on common theoretical foundations.
3) Theoretical concepts must be presented such that the practical implications and the limitations from (idealized) assumptions become apparent.
4) Experimental techniques and results must be critically reviewed, based upon the theoretical concepts presented. Discrepancies and needed work must be pointed out.
5) The volume of the material must remain manageable — and teachable! — in spite of the broad approach taken. Hence historically important contributions cannot be done justice. Only the current state of understanding can be discussed. Also, the presentation of technical details must be limited to selected illustrations of the basic concepts. Yet, efforts must be made to supply up-to-date references to alternative approaches.
6) The presentation must be supplemented with practically relevant ("homework") problems. However, the real world does not present itself as a series of neatly packaged and well-defined exercises the "solution" of which is merely a number that is either correct or incorrect. Hence a good part

of the problems should be phrased so that individual judgment must be exercised and a clear statement of the (always simplifying) assumptions is required.

2. Thermodynamics

This chapter is a concise review of selected topics in classical thermodynamics. Selection and presentation are oriented toward the application of thermodynamic models to crystal growth phenomena in later chapters. Although some of the basics are "brushed up" in the presentation, familiarity with the fundamental concepts of thermodynamics and statistical mechanics is assumed. Supplementary reading in the texts by ANDREWS [2.1,2], MORSE [2.3], and LEWIS and RANDALL [2.4], and in particular, in GASKELL [2.5] is recommended. Additional references are given in the text.

Classical thermodynamics (thermostatics) is concerned with equilibrium states of systems. Equilibrium, most restrictively, can be understood as a state which a *system* takes on when it is "left *long enough* to settle down so that macroscopic quantities such as temperature and pressure become *uniform* throughout" [2.3]. In typical crystal growth situations, however, one has *transients* as well as *gradients* of these macroscopic parameters throughout the bulk phases involved. For instance, during the seeding procedure in a melt pulling process [1.2], seed and melt temperatures undergo considerable changes. But even after steady-state growth is reached, sizeable temperature gradients must be sustained in the solid and liquid in order to drive practically feasible growth rates. Thus in the strict sense, crystal growth and phase transitions in general, are conducted under "non-equilibrium" conditions. Hence the question arises whether a thermostatic, i.e., a macroscopic equilibrium description of crystal growth phenomena can be meaningful.

For an answer to this basic question one must consider the atomic nature of an equilibrium state. The atoms contained in a macroscopically uniform system undergo incessant and rapid changes in their positions and energies (velocities and quantum states). Hence, the system is at any instant in a different *microstate* consistent with the given boundary conditions. If we assume that the transition kinetics among the atomic states is sufficiently effective, then the system passes through a very large number of representative microstates during a time of macroscopic observation. This supplies a sound statistical basis for representing the state of the system by average macroscopic parameters, even during macroscopic transients. As long as the time involved in changes of macroscopic parameters is long as compared to the time constant of the underlying atomic process, the atomic fluctuations in the system will average out, and the equilibrium approach will be an excellent approximation. In other words, a macroscopic or "reduced" description is justified if at no moment the atomic distribution functions of position and energy deviate strongly from their equilibrium form.

This condition concerning the relative time scales for atomic and net macroscopic processes is fulfilled in many crystal growth situations. Times connected with measurable net mass transfer rates between nutrient and growing solid are typically long as compared to times required to establish equilibrium. For example, when a crystal grows from a vapor of the same composition a molecular flux $\underline{N}_i$ impinges on a crystal surface. The higher the temperature of the interfacial system, the more readily will these "adatoms", pass via surface diffusion, through a part of the possible energetic "surface states" in form of various adsorption configurations. At the same time, the higher the temperature, and thus the higher the average kinetic energy of the adsorbed molecules, the larger is also the probability for re-evaporation from the interface into the vapor. At equilibrium the imping flux $\underline{N}_i$ is equal and opposite to the re-evaporated flux $\underline{N}_e$. No net growth occurs. But even off-equilibrium, i.e., when net growth or evaporation prevails, the ensemble of adsorbed molecules is still able to race through essentially the same set of microstates between impingement and re-evaporation if $\Delta N = N_i - N_e \ll N_i$. Then the interface can still be considered, in good approximation, as being "in equilibrium" — in *quasi-equilibrium*. As we will see, N_i has huge values under practical growth conditions. Hence, this approximation can hold for rather large growth rates, i.e., large ΔN's — if the transition mechanism between the various atomic surface states is sufficiently efficient. This condition is frequently neglected, and interfacial equilibrium is taken for granted. Recent results obtained in segregation from solutions and, in particular, in growth from vapors attest to the contrary, however. As we will find in the corresponding sections, considerable deviations from compositional equilibrium between the surface of a growing crystal and the adjacent part of the nutrient must be assumed in a large number of systems.

The above example, dealing with the interface, i.e., a very narrow area of the crystal-nutrient system, allowed us to concentrate on the time scale of the equilibrium concept. Temperature differences across the interfacial layers were tacitly assumed to be insignificant. But this can certainly not be claimed for the bulk nutrient where sizeable temperature differences are required to drive practically satisfactory mass transfer towards the interface. Thus we cannot so easily elude the difficulty imposed by the initial equilibrium definition which demands temperature uniformity throughout the bulk. How can one justify an equilibrium description there? This is done via the same "trick" as for the interfacial area: The bulk phase is thought of as a chain of subsystems which are so small that the temperature differences between their bounds can be neglected. *Local equilibrium* is assumed in each of the subsystems. Mass transfer over macroscopic distances is then envisioned as a sequence of transfer processes through a chain of local equilibria with finite temperature differences. The limitations of this "simulation-model" are obvious. The gradients must not be so steep that the fictitious subdivision, as dictated by the "negligible change" argument, results in subsystems with too few molecules to populate all possible molecular states in a true statistical manner. Then the local equilibrium approach has to break down – no matter how rapidly the exchange between the atomic states proceeds.

From this qualitative discussion we see that "equilibrium" is by no means restricted to systems in which all macroscopic parameters are steady and uniform throughout. Hence the initial simple definition of equilibrium has to be taken only as a *sufficient* condition. Thermodynamic descriptions of steady-state and even transient processes are sufficiently accurate as long as the *necessary* conditions are fulfilled that,

a) processes proceed macroscopically slow as compared to the underlying atomic kinetics, and
b) gradients are not too steep, i.e., the local equilibrium concept is not jeopardized by insufficient size of the resulting subsystems.

Size limitations to a macroscopic description can, of course, simply result from a system's extent even in the absence of considerable gradients. In particular in connection with nucleation phenomena and with highly rarefied gases one has to ask: What is the minimum size of a system, i.e., the number and/or density of molecules in the system needed in order to allow for a macroscopically averaging description. A quantitative answer can again only be given from case to case based on involved statistical considerations that are beyond the scope of this text. Qualitatively, however, one can imagine that with decreasing number of molecules, it becomes increasingly difficult to account for an ensemble with a smooth distribution function — a prerequisite for a macroscopic averaging process. Hence, with decreasing size or density of a system the discrete atomistic structure becomes increasingly significant on account of the accuracy of a continuous macroscopic description.

For instance, one cannot ignore the atomic composition of a gas any more once the density is lowered to the extent that the mean free path length becomes comparable to the container dimensions. For uniform equilibrium systems this may still not impose restrictions on the macroscopic description. Collisions with the walls — given enough time — may establish a Maxwellian distribution, just as it would have resulted at higher densities from collision processes among the molecules. However, efforts to describe "off-equilibrium" processes by coupled local equilibria must fail here.

In condensed phases limitations of the macro-approach will come from the limited size of a system rather than from too low a density. If, for instance, the dimensions of a solid or liquid phase become comparable to the mean free path length of diffusive processes or to the amplitudes of thermal motion, the macroscopic description becomes invalid. We will see that a continuous, thermodynamic model can satisfactorily describe the size of a "critical nucleus" as small as 100 atoms. Yet at smaller sizes, where the addition of each single molecule adds significantly to the dimensions of the cluster, discrete, atomistic models must be employed and fluctuation phenomena, such as Brownian motion, must be taken into account.

With this introduction we have set the boundaries for a meaningful application of equilibrium thermodynamics to transient and non-uniform phenomena. Next, we need to carefully define various basic terms frequently used in thermodynamics.

2.1 Systems

A system is the "part of the universe" that is considered for a specific experiment or calculation. A system must have *walls* (boundaries) with known properties. If it is completely surrounded by *adiabatic* walls, it cannot be influenced from outside (apart from possible effects of exterior fields) except by compressing or expanding its walls. No heat and mass exchange can occur with the surroundings.

Most systems we consider are *closed* with respect to material transport. The number of atoms is thus fixed, but if chemical reactions are involved, the number of molecules need not be fixed. Closed systems may exchange heat across their boundaries.

If the walls allow for heat exchange with the surroundings, one calls the system *diathermal* or non adiabatic. If thermal contact exists with a very large heat reservoir, one has an *isothermal* system. The walls may be mechanically rigid so that no expansion work can be done on or by the system; then the system is called *isochoric*. Or the walls may be set up in such a way that the pressure remains constant; then we have an *isobaric* system.

If a system is *homogeneous* (i.e., if all properties, including the geometrical structure, are the same in all parts), then it can contain only one substance in one phase. A *heterogeneous* system possesses, by definition, internal surfaces of discontinuity which may separate either different phases of one substance and/or different substances. The narrow transition zones between different phases or substances are called *interfaces*. Properties can change with great abruptness across interfaces. The morphology of interfaces and its dependence on growth conditions is one of the keys to crystal growth phenomena and will thus be a major subject in this text.

2.2 State Variables

Quantities such as pressure and temperature, which return to the same value whenever a system returns to the same equilibrium state, independent of the history of the system, are called *state variables* or *state functions*. Heat, thus, cannot be a state function. A system can be brought back to an initial equilibrium state by withdrawing heat and then adding not heat but, for example, mechanical energy. Mechanical energy, therefore, cannot be a state variable either. The quantity that is conserved in this process and which is thus a state variable is U, the total *internal energy* possessed by the system.

The absolute value of the total energy of a system is undetermined, at least for practical purposes. Since we are only concerned about changes in systems, the absolute value of U is not of physical significance to us. The total energy is thus an additive parameter. In principle, $E = mc^2$ defines the total energy of a system precisely. However, c^2 is such a huge scaling factor (3×10^{-8} gram correspond to about 1000 kcal) that even the largest amounts of energy involved in ordinary chemical reactions or phase changes produce changes in mass that are far below the limits of detectability by means of a balance.

The fact that there are equilibrium states which matter in bulk tends to approach *spontaneously* if left to itself and the fact that there are thermodynamical variables — which are uniquely specified by the equilibrium state independent of the history of the system — are not conclusions deduced logically from philosophical first principles. They are conclusions drawn from more than two centuries of experiments.

State variables may be divided into two groups:

Extensive variables X_i, that require a specification of the extent of the system to which they refer. Their value is additive, i.e., it is the sum of the values for all of the constituents. We will specify the extent of a system mostly with the *number of moles*, n, of the contained matter. Examples for extensive variables are the volume V of a system, its internal energy U, and n itself.

Intensive variables Y_i, such as pressure, temperature, and density, are independent of the extent of a system.

Thermodynamics relates energy changes in systems with their changes in state. The amount of energy contained in a system depends on the "quality" (characterized by the intensive variables) as well as on the "quantity" (extensive variables) of the various energy storage mechanisms. Hence thermodynamics deals with pairs of state variables. Each pair relates to a particular way of changing the state of a system. One variable of each pair is extensive, the other is intensive. Their units are advantageously chosen so that the energy change of a system can be specified by one or several terms of the form $X_i dY_i$ or $Y_i dX_i$.

Some examples for pairs of state variables are:

1) The (total) *pressure* P in a system and its extensive counterpart, the *volume* V, are involved in most changes of state.
2) A material may be polarized by a *magnetic field* $\underline{H}$ or an *electric field* $\underline{F}$ impressed upon it. The magnetic or electric work dW done on the system is $\underline{H}d\underline{M}$ and $\underline{F}d\underline{P}$, respectively. The *magnetization* $\underline{M}$ and *polarization* $\underline{P}$ are thus the extensive counterparts of $\underline{H}$ and $\underline{F}$.
3) A system may contain several phases. Changes of state may include changes in the exposed surface area of a phase. The work done on a system by a change in its *surface area* A is $dW = \sigma dA$, where σ is the *surface free energy*, a parameter that will occupy a considerable part of the later discussion on interfacial phenomena. Note that σ is an intensive variable of the system and not of a particular substance. There is no such thing as a sur-

face-free energy of a material *per se*. The value of σ depends on the materials in contact. For example, water has different contact angles on teflon and clean glass and again slightly different values for the same support material depending on the surrounding gaseous atmosphere.

4) The intensive variable related to the mole number n must have the dimension of energy/mole and is usually called the chemical potential μ. It is a very important parameter since it represents the drive for mass transfer in phase transitions and chemical reactions, as we will see in later sections.

5) What is the extensive counterpart to the intensive variable temperature? It must have the dimension of energy/temperature. Heat capacity C could fulfill this requirement. But heat capacity cannot be a state function. Its value depends on the conditions under which a system has been changed. For example, changes of state under constant pressure lead to different values of C than those performed with constant volume — especially with gases. We have not come far enough yet to meaningfully introduce that mystical counterpart of the temperature T. So, let us anticipate later discussion by simply stating that the *entropy* S is the extensive counterpart of T.

2.3 Equations of State

Relationships between the state variables that are necessary to specify the equilibrium state of a system are called equations of state. They are of empirical nature. The first equation of state experimentally determined was the *ideal gas law*

$$PV = nRT \quad , \tag{2.1}$$

where R is the universal gas constant. A compilation of values for R in a large variety of units is given in Appendix A. For a given n (i.e., a closed system) there are here two independent variables and one dependent variable. Real gases exhibit deviations from this simple equation, but all gases approach perfect-gas behavior at low densities if chemical reactions are absent. Hence the ideal gas law is adequate for many discussions, particularly for gases at moderate pressures. Deviations from (2.1) originate physically from two effects in real gases which are neglected in the ideal gas concept: In the first place the individual molecules occupy some finite volume. Hence, the free space in which collisions occur is somewhat smaller than V. Secondly, the molecules of an actual gas attract one another. Hence collisions are not fully elastic. The momentum with which molecules leave one another and hit the wall of the container is somewhat reduced. Hence the pressure measured with a real gas is somewhat smaller than that of an ideal gas at the same temperature and density. Many modifications of the ideal gas law have been proposed, see, e.g., [2.1]. The best known approximation, due to Van der Waals, refines (2.1) with correction terms for pressure and volumes. A theoretically and practically especially interesting equation of state for real gases is the *virial equation* in the form of an infinite power series in the density ρ

$$\frac{PV}{nRT} = 1 + B(T)\rho + C(T)\rho^2 + D(T)\rho^3 + \ldots \tag{2.2}$$

where

$$\rho = \frac{n}{V} \tag{2.3}$$

and B, C, D, ... are called the second, third, fourth, ... virial coefficients. The virial coefficients depend on the temperature and on the particular gas considered, but are independent of density or pressure. For a critical compilation of virial coefficients see [2.6]. The particular importance of (2.2) lies in the fact that it is the only equation of state for gases with a thoroughly sound theoretical foundation. The second virial coefficient represents the deviations from ideal behavior due to interactions between two molecules, the third corresponds to interactions between three molecules, and so on (see [2.7]).

Equations of state for condensed phases (liquids and solids) are typically expressed in terms of the (isobaric) *thermal expansion* coefficient

$$\beta = \beta(P,T,n) \equiv \frac{1}{V}\left(\frac{\partial V}{\partial T}\right)_{P,n} , \tag{2.4}$$

i.e., the fractional change in volume, dV/V, per unit temperature change (at constant pressure); the (isothermal) *compressibility*

$$\kappa = \kappa(P,T,n) \equiv -\frac{1}{V}\left(\frac{\partial V}{\partial P}\right)_{T,n} , \tag{2.5}$$

i.e., the fractional change in volume, dV/V, per unit pressure change (at constant temperature) and the *molar volume*

$$v = \left(\frac{\partial V}{\partial n}\right)_{P,T} \quad . \tag{2.6}$$

Note that in large systems $v \approx V/n$ and hence simply size-independent (constant). If, however, surfaces play a significant role, i.e., if the system is very small, then even the volume may lose its additivity.

With (2.4-6) we obtain from

$$dV = \left(\frac{\partial V}{\partial T}\right)_{P,n} dT + \left(\frac{\partial V}{\partial P}\right)_{T,n} dP + \left(\frac{\partial V}{\partial n}\right)_{P,T} dn \tag{2.7}$$

the equation of state in differential form

$$\frac{dV}{V} = \beta dT - \kappa dP + \frac{1}{V} v dn \quad . \tag{2.8}$$

Frequently β and κ can be treated as nearly constant over a limited range in temperature and pressure. Then, integration of (2.8) between states 1 and 2 yields for a closed, large system ($dn = 0$, $v = \text{const}$)

$$\ln\left(\frac{V_2}{V_1}\right) \approx \beta\Delta T - \kappa\Delta P \tag{2.9}$$

which, if $\Delta V << V$, can be further simplified to the well-known form

$$\frac{\Delta V}{V} \approx \beta\Delta T - \kappa\Delta P \tag{2.10}$$

that is a good approximation for many solids and liquids for moderate temperature and pressure changes. Both β and κ are typically small quantities, of the order of 10^{-6} for metals, for example.

Another example is the equation of state for magnetization of paramagnetic substances in the form of *Curie's law*

$$\underline{M} = \frac{nD\underline{H}}{T} \tag{2.11}$$

where Curie's constant D is proportional to the magnetic susceptibility of the material. Equation (2.11) represents a good approximation for most materials far from saturation, i.e., at low magnetic fields $\underline{H}$.

Fortunately most equations of state involve only one pair of independent state variables. For example, the relation between P, V and T for a paramagnetic gas is independent of H and M and conversely D is independent of P and V to a good approximation. This separation of state variables allows us to neglect most of them when we make a thermodynamic calculation or measurement.

In general we can say: to completely specify the state of a system we must know the value of T plus the value of just one of each pair of variables involved in the problem. To stay with our example: for the specification of a paramagnetic gas we need to know only three independent variables. This determines the macrostate of the paramagnetic gas through (2.1) and (2.11) uniquely.

So far we have tacitly assumed that we deal with homogeneous systems. Heterogeneous systems, of course, require a larger number of independent variables to define an equilibrium state. The exact number depends on the complexity of the system. Internal surfaces (interfaces) can form constraints for some parameters and none for others. For example, large fluid phases in unconstrained mechanical contact cannot sustain any pressure gradients. Hence, specification of the pressure in one phase automatically defines the pressure in all other phases. If, however, some of the phases are very small, i.e., only a few orders of magnitude larger than atoms, sizeable pressure differences can result from surface tension between the phases. Hence, to define the pressure in all components of the system one has to know, in addition to the pressure in the ambient large phases, surface tension and geometry as well as the interfacial equation of state. This problem plays a central role in nucleation theory, for instance.

2.4 First Law of Thermodynamics

There are many ways of phrasing the First Law of Thermodynamics. Particularly useful formulations, in order of increasing complexity, are:

– Energy is conserved. Heat can be converted into other forms of energy.

– There is a "generalized store" of energy contained in a thermodynamic system, called its internal energy U, which can be changed by adding or subtracting energy of any form. The algebraic sum of all these changes is equal to the net change in internal energy.

– Regardless of how many changes a system undergoes, no matter whether reversible or irreversible processes (Sect. 2.7) were involved in these changes, when the system is returned to its initial state, its energy U *is found* to be the same as it was before the changes. Hence the internal energy is a state function. The emphasis on "is found", above, is to remind of the empirical nature of the First Law.

Note that we have already tacitly used the First Law in Sect. 2.2 when the internal energy U was introduced.

Mathematically, the First Law is most simply stated in differential form

$$dU = đQ + đW \quad . \tag{2.12}$$

The change in internal energy dU is the heat *added* to plus the work (of any form) *done on* a system. The bar is used above to characterize "imperfect" differentials, i.e., to indicate that the value of these differentials depends on the path along which the system has been changed. Note the sign convention adopted here. Work and internal energy have the same sign, as recommended by the International Union of Pure and Applied Chemistry (IUPAC) in 1970.

Considering various changes of state on a system, we can rewrite (2.12) in the general form

$$dU = đQ - PdV + \underline{H}d\underline{M} + \underline{F}d\underline{P} + \sigma dA + dE_{chem} + \dots \tag{2.13}$$

If the performance of work on or by a system would be the only way to change a system's energy, then the picture would be much simpler. Not only dU but also dW could be a perfect differential. Whatever work would be performed on a system could eventually be recovered as mechanical (or electrical or magnetic) energy. One would then talk about the heat contained in a system and measure the net balance of heat intake and output simply by the temperature change of a system. The ready convertibility between heat and other forms of energy, however, does not allow for such a simple description. If one can measure ΔQ, the heat absorbed by a system, and ΔW, the work done on the system, the only thing that we know, then, is the change in energy of the system, ΔU. As any form of energy enters a system, it loses its specific "label" and can in principle be withdrawn again in any other form, depending on the path the system takes in the state variable space during the corresponding change of state. The First Law supplies merely an accounting process for the overall energy change.

Let us illustrate the power of the state function concept by applying it to the expansion of a gas. For simplicity's sake the change of state is performed isothermally. This allows for a full presentation of the path in the P-V plane only, as shown in Fig.2.1, if the gas is brought *quasi-statically* from state 1 to state 2. "Quasistatical" implies that the change is performed so slowly that pressure, temperature and hence density remain uniform througout the system.

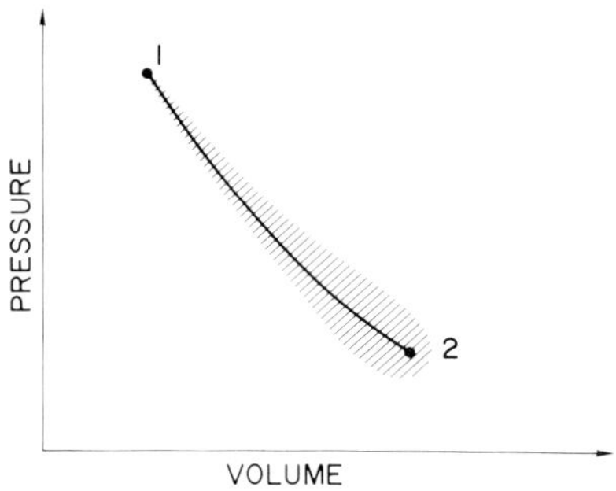

Fig.2.1. Changes of state in the P-V plane

During the process the gas performs expansion work

$$\Delta W = \int_1^2 P(V)dV \quad . \tag{2.14}$$

If we know the equation of state of the gas, e.g., by assuming that it behaves ideally, ΔW can be readily calculated since the changing pressure is uniquely defined at any instant (see Problem 2.1). Furthermore, if we know or can calculate ΔU between the two states then we can deduce from the First Law,

$$\Delta U = \Delta W + \Delta Q \quad , \tag{2.15}$$

the heat exchanged during that process. For instance for an ideal gas, where the internal energy is a function of temperature only, i.e., independent of pressure, ΔU will be zero. Hence the work which the ideal gas performs on the surroundings must be quantitatively balanced by heat taken up from the surroundings. In other words, the heat that is fed into the gas, in order to keep the temperature constant, is completely converted into work.

In contrast to the quasistatic process we will conduct the expansion now under gross off-equilibrium conditions such as a free expansion into a previously evacuated ΔV. During this rapid expansion considerable pressure gradients will exist throughout the gas which are symbolically indicated by the shaded area in Fig.2.1 in contrast to the solid line representing the quasistatic process. However, as soon as the system has settled down, state 2 reached through the free expansion is indistinguishalbe from state 2 after the quasistatic process and is characterized by the same internal energy. Hence for the specific process under consideration where $\Delta U = 0$, we can neglect the intricate kinetic details (turbulence, sound, etc.) of the free expansion and calculate ΔQ as above from (2.15). Since no useful work is performed during the free expansion, i.e., $\Delta W = 0$, ΔQ is also zero.

In general, if one is faced with the task of describing the change of a state function, such as U, during a kinetically complex process one can simply replace the actual path with any computationally convenient path between the same initial and final states. Such a convenient path may, for instance, be a sequence of isobaric and isochoric steps, as used in Problem 2.2.

Another important fact, that is reflected in the above example, is that for *any* off-equilibrium process, both the work done and the heat taken up by the system are smaller than for a quasistatic transition between the same initial and final states.

2.5 Second Law of Thermodynamics

One way of phrasing the Second Law is: The spontaneous tendency of a system to go towards equilibrium cannot be reversed without at the same time changing some "organized energy", work, into "disorganized energy", heat. Let us illustrate this statement with one of the many processes leading towards equilibrium, the flow of heat. Experience shows that heat flows only from hot bodies to cold bodies. Note that heat flow "up-temperature" would not violate the First Law; energy could still be conserved. One can, for example, use a refrigeration machine to further cool a cold body by transporting heat from it to the warmer environment. However, to do this, one needs to perform work and one must devise another dissipative process. We may expand a gas to extract heat from the interior of the refrigerator (colder body), then isolate the gas and compress it, thus heating it up, to facilitate heat exchange with the air in the room (hotter body). On the whole, the system, which here has to include the ambient towards which we cool our coolant, is moving towards equilibrium.

Another formulation of the Second Law is: Under no circumstances, except through the influence of external agencies, will a system change away from the state of equilibrium. Example: A system, already in thermal equilibrium may develop large local differences in temperature due to some chemical reaction, but this will not change the general undirectional flow toward a final state of rest.

Clausius has given an equivalent but more specific statement to our first phrasing of the Second Law: "It is impossible to devise an engine which, working in a cycle, shall produce no effect other than the transfer of heat from a colder to a hotter body." Important here is the clause "working in a cycle". Only if the initial and final state of a system that went through a process are identical, only then is the amount of heat transferred unlimited because we can repeat the process indefinitely. As we have encountered with the refrigerator, one can have non-cyclical processes in which heat is transferred from cold to hot. But, at the end of the process (gas in equilibrium with room temperature air) the gas is not in its original state (equilibrium with refirgerator interior). Thus, the internal energy of the gas must be higher, and it follows from the First Law that more work was necessarily performed on the gas than heat was transported from cold to hot.

Kelvin's formulation of the Second Law is very similar to that of Clausius: "It is impossible to devise an engine which, working in a cycle, shall produce no effect other than the extraction of heat from a reservoir and the performance of an equal amount of mechanical work." If we would drop the clause "working in a cycle", we could actually devise a system which only extracts heat from a reservoir and performs an equal amount of work. For instance, as discussed in Sect. 2.4, a system whose internal energy is a function of temperature only, such as an ideal gas, could be employed here. When placed in contact with a constant temperature bath, it can isothermally transform all the heat it withdraws into useful expansion work. The trouble with such a process is that it cannot continue to do this indefinitely. Sooner or later the pressure in the system becomes too low to be of any use. If we tried to make such an isothermal process cyclic, for example by compressing the gas again, we would find that we must reconvert all the gained work into heat and have no net work left. A cycle, to result in net work done, has to operate between two baths of different temperature.

2.6 Thermal Efficiency

Discussions of general thermodynamic processes can be facilitated by decomposition into a sequence of isothermal and adiabatic processes. In particular, cyclical processes can be analyzed advantageously in terms of a sequence of so-called Carnot cycles. A Carnot cycle, as presented in Fig.2.2 consists of four consecutive branches, two isothermal and two adiabatic processes. It is assumed that all operations in the cycle are performed quasistatically and that the cycle is reversible. Warning: We will encounter processes that are quasistatic, but not reversible, such as heat conduction from a hot body to a cold one. Thus, we will use "reversible" only on the condition that we find a better definition than the coupling to quasi-static. Applying the First Law (energy conservation) to the individual processes in the Carnot cycle (see Fig.2.2) one can write

$$\begin{aligned} \Delta U_1 &= \Delta Q_1 + \Delta W_1 \\ \Delta U_2 &= \Delta W_2 \\ \Delta U_3 &= \Delta Q_3 + \Delta W_3 \\ \Delta U_4 &= \Delta W_4 \quad . \end{aligned} \tag{2.16}$$

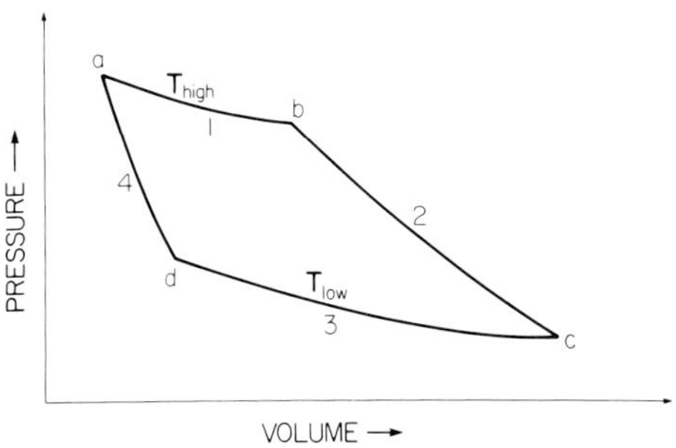

Fig.2.2. Indicator diagram for Carnot cycle

Since $\oint dU = 0$ one obtains from (2.16)

$$\Delta Q_1 + \Delta Q_3 = -(\Delta W_1 + \Delta W_2 + \Delta W_3 + \Delta W_4) \quad . \tag{2.17}$$

One can now express the efficiency η for the conversion of heat into work during the cycle by

$$\eta = \frac{\text{net work performed}}{\text{heat consumed}} = \frac{-(\Delta W_1 + \Delta W_2 + \Delta W_3 + \Delta W_4)}{\Delta Q_1}$$

$$= \frac{\Delta Q_1 + \Delta Q_3}{\Delta Q_1} = 1 + \frac{\Delta Q_3}{\Delta Q_1} \quad . \tag{2.18}$$

Since ΔQ_3 is negative and nonzero, one can see from (2.18) that the efficiency of any heat engine can never reach 1. There is always some ΔQ_3 involved that is given back to the low temperature heat reservoir – even if we would succeed in building a frictionless machine.

For an evaluation of (2.18) one must assume a specific working substance in the system. For an ideal gas, since its internal energy depends only on temperature, one can write for the isothermal branches

$$\Delta Q_1 = -\Delta W_1 = -\left(-\int_a^b PdV\right) = RT_h \ln(V_b/V_a) > 0$$

$$\Delta Q_3 = -\Delta W_3 = -\left(-\int_c^d PdV\right) = RT_l \ln(V_d/V_c) < 0 \quad . \tag{2.19}$$

The required relations between V_b/V_a and V_d/V_c we can obtain from the adiabatic branches. An adiabatic process of an ideal gas is characterized by PV = const (with $\gamma = C_p/C_v$). This can be rewritten as $PVV^{\gamma-1}$ = const. and, employing the ideal gas law, as $TV^{\gamma-1}$ = const. Thus, we get for branches 2 and 4 in Fig.2.2, respectively,

$$T_h V_b^{\gamma-1} = T_l V_c^{\gamma-1}$$

$$T_l V_d^{\gamma-1} = T_h V_a^{\gamma-1} \tag{2.20}$$

and therefore

$$V_b/V_a = V_c/V_d \tag{2.21}$$

which with (2.19) gives

$$-\Delta Q_1/\Delta Q_3 = T_h/T_l \tag{2.22}$$

and hence, from (2.18)

$$\eta = 1 - T_l/T_h \quad . \tag{2.23}$$

Although the above derivation has been based on a Carnot cycle and an ideal gas system, (2.23) is valid for any form of heat engine based upon any working substance. For a general derivation, based upon Sect. 2.7, see Problem 2.3.

Coming back to the practical implications of the thermal efficiency concept, one can deduce from (2.22) and (2.23) that temperature is a measure for the *quality* of heat. Of some ΔQ associated with a higher T_h one can convert more into work than of the same ΔQ at a lower T_h', referring to the same lower temperature T_l.

2.7 Reversibility and Entropy

So far we have been using the adjective "reversible" in a rather undefined manner and more or less tied to quasistatic. Now, based upon the Carnot cycle, we can give a more satisfactory definition and show that quasistatic processes are not necessarily reversible.

Equation (2.22) can be rewritten as

$$\Delta Q_1/T_h + \Delta Q_3/T_l = 0 \quad . \tag{2.24}$$

Since ΔQ_1 and ΔQ_3 represent the total heat exchanged during the Carnot cycle (any other reversible cycle can be simulated by a combination of Carnot cycles) we can conclude that in general for *reversible* cycles

$$\sum \frac{\Delta Q}{T} = 0 \quad \text{or} \quad \oint \frac{đQ}{T} = 0 \quad . \tag{2.25}$$

In a reversible cycle the maximum possible amount of heat ΔQ is taken up by the system and converted with the highest possible efficiency into work ΔW. In an irreversibly conducted cyclic process less heat is taken up by the system and converted with lower efficiency into work than is achievable with a reversible process. Let us illustrate this, in addition to the discussion in Sect. 2.6, with a Carnot cycle in which friction occurs (refer to Fig.2.2). In the isothermal expansion at T_h, instead of taking ΔQ_1 from the bath – less is being withdrawn since part of the energy that could be given off as useful work is used to overcome friction and thus remains in the form of heat within the system. Consequently, $\Delta Q_1^{irr} < \Delta Q_1$. Correspondingly at T_l, more work is required for the compression and more heat must be dumped. Therefore $\Delta Q_3^{irr} > \Delta Q_3$. Overall, less work is performed *by* the system than in the reversibly conducted cycle, less heat is taken at T_h and more heat added to the T_l reservoir. Hence, we can conclude that for irreversibly conducted cyclic processes

$$\oint \frac{đQ}{T} < 0 \quad . \tag{2.26}$$

Note that (2.25) defines a new state function. Although heat is an imperfect or inexact differential, division of the infinitesimal amount of heat absorbed reversibly by the absolute temperature gives a perfect differential. Thus, 1/T can be called an integrating factor because its use allows for integration independent of the chosen path. This new state function is called *entropy* S and is defined in differential form as

$$dS = \frac{đQ_{rev}}{T} \quad . \tag{2.27}$$

The subscript "rev" for reversible is used in order to distinguish it from the actually exchanged heat dQ in real processes, where

$$đQ \leqslant đQ_{rev} \tag{2.28}$$

as discussed above. Entropy is an extensive state variable proportinal to n. In most applications only entropy differences, i.e., values of the entropy measured with respect to some chosen state, are of importance. For these cases (2.27) which allows the calculation of entropy only within an additive constant, is fully adequate. It should be noted in passing that based on the Third Law of Thermodynamics, (see e.g. [2.3]), entropy is a fully calculable entity.

One of the most significant bases of thermodynamics is the fact that entropy, just as the other state functions, depends only on the state of the system and not on the way the system arrived there. The entropy change for a transition between two states is the same, and defined by (2.27), whether the process is conducted reversibly or irreversibly. However, only for reversible processes, where $đQ = đQ_{rev}$ can one calculate ΔS from the actually exchanged heat by

$$\Delta S = \int \frac{đQ}{T} = \int \frac{đQ_{rev}}{T} \quad . \tag{2.29}$$

For irreversible processes, since $dQ < dQ_{rev}$,

$$\Delta S > \int \frac{đQ}{T}$$

or

$$\Delta S = \int \frac{đQ}{T} + \Delta S_{irr} = \int \frac{đQ_{rev}}{T} \quad . \tag{2.30}$$

Equation (2.30) expresses the fact that in an irreversible process an additional amount of entropy ΔS_{irr} is produced ("unnecessarily"), *with respect to* $đQ$. The ΔS_{irr} can be determined from (2.30) by calculating ΔS (that is intrinsic to the transition between initial and final state) along any conveniently chosen reversible path between the two states and subtracting the actual $\int đQ/T$.

In reversible processes entropy is merely conserved and transported from one system (bath) to another one. In irreversible processes less entropy is transported. The difference, ΔS_{irr} in (2.30) is created during the process. Let us illustrate this finding with a concrete example which also shows that "reversible" is more restrictive than "quasistatic". Consider the flow of heat from a body at some higher temperature, T_h, into a reservoir that is big enough so that its temperature, T_R, remains constant. Fig.2.3 indicates the two ways in which we choose to perform this experiment. In (a) we bring the systems in direct contact. Heat will flow spontaneously. A temperature gradient will develop within the body. The small body loses entropy in an undefined way. When $T_h = T_R$, the reservoir has gained entropy

$$\Delta S_R = \frac{C_v(T_h - T_R)}{T_R} \tag{2.31}$$

where C_v is the specific heat of the small body (assumed to be independent of temperature). In (b) we separate the systems by a poor thermal conductor. Consequently, the cooling of the body occurs so slowly that its temperature will stay uniform (although not constant). Hence, the body is at all times

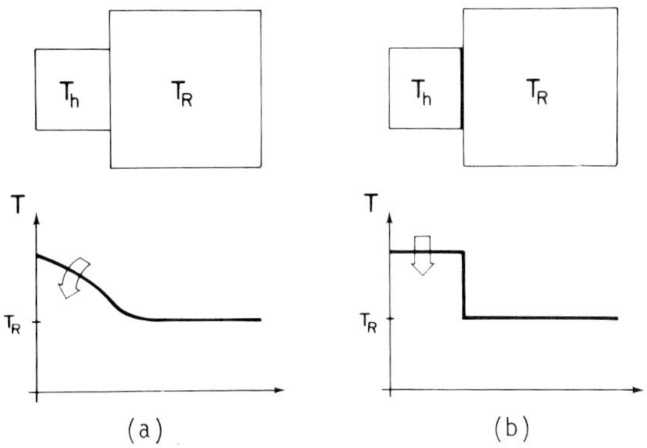

Fig.2.3. Spontaneous and quasistatic heat exchange

near equilibrium. The body thus loses entropy

$$\Delta S_B = C_V \int_{T_h}^{T_R} \frac{dT}{T} = -C_V \ln \frac{T_h}{T_R} \quad . \tag{2.32}$$

The gain in entropy of the reservoir is here just the same as in case (a). Expanding the last term in (2.32) in a Taylor series and comparing it with (2.31) one sees that ΔS_B is always smaller than ΔS_R. Although the entropy of the body decreases during the quasistatic cooling, the *total* entropy of the body and the reservoir increases due to the irreversible nature of the process. Irreversible processes increase the amount of low temperature heat and consequently increase the entropy of the universe.

Note that one can consider the above body and reservoir together as adiabatic, i.e., a completely isolated system. Neither work nor heat are exchanged with the surroundings. Thus, the internal energy cannot change during processes within this system. And still we have encountered a *spontaneous* process (here heat exchange between parts of the system with different temperature) that has increased the total entropy of the system to reach an overall equilibrium. In fact one can show that the Second Law demands, in general, for processes in adiabatic systems

$$dS \geqslant 0 \tag{2.33}$$

depending on the reversibility of the process. Equation (2.33) will be used in Sect. 2.10 as criterion for equilibrium in adiabatic systems.

A last statement of the Second Law may consequently read: The entropy of an adiabatic system can never decrease. It increases in irreversible processes and remains constant in reversible processes.

Entropy is a measure for the *maximum convertibility* of heat into work. If we know the entropy change ΔS connected with the flow of heat into a system (heat engine, etc.) then we can directly calculate that part of the heat taken up that is not available for conversion into work of any kind, even in a reversibly conducted process. The unavailable part of the heat which must be discarded at some lower temperature T_l is

$$Q_{unavail.} \geqslant T_l \Delta S \tag{2.34}$$

where

$$\Delta S = \int \frac{đQ_h}{T_h} \quad .$$

This concept will allow us to define "free energy", i.e., the part of the internal energy of a system that *is* available for work, by subtracting the appropriate entropy-temperature product, for instance, from U.

Besides its macroscopic content, the true significance of entropy lies in its atomistic foundation. Entropy is a measure for the degree of disorder or disorganization of a system. The full meaning of this statement can only be understood on statistical grounds. We will, however, briefly indicate two qualitative implications of the entropy-disorder connection that are of interest for phase transitions.

During a phase change, such as melting, latent heat ΔQ is added to a system at a fixed temperature T_m. Consequently, the entropy of the liquid is higher than that of the solid by $\Delta S = \Delta Q/T_m$. This entropy increase is reflected in a decrease of the structural order in the liquid as compared to the solid.

Later we will see that one can assign thermodynamical functions to interfaces just as well as to bulk phases. The entropy concept will allow us then to predict even the degree of atomic roughness of an interface between a given solid and its fluid from the entropy of transition. The atomic interface roughness determines to a large extent the type of growth mechanism that one encounters in a specific crystal growth situation. Hence, the possibility of predicting it from macroscopic parameters is extremely valuable.

2.8 Chemical Potential

The chemical potential μ was introduced in Sect. 2.2 without definition as the intensive counterpart of the extensive variable n. We will now obtain a definition for μ by discussing the change in internal energy that a system encounters when its mole number n is changed by an infinitesimal amount dn. All molar variables will be denoted with lower case letters.

When dn moles of the same substance in the same state (P, T, ...) are added to a system then the volume is increased by

$$dV = v dn \tag{2.35}$$

where $v = v(P,T)$ is the molar volume of the substance. It follows from the First Law, $dU = đQ - PdV + \ldots$, that a volume increase dV without addition of matter decreases the internal energy by PdV. However, the simultaneous increase in volume *and* the addition of matter dn (at constant P and T) should increase the internal energy of the system U by definition by the amount

$$dU = u dn \tag{2.36}$$

where u is the molar internal energy at the equilibrium values of P and T considered. Consequently, the First Law, including the addition of matter becomes

$$dU = đQ - PdV + (u + Pv)dn + \ldots \tag{2.37}$$

Although derived for constant pressure, (2.37) gives the correct answer even if n is changed but all other independent variables (V, Q ...) are kept constant, i.e., even if the pressure, etc. changes upon addition of dn (see, e.g., [Ref. 2.3, p. 59]).

If we introduce several substances, each characterized by a *partial molar internal energy*

$$u_i = \left(\frac{\partial U}{\partial n_i}\right)_{T,P,n_j \neq n_i} \tag{2.38}$$

and *partial molar volume*

$$v_i = \left(\frac{\partial V}{\partial n_i}\right)_{T,P,n_j \neq n_i} \tag{2.39}$$

it follows that

$$dU = đQ - PdV + \sum_i (u_i + Pv_i)dn_i + \dots \quad . \tag{2.40}$$

For reversible processes these equations can be rewritten entirely in terms of perfect differentials with dQ = TdS. Then the total gain in entropy upon reversible addition of heat to the material already in the system and addition of matter dn is

$$dS = \frac{đQ}{T} + \sum_i s_i dn_i$$

where s_i, the *partial molar entropy* of the i-th component, is

$$s_i = \left(\frac{\partial S}{\partial n_i} \right)_{T,P,n_j \neq n_i} \quad .$$

Thus the gain in heat will not just equal TdS but will be equal to

$$đQ = TdS - T\sum_i s_i dn_i \tag{2.41}$$

for reversible processes and will be smaller for irreversible processes.

Substituting (2.41) for dQ in (2.40) we obtain

$$dU = TdS - PdV + \sum_i (u_i + Pv_i - Ts_i)dn_i + \dots$$

$$= TdS - PdV + \sum_i \mu_i dn_i + HdM + \dots \tag{2.42}$$

where

$$\mu_i = u_i + Pv_i - Ts_i - Hm_i \dots \tag{2.43}$$

is the *chemical potential*.

One should emphasize that the partial molar entities (v_i, u_i, s_i, ...) introduced above are not specific for the respective component anymore and depend on the type and concentration of the other components as well. For instance, the molar volume of a pure bulk material as defined by (2.6) is constant and specific for that material. The partial molar volume, however, according to (2.39) is the infinitesimal volume change upon addition of an infinitesimal amount of that material in presence and hence in interaction with all other components in the system. There are various systems in which at certain concentrations v_i of a component is even negative, i.e., where on further addition of that component the total volume decreases.

2.9 Thermodynamic Potentials

Considering U as an explicit function of the extensive variables $U(S,V,M,n_i,\ldots)$ the perfect differential dU is

$$dU = \left(\frac{\partial U}{\partial S}\right)_{V,M,n_i} dS + \left(\frac{\partial U}{\partial V}\right)_{S,M,n_i} dV + \left(\frac{\partial U}{\partial M}\right)_{S,V,n_i} dM + \left(\frac{\partial U}{\partial n_i}\right)_{S,V,M,n_j} dn_i + \ldots \quad . \tag{2.44}$$

Consequently, by comparison with (2.42)

$$T = \left(\frac{\partial U}{\partial S}\right)_{V,M,n_i} \qquad H = \left(\frac{\partial U}{\partial M}\right)_{S,V,n_i}$$
$$\text{and} \tag{2.45}$$
$$-P = \left(\frac{\partial U}{\partial V}\right)_{S,M,n_i} \qquad \mu_i = \left(\frac{\partial U}{\partial n_i}\right)_{S,V,M,n_j} \quad .$$

If we have determined the internal energy U as a function of the complete set of extensive variables, the various partial derivatives of U are equal to the intensive variables — analogous to force components as derivatives of a mechanical potential with respect to displacement.

But, usually it is not convenient to use only extensive variables as the independent variables. In various experiments we prefer to measure, for example, T instead of S and P instead of V. If U is expressed in terms of these other variables then its partial derivatives are no longer equal to the intensive variables, i.e., U has lost the potential character for these variables. Therefore, it is desirable to find other functions, related to U, that behave like potential functions when expressed in terms of other sets of independent variables. The most convenient potential for the description of a specific process is the one which contains as independent variables the state functions which one chooses to control during the experiment. "Choose to control" can mean, for example, that one keeps these parameters constant. The usefulness of the potentials, however, is by no means restricted to cases where P or T, etc. are kept constant.

In order to define the thermodynamical potentials as generally as possible, yet to obtain compact relations we substitute in the following X for all extensive "mechanical" variables (V, $\underline{M}$, $\underline{P}$, A,...) and Y for all intensive mechanical variables ($-P$, $\underline{H}$, $\underline{F}$, σ,...). Thus, we need to deal only with three pairs of state functions:

— the thermal variables S and T,

— the mechanical variables X and Y, and

— the pair n and μ.

We will now formally introduce thermodynamical potentials for various combinations of three independent variables which have earned special names and discuss their physical significance.

The already mentioned potential for the variables S, X and n is the *internal energy* (sometimes called total energy; also symbol E)

$$U = TS + YX + \mu n \quad . \tag{2.46}$$

The First Law (2.37) using (2.41) and (2.43) can be written as

$$dU = TdS + YdX + \mu dn \quad . \tag{2.47}$$

However, from (2.46) one obtains

$$dU = TdS + SdT + YdX + XdY + \mu dn + nd\mu \quad . \tag{2.48}$$

Thus, we must assume that

$$SdT + XdY + nd\mu = 0 \tag{2.49}$$

which can be rigorously shown based on a computational trick devised by Euler (see, for instance, [Ref. 2.3, pp. 61-62]).

By rewriting (2.49) for one mole one obtains the *Gibbs-Duhem equation*

$$d\mu = -sdT - xdY = -sdT + vdP - mdH - pdF - \ldots \tag{2.50}$$

which we will use frequently in later sections since it conveniently connects changes in T, P and other mechanical variables with the changes in chemical potential.

One can change the independent variable S in (2.47) to T by replacing dS with $đQ_{rev}/T$. Then we see that, holding all the other extensive variables constant, a reversible addition of heat causes an equal increase in U. Therefore, we can call U the heat content at constant n,V,A, etc. (whatever the involved extensive mechanical variables may be). Consequently, we can write the heat capacity at constant volume, n, etc. as

$$C_{V,n,\ldots} = \left(\frac{\partial U}{\partial T} \right)_{V,n,\ldots} \tag{2.51}$$

From the partial differentials of U in (2.45) one can derive another set of useful relations called the *Maxwell relations*. Since a perfect differential must be commutative, i.e., the value of the second partial derivative is independent of the sequence of differentiation, one obtains, for example, from

$$T = \left(\frac{\partial U}{\partial S} \right)_{V,M,n} \quad \text{and} \quad -P = \left(\frac{\partial U}{\partial V} \right)_{S,M,n} \tag{2.52}$$

the relation

$$\left(\frac{\partial T}{\partial V} \right)_{S,M,n} = - \left(\frac{\partial P}{\partial S} \right)_{V,M,n} \quad . \tag{2.53}$$

Each of the new partials is again a state variable. Thus each can be integrated along any convenient reversible path. Since there is a large number of possible Maxwell equations that are of significance only to particular experimental situations, we will, in the following, spell out only a few that are important for later considerations.

The thermodynamic potential for the variables S, Y and n is the *enthalpy*

$$H = U - YX \quad . \tag{2.54}$$

The differential $dH = dU - YdX - XdY$ can be rewritten using (2.47) to the more useful form

$$dH = TdS - XdY + \mu dn \quad . \tag{2.55}$$

Furthermore, employing (2.46) one can write for the enthalpy

$$H = TS + \mu n \quad . \tag{2.56}$$

The enthalpy is obviously useful for describing processes in which the mechanical variables Y are controlled — particularly when they are kept constant — and for adiabatic processes where $dS = 0$ (isentropic). Thus, enthalpy is frequently used for the description of chemical processes and calorimetric measurements ($P = \text{const} = 1$ atm). In the case where no other work than expansion work is involved in a reversible isobaric process,

$$dH = TdS = đQ \quad .$$

Only then is dH a measure for the heat change in a system and can be called the *heat content* of the system. When H is used in connection with other non-isobaric, non-isomagnetic, etc. processes, where $dH \neq dQ$, this name is misleading. So at constant pressure, magnetic field, etc., if heat is exchanged reversibly,

$$\left(\frac{\partial H}{\partial T}\right)_Y = C_Y = C_{P,H,F...} \tag{2.57}$$

is the heat capacity at constant Y. The partial derivatives of H are

$$\left(\frac{\partial H}{\partial S}\right)_{Y,n} = T, \quad \left(\frac{\partial H}{\partial P}\right)_{S,n} = -V, \quad \left(\frac{\partial H}{\partial n}\right)_{S,Y} = \mu \quad . \tag{2.58-60}$$

The potential for the variables T,X and n is the *Helmholtz function*

$$F = U - TS \quad . \tag{2.61}$$

This function is sometimes called work function and/or denoted with A.

The differential $dF = dU - TdS - SdT$ can be correspondingly rewritten to

$$dF = -SdT + YdX + \mu dn \quad . \tag{2.62}$$

Again, with (2.46) for U one obtains also

$$F = YX + \mu n \quad . \tag{2.63}$$

From (2.62) one sees that the Helmholtz function is advantageous for the description of isothermal and/or isochoric processes. In reversible, isothermal processes, ΔF corresponds quantitatively to the work performed by the system (mechanical work if YdX is -PdV only and dn = 0). Or in other words: In any reversible, isothermal process the change in F can be transformed completely into work, such as -PdV. Therefore, F is often called *free energy* or *work content* of the system. Under non-isothermal conditions this name is misleading, however. The partial derivatives of F are

$$\left(\frac{\partial F}{\partial T}\right)_{X,n} = -S, \quad \left(\frac{\partial F}{\partial X}\right)_{T,n} = Y, \quad \left(\frac{\partial F}{\partial n}\right)_{T,V} = \mu \quad . \tag{2.64-66}$$

One of the resulting Maxwell relations

$$\left(\frac{\partial S}{\partial V}\right)_{T,N} = \left(\frac{\partial P}{\partial T}\right)_{V,n} \tag{2.67}$$

will be very helpful for an elegant derivation of the Clausius-Clapeyron equation in Sect. 3.1.

The potential for the variables T, Y and n is the *Gibbs function* (also called free enthalpy, Gibbs free energy, F and Z).

$$G = U - TS - YX \tag{2.68}$$

with

$$dG = -SdT - XdY + \mu dn \tag{2.69}$$

is advantageously used for isothermal, isobaric processes. Corresponding forms are

$$G = H - TS = F - XY = \mu n \quad . \tag{2.70-72}$$

The partial derivatives of G are

$$\left(\frac{\partial G}{\partial T}\right)_{Y,n} = -S, \quad \left(\frac{\partial G}{\partial P}\right)_{T,n,Y} = V, \quad \left(\frac{\partial G}{\partial n}\right)_{T,Y} = \mu \quad . \tag{2.73-75}$$

Equation (2.73) is often called Gibbs-Helmholtz equation. The last differential indicates that for isothermal, isobaric processes the chemical potential is equivalent to the Gibbs function (here Gibbs free energy) per mole.

The First Law in the form of (2.47) can be rewritten as

$$dU = TdS - PdV - dW' \tag{2.76}$$

where dW' stands for all work that is done by (or energy withdrawn from) the system, except for expansion work. Hence,

$$dW' = -dU + TdS - PdV \quad . \tag{2.77}$$

Now for a process under constant P, T and n the right side of (2.77) employing (2.49) and (2.69) is found to be

$$dW' = -d(U - TS + PV) = -dG \tag{2.78}$$

equal to the decrease in Gibbs function of the system. We see that the total work done by a closed system, in addition to expansion work, in an isobaric, isothermal reversible process equals ΔG. In other words ΔG represents the maximum "useful work" that can be obtained in any isothermal, isobaric process from a system. For this particular configuration the name *Gibbs free energy* is appropriate. For the description of processes that are conducted under varying temperature and/or pressure one should use "Gibbs function".

Chemical reactions, phase transitions, etc. always involve "useless" expansion work to a certain extent. For example, in an electrolytic process such as the decomposition of $H_2O \rightarrow H_2 + O_2/2$ electrical work is necessary to generate one mole of hydrogen and half a mole of oxygen. This amount of electrical energy contains, besides the "useful" work that was necessary to dissociate the water molecules, also "useless" expansion work against the atmospheric pressure to accommodate the gases. The change in Helmholtz function ΔF includes both work forms; it characterizes the total work required for this process. Now looking at the work that one could obtain from the back-reaction: ΔF describes the energy that comes from the recombination of H_2 and O_2 as well as the work done by the atmosphere while the gas volume decreases.

With ΔG, however, we obtain only that part of the energy that is used for the intended specific purpose, here the generation of H_2 and O_2. The expansion work is only considered to be an undesirable side effect. In terms of the recombination of the gases: ΔG comes only from the oxygen-hydrogen combustion but not from the compression work that is merely due to the fact that the gases require a larger volume than the water. That part depends on the ambient pressure and is thus not specific for "the system". It would result from volume reduction of any gas.

The same consideration applies to phase changes such as solidification of a liquid. Here also, due to the density difference between solid and liquid, some expansion work is involved in the transition, although at atmospheric pressure typically three orders of magnitude less work than in the gas-liquid transition.

For convenience a compilation of the definitions of the thermodynamical potentials, their symbols and their names as used in other texts is given in Appendix B.

2.10 Equilibrium

Our discussion of thermodynamic concepts began with the most simple definition of equilibrium. Uniformity of the macroscopic parameters, that were later called intensive state variables, was demanded then. Most problems of real interest, however, deal with systems which contain discontinuities and gradients. Phase transitions and transport phenomena, for instance, would be excluded from an equilibrium description from this simple point of view. However, from experience we know that, for example, two phases of the same substance can very well be in equilibrium, in spite of a pressure or density discontinuity at their interface. Fortunately, we found later that uniformity of all intensive variables in a system forms a sufficient, though not necessary criterion for equilibrium. Here we will expand this insight and formulate equilibrium criteria for non-uniform systems. First we will treat discontinuities at the boundaries of homogeneous subsystems. Then, to "feel more comfortable" with transport processes, we will briefly discuss the validity of state variables in the description of non-equilibrium processes.

2.10.1 General criteria

Systems with discontinuities can be most easily described by employing constraints. A constraint is any variable which is part of the specification of the equilibrium state of a system, but which does not occur in the calculation of work done by the system. Let us illustrate this with a "thermal constraint". Consider a completely isolated system, with mechanically rigid walls that are also impervious to heat and matter. Now we assume that another wall of this type is inserted into the system to separate it into two equal parts. If this is done in an ideal way, without work or addition of heat, the equilibrium values for P and T will remain unchanged in both subsystems as compared to the undivided system. Then we suppose that the temperature in the subsystems is changed to $T - \Delta T$ and $T + \Delta T$, that is, to a somewhat lower and higher value than the original equilibrium T. As long as the *constraint*, the inner dividing wall, is thermally impervious, each individual subsystem is in equilibrium, although in different states. The total system remains in an equilibrium compatible with the constraint. As soon as we remove the wall, however, the system will return irreversibly to the original undivided equilibrium state. The internal energy, U, of the isolated system is, by definition, not changed during the equilibration process. But, as shown in Problem 2.4, the entropy is increased because the unconstrained system, the state with uniform P *and* T, has higher probability, characterized by higher S. In connection with (2.30) we have discussed that any *irreversible* process – here any process that occurs *by itself* after removal of a constraint – is connected with an increase in entropy of the system. The resulting unconstrained state has a higher entropy than any constrained state. Hence, we conclude that in general an equilibrium state that is reached through processes during which U, n and all extensive mechanical variables are kept constant, is characterized by a maximum in entropy.

Let us return to the constrained system with $T - \Delta T$ and $T + \Delta T$. Now we bring the system from the constrained to an (internally) unconstrained equilibrium state by running a *reversible* heat engine between the two parts rather than an irreversible process such as heat conduction. In this case the entropy stays constant. Entropy is only transported between the subsystems but not created. Since we know that, with the highest possible efficiency, the engine has converted some ΔU into work and removed it from the system, the internal energy must have reached a minimum in this uniform equilibrium state.

Thus one finds that in general an equilibrium state is characterized by a minimal U if S, n and all extensive mechanical variables are kept constant during all possible processes (such as the removal of an inner constraint) that leads to this state.

Both processes above, adiabatic to the extent that no heat is exchanged with the exterior, are characterized by (2.33), $dS \geqslant 0$, depending on the reversibility, just as we had already obtained in Sect. 2.7 from a similar model experiment.

We see here again that the particular thermodynamic potential appropriate in equilibrium considerations, the potential which is minimized under equilibrium conditions, depends on the thermodynamic variables we choose to control — to keep constant, for example. Constant does not necessarily mean uniform, as we shall see below.

Employing the same trick of relaxing internal constraints in a system, we find from (2.55), $dH = TdS - XdY + \mu dn$, that a process performed under constant S, Y and n is terminated, and thus the system has reached equilibrium when its enthalpy is minimized. Similarly, when T, X and n are kept constant, (2.62), $dF = -SdT + YdX + \mu dn$, reveals that the Helmoltz function has reached a minimum. Finally from (2.69), $dG = -SdT - XdY + \mu dn$, one sees that the Gibbs function is minimized at equilibrium for a system with constant T, Y and n.

Constraints, found useful for the above considerations of equilibrium in uniform systems, are also valuable tools in discussions of equilibrium in *per se* discontinuous systems. For instance if a system contains several bulk phases, the particular function which takes on an extremum value may be calculated for the complete system by summation over all contributing parts that have been calculated independently within the bounds of each part. Yet it may be necessary to consider constraints that contribute themselves to the internal energy of the system. The consequences of such constraints must be carefully considered from case to case. For instance, if a system contains such small subsystems that surface tension effects cannot be neglected anymore, the equilibrium criterion will have to be modified correspondingly. Let us consider, for example, a small drop in equilibrium with its vapor. Due to surface tension the pressure within the droplet is greater than without. One would be in error to assume that the Gibbs function, frequently used in this case as equilibrium criterion, is then simply the sum of the Gibbs functions of the vapor and drop subsystems as will be elaborated upon in the chapter on nucleation.

2.10.2 Interphase mass transfer

All our equilibrium criteria discussed so far referred to closed systems, i.e., $dn = 0$. Transfer of matter, however, plays a key role in equilibria among different phases. Hence, we will in the following derive an equilibrium criterion for mass transfer between phases.

Let us consider the equilibrium between a liquid and its vapor. The constraints we apply are:

1) constant temperature, by immersion of the system into a heat reservoir;

2) constant pressure, by keeping the system open towards atmospheric pressure. Suppose the whole system contains N moles of which in equilibrium n constitute the liquid. Therefore, N-n moles form the vapor phase. Consequently, the Gibbs function of the whole system may be written as

$$G = n\mu_l + (N-n)\mu_v \tag{2.79}$$

and, in equilibrium

$$\frac{\partial G}{\partial n} = 0 = \mu_l - \mu_v \quad . \tag{2.80}$$

From the Gibbs-Duhem equation (2.50), $d\mu = -sdT + vdP$ we see that for the given constraints μ_l and μ_v are independent of n. Hence, we conclude that in an isothermal isobaric system, mass transfer equilibrium will exist *for any ratio* of n to N if $\mu_l = \mu_v$.

Now let us consider the same liquid-vapor system under different constraints. We now close the container (V = const.) and isolate it thermally. We can no longer assume that the chemical potentials are independent of n since any dn between the phases will change pressure and temperature in the completely isolated system. From our above discussion, however, we know that in addition to V also U remains constant. Furthermore, the entropy is maximized in equilibrium. Therefore, we may write, with $n_l + n_v = N$, that

$$V = n_l v_l + n_v v_v \tag{2.81}$$

and

$$\begin{aligned} dV &= n_l dv_l + n_v dv_v + v_l dn_l + v_v dn_v = 0 \\ &= n_l dv_l + n_v dv_v + (v_l - v_v)dn \end{aligned} \tag{2.82}$$

since

$$dn_l = -dn_v = dn \quad .$$

Furthermore,

$$U = n_l u_l + n_v u_v \tag{2.83}$$

$$dU = n_l du_l + n_v du_v + (u_l - u_v)dn = 0 \tag{2.84}$$

and

$$S = n_l s_l + n_v s_v \tag{2.85}$$

$$dS = n_l ds_l + n_v ds_v + (s_l - s_v)dn = 0 \quad . \tag{2.86}$$

Then, multiplying (2.82) with P and (2.86) with -T and adding the results together with (2.84) we obtain

$$\begin{aligned} &n_l(du_l + Pdv_l - Tds_l) + n_v(du_v + Pdv_v - Tds_v) \\ &+ (u_l + Pv_l - Ts_l)dn - (u_v + Pv_v - Ts_v)dn = 0 \quad . \end{aligned} \tag{2.87}$$

The first two terms, according to (2.47), are zero. The last two terms according to (2.43), can be reduced to

$$\mu_l = \mu_v \quad . \tag{2.88}$$

So, under quite different constraints than before we arrive at the same equilibrium condition. One can easily verify that this very same result is obtained when one isolates the container thermally but keeps it open to the external pressure or when a closed container at constant temperature is used. One can show that (2.88) characterizes mass transfer equilibrium for any kind of contact between phases (including semipermeable membranes and very small phases with higher internal pressure) provided that it refers to species that can pass the interface freely (see, e.g., [2.8]). Hence, we arrive at the general condition for mass transfer or distribution equilibrium

$$\mu_i^{(1)} = \mu_i^{(2)} \quad , \tag{2.89}$$

i.e., "two phases cannot be in equilibrium unless the chemical potential for each component i is equal in both phases."

As long as the chemical potential for a component i in a phase labeled 1 differs from that in a phase labeled 2, for instance, if

$$\mu_i^{(1)} < \mu_i^{(2)} \quad , \tag{2.90}$$

there will be spontaneous transfer of matter from phase 2 to 1, since only then will the Gibbs function of the total system decrease and the system will move toward equilibrium. Thus, differences in chemical potential – rather than the concentration – is a measure for the "escaping tendency".

It is erroneous to assume that diffusion proceeds always in the direction of decreasing concentration, as the following examples show:

- A solute can be distributed among two phases of the same solvent with sizeable concentration differences. In Chap. 6 on segregation we will encounter systems with equilibrium distribution ratios of say 100 between solids and their melts and 10^4 between solids and liquid solutions. In spite of such huge concentration gradients across the interfaces, no net mass transfer occurs. Such segregations upon phase change are widely used for the purification of materials and play an important role in connection with many crystal growth processes.
- In some solutions (gaseous, liquid or solid) one can establish significant concentration gradients by imposing a temperature gradient on the system. In most cases, the heavier molecules are concentrated in the colder part of the system. Eventually a steady state develops in which the separating effect of "thermal diffusion" is balanced by the remixing effect of "concentration diffusion". Thermal diffusion, known as the Soret effect in liquids, is also used in separation techniques and plays a role in various crystal-growth mass transfer configurations.

In Sect. 5.2.2 we will express the diffusive mass flux in terms of the driving concentration gradient in the form of Fick's Law. Here we see already that such a law can properly describe diffusion only if all other parameters that enter (2.43) are either uniform throughout the system or in their contributions to μ negligible as compared to the (indirect) effects of concentration.

We have come a long way with the concept of equilibrium. Beyond the most simple definition in the introduction of Chap. 2, requiring uniformity of the state variables we have "accommodated" discontinuities in all sorts of intensive state variables, based upon the various thermodynamical potentials. In some of the examples used to illustrate mass transfer equilibria, we have even dealt with systems that contained gradients of thermal or compositional variables. Systems with gradients and, hence, with flow of heat and/or matter cannot be in equilibrium in a strict sense. Thus, the question arises to what extent can state functions, specifically defined for equilibrium states, be justly used for such systems. We have touched on this question, although in a less specific formulation, already in the introduction of Chap. 2. The idea of "local equilibria" was mentioned there. One divides a non-uniform system into small portions. If it is possible to make these portions large enough to contain a statistically significant number of molecules and yet small enough so that inhomogeneities within a portion are small, then state variables are still considered valid. On first sight this concept appears to be restricted to near equilibrium processes or – as has been believed for some time – to the range of linear rate laws (see Sect. 5.2). However, more recent efforts, in particular by PRIGOGINE [2.9] and his school, have extended thermodynamic methods to processes far from equilibrium. The domain of thermodynamic variables, based upon the local equilibrium criterion, is found to reach considerably beyond the range of linear transport equations. For instance, in chemical reactions between gases, reaction rates have to be very high in order to significantly perturb the Maxwellian distribution of molecular velocity and hence, to jeopardize the local equilibrium concept.

2.11 Chemical Equilibrium

Chemical equilibria and their changes in temperature, pressure and concentration gradients play a crucial role in many crystal growth processes. Whether it is for predictions of growth rates, compatability of containers with melts, dissociation equilibria such as the required oxygen partial pressure above oxides, to name a few, chemical equilibrium considerations belong to the "daily bread" of crystal growers.

In this section we will reformulate the mass transfer equilibrium condition obtained in Sect. 2.10 in a more chemical language. First, (2.89) will be expanded to account for the stoichiometry of chemical reactions. Second, a convenient relation between the yield of a reaction and the chemical potentials of reactants and products will be derived. The presentation concentrates on basic principles and on the pitfalls that some of the conventions in thermochemical discussion present to non-chemists. For details that go beyond the scope of this presentation, for instance, a discussion of reaction equilibria in electrolytic solutions, refer to the classical textbooks by DENBIGH [2.10] and KUBASCHEWSKI et al. [2.11]. Procedures to calculate chemical potentials of components from (sometimes) readily available thermochemical data are discussed in Sect. 2.12.

In chemical reactions atoms are conserved. Hence, one can express a reaction in its most general form as

$$\sum_i \nu_i M_i = 0 \quad , \tag{2.91}$$

where the stoichiometric coefficients ν_i are integers and the M_i's stand for the chemical symbols of the various species involved. For instance, the reaction $H_2 + Cl_2 = 2HCl$ would be replaced correspondingly by $\nu_1 M_1 + \nu_2 M_2 = \nu_3 M_3$ where $\nu_1 = \nu_2 = -1$ and $\nu_3 = 2$. Note the sign convention: minus for the (disappearing) reactants and plus for the products. Stoichiometry, i.e., (2.91) implies also that

$$\frac{dn_1}{\nu_1} = \frac{dn_2}{\nu_2} = \ldots\ldots = \frac{dn_m}{\nu_m} = dr \quad . \tag{2.92}$$

Hence, the change in the number of moles dn_i for each species divided by its stoichiometric constant ν_i is the same for all species at any instant during the reaction. The entity dr can be thought of as a running reaction coordinate.

Suppose that the reaction is conducted under constant pressure and temperature. Under these conditions equilibrium is characterized by a minimum in the Gibbs function of the system, as discussed in Sect. 2.10. The change in G in an isothermal, isobaric reaction is according to (2.69)

$$dG_{P,T} = \mu_1 dn_1 + \mu_2 dn_2 + \ldots + \mu_m dn_m \tag{2.93}$$

and with (2.92)

$$dG_{P,T} = (\nu_1\mu_1 + \nu_2\mu_2 + \ldots + \nu_m\mu_m)dr \quad . \tag{2.94}$$

Now we can readily express the equilibrium condition for the reaction as

$$\left(\frac{dG}{dr}\right)_{P,T} = \sum \nu_i \mu_i = 0 \quad . \tag{2.95}$$

Most chemical reactions are conducted under pressure exchange with the ambient. Hence, the assumption of constant (atmospheric) pressure is reasonable. The constant temperature assumption, however, may be surprising. To justify that, one has to realize that heat transfer proceeds, in general, more rapidly than mass transfer. Thus, a system may be very well in thermal equilibrium without having reached chemical equilibrium yet.

It is important to note that the chemical potentials μ_i in (2.93) and (2.94) are concentration dependent. Although both equations hold for any instant during the reaction, the value of the μ_i's will change with the dn_i's and dr. Due to a continuous change of the composition of the environment with which each species interacts, the μ's change continuously until equilibrium is reached. If the chemical potential of the individual species would be concentration independent, chemical reactions would simply proceed until all species are transformed into the one with the "lowest chemical potential". We will illustrate this point in more detail later.

For an application of (2.95) to specific phases one must connect the chemical potentials to specific equations of state in order to define their concentration or pressure dependence. This is most conveniently done for ideal gases. Although one would expect that the resulting formulation is restricted to gases (with correction factors for deviations from ideality), this is not so. A reaction involving liquids and solids may be regarded, from a thermodynamic standpoint, as a reaction between the vapors of the substances. Equilibrium between the vapors and the condensed substances, in addition to the reaction equilibrium in the vapor phase ascertains reaction equilibrium between the condensed phases themselves. Thus, one can take advantage of the particularly simple form of the ideal gas reaction equilibrium equation for condensed phases as well.

For an evaluation of (2.95) the chemical potentials of the substances involved must be known for the specific equilibrium pressure (i.e., concentration) and temperature. Rather than tabulating experimentally determined chemical potentials for a wide pressure and temperature range one lists them for a *standard state* and supplies functions for computations of the μ's at other P-T combinations (see Sect. 2.12). This convention allows for a condensed data presentation as well as for a smaller number of experiments to characterize a system thermochemically. Thermodynamic variables that refer to a standard state are denoted by a superscript zero.

Suppose the chemical potential μ^0 for an ideal gas is known at the temperature T_0 and the pressure P_0. If the pressure is changed to P, at the same temperature, the new μ can be calculated from the Gibbs-Duhem equation (2.50), $d\mu = -sdT + vdP$, and the molar equation of state for ideal gases (2.1), $Pv = RT$, as

$$\Delta\mu = \int_{P_0}^{P} RT \frac{dP}{P} = RT_0 \ln(P/P_0) \quad . \tag{2.96}$$

Consequently,

$$\mu = \mu^0 + RT_0 \ln(P/P_0) \quad , \tag{2.97}$$

and if P_0 is chosen as unity (e.g., 1 atmosphere), one obtains the convenient form

$$\mu = \mu^0 + RT_0 \ln(P) \quad . \tag{2.98}$$

Note that under this convention the standard chemical potential μ^0 is a function of temperature only. This is often disregarded. In particular, dimension-conscientious physicists have problems with the non-dimensionality of P, that actually stands for the ratio P/P_0.

Analogous, for a mixture of ideal gases, one obtains for each component i from (2.98)

$$\mu_i = \mu_i^0 + RT \ln(p_i) \quad , \tag{2.99}$$

where the partial pressure p_i is defined through the molefraction X_i by

$$p_i = \frac{n_i}{\sum_i n_i} P = X_i P \quad . \tag{2.100}$$

Now, substituting (2.99) for the μ_i's in (2.95), the equilibrium criterion for the ideal gas reaction becomes

$$\sum \nu_i \mu_i^0 = -RT \ln\left[\prod_i (p_i)^{\nu_i}\right] \tag{2.101}$$

which, after defining the *equilibrium constant* or mass action constant as the continued product,

$$k \equiv \prod_i (p_i)^{\nu_i} = p_1^{\nu_1} \times p_2^{\nu_2} \times \dots p_m^{\nu_m} \tag{2.102}$$

can be written as

$$\sum \nu_i \mu_i^0 = -RT \ln(k) \quad . \tag{2.103}$$

Since the left-hand side in (2.103) is a function of T only [see (2.97) and (2.98)], k is also only a function of temperature. As a consequence of the P-convention in (2.98) k is dimensionless. Note also that according to the sign convention for the ν's, (2.91), all partial pressures of the product components appear in the numerator and all reactants in the denominator of (2.102). Remembering that the μ_i^0's are the Gibbs free energies per mole of the pure components at unit pressure one may call the term $\Sigma\, \nu_i \mu_i^0$ the standard change in Gibbs free energy of the reaction at T

$$\Delta G_T^0 = \sum_i \nu_i \mu_{i_T}^0 \quad . \tag{2.104}$$

Substitution of (2.104) into (2.103) leads then to the final condition for chemical equilibrium

$$\Delta G_T^0 = -RT \ln(k) \quad . \tag{2.105}$$

For the derivation of (2.105) we have followed closely a presentation by DENBIGH [2.10]. An alternative route is to substitute (2.99) for the μ_i's in (2.94) which unlike (2.95) is valid for any instant during the proceeding reaction. The resulting change in the Gibbs function of the reaction ΔG_T (as compared to the *standard* change ΔG_T^0) is

$$\Delta G_T = \Delta G_T^0 + RT \ln(k) \quad . \tag{2.106}$$

At equilibrium $\Delta G_T = 0$ and (2.105) results from (2.106).

At this point it may be worthwhile to answer the "naive" question: Why is it that the chemical potentials of ideal gases involved in a reaction change as the reaction proceeds? Ideal gases show no heat of mixing, i.e., forces of interaction are considered negligible; hence, it should not "make any difference" whether some species are replaced by others. In particular, when the number of molecules does not change, as in the reaction between H_2 and Cl_2, even the *total* pressure remains constant during the reaction.

Let us answer this question for the model reaction A + B = 2C. The Gibbs function of this system at any moment is $G = n_a\mu_a + n_b\mu_b + n_c\mu_c$. To illustrate the way in which G changes during the reaction, we express the μ's in terms of (2.99) and obtain

$$G = n_a\mu_a^0 + n_b\mu_b^0 + n_c\mu_c^0 + RT[n_a \ln(p_a) + n_b \ln(p_b) + n_c \ln(p_c)] \quad .$$

Suppose that the reaction is between 1 mole each of A and B and the total pressure (Note: P/P_0 rather!) is unity. Then, replacing the partial pressures through (2.100), with $n_b = n_a$ and $n_c = 2(1 - n_a)$, we get

$$G - 2\mu_c^0 = n_a(\mu_a^0 + \mu_b^0 - 2\mu_c^0) + 2RT\left[n_a \ln\left(\frac{n_a}{2}\right) + (1 - n_a)\ln(1 - n_a)\right] \tag{2.107}$$

which is schematically plotted in Fig.2.4 for some fixed value of $(\mu_a^0 + \mu_b^0 - 2\mu_c^0)$ and with n_a ranging from 1 (beginning of reaction) to 0 (fictitious

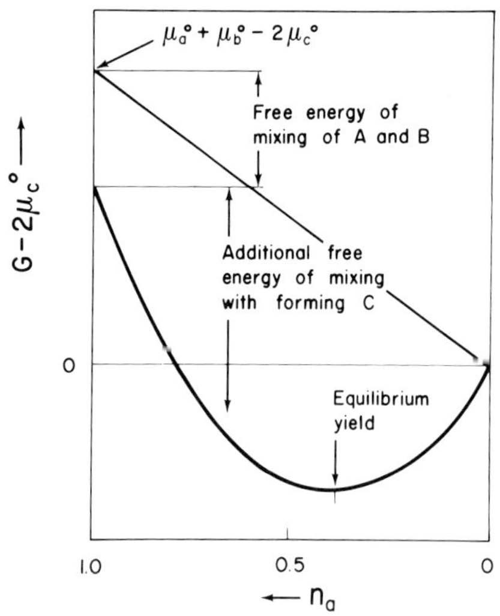

Fig.2.4. Gibbs function of a reacting system. After [2.10]

complete reaction of A and B to C). The term $G - 2\mu_c^0$ represents the amount by which the Gibbs function of the system exceeds that of two moles of C when there are still n_a moles of A (and B) left in the system. At complete conversion to C ($n_a = 0$) this excess vanishes. An interpretation of the various terms in (2.107) reveals two interesting features.

First, one sees that at $n_a = 1$, hence, before any C is formed, the Gibbs free energy of the A-B mixture is lower than the sum of the free energy of the separated substances. This results from an increase in entropy due to the irreversible nature of the mixing process. As shown in Problem 2.5, the second term on the right-hand side of (2.107) represents the entropy of mixing of identical amounts of two ideal gases. From (2.70), ΔG at constant temperature is $\Delta G_T = \Delta H_T - T\Delta S_T$. Since $\Delta H = 0$ for the mixing of ideal substances, any increase in entropy lowers the Gibbs function of the system. Note that this is not necessarily true for non-ideal substances where a positive ΔH can override the $T\Delta S$ term so that mixing of A and B does not occur. Such systems with "miscibility gap" will occupy a major part of later discussions on phase equilibria.

Second, one sees that with proceeding reaction (decreasing n_a) the second term decreases as A and B are more and more converted to C. This decrease, however, is opposed by the growing third term which represents the entropy from mixing of C into A and B. Since both terms vanish at $n_a = 0$, the function $(G - 2\mu_c^0)$ must exhibit a minimum somewhere between $0 < n_a < 1$. Hence, there is necessarily some compositional range in which the free energy of the mixture is even smaller than if there were complete conversion to C. The equilibrium yield in C is determined by the minimum in Fig.2.4 (see Problem 2.6). Note that without the entropy of mixing terms the reaction would proceed along the straight line in Fig.2.4 to $n_a = 0$ where all A and B are converted to C, and this is because the chemical potentials μ_i are artificially kept constant in the form of the μ_i^0's.

With (2.105) the equilibrium yield of a chemical reaction between ideal gases can be predicted, provided the standard chemical potentials are known at the temperature of interest. Since, by the above conventions, the equilibrium constant for this idealized group of reactions depends on temperature only, one may be tempted to assume that the concentration of the products, i.e., the yield, is also pressure independent. This, however, is only true for reactions in which the total number of molecules does not change. For reactions with $\Sigma\nu_i \neq 0$ one obtains through (2.100) and (2.102) unequal powers of P in the numerator and denominator of k and a pressure-dependent yield (see Problem 2.12). Reactions which result in more molecules are favored by lower pressures, whereas reactions with $\Sigma\nu_i < 0$ are aided by higher pressures. As illustrated in Problem 2.10, such a pressure dependence of the yield can be used to overcome the poor yield that stems at ambient pressures from an unfavorable value of ΔG_T^0.

The pressure in a system can be increased either directly by a piston compressing the reactants and products or indirectly by addition of an inert gas. Note, however, that an inert gas does not only affect the total pressure P but it must also be included in calculating the mole fractions. As one can deduce from a discussion similar to Problem 2.10, pressure increase via addition of an inert gas is typically a less efficient means of increasing the yield than direct pressurization. Besides this mere equilibrium argument we will encounter other kinetic reasons against employing high "inert" gas pressures in reactive crystal growth processes.

After this discussion of reaction equilibrium between ideal gases, let us turn to some conventions that are advantageous for applications of (2.103) – (2.105) to *real gases* and *condensed phases*.

Instead of defining an ideal gas by its equation of state, (2.1), one can base the definition on (2.98) which has the same physical content. Thus, one can define a real gas by an analogous relation for its chemical potential at T and P

$$\mu = \mu^0 + RT \ln(f) \tag{2.108}$$

with $f/P \to 1$ as $P \to 0$, where the *fugacity* f is a fictitious pressure ("fudge factor") that is introduced in order to preserve the simple form of (2.98) even for real gases. The limiting relation, as part of (2.108), makes the fugacity equal to the pressure under conditions where the gas approaches ideal gas behavior, i.e., at low P and/or high T. Note that the fugacity is a dimensionless entity just as the "pressure" in (2.98) that is obtained by dividing P by unit pressure. For most gases the ratio f/P does not deviate appreciably from unity except at pressures significantly greater than atmospheric. Hence, the value of μ^0 in (2.108) is practically the same as that of the chemical potential at unit pressure. Strictly speaking, of course, μ^0 here is the chemical potential of the gas at unit fugacity and is only a function of temperature.

Fugacities of pure gases are functions of temperature and pressure. If the deviation from ideal gas behavior has been determined, its fugacity can be computed. If, for instance, the *virial coefficients* B(T) and C(T) in

$$PV = nRT\left[1 + B(T)\frac{n}{V} + C(T)\left(\frac{n}{V}\right)^2 + \dots\right] \tag{2.2}$$

are known at the temperature of interest (see, e.g., [2.6]) the Gibbs-Duhem equation, (2.50) can be readily integrated at constant T between the reference state pressure P_0 and the actual P employing the virial equation, (2.2), for the molar volume. One obtains thus for the chemical potential of a non-ideal gas

$$\mu = \mu^0 + RT\ln(P) + BP + \frac{C - B^2}{2RT}P^2 + \dots , \tag{2.109}$$

where B and C are the second and third virial coefficients, respectively. Comparison with (2.108) gives then

$$\ln(f) = \ln(P) + \frac{BP}{RT} + \frac{C - B^2}{2(RT)^2}P^2 + \dots . \tag{2.110}$$

Mixtures of imperfect gases are treated analogously. Hence, from (2.108) one obtains for each individual component

$$\mu_i = \mu_i^0 + RT\ln(f_i) \tag{2.111}$$

with $f_i/p_i \to 1$ as $p_i \to 0$. Since fugacity has been defined in such a way that μ_i^0 is a function of temperature only, μ_i^0 has the same value as for the pure component i at the same temperature independent of the partial pressure p_i. The fugacities f_i, however, depend on the composition (partial pressures) as well as on T. Their characterization as function of P, T and composition requires such a large amount of experimental data that it has been carried out for only a small number of mixed gas systems. In practice fugacities of the components are usually approximated, as described in DENBIGH [Ref. 2.10, pp. 128-130].

The chemical potentials as defined with fugacities in (2.111) can now be substituted in the chemical equilibrium criterion (2.95) to obtain the yield equation for real gas reactions

$$\Delta G_T^0 = -RT \ln\left|\prod_i (f_i)^{\nu_i}\right| = -RT \ln(k) \qquad (2.112)$$

with $\Delta G_T^0 = \sum_i \nu_i \mu_{i_T}^0$ as defined in (2.104).

Now let us expand this concept to reactions that include condensed phases. As mentioned before one can always treat, from a thermodynamic point of view, such reactions as if they would occur only in the vapor phase above the condensed phase. However, it is more convenient to express the equilibrium constant of a condensed phase reaction directly in terms of the actual composition. In order to do this, one has in principle only to substitute "appropriate" expansions for the chemical potentials of all components in (2.95). "Appropriate" is put in quotation marks to anticipate various pitfalls that are hidden in the conventions which have yet to be introduced in order to apply (2.112) to condensed phases.

The method usually adopted in dealing with condensed phases is to find the magnitude of the pure number γ_i which, when multiplied by the mole fraction X_i of the particular component, preserves the simple form of (2.98) for the chemical potential

$$\mu_i = \mu_i^* + RT \ln(\gamma_i X_i) \quad , \qquad (2.113)$$

no matter how large the actual deviation is from ideality. The standard state chemical potential in (2.113) has an asterisk as superscript in order to distinguish it from the μ_i^0 of gases. In contrast to the μ_i^0's that were defined such that they depend on temperature only, the μ_i^*'s depend on temperature *and* pressure. Typically, however, the PdV terms involved in reactions of condensed phases are insignificant, particularly when compared to the uncertainty in thermochemical data, unless the pressure is greatly in excess of one atmosphere.

The product of the *activity coefficient* γ_i and the mole fraction X_i is commonly called the *activity* of the component i,

$$a_i = \gamma_i X_i \quad . \qquad (2.114)$$

The activity a_i can be imagined as an effective relative concentration.

Only for very few systems, such as the liquid pairs CCl_4-$SnCl_4$, chlorobenzene-bromobenzene, and n-hexane-n-heptane, is $\gamma_A \approx \gamma_B \approx 1$ over the whole composition range. For most systems, particularly when they contain comparable amounts of all components, the γ_i's deviate from unity. For instance, in a binary system in which the intramolecular attraction is stronger than in pure A or B, the components will appear "less active" than in their pure state. This can be reflected in lower vapor pressures, boiling points, etc. In order to simulate such a real system with an ideal one, as is done by (2.113), one has to reduce the concentration in the simulation system ($\gamma_i < 1$) to obtain the properties of the real system. On the other hand, systems with increased intramolecular repulsion display, for instance, higher vapor pressures than the pure substances. Hence, a "correction factor" $\gamma > 1$ is required in the idealized description of (2.113). Note also that the activity coefficients are functions of P, T and the mole fractions of all components in the system.

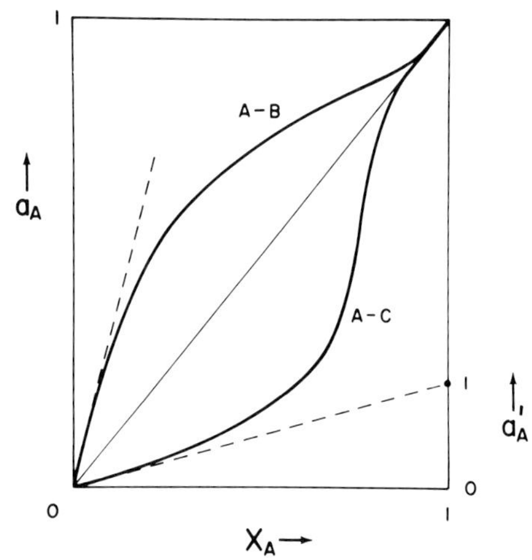

Fig.2.5. Typical activity – molar fraction relationships in real solutions

Equation (2.113), as it stands, does not provide a complete definition of γ_i, unless the standard state for μ_i^* is also specified. Hence, one must define, for instance, as in connection with (2.108), under what conditions γ_i becomes equality unity. Here one can take advantage of the fact that components in real solutions are typically found to approach ideal behavior both as $X_i \to 1$ (Raoult's law) and $X_i \to 0$ (Henry's law). With ideal behavior we mean here a linear relationship between activity and mole fraction of a component. This behavior is schematically indicated in Fig.2.5 for two binary solutions, AB and AC, with a common component. One sees that as $X_A \to 1$, the activity coefficient for A approaches $\gamma_A = a_A/X_A = 1$ (diagonal line) in both systems considerably before $X_A = 1$. Small amounts of other components (solutes) do not change the chemical potential of a component (solvent) significantly. Thus, it appears advantageous to choose for larger X_A the chemical potential of the pure A at the same P and T as standard state μ_i^*.

On the other hand, one sees that with increasing dilution of A, a_A approaches also linear behavior as represented by the dashed tangents. In order to define a $\gamma_A = 1$ also for this lower range we have to define a new reference state and cannot refer to the μ_i^* for $X_A \to 1$ anymore. To take advantage of this linear range we *make* $\gamma_A = 1$ as $X_A \to 0$ by referring to a hypothetical ideal solution, that, e.g., for A in C is characterized by the lower dashed line in Fig.2.5. The new μ_i^* corresponds then to the hypothetical ideal solution that obeys Henry's law at all mole fractions up to $X_A = 1$, where its activity a_A' reaches unity. The shrinkage of the a_A'-scale as compared to the a_A-scale reflects the fact that the fictitious ideal solution used to replace the real solution in (2.113) contains less A, as discussed above.

The same considerations apply to the $X_A \to 0$ end of the A-B solution. Note, however, that the upper dashed line defines again a different μ_i^*. Hence, the statement made earlier that the standard chemical potential μ_i^* depends only on P and T applies only to components, the standard states of which refer to the pure substances. At all compositions where the activity cannot be approximated by either of the straight lines, the fugacity f, by way of its deviation from unity, has to account for the non-ideality of the system.

The actual choice of the reference state, and hence of the value of μ_i, will be made from case to case on practical grounds. Obviously, for solvents in solutions one will choose the pure component as the standard state. For components for which, for instance due to solubility limits, X_A cannot be varied up to unity, one will more advantageously determine the activity for the lower range of X_A and then extrapolate the tangent at $X_A = 0$ to $X_A = 1$. For various possibilities of measuring activity coefficients see the corresponding chapters in the monographs of LEWIS and RANDALL [2.4], DENBIGH [2.10], and KUBASCHEWSKI [2.11].

Finally by substituting (2.113) into (2.105) we obtain an expression for the equilibrium yield of reactions in condensed phases in the form of

$$\Delta G^0_{T,P} = -RT \ln \left[\prod_i (a_i)^{\nu_i} \right] = -RT \ln(k) \quad , \tag{2.115}$$

with

$$\Delta G^0_{T,P} = \sum_i \nu_i \mu^*_{i_{T,P}} \quad . \tag{2.116}$$

Now let us summarize the various conventions by considering a hypothetical reaction in which a solid is precipitated from a reaction between liquid components under formation of a gas, e.g., $2A(l) + 3B(l) = C(s) + 4D(g)$. For this case one obtains

$$\Delta G^0_T = -RT \ln \frac{1 \cdot f^4_D}{a^2_A \cdot a^3_B} \quad .$$

The activity of the solid is unity here, by definition, since the solid phase consists of the pure C only ($\mu_i = \mu^*_i$). Note that this is not specific for solids. If, in a similar reaction, a monocomponent (pure) liquid would be produced, its activity would be unity as well. Of course, the fact that the gas in the above reactions may dissolve to some extent in the liquid phase has to be reflected in the value for its fugacity as well as in the activities of the liquid components.

For reactions in electrolytic solutions the reaction yield criterion must be expanded to account for overall charge neutrality – just as the general mass transfer equilibrium criterion had to be modified to account for the stoichiometry of reactions. For a very detailed presentation of this subject see the monograph of DENBIGH [Ref. 2.10, pp. 302-320].

2.12 Thermochemical Calculations

2.12.1 General discussion and approximations

In order to compute the equilibrium constant k from (2.112) or

$$\Delta G^0_{P,T} = -RT \ln(k) \tag{2.115}$$

one must know the change in standard Gibbs free energy of the reaction. The conventions discussed in Sect. 2.11 free one from specifying the actually concentration dependent chemical potentials of the reactants and products. For a calculation of the equilibrium constant it suffices to know the chemical potentials of the components "in their standard states". Remember that the standard change in Gibbs free energy, as the sum of the stoichiometry-factor weighted standard chemical potentials, depends only on T for gaseous reactions, see (2.104). For reactions between condensed phases it is, in principle, also pressure dependent. As discussed in connection with (2.113), however, this pressure dependence can be neglected with *condensed phases* unless very large pressures are used. Hence, we will in the following tacitly assume that given reactions occur in a pressure range where standard state chemical potentials for 1 atm can be used with reasonable accuracy. Note also that the pressure dependence of *gaseous* reaction equilibria is not ignored in (2.115). It is only hidden in the form of fugacities in k itself, see (2.112), and one is able to refer to pressure independent μ_i^0's.

It should be emphasized again that (2.115) allows us to completely ignore the kinetic details of a reaction. No matter how irreversibly the reaction proceeds, i.e., how strongly pressure and temperature deviate during the reaction from P and T in (2.115), once the system has "settled down" and reached equilibrium, the yield will be described correctly. This is one of the great profits of the state function concept.

The above discussion might suggest that an equilibrium yield can only be meaningfully calculated for a system that has, for instance, reached temperature uniformity throughout. In the sense of our expanded understanding of equilibrium, however, this is not a necessary condition (see Sect. 2.10). All that we have to ask for is whether the chemical reaction kinetics proceeds more rapidly than any transport kinetics that might carry the components into areas with significantly different temperature. As long as this is the case one can treat the system in terms of local equilibria and hence, obtain local yields. This assumption is rather important in chemical transport crystal growth processes to be discussed later. There one makes actual use of the shift of a chemical equilibrium in a temperature gradient. Quantitative discussions of transport rates are then tacitly based upon local equilibrium. The validity of this assumption must be checked from case to case. As we will see, various exceptions have indeed been observed.

For such "isothermal" reactions (isothermal in a uniform or local sense) it is advantageous to express the standard free energy change, from (2.70), as

$$\Delta G^0_T = \Delta H^0_T - T\Delta S^0_T \quad , \tag{2.117}$$

since both enthalpies and entropies of the components can directly be calculated from experimental heat capacity data, as shown below. In principle, enthalpies and entropies can be determined with reference to their values at absolute zero. But, as we are here only concerned with their changes as a result of a chemical reaction, rather than with their absolute value, we can choose any convenient, i.e., experimentally readily accessible, reference state. Since atoms are conserved in chemical reactions one chooses advantageously the elements as reference. One defines that the Gibbs free energy *of formation* of the elements in their stable state at any given temperature and when P = 1 atm, i.e., in their standard state, is zero. The standard states of compounds (also taken at 1 atm) are then characterized by the free energy of formation from the elements in their standard state. It is customary to use the symbols $\Delta G^0_{f,T}$, $\Delta H^0_{f,T}$ and $\Delta S^0_{f,T}$ for the standard free energy, enthalpy and entropy of formation, respectively. The subscript f is used to indicate *formation from elements*. It is not used with ΔG^0_T, ΔH^0_T or ΔS^0_T for reactions between compounds. Sometimes this may cause confusion. Note that enthalpy of formation and heat of formation become synonymous here since the standard states of compounds are formed isobarically from those of their elements. For clarity's sake it should also be pointed out here that in addition to $\Delta G^0_{f,T}$, $\Delta H^0_{f,T}$ and consequently $\Delta S^0_{f,T}$ for elements *are defined* to be zero simply as a convenient reference. This, of course, does not imply that heats and entropies (rather than heats and entropies *of formation*) of elements are zero.

Thermochemical data are presented in various ways. (A brief discussion of sources is given in Sect. 2.12.2). One way is to tabulate the standard heats and entropies of formation for each compound in certain increments as function of temperature. Another method is to plot free energies of formation versus temperature, typically in groups of similar compounds. Both of these approaches are very convenient for the user. $\Delta G^0_{f,T}$ values for the temperature of concern can be retrieved either directly or with some interpolation from the compilation. Yet another very common way is to tabulate the heat of formation, $\Delta H^0_{f,T_0}$, and free energy of formation, $\Delta G^0_{f,T_0}$, typically for the standard state temperature T_0 of 25°C ≡ 298.15 K and to list in addition values for the coefficients a, b, c ... in a polynomial approximation for the heat capacity at different temperatures. This requires only one line in a multiple column table in order to characterize one compound over a certain temperature range. However, it also requires the computation of ΔG_T^0 at the temperature of interest from $\Delta G_{T_0}^0$ of the reaction, and the heat capacity data of the compounds. This can be done in a rather straightforward manner, although the procedure is prone to computational errors because of the many terms that have to be handled.

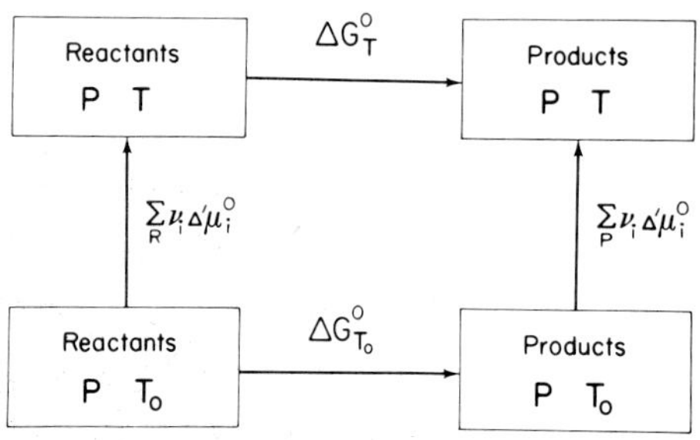

Fig.2.6. Calculation of ΔG_T^0 from $\Delta G_{T_0}^0$

Let us illustrate the procedure and employ thereby one of the commonly used forms for the $C_p(T)$ expansion. The problem is illustrated in Fig.2.6. One sees that two categories of changes are involved which we will distinguish by different symbols to avoid confusion. Firstly, there is the given $\Delta G_{T_0}^0$ and the ΔG_T^0 to be computed for the reaction. These isothermal changes, denoted by Δ, are obtained at T_0 and T, respectively, from the standard free energies of reactants and products. Sedondly, there are changes in the values of the standard free energies of reactants and products due to the transition from T_0 to T. These changes, denoted by Δ', are obtained from the temperature dependence of the heat capacity data. Rewriting (2.117) for the standard state parameters one obtains for the reaction at a given T

$$\Delta G_T^0 = \Delta H_T^0 - T\Delta S_T^0 = \Delta' H^0 - T(\Delta S_{T_0}^0 + \Delta' S^0) \quad . \tag{2.118}$$

Now for an isobaric process, one can calculate the $\Delta' H$ and $\Delta' S$ with Kirchhoff's relation

$$\Delta' H^0 = \int_{T_0}^{T} \Delta C_p^0 dT' \quad , \tag{2.119}$$

which is an integrated form of (2.57). Analogously one gets for the entropy change, with (2.27)

$$\Delta' S^0 = \int_{T_0}^{T} \frac{dQ_{rev}}{T} = \int_{T_0}^{T} \frac{\Delta C_p^0}{T'} dT' \quad . \tag{2.120}$$

Suppose the heat capacity of each component in its standard state is given by the frequently used approximation

$$C_p^0(T) = a + bT + cT^2 + \frac{d}{T^2} \tag{2.121}$$

with a, b, c and d the tabulated empirical constants. Then $\Delta C_p^0(T)$ for the reaction is

$$\Delta C_p^0(T) = \Delta a + \Delta bT + \Delta cT^2 + \frac{\Delta d}{T^2} \tag{2.122}$$

where $\Delta a = \sum_i \nu_i a_i$, $\Delta b = \sum_i \nu_i b_i$, etc. since each component enters weighted with its stoichiometric factor [see (2.91)]. Substitution of (2.122) into (2.119) and (2.120) and integration and collection of the terms in (2.118) leads finally to (see Problem 2.7)

$$\begin{aligned} \Delta G^0(T) = {} & \Delta H_{T_0}^0 - T_0 \Delta a - T_0^2 \frac{\Delta b}{2} - T_0^3 \frac{\Delta c}{3} + \frac{\Delta d}{T_0} \\ & + T \left| \Delta a(1 + \ln T_0) + T_0 \Delta b + T_0^2 \frac{\Delta c}{2} - \frac{\Delta d}{2T_0^2} - \Delta S_{T_0}^0 \right| \\ & - T^2 \frac{\Delta b}{2} - T^3 \frac{\Delta c}{6} - \frac{\Delta d}{2T} - T\Delta a \ln(T) \quad . \end{aligned} \tag{2.123}$$

If any of the components, or of the elements from which the components were formed, undergoes a phase transition in the integration range $T_0 \rightarrow T$, its effect on heat and entropy of formation has to be taken into account. Assuming that only one of the components, say j, undergoes a phase transformation at T_t with a (latent) molar heat of transition $\Delta'h$, (2.119) becomes

$$\Delta'H^0 = \int_{T_0}^{T_t} \Delta C_p^0 dT' + \nu_j \Delta' h_j + \int_{T_t}^{T} \Delta C_p^0 dT' \quad . \tag{2.124}$$

Note that the heat of transition weighted with the stoichiometric factor is subtracted for reactants and added for products. The summation is interrupted at T_t since due to the change in heat content the coefficients in C_p^0 of the component and, hence, in ΔC_p^0 have different values above T_t. Equation (2.120) is changed correspondingly to

$$\Delta'S^0 = \int_{T_0}^{T_t} \frac{\Delta C_p^0}{T'} dT' + \frac{\nu_j \Delta' h_j}{T_t} + \int_{T_t}^{T} \frac{\Delta C_p^0}{T'} dT' \quad . \tag{2.125}$$

These new relations lead to a considerable increase in complexity of (2.123) particularly when several phase transitions have to be accounted for.

In order to reduce this cumbersome complexity it is customary to tabulate heats and entropies of formation at 298 K, along with C_p coefficients, not only for the real (most stable) standard state but also for the (extrapolated) liquid and/or gaseous state. Based on these adjusted data one can obtain ΔG_T^0 values for phases that are not stable at 298 K by uninterrupted integration from 298 K to a temperature of actual existence of thos phases.

Equation (2.123) might give an impression of a high degree of accuracy for thermochemical calculations that is rather unjustified. In most cases many of the terms can be neglected without exceeding the error that results from the limited accuracy of the experimental data. As drastic a measure as replacing (2.123) by

$$\Delta G_T^0 = \Delta H_{T_0}^0 - T \Delta S_{T_0}^0 \tag{2.126}$$

which, as can be seen from (2.119) and (2.120), implies only that $\Delta C_p^0 = 0$, is often justifiable. However, if phase transitions occur in the considered temperature range, their heats should be accounted for even in such a "crude" approximation since they lead to terms that are typically large as compared to the C_p-terms.

A somewhat less drastic measure for simplifying (2.123) is to assume that the heat capacities of the components are temperature independent. For gases this is often a good approximation since most of the motional degrees of freedom are fully excited at moderate temperatures. But even with condensed phases, particularly at elevated temperatures, it may be justified in *estimates* of reaction equilibria. As one can see from (2.123) this assumption adds only one term proportional to T ln(T) to (2.126) [see (2.127)].

For a comparison of the accuracy of various approximations as applied to some systems, refer to [Ref. 2.11, Chap. 5].

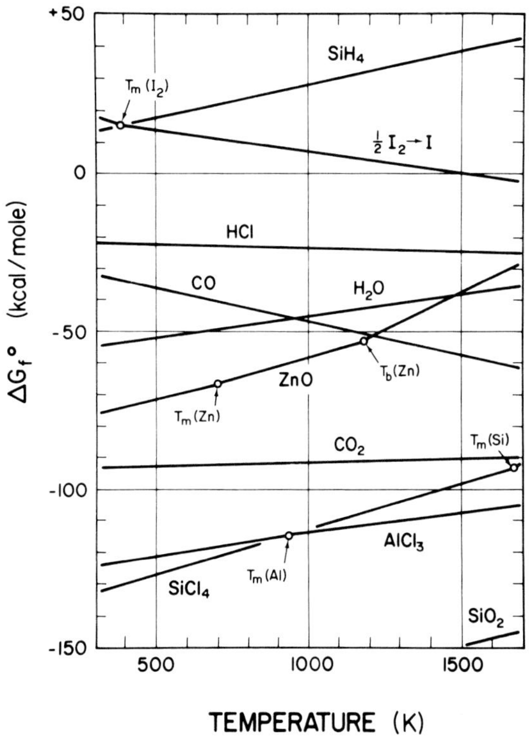

Fig.2.7. Free energy of formation vs. temperature for some compounds relevant to crystal growth. T_m and T_b are melting and boiling points of the elements indicated. Based on data from [2.11 and 13]

In general one observes that plots of ΔG_f^0 versus T, known as Ellingham diagrams [2.12], can be rather well approximated with straight lines between points of phase transitions. [Note that the latent heats of phase transitions of the compounds involved as well as of their elements result in "kinks" in

$\Delta G_f^0(T)$.] This linearity is, for instance, reflected in Fig.2.7. There the standard free energies of formation (i.e., of reaction from the elements), although presented as straight segments, deviate at most 1 kcal/mole from the actual $\Delta G_f^0(T)$ functions. This is negligible as compared to an average uncertainty of about 5 kcal/mole for typical thermochemical data. It is also less than the intrinsic inaccuracy of small graphs that results from the drafting process, linewidth, etc. (see also Problem 2.9).

Upon closer inspection of (2.118), (2.119) and (2.120) the "straight character" of $\Delta G_f^0(T)$ functions becomes less surprising. Changes in C_p with temperature are not very pronounced *per se*. In addition, they lead to changes in S and H that are in first order proportional to T^{-1} and to T, respectively. Hence, changes in heat capacity lead to changes in ΔH^0 and ΔS^0 in such a way as to counterbalance in $\Delta G^0(T)$.

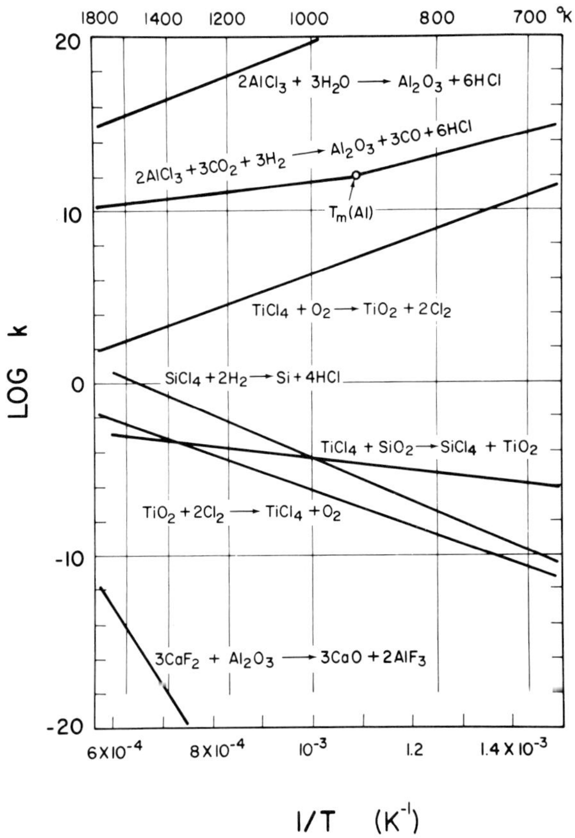

Fig.2.8. Logarithm of reaction constant vs. inverse temperature for some reactions occurring in crystal growth processes. Computed from data in [2.13 and 2.20]

The negligible curvature of $\Delta G_f^0(T)$ functions is also reflected in plots of ln k for reactions versus 1/T that resemble straight lines for many reactions as shown in Fig.2.8 for a few systems (see Problem 2.8).

So far we have based the computation of chemical yields on clearly specified, single chemical reactions. In practice, however, the specific reactions that underly the (partial) transformation of certain reactants into certain products are often not too well known. Frequently there are several reactions of different weight proceeding in parallel. Simplifying statements of *the* underlying reaction, for instance in the chemical vapor transport literature (see [1.2]), should therefore be taken with a grain of salt. Fortunately, there is actually no necessity to selectively invoke (possibly erroneous) chemical reactions as ties between reactants and products. It suffices to specify ΔG_f^0 for "all" molecular species possibly participating in a chemical transformation, without being restricted to a particular set of stoichiometry coefficients. No matter how many "reaction paths" there are and how complexly interwoven they may be, minimization of the *total* Gibbs function constrained by the mass balance conditions will give the correct yield. The enormous data handling and complex computational chores that may arise from this (less biased) approach were obviously difficult to tackle before high speed electronic computers became available. An introduction to computational methods for the numerical solution of complex chemical equilibrium problems is presented in an excellent monograph by VAN ZEGGEREN and STOREY [2.14]. Applications of the "free-energy minimization technique", recently introduced to vapor-solid crystal growth reactions by NOLÄNG and RICHARDSON [2.15] will be discussed in [1.2].

2.12.2 Sources of thermochemical data

Until recently thermochemical data were scattered over a large number of specialized sources. Fortunately this situation has improved. BARIN and coworkers have put forward an extensive survey [2.16, 17] which combines most data from the earlier compilations with more recent results. Vapor pressure data are tabulated based upon the relation

$$\log(p) = aT^{-1} + b\log(T) + cT + d \quad . \tag{2.127}$$

Molar heat capacities are given as coefficients for the approximation

$$C_p = A + B\ 10^{-3}T + C\ 10^{5}T^{-2} + D\ 10^{-6}T^{2} \quad . \tag{2.128}$$

Heats, entropies and free energies of formation are tabulated in increments of 100 K.

The interest in chalcogenides (Group VIA compounds) for solid-state devices is reflected in a recent compilation by MILLS [2.18]. Most valuably, for each data critical statements as to the employed measurement technique and/or theoretical method are made. The compilation contains structure and vapor pressure information together with the usual thermochemical parameters.

The most convenient form of thermochemical data presentation are graphs of free energies of formation versus temperature. Among recent compilations of this type, probably the most versatile for crystal growers has been put forward by REED [2.13]. Free energies of formation for a large number of hydrides, nitrates, oxides, sulfides, selenides, tellurides and halides are given. Furthermore, the collection contains charts on free energies of formation of elemental gases (of elements that are solid at 298 K), free energies of dissociation of diatomic gases and of ionization of some elements. Particular attention should be given to the tabulation and graphs of oxygen affinity of metals and oxides. Also this compilation contains a series of very useful overlay nomographs that facilitate the evaluation of the graphs considerably. Graphs of free energy of formation for many gaseous halides have been presented in a paper by JEFFES [2.12].

Last, but not least, a very useful service should be mentioned that is provided by the National Bureau of Standards. Up-to-date thermochemical data and data references for organic and inorganic compounds can be obtained by phone from Dr. D.D. Wagman. Chemical Thermodynamic Data Center, Physical Chemistry Division, Institute for Basic Standards, NBS, Washington, DC 20234, USA.

When all these sources fail, one is left with the necessity of estimating the missing data. "This task is not as difficult as it might first appear as most of the principles which are applied have been well established for a considerable time." This verbatim quotation is taken from a helpful chapter of KUBASCHEWSKI et al. [2.11] on the estimation of thermochemical data.

For estimates of enthalpy and entropy values for gaseous inorganic compounds one may resort to interpolation methods pointed out by SIRTL [2.19], which are based on some regularities observed within homologous series of compounds.

In concluding one should point out that there is an often confusing and time consuming nonuniformity in the presentation of thermochemical data. Some authors present the data "per mole of X_2", i.e., per mole of gas consumed in the reaction of the elements that led to the compound. X_2 stands here, e.g., for H_2, O_2, Cl_2, etc. Hence, to obtain the proper permole value, for instance, for AsH_3, H_2O, TiO_2, and $TeCl_4$ the shown values have to be multiplied by 3/2, 1/2 (with respect to O_2), 1 and 2, respectively. Other data, however, are presented in units per gram atom of the gaseous element in the compounds. Here, for instance, the data for H_2O can be taken uncorrected, whereas those for TiO_2 and $TeCl_4$ have to be doubled and quadrupled, respectively. Tables of yet different authors are given in units per mole.

Problems

2.1 How much work is performed *on* an ideal gas system that is isothermally and quasistatically expanded from volume V_1 to V_2?

2.2 Consider an ideal gas system, the state of which is to be changed quasistatically from point 1 to point 2 in the PV plane (see the figure below).

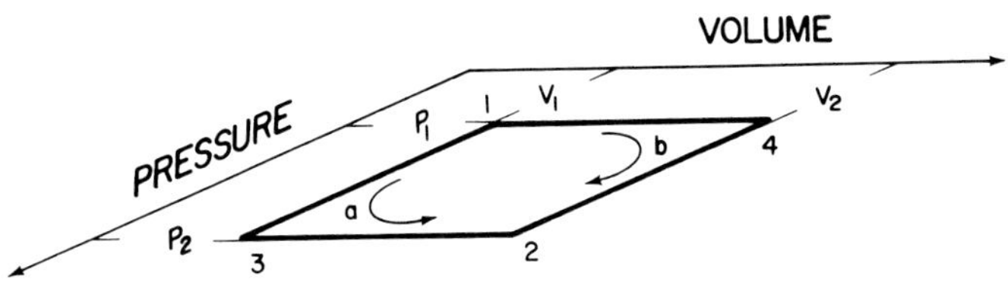

Two paths, a and b, are chosen. They differ in the sequence of isobaric and isochoric subprocesses. Calculate the work ΔW and ΔQ exchanged between 1 and 2. Show that ΔW and ΔQ depend on the path, but that ΔU between 1 and 2 is path independent. Hint: The molar heat capacity for an ideal gas in isochoric processes is $C_V = 3/2R$ and in isobaric processes is $C_P = C_V + R$.

2.3 Derive (2.23), the maximum efficiency of a heat engine (see figure on next page), without any specific assumptions about the type of under-

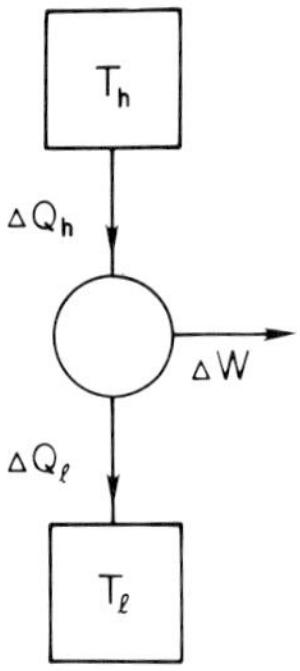

lying cycle and working substance. The discussion should be merely based upon the First and Second Laws of Thermodynamics. Hint: Consider the high and low temperature reservoirs together with the heat engine as an adiabatic system.

2.4 Two bodies of same composition and mass, initially at the temperatures $T + \Delta T$ and $T - \Delta T$, respectively, are brought into thermal contact. Show that the overall entropy change, that occurs during thermal equilibration between the bodies, is positive. Assume the specific heat of the material to be temperature independent.

2.5 (a) Show that upon isothermal, isobaric mixing of two ideal gases A and B the entropy of the system increases by $\Delta S = -X_A R \ln(X_A) - X_B R \ln(X_B)$ per mole of A and B. The X_i's are the mole fractions. For instance $X_A = n_A/(n_A + n_B)$.

(b) Show, without assuming a specific equation of state, merely on statistical grounds, that the above relation for the configuration entropy of mixing applies to any molecular solution between two components, A and B.

2.6 Derive (2.107) under the assumptions given in the text, and show that at equilibrium (i.e., for the minimum of the function in Fig.2.4)

$$\frac{n_a}{2(1 - n_a)} = \exp[(2\mu_c^0 - \mu_a^0 - \mu_b^0)/2RT] \quad .$$

2.7 Derive (2.123), i.e., a relation for the change in the standard state Gibbs potential ΔG_T^0 of a reaction at T in terms of its standard state enthalpy and entropy at T_0 and the coefficients a, b, c, and d of a polynomial approximation for $C_p^0(T)$, (2.121), as indicated in the text.

2.8 One method of producing epitaxial layers of silicon employs the reduction of silicon tetrachloride with hydrogen in the form

$$SiCl_4(g) + 2H_2(g) \rightleftarrows Si(s) + 4HCl(g) \quad .$$

Verify the plot of ln(k) vs 1/T for this reaction in Fig.2.8 from the data given in Fig.2.7. What temperature range and relative concentrations of $SiCl_4$ and H_2 would you recommend for the conduction of this process in an open-tube flow-through system, i.e., $P \approx 1$ atm. Give a detailed discussion of your choice.

2.9 Estimate the uncertainty in the equilibrium constant of a reaction at 1000 K that results when its Gibbs free energy change is only known to within ± 5 kcal/mole. Hence what uncertainty in the concentrations of the products does one have to face for a simple reaction of your choice?

2.10 Consider the gaseous reaction A + 2B = C which will be supposed to have a very unfavorable value of ΔG_T^0, say 10 kcal/mole at 500 K. Assume that the reactant mixture contains originally one mole of A and two moles of B. Calculate the yield of the reaction for a total pressure (at equilibrium) of 24, 82, and 316 atmospheres, respectively. Note that the number of molecules decreases during the above reaction. Hence, qualitatively, what pressure dependence would you expect for a gaseous reaction in which the number of molecules increases? Answer: The resulting number of moles of C is 0.01, 0.1 and 0.5, respectively. [2.10].

2.11 At elevated temperatures, gaseous iodine dissociates to atoms to such an extent that observed pressures P deviate measurably from expected pressures P_0. If the total pressure difference is due to dissociation, calculate the equilibrium constant k_p for the process $I_2(g) \rightleftarrows 2I(g)$ at 800°C if $P_0 = 0.3153$ atm and if $P = 0.3429$ atm.
Answer: $k_p = 1.060 \times 10^{-2}$ atm [2.21].

2.12 At what pressure (in torrs) is only 1.0% of I_2 not dissociated to atoms at 800°C if $k_p = 1.060 \times 10^{-2}$ atm for the reaction $I_2(g) \rightleftarrows 2I(g)$?
Answer: $P = 4 \times 10^{-2}$ torr.

3. Phase Equilibria and Phase Diagrams

Crystal growth can be called, in a wide sense, the science and technology of controlling phase transitions that lead to a (single crystalline) solid. Hence thorough knowledge of the thermodynamics of phase transitions and skill in deducing from phase diagrams the consequences for crystal growth are among the keys to efficient work in that area. In this chapter we will familiarize ourselves with the thermodynamic basis of phase diagrams. The discussion will first center upon liquid-solid equilibria and later expand the concepts to vapor-condensed phase equilibria. This general presentation is then illustrated in Chap. 4 with a broad discussion of specific systems and the consequences of their phase equilibria for the selection of specific crystal growth techniques. Large parts of this chapter are structured after a monograph on principles of phase diagrams by GORDON [3.1]. The presentation has also benefited from recent reviews by YEH [3.2] and KRÖGER [3.3]. For supplementary reading a more advanced presentation by PELTON and THOMPSON [3.33] is particularly recommended.

The basis for a treatment of any phase change is contained in the general condition for mass transfer equilibrium (see Sect. 2.10.2)

$$\mu_i^{(1)} = \mu_i^{(2)} \quad . \tag{3.1}$$

Two phases can only coexist stably if the chemical potential of each of their components is equal in each phase. This points to the "central trick" in all crystal growth processes:

The state variables of a system have to be manipulated such that the chemical potential of the desired solid phase becomes lower than that of all other phases present in a system. The Gibbs-Duhem equation, (see Sect. 2.9)

$$d\mu = -sdT - xdY = -sdT + vdP - \underline{m}\cdot d\underline{H} - \underline{p}\cdot d\underline{F} - \dots \tag{3.2}$$

tells us that the required lowering of the chemical potential can be achieved *in principle* by various means.

The most common way to lower the chemical potential of the solid below that of a nutrient phase is by lowering of the temperature. For instance, for a monocomponent liquid-solid system we can, according to $d\mu/_{Y} = -sdT$ with $s_l > s_s$, schematically plot the chemical potential for liquid and solid phase around the equilibrium temperature T_m as in Fig.3.1. Withdrawal of heat from the liquid ("cooling") at the "melting point" T_m will drive growth of the solid at the expense of the liquid. Of course, this statement will have to be refined. In the discussions of nucleation and interfacial growth kinetics [1.1] we will find that it takes a certain finite ΔT (supercooling) to stimulate the transition and to make the solid grow with a finite rate. This, however, will be ignored in the current thermodynamic consideration.

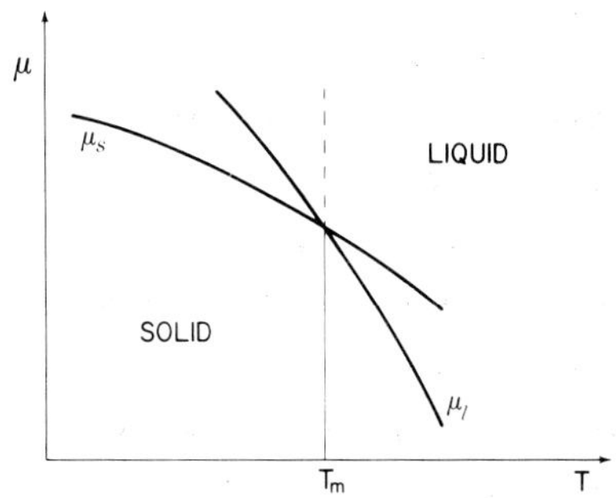

Fig.3.1. Temperature dependence of the chemical potentials of a solid and its liquid near the melting point

Change of pressure is another means of stimulating solidifications. Depending on the sign of the density change upon freezing, i.e., whether $v_l > v_s$ (which is considered "normal") or $v_l < v_s$, (as in H_2O, Bi, Si, Ge, Hg, and III-V compounds), a pressure increase or decrease, respectively, is required. Due to the typically small density difference between condensed phases, changing the pressure is a rather inefficient, although sometimes necessary, way of achieving solidification. Note that the above discussion does not include a pressure dependence of v yet and is simply based upon a difference in v_l and v_s at "normal" pressures.

In principle, we could also expose the solid-liquid system to an electric or magnetic field to drive a phase transition. If, for example, the polarizability of the solid is larger than that of its liquid, we will stimulate growth of the solid phase. In practice, however, the change in chemical potential that can be achieved with reasonable fields is typically too small to make this an attractive means of controlling bulk solidification. Yet a similar phenomenon, the Peltier effect, though restricted to the interface, has been advantageously used as means of controlling solidification in semiconductor production and basic melt growth studies, as we will see in Sect. 6.5.

In multicomponent systems we will encounter other means of stimulating phase transitions. Though the formulation will be more complex, the key criterion for equilibrium among two phases for a specific component will remain the condition of equal chemical potential. Spontaneous transfer of matter will occur as long as this equality is not established. Note that "spontaneous" is here used in the sense of "by itself" and does not imply any rate considerations, as these are beyond the scope of "thermostatics".

For a one-phase system we found in Sect. 2.3 that an equation of state related three state variables. Thus, for the characterization of an equilibrium state we had the choice of two independent variables (P and T, for instance) which then determined the value of the third, dependent variable (V for this case). In the above examples for two-phase monocomponent systems, however, we have varied only one state variable independently of the others and the result was that we perturbed equilibrium. From the Gibbs-Duhem equation (3.2), one can see that in order to maintain equilibrium between two phases upon change of one variable, one must change an additional state variable in a very specific, i.e., dependent way. This insight will be formulated in the next two sections in two important relations: The Clausius-Clapeyron equation and the Gibbs phase rule.

3.1 Clausius-Clapeyron Relation

During a transition to another equilibrium state two phases remain in equilibrium only if for each component

$$d\mu^{(1)} = d\mu^{(2)} \quad . \tag{3.3}$$

Keeping all the other intensive variables constant we can thus write for small variations in P and T

$$\left(\frac{\partial\mu^{(1)}}{\partial T}\right)_P dT + \left(\frac{\partial\mu^{(1)}}{\partial P}\right)_T dP = \left(\frac{\partial\mu^{(2)}}{\partial T}\right)_P dT + \left(\frac{\partial\mu^{(2)}}{\partial P}\right)_T dP \quad . \tag{3.4}$$

From the Gibbs-Duhem equation (3.2), we get for $d\mu \approx 0$ (i.e., close to an equilibrium state)

$$(s^{(2)} - s^{(1)})dT = (v^{(2)} - v^{(1)})dP \quad . \tag{3.5}$$

Substituting (3.5) into (3.4) we obtain then the *Clausius-Clapeyron* equation

$$\frac{dP}{dT} = \frac{\Delta s}{\Delta v} \tag{3.6}$$

with $\Delta s = s^{(2)} - s^{(1)}$ and $\Delta v = v^{(2)} - v^{(1)}$. Equation (3.6) determines the pressure change dP, which is required to maintain phase equilibrium when there is a temperature change dT, or vice versa. Similarly, one can relate other two terms of the chemical potential [see (3.2)]and thus, for instance, determine the change in magnetic field required upon a change in pressure, etc. In this formulation (3.6) holds only for monocomponent systems since it was assumed that the chemical potentials depend on T and P only, and not on the presence of other components.

The above derivation of the Clausius-Clapeyron relation was chosen for a clearer presentation of its physical basis. Yet (3.6) can also be shown to directly result from the Maxwell relation (2.67) that was obtained from partial differentials of the Helmholtz function [Ref. 3.10, p. 55].

In the following we will discuss several implications of the Clausius-Clapeyron equation. The difference in entropy, Δs, between a mole of phase 1 and a mole of phase 2 is the latent heat L connected with a transition, divided by the transition temperature T_{tr}. Hence, we can rewrite (3.6) to

$$\frac{dT_{tr}}{dP} = \frac{T_{tr}}{L}(v^{(2)} - v^{(1)}) \quad , \tag{3.7}$$

which relates the pressure change in (or on) a system with the consequent change in transition (equilibrium) temperature.

For equilibria between vapors and their condensed phases, where typically $v_l \approx v_s \approx 10^{-3}v_v$, one can expect a significant pressure dependence of the equilibrium temperature from (3.7). With $v_v >> v_s \approx v_l$ one can set, to a good approximation, for $\Delta v = v^{(2)} - v^{(1)} = RT/P - v^{(1)} \approx RT/P$, and (3.7) becomes

$$\frac{dP}{dT} = \frac{PL_v}{RT^2} \quad . \tag{3.8}$$

This can be readily integrated to obtain the vapor pressure of a substance as function of temperature if L_v the (latent) heat of evaporation or sublimation is known as a function of temperature. For instance, of one assumes L_v to be pressure and temperature independent, one obtains from (3.8)

$$\ln(P) = -\frac{L_v}{RT} + \text{const} \quad . \tag{3.9}$$

As a comparison with the vapor pressure data represented in Fig.3.2 shows, constancy of L_v over relatively large changes in P and T is indeed a good approximation for many materials (see also Problem 3.1).

For liquid-solid systems with $v_l \approx v_s$ and $L_m \approx (1/10 \rightarrow 1/50)L_v$, one will expect that the melting point depends much less on the pressure. In addition, since there are solids with higher and some with lower density than their melts, the sign of the change in transition temperature depends on the specific system. For instance, ice tends to melt under increased pressure, whereas "normal" materials do not. This effect is at least partly responsible for the low friction encountered in ice skating; for a quantitative estimate see Problem 3.2.

The significance of the pressure dependence of transition temperatures is often underestimated in material preparation. Since it takes in general several thousand atmospheres to change the melting point of a material by a few percent, one is tempted to ignore such shifts (and their consequences) that may arise under normal experimental conditions. This may be justified, for instance, for the solidification of monocomponent systems, i.e., for very pure materials. But with multicomponent systems, e.g., growth of "doped" crystals, the situation is quite different. There, as will be discussed in Sect. 6.5, very stable growth rates are required to obtain homogeneous composition in the growing solid. The growth rate, however, depends on the difference between transition temperature and actual temperature of the interface [1.1]. Hence, changes in either or both causes compositional changes. Recently rather modest pressure variations that occurred during solidification of tin-doped InSb have been identified as the origin of composition variations in this material [3.5]. InSb expands 14% upon solidification. Its melting point changes by -0.008°C/atm.

Another recent illustration of (3.7) in connection with solid-liquid phase transitions is the "overheating" of ice [3.6]. Ice, when exposed to high intensity optical radiation, begins to melt internally at points where,

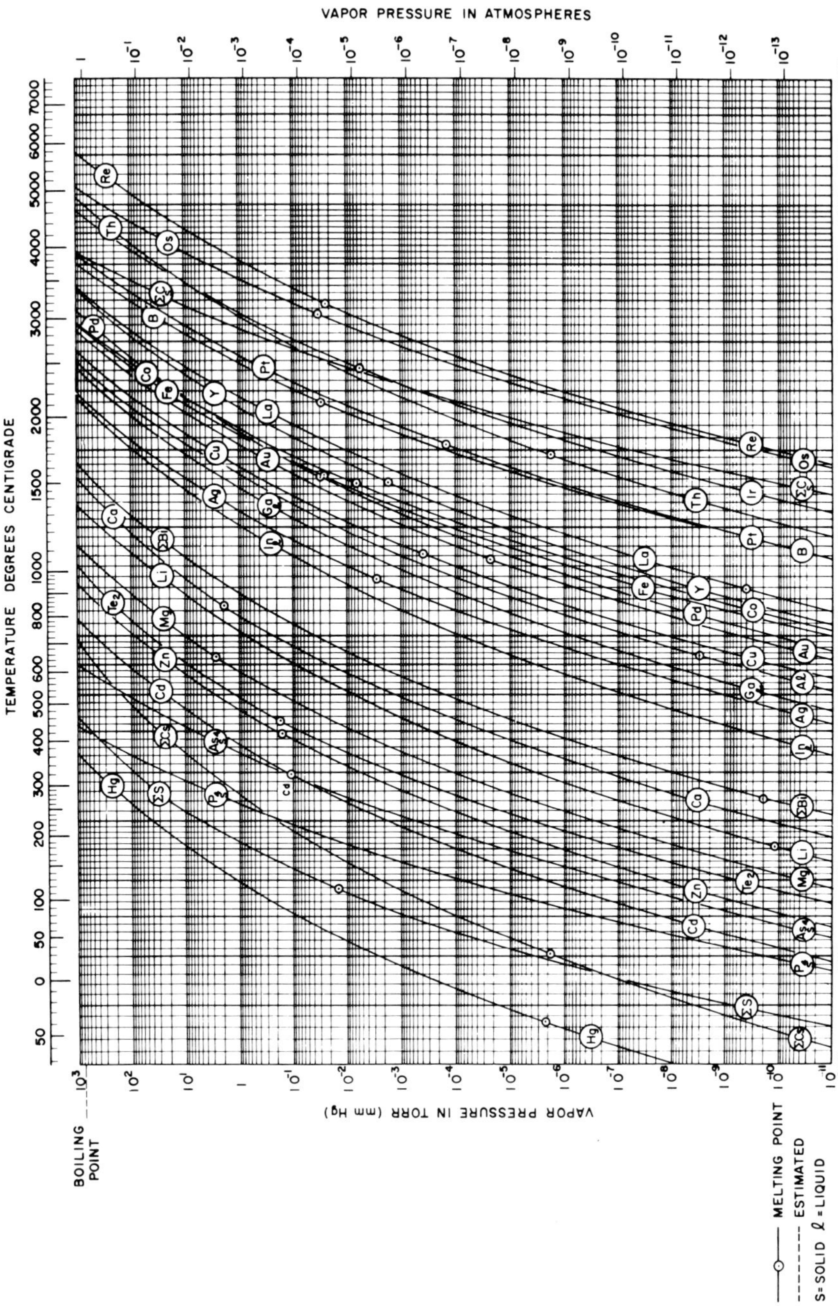

Fig.3.2. Vapor pressure data. From [3.4] by permission of Wiley and Sons

due to occlusion of solid particles, or similar "centers", more energy is absorbed than in the rest of the crystal. If this local melting process is conducted such that no vapor bubbles form, the negative pressure produced by melting the ice internally can raise the melting point by several degrees.

The Clausius-Clapeyron equation can also be used (after slight transformation, see [Ref. 3.7, pp. 122-124]) to calculate the change in latent heat with pressure. And in turn, if one knows L as a function of P, then one can compute the difference in specific heat of the two phases, etc.

3.2 Gibbs Phase Rule

So far we have only considered two-phase systems. In the following we will expand the equilibrium criterion to more complex systems and determine how many independent variables are required for the characterization of equilibrium states of such complex systems.

Suppose a system formed by C components is distributed between P homogenous phases. By extrapolation, from n to n + 1, one can easily devise the equilibrium condition for this system to be

$$\begin{aligned}
T^{(1)} &= T^{(2)} = \dots = T^{(P)} \\
p^{(1)} &= p^{(2)} = \dots = p^{(P)} \\
\mu_1^{(1)} &= \mu_1^{(2)} = \dots = \mu_1^{(P)} \\
&\vdots \\
\mu_i^{(1)} &= \mu_i^{(2)} = \dots = \mu_i^{(P)} \\
&\vdots \\
\mu_C^{(1)} &= \mu_C^{(2)} = \dots = \mu_C^{(P)}
\end{aligned} \tag{3.10}$$

(Note that superscripts designate the phases and subscripts stand for the components) Put into words, (3.10) reads as follows: The temperatures of all phases of a system must be alike at equilibrium. When there are no surface effects, membranes, shear stresses, etc., the pressure must be the same in all phases at equilibrium. If each phase is open to a component so that it may enter or leave, the chemical potential of that component must be the same in all phases at equilibrium.

The overall composition of the above system is completely defined by C-1 concentration terms since the concentration of the last component is equal to the remainder. To define the composition of all P phases, we need, therefore, P(C-1) concentration terms or independent concentration variables. Hence, together with temperature and pressure, in the absence of other intensive variables, [P(C-1) + 2] parameters are required to characterize a system at any moment, for instance, as long as there is net mass transfer among phases. In equilibrium, however, the conditions (3.10) fix automatically (P-1)C variables, leaving P(C-1) + 2 - (P-1)C - P + 2 undetermined. Consequently, we can say that in equilibrium there are only

$$f = C - P + 2 \tag{3.11}$$

variables to be specified to uniquely define the state of a system; f standing here for "degrees of freedom", i.e., the number of intensive state variables that can be independently varied without changing the number of phases.

For example, in a one-component two-phase system we obtain f = 1, in agreement with the earlier finding that, e.g., either the pressure or temperature are enough to fix the state of such a system. For a one-component three-phase system we find f = 0, which means that at the "triple point" the state of a system is uniquely specified. No state variable can be changed without removing the system from its triple state.

The "number of components" is not always as easy to decide upon as in the above simple examples. Chemical reactions can complicate the choice of C considerably. MOORE [3.8] has given a general definition that can hardly be improved in its clarity: "The number of components is equal to the total number of independent chemical constituents in the system minus the number of distinct chemical reactions that can occur in the system between these constituents. By number of independent constituents we mean total number minus the number of any restrictive conditions, such as material balance or charge neutrality. By a distinct chemical reaction we mean one that cannot be written simply as a sequence of other reactions in the system."

Consider, for example, the system consisting of calcium carbonate, calcium oxide and carbon dioxide. There are three distinct chemical constituents, $CaCO_3$, CaO and CO_2. One reaction occurs between them: $CaCO_3 \rightarrow CaO + CO_2$. Hence the number of components C = 3 - 1 = 2. A more complex example is the system discussed in Problem 3.3.

Reactions that are included in calculating the number of components must be actually occurring with measurable rates in the system. For instance, a mixture of water vapor, hydrogen and oxygen must be considered as a three-component system as long as the reaction $H_2 + O_2/2 \rightarrow H_2O$, due to absence of a suitable catalyst, is hindered.

At this point we have at our disposal all basic thermodynamic concepts that are essential for an efficient discussion of phase diagrams.

3.3 Monocomponent Systems

A pure substance represents by definition a monocomponent system. Suppose that the state of such a system is determined by P and T only. Each phase can then be represented by a μ-surface in the μ-P-T space. A simple substance which occurs in the three phases, "solid, liquid and gaseous", only is thus described by three μ-surfaces. Let us schematically construct two of these surfaces, say the ones for the solid and vapor phase, irrespective of whether the phases are observable at all P's and T's or not. The general shape of these surfaces can be easily deduced from the Gibbs-Duhem equation, (3.2), and is schematically shown in Fig.3.3. Since $v_V > v_S$ and $s_V > s_S$, the "vapor surface" has higher μ values at low P and T than the "solid surface" and has, in general, a steeper inclination with respect to the P-T plane.

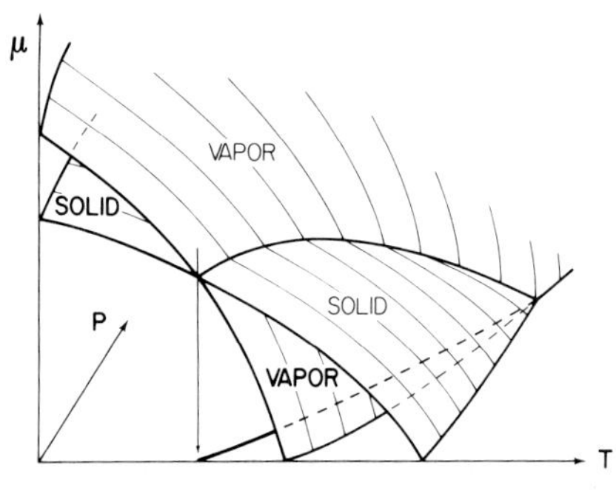

Fig.3.3 Intersecting chemical potential surfaces for a solid and its vapor in the μ-P-T space (schematic)

Any two of the three possible surfaces intersect along a *coexistence line*. In Fig.3.3 it is the sublimation curve. Along each line the two neighboring phases have the same chemical potential. Each coexistence line specifies the pressure and temperature combinations at which the two phases are in equilibrium. An important feature of equilibria between macroscopic phases is that they are independent of the actual amounts of the bulk phases.

At points removed from the coexistence lines one of the μ-surfaces lies lower than the others, as can readily be seen from an isobaric cut through the μ-P-T space in Fig.3.4a. The $\mu(T)$ functions have been drawn beyond the intersections. This is to indicate that a component may not be necessarily in its lowest μ configuration nearby the coexistence lines and may be in a *metastable* state instead. Vapors, for instance, may stay uncondensed at pressures equal to several times their vapor pressure at a given temperature as in the

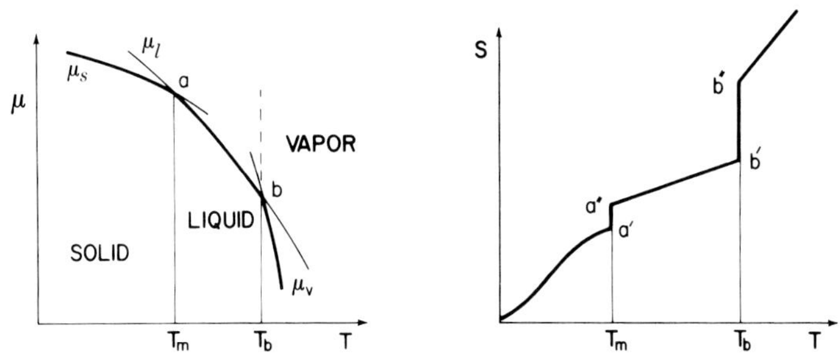

(a) (b)

Fig.3.4. (a) Traces of the μ-surfaces, resulting from an isobaric cut through the μ-P-T space
(b) Corresponding S-T dependence

Wilson cloud chamber. Liquids can also be superheated as in a bubble chamber particle detector. In the absence of a corresponding condensed phase, both vapors and liquids can be considerably *supercooled*. Whether superheating of solids exists has been questioned for a long time. However, most recent experimantal data obtained on dislocation-free Ga and some earlier observations [3.9] seem to establish such a metastability, i.e., a kinetic barrier for melting of certain solid material. Hence, the statement that only the phase with the lowest μ-value at a given P-T combination will exist has to be taken as a mere equilibrium argument. Whether a transition from a higher lying μ-surface to a lower one occurs in reality with a perceptible rate depends on the *kinetic* conditions during nucleation and growth, respectively [1.1]. Note that the mere passage of atomic particles through metastable phases, e.g., through vapor in the cloud chamber and liquid in the bubble chamber, may suffice to initiate phase transitions.

Figure 3.4b shows schematically the temperature dependence of the entropy of a monocomponent system as it is brought reversibly over the three stable parts of the μ-surfaces, such as depicted in Fig.3.4a. We see that as P is kept constant, S increases steadily with T until a phase change occurs. There S takes a sudden jump of the amount L/T_{tr} and continues steadily again in the new phase. Note that the heat capacity $C_P = T\partial s/\partial T_P$ becomes infinite at T_{tr}. Phase changes that are connected with such discontinuous changes in volume, entropy, magnetization, etc. are called *first-order* phase changes.

There are, however, other types of phase changes, such as the initiation of a different kind of ordering in a lattice or the appearance of superfluid helium II. Some of these changes involve only a change in slope of S rather than a discontinuity in its value. The heat capacity changes discontinuously here but does not become infinite. Changes of this type are called phase changes of second order. For a more detailed discussion of phase transitions of types higher than first order, which goes beyond the scope of this course, see, for example, [Ref. 3.10, pp. 136] and [3.11].

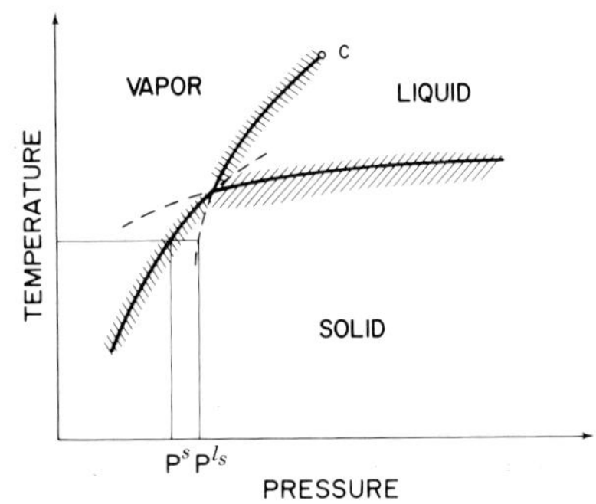

Fig.3.5. Schematic T-P phase diagram for a monocomponent material that expands upon melting ("normal material")

Projection of the coexistence curves into the P-T plane yields the two-dimensional P-T phase diagram of a monocomponent system (Fig.3.5). Sublimation pressure, vapor pressure and melting curves intersect at the *triple point*. Assuming no further phase transitions exist, the melting curve steadily proceeds, theoretically, to infinite values of P and T. The vapor pressure curve ends at the *critical point* C where liquid and vapor become indistinguishable. This is a consequence of the higher compressibility of gases. At high enough pressures v_v approaches v_l and the gas becomes structurally so similar to the liquid that $s_v \to s_l$ and consequently *at* C, $\Delta\mu$ and the latent heat, L_v, become zero.

For normal substances the slope of the melting curve is positive, as can be deduced from the Clausius-Clapeyron equation. Hence "anormal" substances like water and III-V compounds, that expand upon solidification, have a (slightly) negatively inclined melting point curve.

The shaded areas in Fig.3.5, that extend the phase stability ranges beyond the coexistence lines, are to indicate the metastable states. Note also that the coexistence lines in Fig.3.5 have been continued (dashed) beyond the triple point. This is to account for the fact that two coexisting phases can be brought to a certain extent into these unstable configurations where, for instance, a *supercooled* liquid coexists with its vapor. As shown by the phase diagram, the supercooled liquid has then a vapor pressure P^{ls} which is higher than the vapor pressure P^s of its incipient solid at the same temperature. Hail formation may be based on this phenomenon. Ice crystals falling through a cloud which contains supercooled water droplets tend to grow rapidly from the vapor phase.

The data of a P-T phase diagram apply strictly speaking only to the corresponding monocomponent equilibria. In reality, however, one encounters mostly systems where at least one additional *inert* gas component is present. To what extent can monocomponent phase equilibria data still be used meaningfully in such situtations? Let us illustrate this with the case of pure H_2O versus H_2O in air. The triple point data for H_2O are 0.01°C and 4.58 mmHg. If one adds an inert gas to the three coexisting water phases, then according to the phase rule (3.11), $f = C - P + 2 = 2 - 3 + 2 = 1$, the triple point has lost its uniqueness. Hence solid, liquid and gaseous H_2O can in principle coexist at any other temperature as long as the (dependent) pressure is chosen appropriately, and vice versa. For instance, at 1 atmosphere solid and liquid water coexist at 0.00°C with water vapor of 4.58 mm Hg. One sees that, although the pressure in the phase diagram represents, strictly speaking, the vapor pressure of the pure material, the data are still rather accurate in the presence of inert gases of considerably higher pressures. This, as discussed earlier in connection with the pressure dependence of melting points, is mostly because of the low compressibility of condensed phases. The emphasis on "inert" above is to stress the fact that interactive gases may influence phase equilibria drastically. For instance, HCl, which dissolves readily in H_2O, lowers the freezing point of the aqueous solution by 0.5°C at a concentration as low as 0.5 wt. %.

At high inert gas pressures the resulting vapor pressure change of liquids can not be neglected anymore. Assume that an external hydrostatic pressure

P_e is imposed on a liquid of molar volume $v^{(l)}$. An estimate of the vapor pressure change can be based upon the equilibrium condition $d\mu^{(v)} = d\mu^{(l)}$, see (3.3). Hence, for an isothermal pressurization

$$v^{(v)}dP = v^{(l)}dP_e \tag{3.12}$$

where $v^{(v)}$ and P are the molar volume of the vapor and (changed) partial pressure of the vapor, respectively. If the vapor behaves like an ideal gas, $v^{(v)} = RT/P$, and (3.12) becomes

$$\frac{dP}{P} = \frac{v^{(l)}}{RT}\, dP_e \quad . \tag{3.13}$$

Since liquids have low compressibilities, we assume constant $v^{(l)}$. Equation (3.13) can then be integrated to

$$\ln \frac{P_1}{P_2} = \frac{v^{(l)}(P_{e1} - P_{e2})}{RT} \quad . \tag{3.14}$$

Applying (3.14), for instance, to mercury at 100°C one estimates that the vapor pressure of 0.273 mm is raised by a factor of 1.6 through pressurization of the liquid from 1 atm to 1000 atm (Problem 3.4).

In a pure material only one gaseous and/or liquid phase can exist since – loosely speaking – with indistinguishalbe particles only one random arrangement is possible. If we add, as in a solid, symmetry as an element of distinction, then one expects that one component may occur in symmetric crystalline arrangements of many types. The various possible crystalline forms of a given pure substance are called allotropes of polymorphs. Polymorphic transitions can occur as a result of temperature and pressure changes. The stability criterion for a certain phase follows again from the Gibbs-Duhem equation and is illustrated in Fig.3.6. Of two possible phases at constant pressure, the one with the higher entropy will be the stable one at higher temperatures (assuming that the transition temperature, T_{tr}, is below the melting point, T_m, of the material, of course). Changes of pressure as compared to temperature changes have only small effects on the chemical potential of solids. Since polymorphs have only small differences in molar volume, excessive pressures are required to obtain any sizeable $\Delta\mu$ and, hence, to drive a solid-solid phase transition.

One of the most important examples of polymorphism is the γ-iron to α-iron transition which plays an essential role in steel hardening. One of the "most expensive" polymorphisms is that of carbon shown in Fig.3.7. One sees that heating of carbon at atmospheric pressure results in melting at around 4200 K, whereas pressurization to sufficiently high pressures leads to diamond. (Remember that 1 kilobar = 987 atm.) The phase diagram seems to indicate that graphite can be converted to diamond at room temperature and relatively low pressures. This expectation, however, ignores the actual kinetic factors. There is a significant *activation barrier* [1.1] impeding the spontaneous transformation of graphite to diamond, and (fortuantely) vice versa. Temperatures around 1500°C and pressures of 50-60 kbar are required for diamond synthesis from graphite even when performed in molten metal fluxes

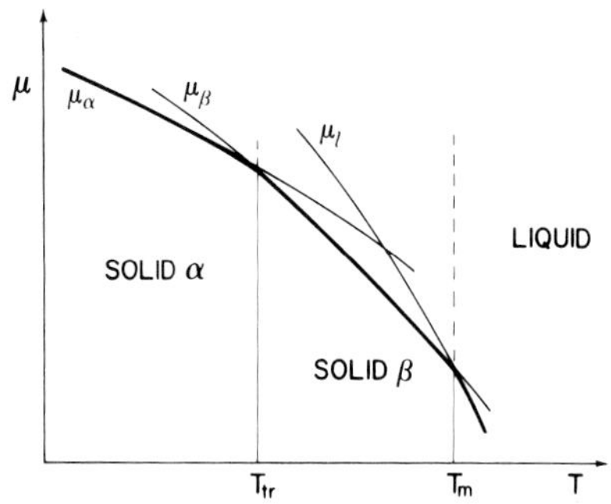

Fig.3.6. Stability range of two solid polymorphs in a μ-T plot

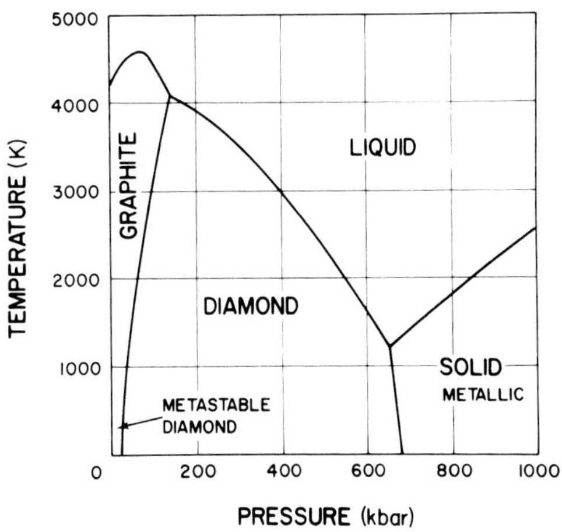

Fig.3.7. T-P phase diagram for carbon. After [3.1]

(solution growth [1.2]) which enhance the transformation greatly*. Flawless diamonds of up to 5 mm size (equivalent to 1 carat or $ 3,500) have been grown at General Electric Company [3.12]. The importance of kinetic factors is even more emphasized by the fact that diamond has recently been grown from carbonaceous vapors onto diamond seeds at around 1050° C and pressures as low as 10^{-4} atm [3.13-16]. This is clearly in a range where diamond is thermodynamically unstable with respect to graphite and thus the chemical potential of graphite is lower than that of diamond. Hence, during growth, when the atmosphere about the diamond seeds is supersaturated with respect to diamond, then it will be even more supersaturated with respect to graphite. Thus there is also a trend for graphite formation together with the diamond growth. Details of the synthesis of diamond which forms a particularly instructive example for various kinetic principles in flux and vapor crystal growth will accompany us throughout various sections of this text.

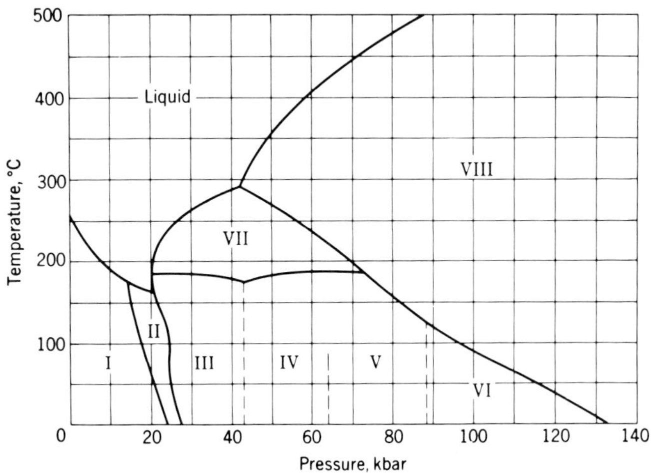

Fig.3.8. Equilibrium phase diagram for bismuth. From [3.1]

One of the most complex examples for polymorphism is that of bismuth, where stability ranges for eight different solid phases have been found, as shown in Fig.3.8. Note that the slope of the melting curve of Bi at low pressures is negative, in contrast to carbon. As already discussed, the explana-

---

* "A diamond is a piece of coal that stuck to the job". ("Today's Chuckle" in Salt Lake Tribune of 10-16-73).

tion for this result can be based upon the Clausius-Clapeyron equation and the "anormal" density change upon melting.

3.4 Binary Systems

3.4.1 Definitions and general remarks

The components we will be dealing with in two-component and multi-component systems are understood not to combine chemically, but to "mix" with one another. Mixing can occur in three different forms:

1) *Mechanical mixtures* are intimate associations of the components in a state of subdivision large enough so that the fraction of atoms in the interface is negligible compared to the atoms within the phases. This requirement puts a lower limit on the allowed particle size in a mechanical mixture. Since the effective thickness of interphase boundaries is a few atomic diameters, particles must not be smaller than a few hundred atomic dimensions.
2) *Solutions* are mixtures of the components on an atomic scale.
3) *Mechanical mixtures of solutions* have particles of the above minimum dimensions that consists of solutions.

In order to decide which of these mixing forms is the thermodynamically preferred and stable one for a given overall composition at certain values of P and T, one must decide which one has the lowest Gibbs function. Binary phase diagrams, as we will see, contain this information for all possible combinations of two substances at some given P.

In principle, binary phase diagrams could be presented in the form of coexistence surfaces in a C-T-P space, where the relative concentration C ranges from 0 to 1 for each component. But, even for the simplest case, where both pure components have a simple phase diagram, such as Fig.3.4, and are isomorphic (same crystal structure) as well as mutually soluble in all proportions in all phases, one obtains a complex three-dimensional representation as indicated in Fig.3.9. Even that figure is greatly oversimplified, as we will soon see. Actualy there are two surfaces required per two coexisting phases since, in general, their compositions are not the same. This complexity (but also the information contained) is grossly reduced by presenting isobaric sections (typically at 1 atm) through the three-dimensional diagram. Most presentations show only the solid-liquid range since this contains the transitions which were traditionally of most common interest. We will first follow this customary approach. In modern materials preparation, however, vapor techniques have been gaining considerable importance. Hence in Sect. 3.4.7 on vapor-liquid-solid equilibria we will expand the discussion to three-dimensional (C-T-P) phase diagrams.

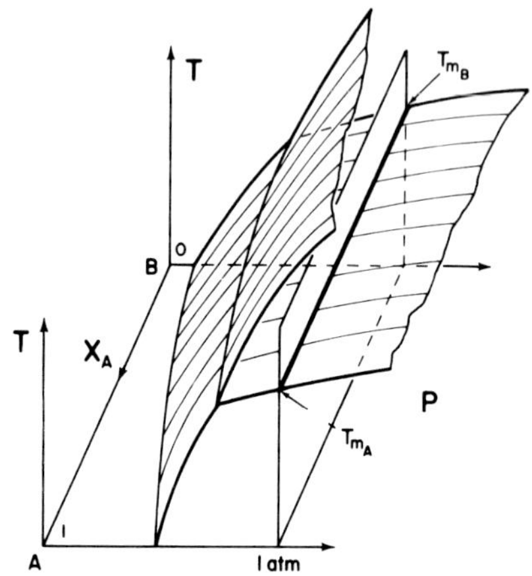

Fig.3.9. Obersimplified three dimensional phase diagram for a binary system

The change in Gibbs function for a phase α is, according to (2.69)

$$dG^{(\alpha)} = -S^{(\alpha)}dT + V^{(\alpha)}dP + \sum_i \mu_i^{(\alpha)} dn_i^{(\alpha)} \tag{3.15}$$

where the summation is over all components i which are present in α. We see that for a two-component system $G^{(\alpha)} = G(P, T, n_A, n_B)$. How many of these four variables are independent? From the phase rule, f = C - P + 2, we know that this depends on the number of phases present.

In a *single-phase binary* system f = 3. Three independent variables characterize here an equilibrium state. This is because in a closed single-phase binary system the specification of one concentration also fixes the other one. Expressing the concentrations in terms of the mole fractions,

$$X_A^{(\alpha)} = \frac{n_A^{(\alpha)}}{n_A^{(\alpha)} + n_B^{(\alpha)}} \quad \text{and} \quad X_B^{(\alpha)} = \frac{n_B^{(\alpha)}}{n_A^{(\alpha)} + n_B^{(\alpha)}} \quad , \tag{3.16}$$

the relation between the two concentrations is

$$X_A^{(\alpha)} + X_B^{(\alpha)} = 1 \quad . \tag{3.17}$$

The molar Gibbs function of phase α can thus be written in the form

$$g^{(\alpha)} = X_A^{(\alpha)}\mu_A^{(\alpha)}(P, T, X_A) + (1 - X_A^{(\alpha)})\mu_B^{(\alpha)}(P, T, X_A) \quad . \tag{3.18}$$

Hence, the three independent variables are P, T, and X_A.

In a *two-phase binary* system f = 2. This reduction comes from the condition that in equilibrium the chemical potential of each individual component must be equal in both phases, i.e.,

$$\mu_A^{(1)}(P, T, X_A^{(1)}) = \mu_A^{(2)}(P, T, X_A^{(2)}) \quad \text{and}$$

$$\mu_B^{(1)}(P, T, X_B^{(1)}) = \mu_B^{(2)}(P, T, X_B^{(2)})$$

so that only two independent variables, either P and T or P and X_A or T and X_A are left.

The molar Gibbs function for a certain phase of a binary *mechanical mixture*, in the form of (2.70) and in terms of mole fractions, is

$$g^m = X_A h_A + X_B h_B - T(X_A s_A + X_B s_B) = X_A \mu_A^* + X_B \mu_B^* \tag{3.19}$$

where the h_i and s_i are molar entities, i.e., not influenced by the (remote) presence of the other component. Hence, the Gibbs function is simply additive with respect to pure components. (Note that phases have no chemical potential, only components do).

In *solutions* the Gibbs function is not additive. In pure A all neighboring atoms are also A. In solutions, however, as X_B increases, more and more neighbors are B. Hence, the changing neighbor interactions will change the potential and kinetic energies of A and B in the solution as compared to their pure states. Both potential and kinetic energy changes are reflected in changes of the internal energies of the components. In addition, the kinetic energy change (change in vibrational and rotational frequencies and hence, change in motional randomness) also causes a change in the entropy of the components. Therefore, the chemical potentials μ_A and μ_B depend on the composition and the other intensive state variables such as P and T as anticipated in (3.18). Consequently in solutions one has to replace the molar parameters [see, e.g., (2.6)] by partial molar quantities (see Sect. 2.8):

$$\begin{matrix} h_A & & s_A & & \bar{h}_A & & \bar{s}_A \\ & \text{and} & & \text{by} & & \text{and} & \\ h_B & & s_B & & \bar{h}_B & & \bar{s}_B \end{matrix}$$

where the bar is to indicate the dependence on the other component, i.e., the non-ideality.

But these are still not all the changes which occur upon dissolution of A in B (or vice versa). In addition there is the *configurational entropy of mixing* connected with the formation of a solution. Even if atoms A and B were virtually identical, so the motional entropy part of the atoms in the pure substances and in the solution were approximately the same, the bringing of the pure substances into states of lower order would cause an increase in entropy. So instead of (3.19) we can write for the molar Gibbs function of a certain phase of a solution

$$g^s = X_A\bar{h}_A + X_B\bar{h}_B - T(X_A\bar{s}_A + X_B\bar{s}_B) - T\Delta s_m \tag{3.20}$$

where Δs_m is the configurational entropy of mixing. Sometimes it is useful to include the configurational entropy term into $\bar{s}_A$ and $\bar{s}_B$. At this point, however, we list it separately to emphasize that the configurational entropy is not associated with the immediate interaction between the components, but rather with the solution as a whole.

Based upon (3.19) and (3.20) one can now decide whether the mechanical mixture or a solid solution will form under given conditions. In general we see already that even if $\Delta h = \sum_i (h_i^m - h_i^s)$, the heat of dissolution is positive, it is possible that the entropy terms overcompensate so that $\Delta g = g^m - g^s < 0$, i.e., the components form a mechanical mixture (separation of the components) rather than a solution. Furthermore, since the Gibbs function depends on P, T *and* on composition, separation may occur in certain parts of the composition range ($0 < X_A < 1$) whereas solutions may be more stable in other parts. Most of the remainder of this chapter is concerned with the conditions that lead to the various combinations of solution and mixture formation in a system.

3.4.2 Ideal solutions

A solution is called ideal if

$$\bar{h}_A = h_A,\ \bar{h}_B = h_B \quad \text{and} \quad \bar{s}_A = s_A,\ \bar{s}_B = s_B \quad . \tag{3.21}$$

These conditions do *not* imply that there are "no interactions" between atoms of A and B. They mean that the interactions are the same as in the pure components, although A and B can be distinguished. Real solutions can therefore only approximate this idealized model. Mixtures of isotopes and isomorphs come the closest to ideal conditions.

The molar Gibbs function for an ideal solution is consequently

$$g^{ids} = g^m - T\Delta s_m \tag{3.22}$$

with

$$\Delta s_m = -R[X_A \ln(X_A) + X_B \ln(X_B)] \tag{3.23}$$

as shown in Problem 2.5. Hence, one can rewrite (3.22) in analogy to (3.19) to

$$g^{ids} = X_A\mu_A + X_B\mu_B \tag{3.24}$$

where the chemical potentials are

$$\mu_i = \mu_i^* + RT \ln(X_i) \quad . \tag{3.25}$$

Note that (3.25) is identical with (2.113) if the activity coefficient γ_i is unity. The μ_i^* is correspondingly the chemical potential of the pure component at the pressure and temperature of the solution. It is worthwhile to retrace the "different" origin of (3.25) and (2.113). Equation (2.113) was derived from a specific, phenomenological equation of state, the ideal gas law [see also (2.98)]. Beyond that, the derivation of (3.25) beautifully illustrates the statistical meaning of the RT $\ln(X_i)$ term that resulted from the configurational entropy contribution to the chemical potential.

The entropy of mixing term (3.23) can be shown to be always positive for all compositions and finite temperatures. Therefore $g^{ids} < g^m$, hence, under the idealized assumptions (3.21), a single solution is always more stable than a mechanical mixture. Similarly, it can be shown that the single solution is more stable than any mechanical mixture of solutions.

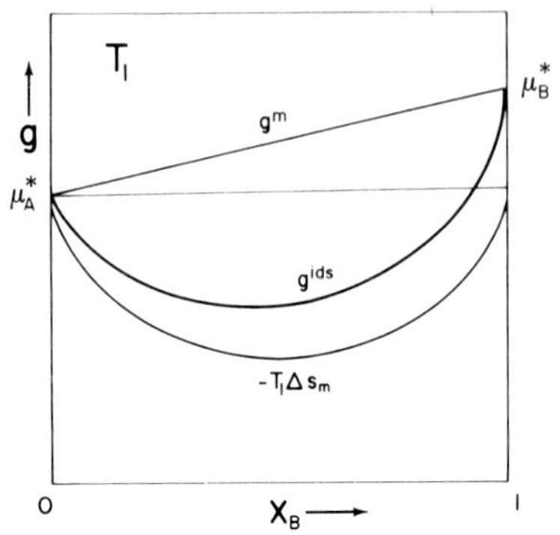

Fig.3.10. Gibbs function for a mechanical mixture, entropy of mixing and Gibbs function for an ideal solution as function of composition (schematic). After [3.1]

In the following we will construct a typical phase diagram for an ideal solution. As a first step it is instructive to visualize the concentration dependence of the configurational entropy term and hence, of $g(\equiv g^{ids})$. Together with g^m; both have been schematically plotted in Fig.3.10 for a given temperature. As for the temperature dependence, one can see from (3.15) that g decreases with increasing temperature, at fixed composition and pressure. Furthermore, due to the increasing weight of the $T\Delta s^m$ term the g-curves become more convex with increasing T. Thus, we can schematically illustrate the temperature dependence of the free energy of an ideal solution as shown in Fig.3.11.

For a system in which the liquid as well as the solid state form ideal solutions, both states have free energy curves of the shown form. At a high temperature (T_6 in Fig.3.12) the free energy for the liquid solution is lower over the whole composition range. Thus only the liquid solution is stable here. As the temperature is lowered, g increases for both liquid and solid solution [see (3.15)]. In general, at a given temperature, the liquid form of a substance has a higher entropy than the solid. Consequently, the free-energy function for the liquid will be raised more than that for the solid. Eventually the two functions begin to intersect upon lowering of T. At the melting point of pure A, T_5, liquid A and solid A are in equilibrium. For all other compositions at T_5 the stable state is a single liquid solution. As we lower the temperature further, for instance, to T_4, $g^{(s)}$ and $g^{(l)}$ intersect at some intermediate composition X_1. What is the stable state for an alloy of *overall* composition X_1? As we will show later, the state with the lowest g possible is a mixture of solid and liquid solutions with compositions

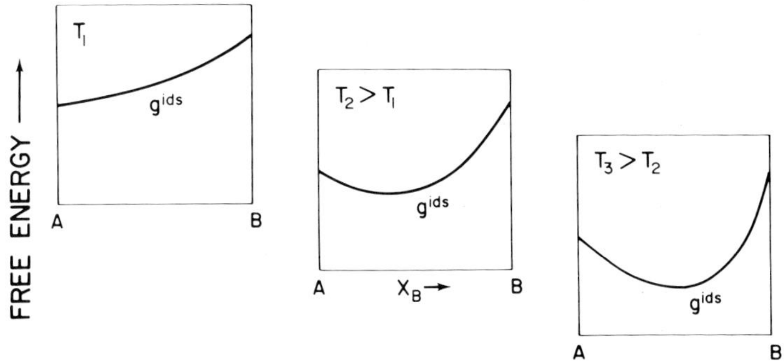

Fig.3.11. Gibbs function (free energy) curves for ideal solution at various temperatures. $T_1 < T_2 < T_3$

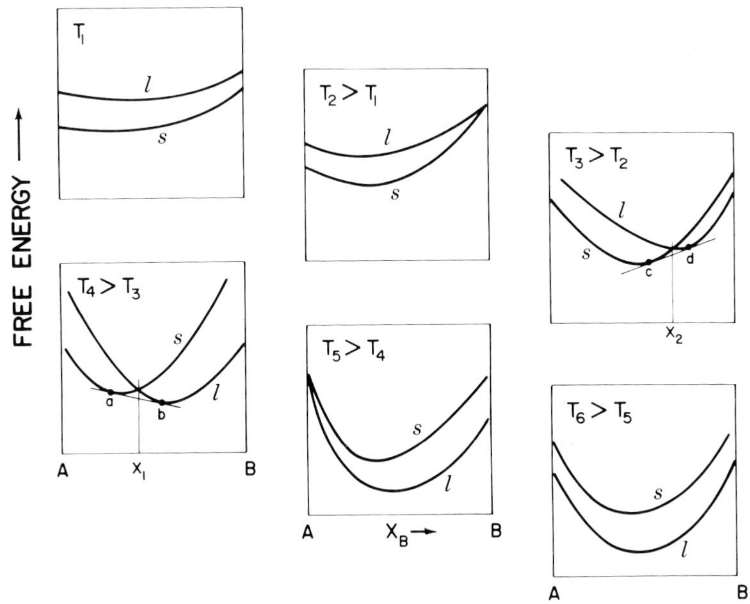

Fig.3.12. Free-energy curves for the liquid and the solid phase in an ideal system with temperature increasing from $T_1 \rightarrow T_6$

given by the points a and b, which are given by the common tangent of $g^{(s)}$ and $g^{(l)}$. This applies to any overall compositions between X_a and X_b. Within these boundaries, regardless of the overall composition, the liquid phase always has the composition X_b and the solid phase has X_a at the given T for which we have plotted $g^{(l)}(X)$ and $g^{(s)}(X)$. The relative amounts of solid and liquid solution, however, depend on the overall composition as we will discuss below in connection with the lever rule. Systems with overall composition to the left of X_a form a single solid solution at T_4. Systems with overall composition to the right of X_b form a single liquid solution at T_4. This "thought-experiment" can be continued through temperatures T_3 and T_2 until at T_1 the whole system consists of a single solid solution. Based upon the various steps in Fig.3.12 one can construct an equilibrium phase diagram for the system A-B as depicted in Fig.3.13. Each temperature determines uniquely the composition of a solid and a liquid solution which can be in equilibrium with each other. The locus of all liquid solution freezing points is called *liquidus*; that for the solid solution melting points is called *solidus*. Constant temperature lines between solidus and liquidus are descriptively called *tie-lines*. Liquidus and solidus separate the phase diagram into three regions:

- a one-phase liquid region
- a one-phase solid region
- a two phase liquid-solid region

Only the first two have physical significance. The third (here lens shaped) region is merely an "information storage space". Equilibria between solid and liquid solutions of the same composition, say X_1, are excluded by this area.

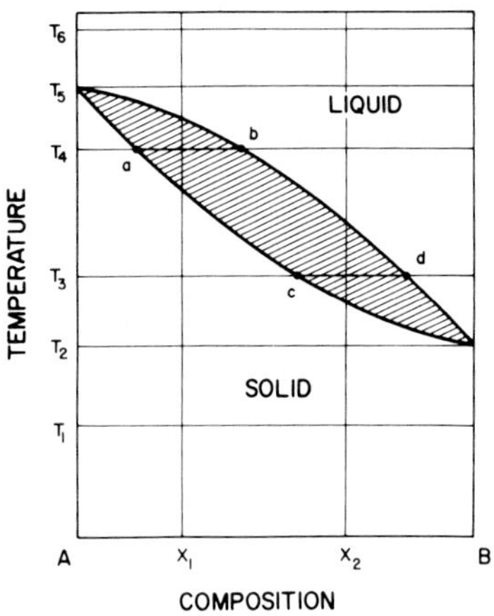

Fig.3.13. Ideal-system equilibrium phase diagram resulting from the free-energy curves in Fig.3.12

The *lever rule*, which can be derived from simple mass conservation considerations (see Problem 3.5), states the following: The amounts of the coexisting phases are inversely proportional to their "distances" from the overall composition of the system. This means, referring to Fig.3.14a, that

$$\frac{n_A^{(s)} + n_B^{(s)}}{n_A^{(l)} + n_B^{(l)}} = \frac{\overline{OL}}{\overline{OS}} \tag{3.26}$$

where $n_A^{(s)} + n_B^{(s)}$ and $n_A^{(l)} + n_B^{(l)}$ are the total number of moles in the solid and liquid, respectively.

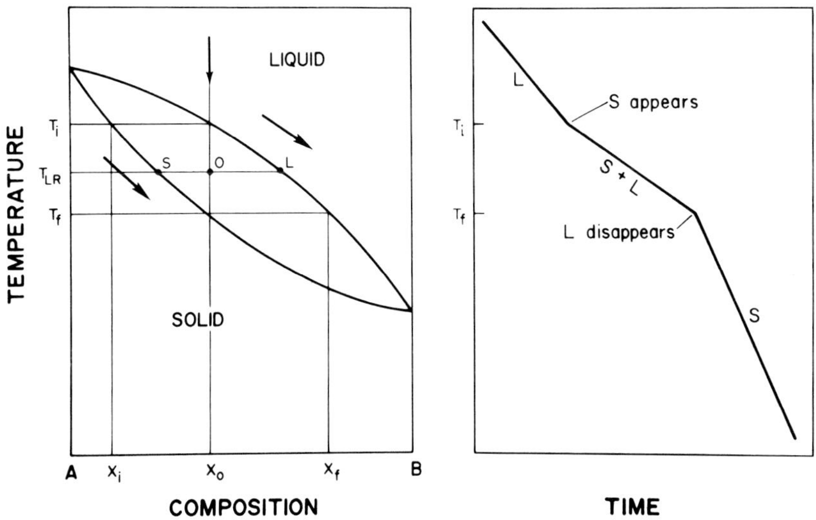

Fig.3.14. Equilibrium solidification of an ideal binary liquid. a) Phase diagram, lever rule. b) Cooling curve for constant heat flux from system (schematic)

Let us follow the compositional and thermal changes that occur upon progressive *equilibrium solidification* of a melt in a system corresponding to Fig.3.13. We can, for instance, arrange for cooling with constant heat flux from the system. As we lower the temperature of the melt (Fig.3.14b) of, say, initial composition X_0 (Fig.3.14a) the first minute crystal appearing at T_i has the composition X_i. Further withdrawal of heat, in order to solidify a larger portion of the system, results in changes of the composition of the diminishing melt toward X_f. Concurrently, the composition of the whole growing solid is shifted toward X_0. Note that this requires, superimposed on to the solidification flux from the liquid to the solid phase, a diffusion flux of B into the already solidified portion of the system. Note also the reduced gradient of the cooling curve, Fig.3.14b, that stems from the liberated heat of solidification. Close to T_f, the remaining melt (now a small droplet) contains a bit less B than X_f and the solid counterpart a bit less than X_0. Finally *at* T_f the whole melt is solidified, the system has again a uniform composition X_0 (see Problem 3.6) and the cooling curve resumes a steeper gradient.

This consideration is based on the assumption that diffusion can establish equilibrium between the bulk phases throughout the whole progressing process. Since diffusion is a slow process, particularly in solids, very low crystallization rates are required to avoid concentration gradients in the forming solid. On the other hand, one can conduct such a freezing process intentionally rapidly enough to override the equilizing diffusion in order to achieve separation and purification, as will be discussed in Sect. 6.3.

We still need to prove that the composition of coexisting solid and liquid solutions in equilibrium are characterized by the common tangent to $g^{(s)}(X)$ and $g^{(l)}(X)$. According to (3.24) and (3.17), (dropping the parenthesis around the phase superscripts),

$$g^l = X_A^l \mu_A^l + (1 - X_A^l)\mu_B^l \quad \text{and} \quad g^s = X_A^s \mu_A^s + (1 - X_A^s)\mu_B^s \tag{3.27}$$

and hence

$$\frac{\partial g^l}{\partial X_A^l} = \mu_A^l - \mu_B^l \quad , \quad \frac{\partial g^s}{\partial X_A^s} = \mu_A^s - \mu_B^s \quad . \tag{3.28}$$

In equilibrium, however, $\mu_A^l = \mu_A^s$ and $\mu_B^l = \mu_B^s$. Therefore, from (3.28)

$$\partial g^l/\partial X^l = \partial g^s/\partial X^s \quad . \tag{3.29}$$

This common tangent criterion does not prove that there are no systems possible in which coexisting solid and liquid phases have the same composition. In fact, we will encounter such systems that "melt congruently" at certain compositions X. For the ideal binary system, however, one can show that (3.29) cannot be fulfilled for any $X^l = X^s$ except for $X^l = 0$ and $X^s = 1$, i.e., at the melting points of the pure components.

Although Fig.3.13 has been derived for ideal systems, there are some real systems that possess lens-shaped phase diagrams. Figures 3.15a and b represent examples for such binary systems. Note that germanium is isomorphous with silicon and copper with nickel, i.e., both components have the same crystal structure in their pure solid state (diamond structure in the first

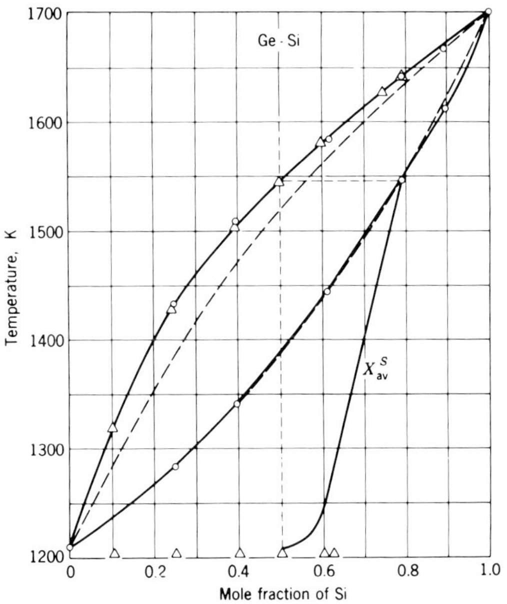

(a)

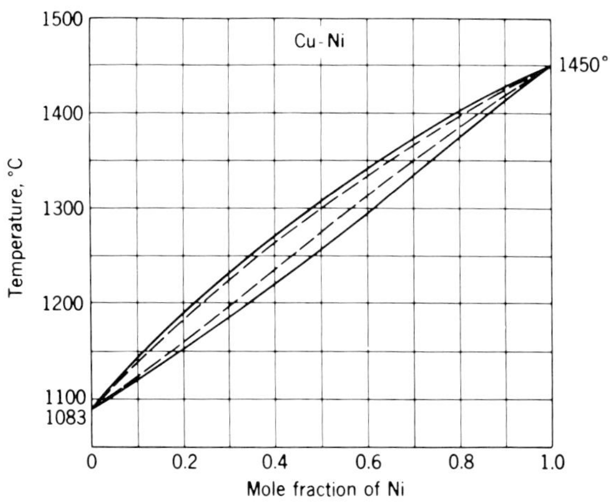

(b)

Fig.3.15. Phase diagrams for (a) germanium-silicon; (b) copper-nickel. Full lines are experimental, dashed lines were calculated. From [3.1]

case and face centered cubic in the latter). In a following paragraph we will see that only isomorphous components show complete intersolubility if certain thermodynamic requirements are met. This has led to the custom of calling lens-shaped phase diagrams isomorphous.

The goal of this introduction to phase diagrams is familiarity with interpretation of phase diagrams and application to the control of phase transitions. Hence, we will here only briefly mention that one can actually calculate the solidus and liquidus for quasi-ideal systems from the melting temperatures and enthalpies of fusion (heat of fusion at constant P) of the pure components. (See, for instance, [Ref. 3.1, Sect. 5.12]). The dashed lines in Figs.3.15a and b are the results of such calculations, which for the Ge-Si and Cu-Ni systems were carried out by THURMOND [3.17] and SELTZ [3.18], respectively. For more recent work on calculations of the liquidus-solidus gap in homogeneous, monotonic alloy systems see STEININGER [3.19].

3.4.3 Real solutions

The Gibbs function for real solutions can be written, combining (3.20) and (3.23), as

$$g^s = X_A\bar{h}_A + X_B\bar{h}_B - T(X_A\bar{s}_A + X_B\bar{s}_B) + RT[X_A \ln(X_A) + X_B \ln(X_B)] \quad . \tag{3.30}$$

It is, however, advantageous to treat real solutions in the form of

$$g^s = g^{ids} + \text{correction term} \quad ,$$

where the correction term accounts for the changes in entropy and enthalpy due to the presence of the second component in the solution. The correction term is thus itself a function of the components and their ratios (relative concentrations).

Hence, we will rewrite (3.30), employing (3.19) and (3.22), to

$$g^s = g^m + X_A(\bar{h}_A - h_A) + X_B(\bar{h}_B - h_B) - T[X_A(\bar{s}_A - s_A) + X_B(\bar{s}_B - s_B)] + RT[X_A \ln(X_A) + X_B \ln(X_B)] \quad . \tag{3.31}$$

By regrouping one obtains

$$g^s = g^m + \Delta g^{xs} - T\Delta s_m \tag{3.32}$$

where the correction term

$$\Delta g^{xs} = X_A(\bar{h}_A - h_A) + X_B(\bar{h}_B - h_b) - T[X_A(\bar{s}_A - s_A) + X_B(\bar{s}_B - s_B)] = \Delta h^{xs} - T\Delta s^{xs} \tag{3.33}$$

is commonly called the "excess free energy of dissolution". In ideal solutions $\Delta g^{xs} = 0$; hence, Δg^{xs} characterizes the difference between the actual Gibbs function of the real solution and the value of g if the solution were ideal.

Expressed in terms of the chemical potentials of the components in the solution μ_i, the molar Gibbs function is also

$$g^s = X_A\mu_A + X_B\mu_B \quad . \tag{3.34}$$

The chemical potentials of the components in real solution are, according to (2.113), $\mu_i = \mu_i^* + RT\ln(\gamma_i X_i)$ where, as discussed in Sect. 2.11, the standard chemical potential μ_i^* is not necessarily that of the pure component i in its standard state. In general, one can only obtain the activities $a_i = \gamma_i X_i$ and thus the Δg^{xs} for specific solutions experimentally. As an example, Table 2.1 lists the activities as they result in liquid zinc-tin solutions for both components, determined by an electromotive force technique.

Table 3.1. Activity data for the liquid Zn-Sn system at 431°C. From [3.20]

X_{Zn}	a_{Zn}	a_{Sn}
0.0	0.000	1.000
0.1	0.204	0.904
0.2	0.380	0.815
0.3	0.530	0.730
0.4	0.655	0.655
0.5	0.750	0.585
0.6	0.821	0.524
0.7	0.875	0.465
0.8	0.914	0.405
0.9	0.950	0.320
1.0	1.000	0.000

There are several theoretical approaches to calculate Δg^{xs} under specific idealized assumptions. One of these concepts is based on assuming that $\Delta s^{xs} = 0$. Hence, in these so-called *regular solutions*

$$g^s = g^m + \Delta h^{xs} - T\Delta s_m \quad \text{or} \quad \Delta g^{xs} = \Delta h^{xs} \quad , \tag{3.35}$$

i.e., at constant pressure the excess Gibbs function consists of the *heat of dissolution* (↔ solution) only. Δh^{xs} is, in general, experimentally readily accessible. This concept holds for many real solutions rather satisfactorily. If the change in entropy due to solution formation is not known for a system, then the regular solution concept, in which only the configurational entropy change is accounted for, often supplies a good approximation and allows for semi-quantitative predictions. Calculations of Δh^{xs} for specific systems will be referred to in Sect. 6.2 on segregation. One should keep in mind, however, that in many systems the vibrational entropy-of-dissolution term is not small as compared to Δh^{xs}.

There are many different ways the free energy of real solutions may deviate from ideality, but there are only a small number of *distinctly* different types of binary phase equilibria. In the following we will be using the sign and magnitude of Δg^{xs} (and Δh^{xs}) as a guideline for a discussion of the various resulting types of phase diagrams.

Regular solutions with $\Delta h^{xs} < 0$

The theory of regular solutions shows that $\Delta h^{xs}(X)$ is a symmetric parabolic function (see, for instance, [Ref. 3.1, pp. 79]). Consequently, the Gibbs function curves, at constant temperature, have the same convex downward shape as those for ideal solutions; as schematically indicated in Fig.3.16.

Now going beyond regular solutions, for all systems in which both liquid and solid phases have a moderate negative Δh^{xs}, we can thus expect Gibbs function curves that are convex downward at all temperatures. However, depending on whether

$$|\Delta g^{xs(s)}| \lessgtr |\Delta g^{xs(l)}| \quad , \tag{3.35a}$$

one can obtain three different types of phase diagrams as depicted in Fig. 3.17. Solidus and liquidus do not necessarily simply descend anymore from $T_{m\ high}$ to $T_{m\ low}$ as in ideal systems. Extrema can appear at intermediate

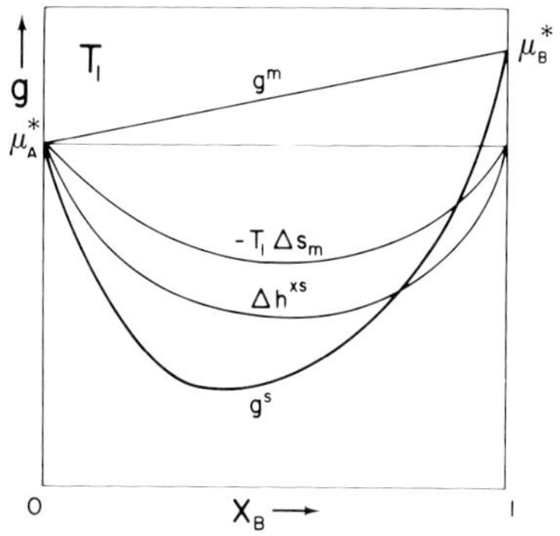

Fig.3.16. Energy composition curves for negative Δh^{XS}. After [3.1]

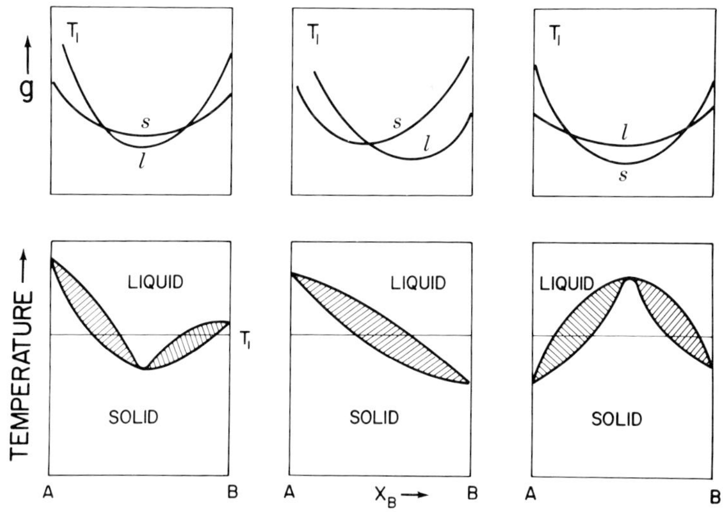

Fig.3.17. Schematic phase diagrams for negative Δg^{XS}. (a) $\Delta g^{XS(l)}$ more negative than $\Delta g^{XS(s)}$. (b) comparable excess Gibbs function for solid and liquid solutions. (c) $\Delta g^{XS(l)}$ less negative than $\Delta g^{XS(s)}$. After [3.1]

compositions. At these extremum points, where solidus and liquidus becomes identical, solid and liquid solutions of the same composition can coexist in equilibrium. Note that even "very real" solutions can have a lens-shaped phase diagram if the excess Gibbs function of dissolution is comparable for solid and liquid phases. Lens shape is thus no criterion for the ideality of a solution. For calculations of the liquidus-solidus gap in such "monotonic" real systems, see STEININGER [3.19].

An important point to be made here is that real systems with negative excess heat of dissolution Δh^{xs} have no miscibility gap, i.e., they form solutions over the whole composition range. Ordering and formation of intermediate phases that are typical for this group will be discussed later.

Regular solutions with $\Delta h^{xs} > 0$

With solutions possessing positive Δh^{xs} the situation is more complex. As illustrated in Fig.3.18 for one phase, the enthalpy terms now compete with the entropy of mixing term. At high temperatures, where the $T\Delta s_m$ term is

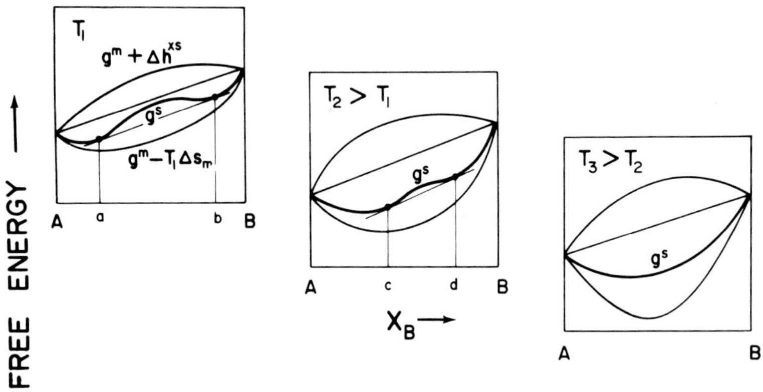

Fig.3.18. Free-energy composition curves for positive Δh^{xs} at several temperatures. $T_1 < T_2 < T_3$. After [3.1]

large, g^s will still be convex downward. Upon lowering of T, however, $T\Delta s_m$ will eventually become comparable with Δg^{xs}, and $g^s(X)$ will become inflected (see points c and d in Fig.3.18). In a later part of this section we will encounter arguments that for many systems the slope of g^s must be negative at the pure component ends of the composition coordinate. Hence, a g^s-curve remains inflected upon further lowering of T, but the inflected part broadens. When an inflection is present, then, using again the common slope argument of (3.20), systems with overall composition X between points c and d (Fig.3.18) will exist in the form of a mixture of two solutions as the state of lowest g^s rather than forming one solution of composition X. The composition of the two solutions is given by the points of common tangency.

By scanning in this manner through the entire temperature range in which the considered phase exists, one can construct the *solvus* line in phase diagrams, as schematically shown in Fig.3.19. The solvus separates the equilibrium existence range of one solution from the temperature-composition combinations for which two coexisting solutions of different (but same overall) composition are more stable. Hence, the area between solvus and composition coordinate can be thought of as an instability range for single solutions. This miscibility gap is similar to the areas between solidus and liquidus encountered earlier. Note, however, that there the two different solutions were in equilibrium in different phases, whereas here we have two different solutions of the same phase. The relative amounts of the solutions can again be obtained with the lever rule (3.26) from the corresponding sections of the tie-line.

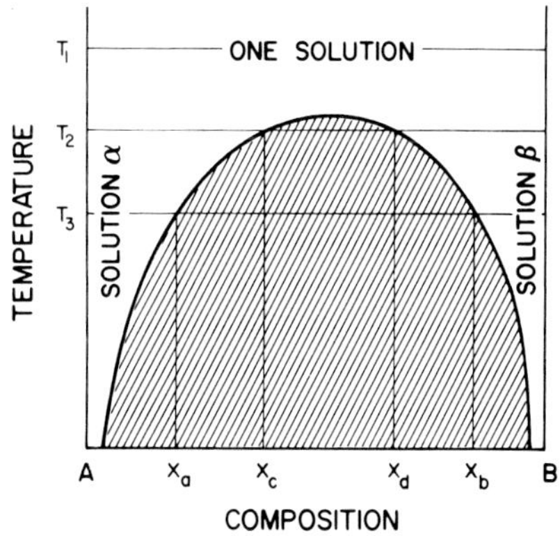

Fig.3.19. Miscibility gap for a phase corresponding to the g^s-curves in Fig.2.18. After [3.1]

For moderate Δg^{xs} in *liquid systems* there are usually no miscibility gaps since the entropy of mixing term is typically high enough to prevent inflections in the g^s-curves at the temperatures of liquid stability.

Now proceeding to equilibria between two phases, the relative magnitudes of $\Delta g^{xs(l)}$ and $\Delta g^{xs(s)}$ determine again the nature of a phase diagram for systems with positive Δg^{xs}. Fig.3.20 depicts schematically the three distinctly different types one can obtain. Each type shows a miscibility gap at low temperatures.

Let us outline the compositional and thermal changes that occur upon cooling in such systems with miscibility gap in the solid phase. We pick for the discussion Fig.3.20c which is shown on a larger scale in Fig.3.21a. The corresponding cooling curve for constant heat flux from the system is schematically presented in Fig.3.21b. When a liquid solution of composition X_0 is cooled down, at T_1 it becomes saturated with respect to a solid solution of

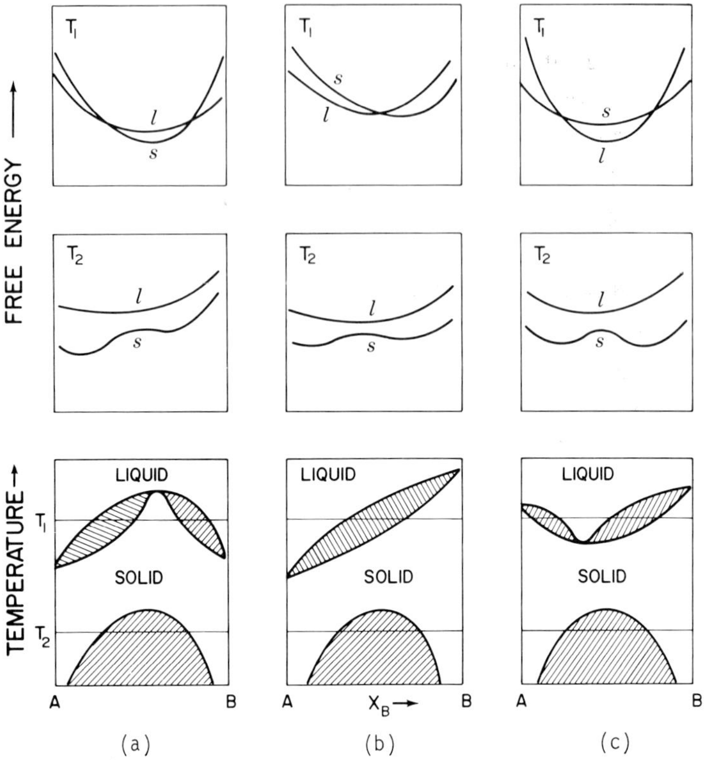

Fig.3.20. Schematic phase diagrams for positive Δg^{XS}. (a) liquid more positive than solid; (b) comparable; (c) liquid less positive than solid. After [3.1]

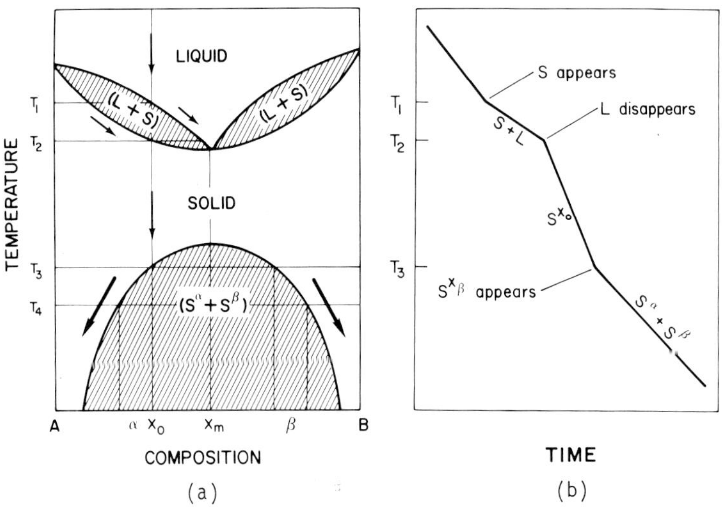

Fig.3.21. (a) Cooling composition path in a binary phase diagram with solid miscibility gap. (b) Corresponding cooling curve for constant heat flux from system (schematic)

a composition given by the intersection of the tie-line with the solidus. Upon quasistatic lowering of T, the whole system will be solidified at T_2 where it regains its homogeneous composition X_0. Now, if we cool the system further down, at T_3, at the solvus line, the solid solution X_0 becomes saturated with respect to a second solid solution of composition X_β. The composition is given again by the tie-line intersection with the solvus. As we cool down even further to T_4, slowly enough to allow for maintenance of equilibrium conditions, the mother phase α will continuously decrease its B content, thereby remaining saturated with respect to a β phase of continuously increasing B content. Their relative amounts are again proportional to the inverse distances on the tie-line as divided by the unchanged overall composition X_0. At each point of appearance or disappearance of a phase the cooling curve (Fig.3.21b) shows a characteristic change in slope (Problem 3.7a). (See also Problem 3.7b.) It should be emphasized here again that this discussion is restricted to equilibrium conditions. With practical solidification rates, however, solution separation of this type is often not found due to the slow diffusion rates in solids.

In the literature it is customary to label the "non-existence" areas of phase diagrams, that we have shaded for better distinction, rather than the existence areas for the phases. For instance, as indicated in brackets in Fig.3.21, the area between solidus and liquidus would read L + S and the area between solvus and composition coordinate axes would show $S^\alpha + S^\beta$. The L's and S's stand simply for the liquid and/or solid solutions that coexist *at the boundaries* of these areas. This may be somewhat confusing to the novice. Hence, as long as the existence areas are large enough in our simple schematic diagrams, we will label them directly. With most real phase diagrams, this convention is practially not feasible since existence areas frequently shrink to lines.

In connection with the inflection of the g^s-curve in Fig.3.18 we have used a "negative gradient" argument for the pure ends of the phase diagram. This remains to be proved. From the general expression for the Gibbs function of a real solution, (3.30) and $X_A + X_B = 1$, the change in g^s with, say, impurity concentration X_B is

$$\frac{\partial g^s}{\partial X_B} = (\bar{h}_B - \bar{h}_A) - T(\bar{s}_B - \bar{s}_A) + RT \ln \frac{X_B}{1 - X_B} \quad . \tag{3.36}$$

For the change in g^s with X_A a symmetric expression would be obtained. The argument for a negative gradient, as commonly made, goes then as follows: As X_B goes toward zero, the ln term in (3.36) goes toward minus infinity. Consequently, with "finite values" for $\bar{h}_B$, $\bar{h}_A$, $\bar{s}_B$ and $\bar{s}_A$, the gradient must become negative at some small value of X_B as one moves toward $X_B = 0$. Although this argument may be helpful for the design of phase diagrams with not too excessively different enthalpies and entropies of the components in a system, it should be used very cautiously. One should keep in mind that the parameters in (3.36) require a certain minimum number of atoms in an ensemble so that it can be treated statistically in a meaningful way (see brief discussion in the introduction of Chap. 2). If one of the components becomes too dilute (i.e., $X_B \to 0$) its $\bar{h}$ and $\bar{s}$ terms and even the configurational entropy term in (3.36) become undefined. (Note also that, as in Problem 2.5, the Δs_m expression is typically derived based upon Stirling's approximation that is only valid for large numbers.) Hence, the above "$-\infty$ argument" is questionable. Various authors base conclusions such as "any substance is soluble in any other substance to some extent" or "pure substances are thermodynamically impossible" upon (3.36) and the negative gradient argument. Conclusions of this generality, that neglect the size limitations of an ensemble for which macroscopic state functions are still meaningfully defined, appear overly ambitious. For further arguments and for references on this disputed subject see [Ref. 3.21, p. 30].

3.4.4 Invariant transitions

In most binary systems encountered so far the composition changed continously during a liquid-solid transition. Consequently, the temperature had to be lowered continuously to complete solidification. Now, from the phase rule, $f = C - P + 2$, one would expect that even this last degree of freedom (here T, after P is fixed) is lost when:

1) the effective number of components is reduced by one. For instance, when solid and liquid have the same composition and hence become thermodynamically indistinguishable. Then C = 1 even in a binary system with two chemically distinguishable species in two phases. Such "congruently melting" systems, although tacitly encountered already in Figs.3.17 and 3.20, will be discussed in Sect. 3.4.6.

2) the number of phases of different composition is increased by one. This can occur when two phases react to form a new, third one or when one mother phase decomposes into two other new phases.

In an invariant equilibrium situation ($f = 0$) the relative amount of the phases is only determined by the heat flow to or from the system since the temperature, here invariant, has lost the regulatory function that it had concerning composition.

In the following we will consider only invariant reactions among condensed phases. One should, however, keep in mind that invariant reactions including gases can play an essential (often detrimental) role in crystal growth processes. For instance, gases can have a higher solubility in melts than in the corresponding solids. This leads to gas bubble formation at solid-liquid interfaces, which is generally very undesirable for heat and mass transfer reasons.

The most commonly occurring invariant reaction in binary solution systems, the *eutectic reaction*, consists in the separation of two different solid solutions from a liquid phase or, more generally, in

$$L^{(1)} \underset{\text{heating}}{\overset{\text{cooling}}{\rightleftharpoons}} S^{(1)} + S^{(2)} \quad .$$

To see when eutectic reactions can occur, let us look at the g(X) scheme of a system that contains only two final solid solution phases and one liquid phase. A positive Δg^{xs} is assumed. If $\Delta g^{xs(s)}$ is sufficiently larger than $\Delta g^{xs(l)}$ so that at a certain temperature there is a common tangent to $g^{(l)}$ *and both* free-energy functions g^{α} and g^{β}, as depicted in Fig.3.22, a eutectic reaction occurs. The eutectic composition then determines the lowest possible solidification and melting temperature for the system. The root for the term eutectic lies in the Greek word eutectos which means "easily fused".

We will now discuss the three significantly different freezing paths that one can choose within a binary eutectic phase diagram. For this purpose Fig. 3.23 has been drawn oversimplified with vertical solvus lines.

Alloy 1: Cooling of the eutectic liquid results in solidification with respect to solid α and β at the eutectic temperature T_e. Then, under equilibrium conditions the temperature does not change any more while the transformation goes on – as long as three phases are present. Extraction of heat does not lower T. A microstructure of an intimate mixture of phases α(25% B) and β(80% B) forms. The actual structure depends on the kinetic details of the solidification pro-

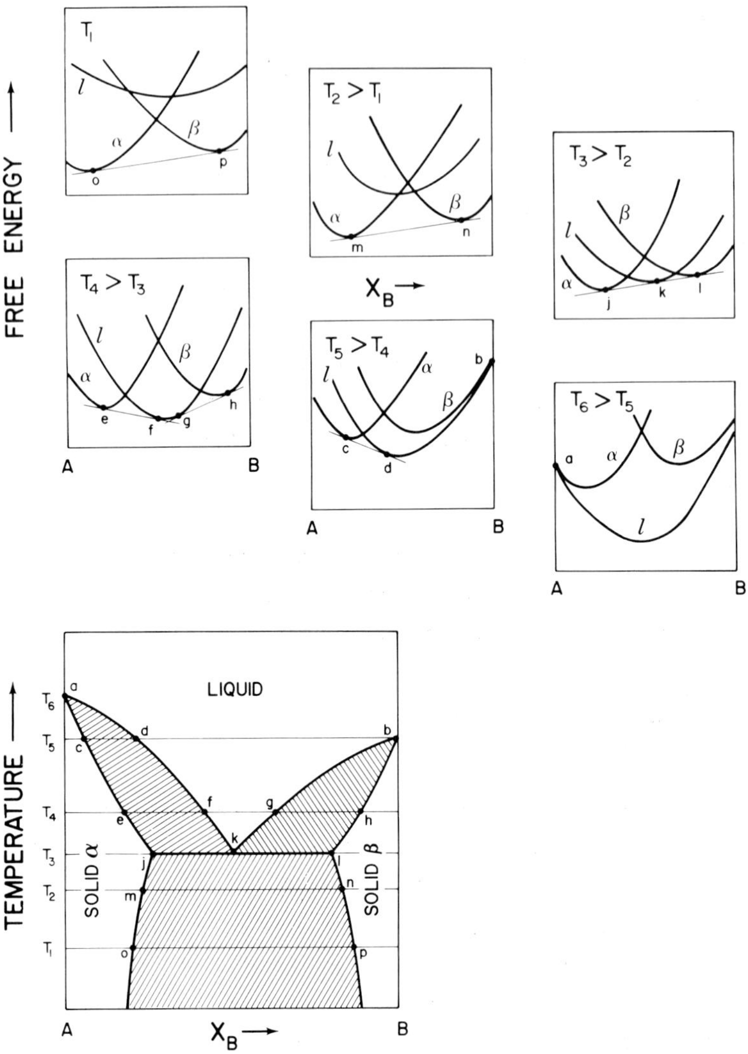

Fig.3.22. Free-energy composition curves and the resulting temperature-composition equilibrium diagram for a eutectic system. After [3.1]

cess, i.e., the off-equilibrium conditions. Figure 3.24 indicates schematically the various types of microstructures which can occur in eutectic solidification. Examples and applications will be discussed in later chapters.

Alloy 2: In this system the overall composition line does not intersect the eutectic line. Consequently, the system behaves like one with complete solid solubility range. For instance, at T_1 solid particles of phase α(7% B) are in

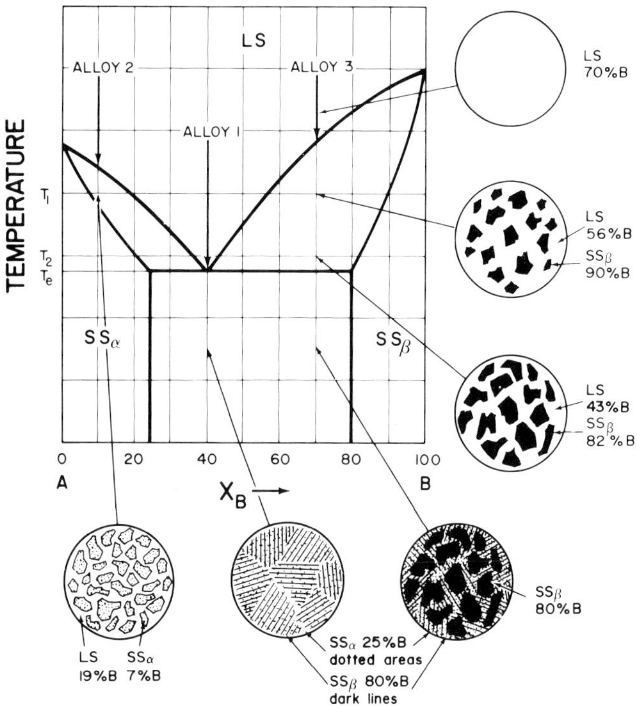

Fig.3.23. Three different solidification processes in a simplified eutectic system (solvus lines drawn vertically) and their structural results. (LS: liquid solution SS: solid solution)

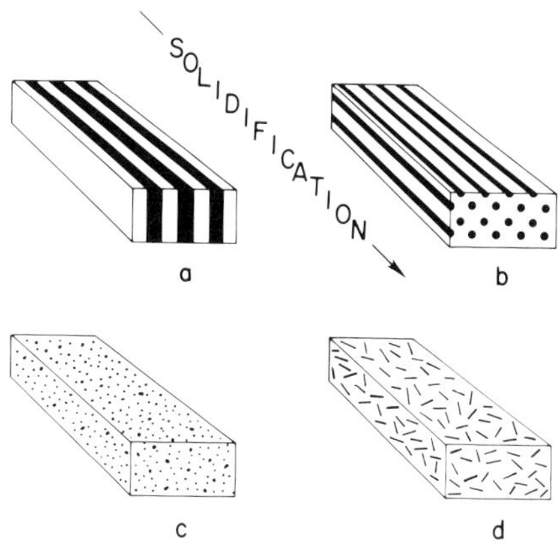

Fig.3.24. Schematic illustration of various eutectic structures: (a) lamellar; (b) rodlike; (c) globular; (d) acicular. After [3.22]

equilibrium with a melt of 19% B in A. Their relative amounts follow from the lever rule, (3.26).

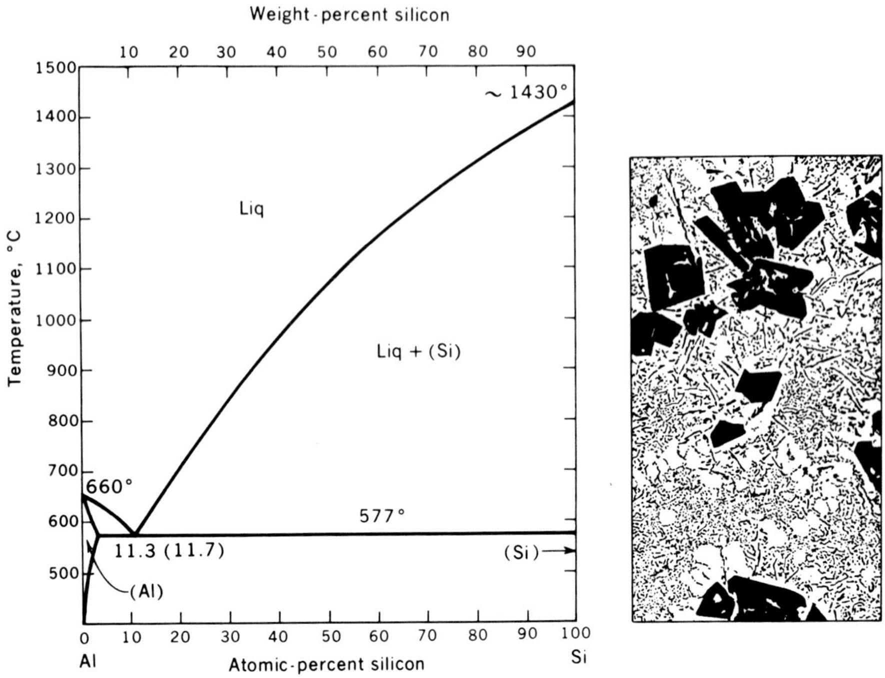

Fig.3.25. Phase diagram of the Al-Si system and structure of solid Al-20% Si alloy, about 100 X. The large angular phase is primary silicon. Note the essentially random orientation of the silicon in the eutectic matrix. This matrix would be classified as an acicular eutectic. From [3.23] by permission of McGraw-Hill Book Co.

Alloy 3: At the liquidus the melt becomes saturated with respect to a B-rich solid solution that begins to solidify. Upon further cooling, the B-rich particles in the melt grow and simultaneously drive some B back into the melt (equilibrium consideration). When the eutectic composition of the residual melt is reached, the same microstructure as in the case of alloy 1 forms in between the large β particles. The relative amount of β particles and residual melt at T_e is again determined by the lever rule. Figure 3.25 shows the structural result for such a solidification process in the Al-Si system.

In reality, the solvus line is not perpendicular. Hence, all compositional results of the above discussion describe only the "primary solidification". In real systems, due to the tilt of the solvus lines, the composition of the solidified parts of the system must change upon further lowering of the temperature as discussed already in connection with Fig.3.21. For instance, an

A-rich solid solution α (alloy 2 in Fig.3.23) will become saturated with respect to a B-rich β solid solution upon reaching the solvus line. The composition of the new B-rich particles forming is given by the intersection of the tie-line with the β-solvus. Further equilibrium cooling shifts the newly forming β as well as the primarily formed α phases along the corresponding solvus lines. For this to occur B must diffuse out of α and into β. Yet in practice with finite cooling rates, the degree of supersaturation for β within α can become high enough to drive nucleation of more β particles within α, and the already existing β zones will not grow further. Note, however, that then the newly nucleating internal β zones have a composition different from the external β zones. Since processes of this type are typically undesirable in single crystal growth, we will leave it at this example which was only discussed to indicate the complexity in structure and composition that can result from actual, non-equilibrium conditions.

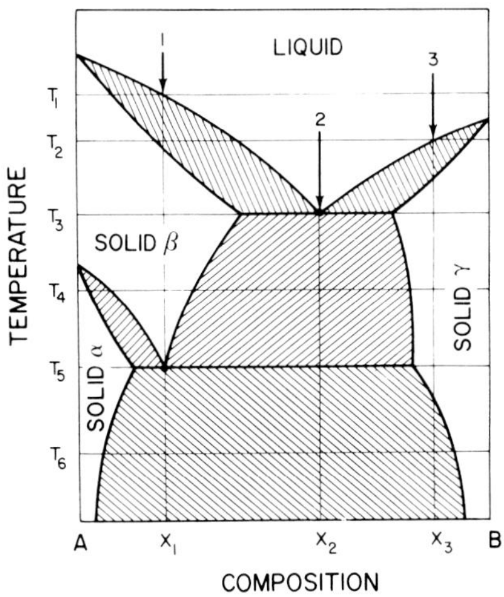

Fig.3.26. Phase diagram with an eutectic and eutectoid invariant reaction. After [3.1]

Many phase diagrams show more than one eutectic reaction. Secondary eutectic reactions of the type indicated in Fig.3.26 are called *eutectoids*. This reaction occurs when a solid solution resulting from a eutectic transition, becomes unstable with respect to a distinctly different solid phase (see Problem 3.8).

Systems with a miscibility gap in the liquid phase show a *monotectic* invariant transition characterized by

$$L^{(1)} \underset{\text{heating}}{\overset{\text{cooling}}{\rightleftharpoons}} L^{(2)} + S^{(1)} \quad .$$

Whereas eutectics appear frequently without other invariant processes in the same system, this is rarely the case with monotectics. Typically they form combinations with other transitions, as shown in Fig.3.27. Usually there is enough difference in density between $L^{(\alpha)}$ and $L^{(\beta)}$ that if the liquids are kept above T^M and below T^C, they form separate layers in a container. At $T > T^C$, however, the compositional difference between the two liquids disappears, and they begin to form one liquid solution.

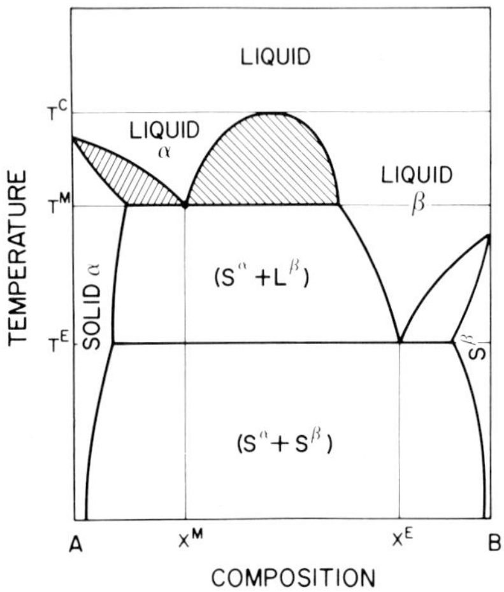

Fig.3.27. Phase diagram incorporating a monotectic reaction. After [3.1]

As one cools a liquid alloy of the monotectic composition X^M, at T^M $L^{(\alpha)}$ begins to decompose into $L^{(\beta)}$ and $S^{(\alpha)}$. This invariant reaction continues upon extraction of heat until $L^{(l)}$ has disappeared. Further lowering of T (see Fig.3.27) will shift the composition of α towards A and of L towards B. When the eutectic temperature T^E is reached the composition of L is X^E.

The third type of the three-phase invariant reactions is the *peritectic* reaction

$$L^{(1)} + S^{(1)} \underset{\text{heating}}{\overset{\text{cooling}}{\rightleftarrows}} S^{(2)} \quad .$$

The g(X) curves that lead to the peritectic reaction are rather similar to those of eutectic systems (see Problem 3.9). In eutectic systems (Fig.3.22) before reaching the common tangent situation for all three phases (T_3) upon lowering of T, a two tangent situation (T_4) is encountered. In peritectic systems this sequence is reversed. This is due to the relative position of the g(X) minima along the composition axis. In a eutectic the "liquid minimum" lies between those of the two solid phases. In the peritectic case, however, it lies outside the "solid minima". This situation is favored by large differences in the melting points of the two components. (Example: FeO and MnO with T_m = 1365° and 1785°C, respectively, T_p = 1430°C.)

Referring to Fig.3.28, we will now discuss the result of equilibrium cooling on various alloys in a peritectic phase diagram.

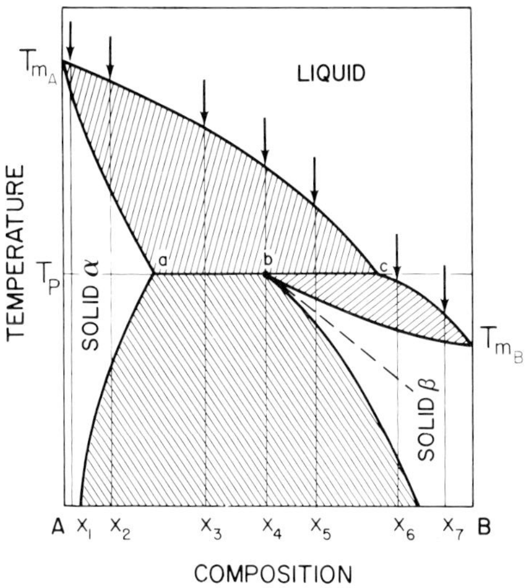

Fig.3.28. Peritectic phase diagram with alloys of various overall compositions

Alloys 1 and 7 undergo the previously discussed compositional changes upon freezing, and the solid phases remain unchanged upon further cooling. *Alloys 2 and 6*, in addition, encounter solid phase transitions, as discussed in detail in connection with the eutectic phase diagram. *Alloys 3, 4 and 5* all go through the peritectic transition. *Alloy 4* precipitates out α until the composition of the liquid phase corresponds to c. Then, at T_p, L and α react completely to β. The slightest further withdrawal of heat causes some α to separate from β again according to the solvus line. *Alloy 3*, just as alloy 4, precipitates out α, and L becomes shifted toward c. After the invariant peritectic reaction, however, some α is left while L is completely used up for β formation. Upon further lowering of T, this α is supersaturated with respect to β and consequently more β is formed with the corresponding composition change in both solid phases along the solvus lines. In *alloy 5*, after α and

L react to β, some L remains. Upon lowering of T, L is frozen completely to β. Then at the intersection with the solvus α starts forming within β.

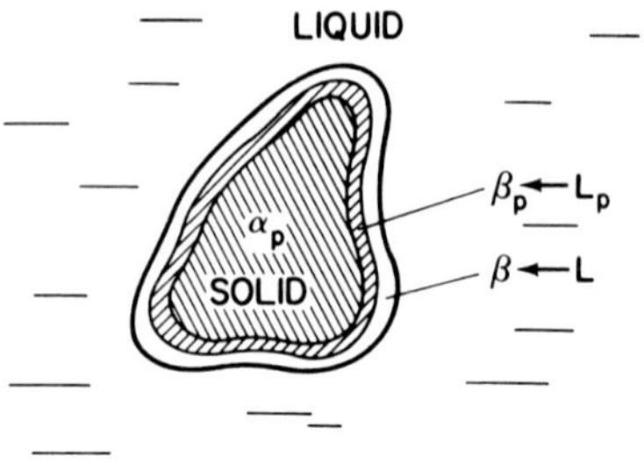

Fig.3.29. Coring in non-equilibrium peritectic reactions

There is a significant difference in mass transfer between eutectic and peritectic reactions. In a eutectic solidification the compositional readjustment occurs mostly by diffusion through the liquid phase. In peritectic solidifications, however, diffusion in the solid phase is the rate limiting parameter. The peritectic reaction, i.e., $L_p + \alpha_p \rightarrow \beta_p$, takes place at the interfaces between the α-phase particles and the surrounding melt. After formation of the first layer of β_p onto the α_p particles, the reaction rate decreases drastically. So, under any practical circumstances α_p will more or less stay as is while L continues to form layers of β that are increasingly more rich in B than β_p. This non-equilibrium primary solidification to β will follow some other lower lying solidus (dashed line in Fig.3.28; see Problem 3.6). The product of the reaction is thus a mixture of α_p and "cored" β.

3.4.5 Ordered and intermediate phases

In the general discussion of real systems with $\Delta h^{xs} < 0$ it was indicated that this group shows disorder-order solid state transitions and formation of intermediate phases. We will now briefly discuss these phenomena. Ordering and intermediate phase formations are by no means restricted to negative heat of dissolution systems and do occur in solutions with $\Delta h^{xs} > 0$ as well, although less frequently. Intermediate phase formation applies to many important semiconducting compounds, such as the III-V and II-VI systems.

Examples for order transitions are given in Fig.3.30 for the Cu-Pt system. A detailed description of disorder-order transition is beyond the scope

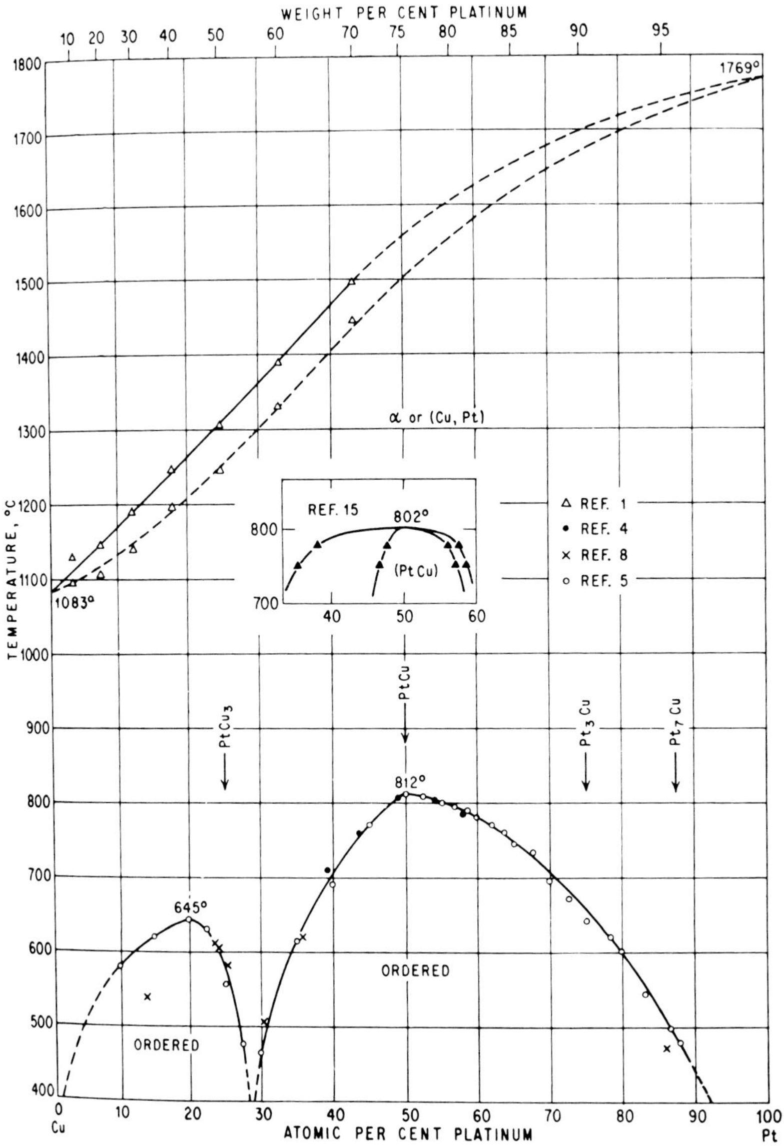

Fig.3.30. The copper-platinum equilibrium phase diagram displaying formation of ordered subsystems. From [3.24] by permission of McGraw-Hill Book Co.

of this text. We will only indicate structural changes which occur. In a typical substitutional solid solution (in contrast to interstitial solid solutions), solute atoms occupy regular host lattice sites at random. In an ordered solution, however, solute atoms occupy a certain set of lattice sites, forming a regular sub- or superlattice within the set of lattice sites that are occupied by the solvent (or host) atoms. For instance, in the classical case of $AuCu_3$, both metals form face-centered cubic structures in their pure state. In the alloy $AuCu_3$ at high temperatures the gold atoms are more or less randomly located substitutionally on all sites of the cubic face--centered lattice. In the ordered phase range, however, gold occupies only cube corner sites and copper only face-centered positions. (Note that there are three face-centered sites and only one cube corner site within a primitive cell in the cubic face-centered lattice; hence, the stoichiometric ratio Cu_3Au.) The gold atoms thus form a simple cubic superlattice within the ordered alloy.

Ordered solutions will obviously only occur with components that have very similar structures, i.e., that are isomorphous. Components that differ strongly in properties such as atomic size, electronic and crystal structure are unlikely to form ordered solutions. However, because of these differences, stable intermediate phase formation may be favored. Let us outline two extreme cases of the "spectrum of reasons" that can lead to intermediate phase formation. In reality, of course, combinations are encountered as well.

If there are stronger attractive forces between unlike atoms, there is a tendency to form phases with highly ordered patterns that differ from ordered solutions mainly in the underlying type of bonding. In intermediate phases the bonds have a much stronger ionic or covalent character than in the metallic ordered solutions. As a "rule of thumb", the more dissimilar the electronic structure of the components, the less metallic and the more ionic or covalent are the bonds. Because of the directed nature of the bonds and the high order in intermediate phases, their entropy (vibrational and configurational) as well as enthalpy and consequently the free energy, are typically low. Hence, these *compound-like intermediate phases* show high stability.

Figure 3.31 shows schematically a set of g(X)-curves for the stable solid phases in such a compound forming binary system. Typically the two components have a somewhat different "crystal structure", (i.e., actual structure and/or lattice parameters). Consequently, $g(\alpha)$ and $g(\beta)$ intersect sharply in the central portion, rather than joining smoothly as the g(X) for isomorphous components do. We assume that A and B are quite different with respect to their electronic structure and that a stable, compound-like intermediate phase γ does form. From the bond argument one could expect that there are sharp $g(\gamma)$ minima at simple ratios such as AB_2, AB and A_2B. At small stoichiometric ratios there is a maximum of bonds between unlike atoms. However, as is typical for solid solutions, the free energy does not vary as rapidly with composition as one would anticipate from, for instance, gaseous reactions. Hence, g(γ) has a comparably broad minimum, though much narrower than g(α) and g(β). Intermediate phase formation has a drastic effect on the intersolubility of the two components, as indicated in Fig.3.31. Without γ-formation the miscibility gap would extend from X_a to X_b. The intermediate phase formation shifts these limits to X_a' and X_b'. Hence, a broader miscibility gap arises than is typical for systems with $g^{XS} > 0$. Note also the relatively narrow stability range for the compound-like phase itself.

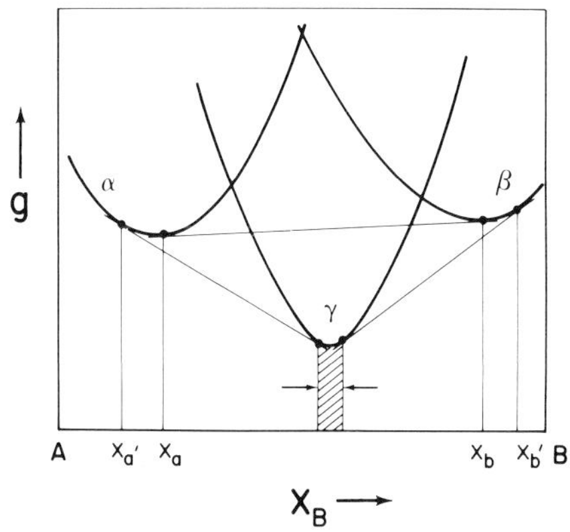

Fig.3.31. Typical free-energy curves for A- or B-rich phases and an intermediate phase in a system in which the electrochemical dissimilartiy is large

Intermediate phases may even form when the interatomic forces between like atoms are equal or even stronger than between unlike ones. Here the basis is a sizeable difference in crystal structure of the pure components. The free energy of an intermediate phase may then be lower than for either of the two parent structures. For instance, a lower strain term may result compared to the one which would be introduced in either α or β due to the large difference in atomic sizes between A and B. In addition, since these intermediate phases tend to be random solid solutions, the entropy of mixing term may also favor the new phase. Fig.3.32 displays typical g(X)-functions for such a *solid solution type intermediate phase* forming binary system. Note that the new γ phase, due to its solid solution character (where stoichiometric ratios are of minor significance) has a much broader minimum than in the compound-like case. Consequently, the γ stability range is also wider and the intersolubility restrictions less drastic.

Although the above discussion on intermediate phase formation was "graphically based" upon parabolic g(X)-function, it should be emphasized that intermediate phase formation is not restricted to systems with $\Delta h^{xs} < 0$ and does also occur for solutions with positive heat of dissolutions, i.e., for g(X) with inflection.

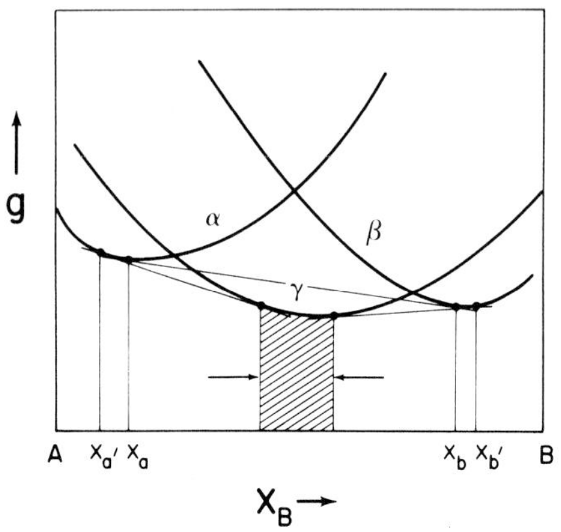

Fig.3.32. Typical free-energy curves in a system in which a solid-solution type intermediate phase forms

3.4.6 Congruent and incongruent melting

Solid and liquid phases of the same composition in equilibrium at a certain temperature, i.e., congruent melting, was seen already in Figs.3.17 and 3.20. A discussion was postponed up to this point because invariant reactions as well as intermediate phase formation frequently play an essential role in the occurrence of congruent melting in phase diagrams.

The g(X)-curves and the resulting phase diagram for a system with two eutectics in which an intermediate phase γ forms directly from a melt of same composition are shown in Fig.3.33. For instance, the system Bi-Te with the intermediate compound Bi_2Te_3 falls into this category. A "streamlined" version of the phase diagram (Fig.3.33) as it often occurs in the literature, is depicted in Fig.3.34. Usually the stability range for γ is quite narrow and, hence, is represented by a single vertical line. The stability range for α and β is also frequently so narrow that the solvus lines (dashed in Fig.3.34) more or less coincide with the vertical pure component lines and consequently are not shown in the phase diagram. Solidification from alloys of overall compositions to the left and right of the eutectic points thus leads initially to almost pure A and B, respectively. Fig.3.34 indicates that "pure" γ can be obtained in two basically quite different ways:

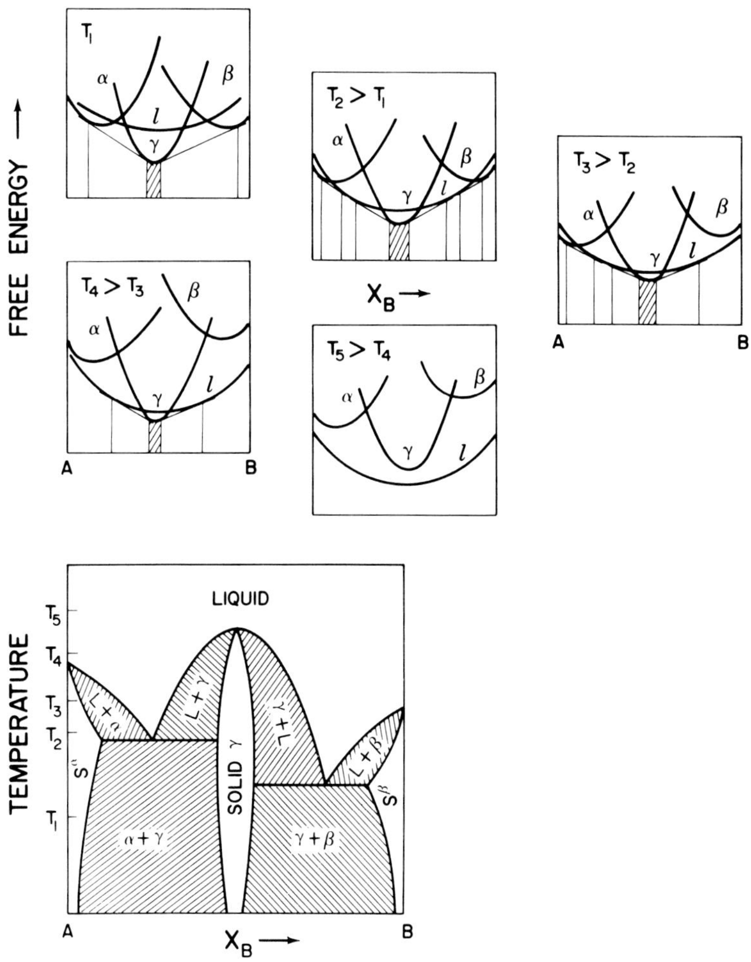

Fig.3.33. Free-energy functions and equilibrium phase diagram with intermediate phase formation directly from melt. After [3.25]

1) Melt of composition X_m leads directly and quantitatively to solid γ. This process, though conducted at the highest temperature, T_m, has the advantage that transport problems due to different melt and solid composition are avoided.
2) Any melt with a composition between the two eutectic concentrations can be partly converted to γ. The more the composition deviates from X_m the lower is the solidification temperature. However, since either A or B is rejected at the interface, complex transport phenomena do occur in the melt. The melt can only be converted to γ as long as $X_{e_1} < X(L) < X_{e_2}$.

The consequences of these two cases, 1) and 2), for crystal growth processes will be discussed in the chapter on applications of phase diagrams based upon systems of general interest.

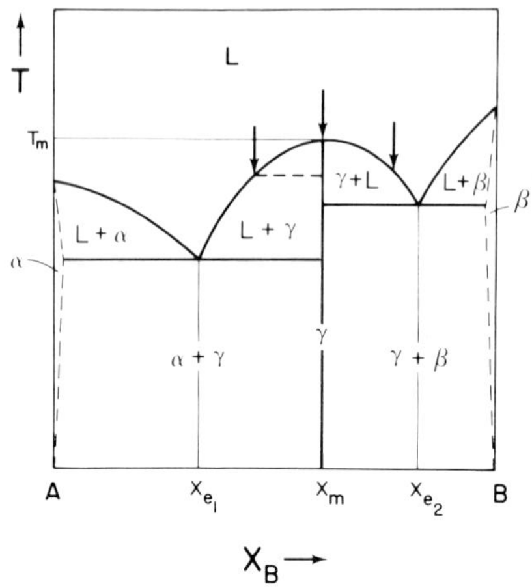

Fig.3.34. Congruently melting compound (composition X_m) in binary phase diagram with two eutectics

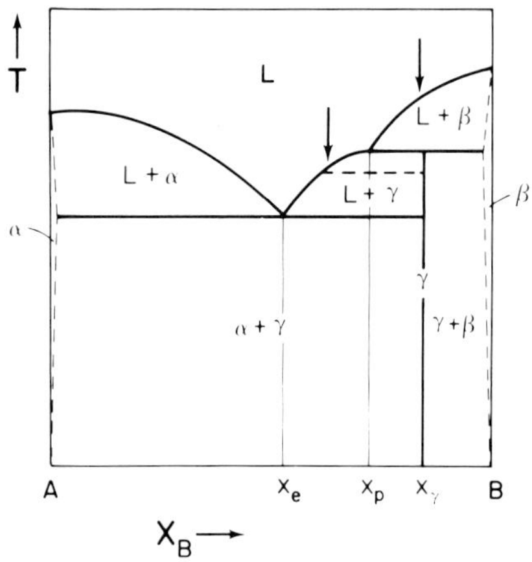

Fig.3.35. Incongruently melting intermediate phase γ in binary system

When a compound decomposes into components upon heating before its "melting point" is reached, it is said to melt incongruently. In this sense a solid with the peritectic composition X_4 (see Fig.3.28) melts incongruently at T_p since it decomposes into liquid of composition X_c and solid of X_a.

Figure 3.35 represents a system in which an intermediate compound melts incongruently due to a peritectic transition. The solid solution γ begins to "melt" by separating into a liquid solution of X_p and more or less pure solid B. Here again, pure γ can be obtained in two ways:

1) Liquid solutions of compositions between X_p and B lead through the peritectic reaction to γ. This is not a very practical way since γ forms at the interface between β and L, thus drastically impeding further reaction. In principle, of course, under quasistatic conditions a melt of X_γ can completely be converted into γ.
2) Any melt of composition between X_e and X_p can be partly converted into γ without involving another solid phase. These processes occur at lower temperatures than 1).

3.4.7 Vapor-liquid-solid equilibria

In the interest of simplicity, we have centered the foregoing discussion of binary systems upon solid-liquid equilibria. Yet for modern crystal growth, a thorough understanding of vapor-condensed phase equilibria is equally important. Vapors may constitute the nutrient (or part of it) in a crystal growth system, or, less desirably, evaporation from liquid nutrients may result in intolerable material losses. In both cases, one must understand the coexistence conditions between the vapor and condensed phases in the system over a considerable pressure range in order to be able to choose advantageous growth conditions. Hence, one is frequently faced with the necessity of reading or determining complex sections through the V-L-S stability ranges of a T-P-X phase diagram. The *thermodynamic description* of phase equilibria presented earlier in this chapter applies to equilibrium between all states of matter. Therefore, in the following we will mainly concentrate on the *graphical presentation* of the full three-dimensional phase figure of binary systems as it has been introduced by BAKHUIS ROOZEBOOM in 1904 [3.26]; see also the monograph by ZERNIKE [3.27]. An understanding of the complex conditions in three dimensions will allow us to better "see" the origin and connections of the various T-P, T-X, P-X projections and sections used in the literature for presentations of vapor-liquid (V-L), vapor solid (V-S) and vapor-liquid-solid (V-L-S) equilibria. It will also form a valuable supplement to the earlier discussion of isobaric solid-liquid equilibria.

Solid-liquid equilibria in systems with vapor pressures below one atmosphere in the temperature range of interest are typically determined at a *total pressure* of P = 1 atm of inert gas and vapor rather than at atmospheric pressure of the vapor of the material alone under a piston. Hence, most binary phase diagrams are only "pseudo-binaries". However, as we have discussed in Sect. 3.3, very little accuracy is lost this way. The presence of a partial pressure of inert gas, that may be considerably higher than the intrinsic vapor pressure at a given temperature and composition, has little effect upon vapor-condensed phase equilibria. The data in a T-P-X space figure must be understood in the same sense. Strictly speaking, they are supposed to represent a true "piston-pressured" system, although they may not have been determined that way. We do understand, however, that in first order the pressure of an inert gas does not shift the V-L-S equilibria. This, of course, is only true as long as the partial pressure of the inert gas is low enough so that the changes in Gibbs function of the condensed phases (by compression) and in activity of the equilibrated vapor (non-ideal behavior) can be neglected. Beyond that, if real gases are used for pressurization, one will have to expect that "inert" gases are dissolved in the condensed phases, which changes their Gibbs functions. More significant interactions between inert gases and vapors concern evaporation and condensation *rates*. These are non-equilibrium phenomena and, hence, will be discussed at a later point.

The least complex binary T-P-X phase figure is obtained for systems with continuous solid solution range. Such a system is also bound to have full miscibility in the liquid phase. Consequently, as indicated in Fig.3.36, the phase figure consists of three bodies:

1) the vapor-liquid equilibrium body which is bounded by a *chaoticus** surface V (the dew point surface) and a *liquidus* surface L (the boiling point surface) which intersect the T-P planes at X = 0 and X = 1 along the boiling point curves of the pure components;
2) The liquid-solid equilibrium body, which is bounded by a *liquidus* surface L (the freezing point surface) and a *solidus* surface S (the melting point surface) which intersect the pure component planes along the melting point curves for A and B, respectively;
3) the vapor-solid equilibrium body, which is bounded by a *chaoticus* surface V (the condensation-to-solid surface) and a *solidus* surface S (the sublimation-pressure surface) which intersect along the sublimation pressure curve for pure A and B.

To facilitate the graphical presentation in Fig.3.36, we have assumed a system in which both the triple-point pressure and temperature of B are considerably higher than for A. Furthermore, we have assumed that the equilib-

---

* from the Greek word chaos from which the international word gas derives.

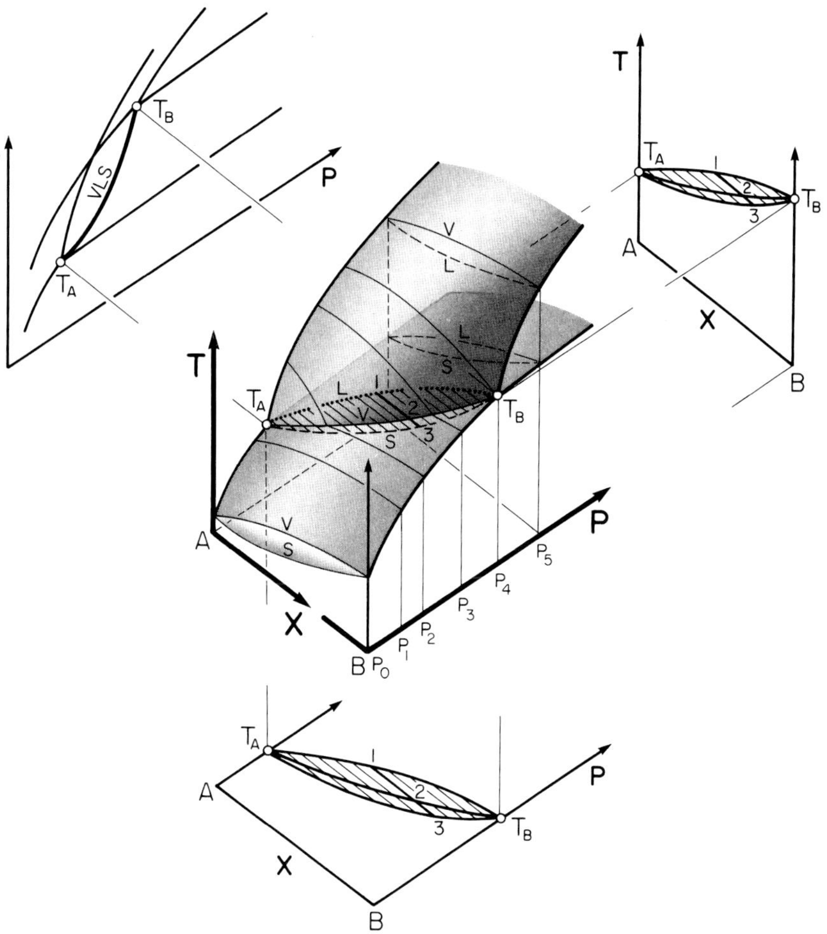

Fig.3.36. T-P-X phase diagram without critical region for a binary system without miscibility gaps. Three-dimensional phase figure and corresponding T-P, T-X and P-X projections

rium surfaces (two each) monotonically connect the sublimation pressure, boiling point and melting point curves of A and B. To indicate the relative inclinations of the equilibrium bodies, their upper surfaces have been shaded and show some lines of intersection with planes of constant pressure, P_1 - P_4. The lens-shaped cross section of the bodies is indicated in the isobaric sections at P_0 and P_5, respectively. Visualization of the phase figure may be aided by the imagination that each of the three bodies is generated by sliding its lens-shaped vertical cross section along the corresponding pair of lateral curved "rails" (the melting point curves, boiling point curves and sublimation pressure curves).

These three equilibrium bodies intersect along three curves. The two chaoticus surfaces meet in curve V (solid line), the two liquidus surfaces intersect along curve L (dotted line) and the lowest lying curve, S (dashed line) is common to the two solidus surfaces. These three curves from the *three-phase strip* (here sickle shaped; also called three-phase blade) which connects the triple points T_A and T_B. Since all bulk phases in equilibrium have the same pressure and temperature, coexisting phases lie on straight lines, *three-phase axes*, normal to the T-P plane, as indicated by the par-alled shading in the T-P-X figure of Fig.3.36. These three-phase tie-lines rule the single-curved three-phase strip. (The reader may be reminded that ruled surfaces are generated by a moving straight line, in contrast to double-curved surface — such as the bounding surfaces of the three two-phase bodies — that are generated by curved lines.) The ruling of the three-phase strip parallel to the X-axis results in a single (curved) line projection in the T-P plane which for the assumed case lacks an extremum between T_A and T_B. The other lines in the T-P projection are simply the sublimation pressure, melting point and boiling point curves of the pure components. The projections of the three-phase strip into the P-X and T-X planes are lenses which monotonically descend from T_B to T_A. The equilibrium curves for the pure components project into the P- and T-axis, respectively, and hence do not show up in these views.

It should be noted that in this case of continuous solubility with monotonically descending equilibrium surfaces all coexisting phases have different compositions, or in other words, all possible condensed phases melt or vaporize incongruently. A variation of this case, with a congruently melting solid solution composition is assigned in Problem 3.10.

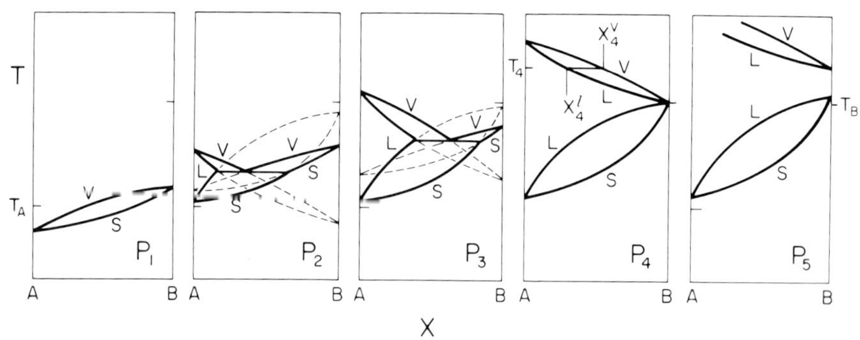

Fig.3.37. Isobaric sections P_1 - P_5 through the foregoing three-phase space figure (schematic)

Isobaric sections at pressures P_1 - P_5 through the three-dimensional T-P-X space figure are given in Fig.3.37. The reader is urged to correlate the various coexistence lines in these T-X sections with the corresponding parts of the T-P-X figure. Here, we will only compare the isobaric binaries at P_3 and P_4. At P_4, which corresponds to the triple point pressure for B, the "lens" for coecisting vapor mixtures and liquid solutions meets the lens for liquid-solid solution equilibria in T_B. Hence, at this pressure, only pure B can simultaneously exist in all three phases. All others are two-phase equilibria. The composition of coexisting vapors and liquid solutions, for instance, can be deduced from the upper lens in the very same manner as one determines liquid-solid data from tie lines between liquidus and solidus. For example, at T_4 one finds that vapor of compositions X_4^v, on the dew or condensation line, is in equilibrium with liquid of composition X_4^l, on the boiling point line. As one lowers the pressure to P_3, the two lenses penetrate each other. Moreover, on the B-side, the solid-vapor equilibrium lens begins to appear. All three lenses terminate on the three-phase axis at T_3 where liquid of composition X_1 is in equilibrium with vapor of X_2 and solid of X_3. These points are also shown in the space figure and its P-X and T-X projections in Fig.3.36.

Isothermal sections of a space figure, i.e., P-X diagrams largely resemble the corresponding T-X diagrams turned round over the X-axis. Hence, the presentation of a typical sequence of isothermal sections is left to the reader in Problem 3.11. P-X diagrams are of particular value in materials preparation. In many occasions, one will choose the temperature of a process mainly for reasons of kinetics, apparatus capability, material properties, etc. A P-X diagram facilitates then the adjustment of the composition of the desired condensed phase through the vapor phase composition, i.e., through adjustment of the partial pressure ratio that can frequently be controlled by exterior means.

The vapor pressure curves in P-X diagrams represent the total vapor pressure above the condensed phase (solid or liquid) as a function of composition at the considered temperature. The shape of these vapor pressure curves allows for conclusions with respect to the intermolecular forces between the components in the condensed phase solution. Above ideal solutions the concentration dependence of the partial vapor pressures of the components obeys *Raoult's law*

$$p_i = X_i p_i^0 \quad , \tag{3.37}$$

i.e., the partial pressure of each component is proportional to the vapor pressure of the pure condensed component with the mole fraction as proportionality constant. The partial pressure is understood as the pressure that would exist in a container if only the considered component were present with the same concentration. Hence, as illustrated in Fig.3.38, the partial pressures above ideal solutions can be represented by straight lines in a P-X plot.

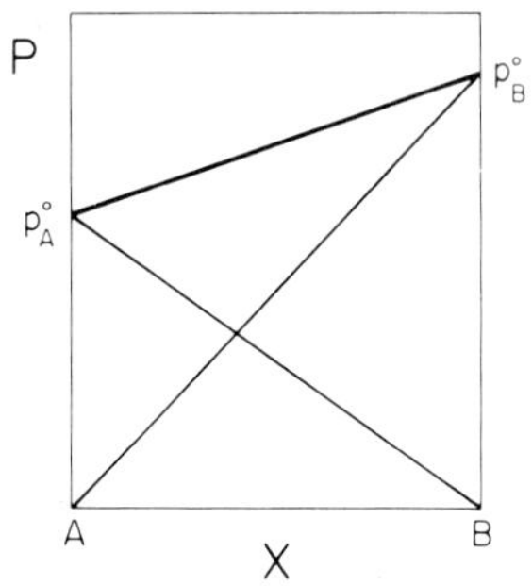

Fig.3.38. Illustration of Raoult's and Dalton's law: P-X diagram of total vapor pressure and partial pressures above an ideal solution

If, in addition, the vapors above the originating phase behave ideally, *Dalton's law*

$$P = \sum_i p_i \tag{3.38}$$

applies, i.e., the total pressure is simply the sum of the partial pressures and the P(X)-function is also a straight line. Dalton's law applies with reasonable accuracy to real gases as long as the vapor density is not too high and/or the vapor does not contain a significant concentration of polarized species.

When the pressure on a single-phase solution is lowered to the point where the composition coordinate intersects with the vapor pressure curve (e.g., at X_1^l in Fig.3.39b) the underlying liquid or solid begins to vaporize. The vapor that is formed is richer than the condensed phase in the more volatile component, as proven in Problem 3.12. This compositional change (to $X_1^v \equiv Y_1$) places the chaoticus below the vapor pressure curve over the whole X-range, as indicated in Fig.3.39. As long as ideal behavior can be assumed in the vapor, the mole fraction of a vapor component is simply

$$Y_i = \frac{p_i}{P} \quad , \tag{3.39}$$

from which one obtains the useful relation for a binary vapor in which A and B exist in the same molecular form (see also Sect. 4.4)

$$\frac{p_A}{p_B} = \frac{Y_A}{Y_B} \quad . \tag{3.40}$$

For non-ideal solutions the partial vapor pressures are no longer directly proportional to the mole fractions of the components in the solution. If the A-B cohesive interactions are stronger than the A-A and B-B interactions, the escaping tendency from the solution will be reduced. The resulting negative deviation from Raoult's law for such a system is schematically indicated in Fig.3.39a.

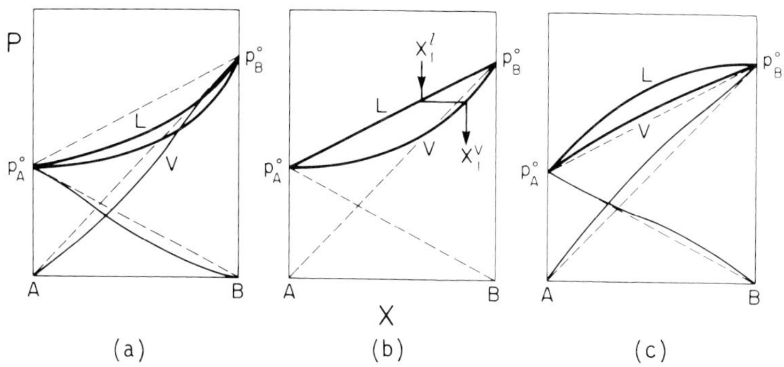

Fig.3.39. Schematic P-X diagrams for (a) solutions with negative deviations from Raoult's law, (b) ideal solution, (c) positive deviation. Thin lines: partial pressures. Dashed lines: ideal behavior

Positive deviations from Raoult's law (Fig.3.39c) will result when the attractive interaction between A and B molecules is weaker than between likewise molecules. For example, positive deviations are often observed in aqueous solutions. Pure water is itself strongly associated. The addition of a second component may partially "depolymerize" the water, leading to an increased partial water vapor pressure. Figures 3.39a and c reflect also the fact that for real solutions Raoult's law always applies to a component in the limit as its mole fraction approaches one. Furthermore, it is noted that in both cases at low concentrations ($X_i \to 0$) the partial pressures of the components are again proportional to their mole fraction; the proportionality factor K for real solutions, however, deviates from p_i^0. This behavior is expressed by *Henry's law*

$$p_i = X_i K(T) \quad , \tag{3.41}$$

where K(T) must be experimentally determined for each binary. This "ideal" behavior of real solutions, i.e., of the activities, in the lowest and highest part of the composition range has already been indicated in connection with Fig.2.5.

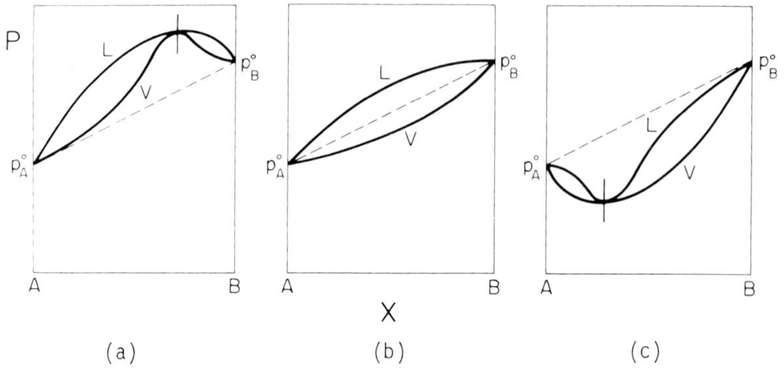

Fig.3.40. P-X phase diagrams for real solutions; (a) and (c) show congruently vaporizing solutions

Sufficiently great deviations from ideality (i.e., from Raoult's law) may lead to extrema in the P-X diagram, as depicted in Fig.3.40. Extremum formation is facilitated by comparable vapor pressures of the pure components. At a maximum or minimum in the vapor pressure curve, the vapor and condensed phase must have the same composition, i.e., the condensed phase *vaporizes congruently*. A liquid solution with the composition of an extremum in the vapor pressure curve is called an *azeotrop*. A maximum in the vapor pressure curve (P-X plot) implies a minimum in the boiling point curve (T-X plot), and vice versa. Note the analogy to the T-X diagrams for fully miscible liquid-solid binaries in Fig.3.17.

For the next case to be presented in a T-P-X binary (Fig.3.41) we restrict solid solution formation to narrow ranges at the pure component ends of the composition axis, with a broad miscibility gap in between. For the liquid we maintain continuous miscibility. As for the foregoing case we will first discuss the individual equilibrium bodies and then consider their intersection. For the vapor-liquid equilibrium body we choose again, for the sake of graphical simplicity, a lens-shaped cross section. In reality, of course, it may possess any of the cross sections shown in Figs.3.39 and 3.40. The cross section of the liquid-solid body is that of a eutectic binary, as indicated in the isobaric section at P_5. The tie-lines between solid S_A (solid solu-

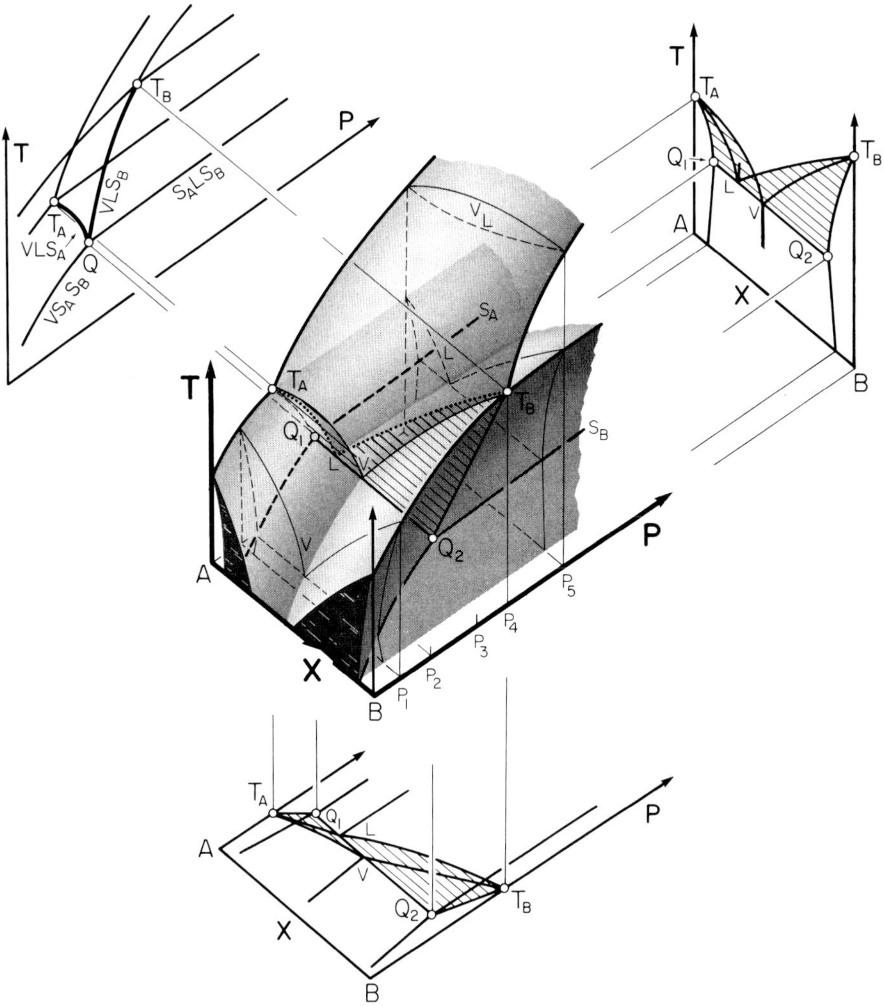

Fig.3.41. T-P-X phase diagram without critical region for a binary system with solid solubility gap. Three-dimensional phase figure and corresponding T-P, T-X and P-X projections

tion of some B in A), S_B and L form a three-phase strip which is bound by the straight line Q_1Q_2 and the slightly curved lines Q_1S_A and Q_2S_B. Hence, the three-phase strip, which forms also the bottom of the eutectic trough, does not extend all the way to the pure component T-P planes. Since the strip is ruled by straight lines parallel to the X-axis, by the tie-lines, its projection into the T-P plane occurs as a line, S_ALS_B. Note that this three-phase projection line has a positive gradient for normal materials and slopes negatively for materials that contract upon melting. Due to the restriction on the solid miscibility, the V-S equilibrium consists also in the coexistence of two different solid phases, S_A and S_B and vapor. Correspondingly the V-S equilibrium body has also a eutectic-shaped cross section as revealed by the isobaric section at P_1. The corresponding three-phase strip (again ruled with tie-lines normal to the T-P plane) is supended between the steeply downward sloping lines that originate in Q_1 and Q_2. In the T-P projection this strip, which forms the bottom of the eutectic trough that descends from point V, occurs as line VS_AS_B.

These two three-phase strips intersect at a certain temperature and pressure. Along the straight line of intersection, Q_1Q_2, equilibrium prevails between two solids, a liquid and a vapor. Q_1Q_2 is called the *quadruple axis*. As one can verify from the phase rule, this four-phase equilibrium situation is invariant (f = 0). Hence, the projections of Q_1Q_2 results in a point Q in the T-P plane; or in other words, Q_1Q_2 lies parallel to the X-axis in the space figure. This, of course, can also be deduced from the fact that both three-phase blades that intersect along Q_1Q_2 are ruled parallel to the X-axis. It is essential to note that the endpoints of the quadruple axis, Q_1 and Q_2 approach only the pure component planes in the limit as the solid solubility ranges converge towards zero compositional width. The situation of the quadruple axis is further illustrated by P-X sections at the quadruple temperature in Fig.3.42.

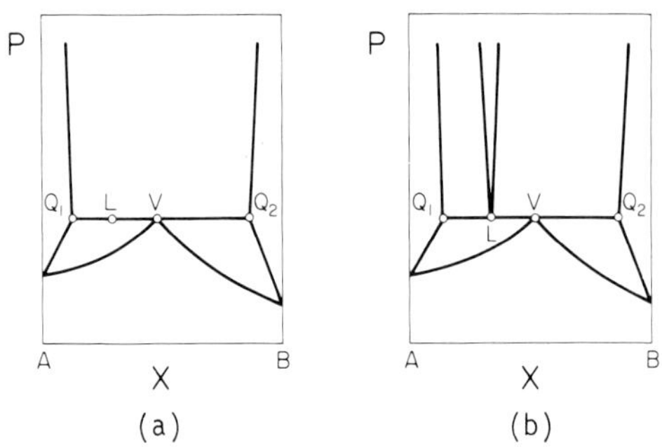

Fig.3.42. P-X diagrams through the quadruple axis, (a) for normal substances, (b) for materials that contract upon melting (H_2O-type)

In addition to the VS_AS_B and LS_AS_B equilibria we have to accomodate two more three-phase equilibria, namely VLS_A and VLS_B. The corresponding three-phase strips in the space figure are spanned between the triples of curved lines T_AQ_1 - T_AL - T_AV and T_BL - T_BV - T_BQ_2, respectively. These two three-phase strips, in contrast to the ones discussed before, have each a point in the corresponding pure component planes: the triple points T_A and T_B where solid A and B, respectively, coexist with their pure melt and vapor. These two strips, again ruled parallel to the X-axis, occur as curved lines VLS_A and VLS_B in the T-P projection.

The above description of the space figure is supplemented with a schematic presentation of isobaric sections P_1 - P_4 in Fig.3.43 (see also Problem 3.13) and a sequence of isothermal sections in Fig.3.44. The reader is urged to test his understanding of the three-dimensional figure by correlating the various lines and their intersections in the T-X and P-X diagrams with the corresponding equilibrium surfaces. In particular, we want to draw attention to the narrow coexistence ranges S_A + L, L + S_B and stability ranges S_A and

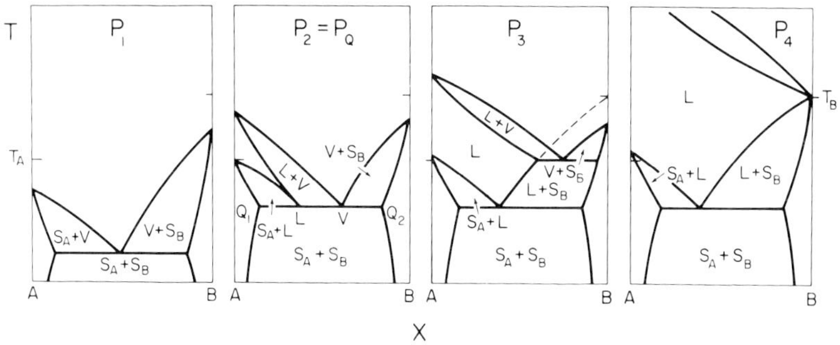

Fig.3.43. Isobaric sections (P_1 - P_4) through Fig.3.41 (schematic)

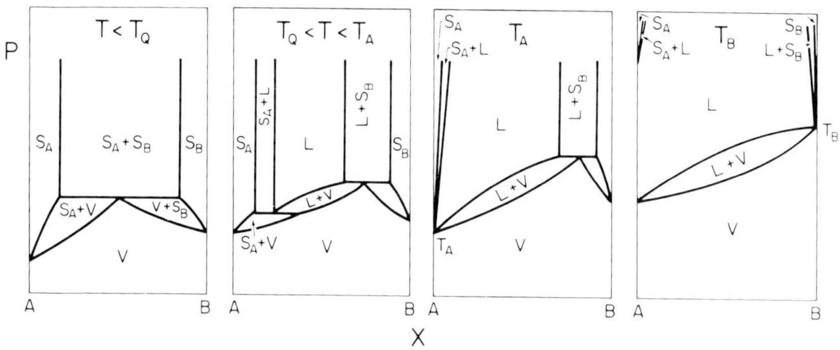

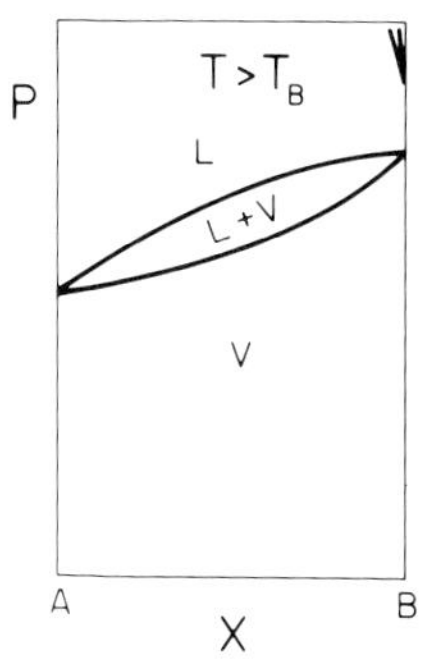

Fig.3.44. Isothermal sections through Fig.3.41 (schematic)

S_B in the P-X diagrams that result for materials which expand upon melting (normal materials).

Many of the technologically important crystal growth materials consist of compounds rather than solid solutions of components. Hence, in the following, we will add to the preceding case the formation of a compound A_nB_m at a certain stoichiometric ratio n:m. Such a system may be described as a combination of the binary systems of each of the components with the compound. The space model can, in essence, be constructed by placing side by side the phase figure of the two systems A - A_nB_m and A_nB_m - B. The composition axis must be divided by the two subsystems in the ratio n:m, for instance 1:1 if the compound is AB, as assumed for simplicity's sake in the following graphical representations.

Strictly thermodynamically speaking, a compound cannot have a triple point. For only if the Gibbs free energy of formation of a compound were infinite could the solid be "non-dissociating" at elevated temperatures, i.e., solid of strictly stoichiometric composition could coexist with stoichiometric vapor and/or liquid (see Problem 3.14). Hence, any real compounds are dissociated to some extent, though the thermally induced composition changes may be too insignificant (i.e., one percent or less) to be represented on the conventional scale of a phase diagram. Hence the traditional distinction between "dissociating" and "non-dissociating" compounds must be understood as a rather crude quantitative differentiation, which, in particular, in the light of modern solid state technology has even lost its practical justification. As we will see in Chap. 4 with compound semiconductors, thermally induced deviations from strict stoichiometry in a range way below one percent can result in drastic changes of (electrical) properties.

If a compound remains approximately stoichiometric, i.e., is "non-dissociating" and can, for the appropriate T-P-X combination, be graphically presented by a triple "point", the phase figure is simply a side-by-side presentation of the figures of the subsystems. For instance, for a liquid-solid binary with a compound AB, one may obtain a T-X diagram as depicted in Fig. 3.45a. The liquidus curves around X = 0.5 are clearly two different curves meeting each other under an angle. This angle is a measure for the melting point depression of AB upon addition of A or B.

Yet many compounds show dissociation that is significant on the macroscale. The non-stoichiometric compound that is obtained on heating can be regarded as the fictitious stoichiometric compound with some excess A or B dissolved in it. Addition of another small amount of either A or B can be expected to leave the melting point of the real compound relatively unaffected since there are already finite amounts of these excess components present. Hence, with increasing deviation from stoichiometry in different dissociating compounds one will have a transition from the cusp-shaped liquidus maximum in Fig.3.45a to a continuous curve with horizontal tangent, as depicted in Fig.3.45b. Hence, the sharpness of the melting point maximum can be regarded as some measure for the degree of dissociation of a compound. Unfortunately, many phase diagrams have not been determined with sufficient accuracy to make this a reliable argument.

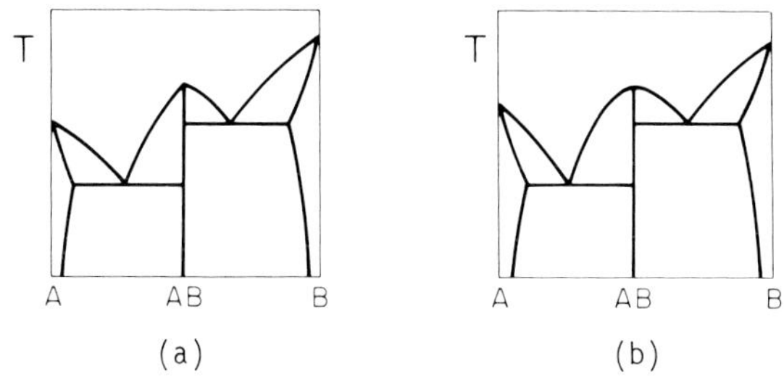

Fig.3.45. T-X phase diagram for binary system in which a compound AB is formed, (a) AB does "not dissociate" (b) AB dissociates at high temperatures

For the three-dimensional presentation of the phase diagram of a binary system which forms a dissociating compound, Fig.3.46, we chose the case in which the vapor pressure of the compound is intermediate between those of the components. This choice is quite arbitrary and was made for graphical reasons. Cases with compound vapor pressures being lower or higher than that of either component are found as well. Since we have again assumed narrow solid solubility ranges at the pure component ends of the composition axis, the space figure, except for the center part, is a side-by-side arrangement of two Figs.3.41. The center, i.e., the coexistence curves around the composition plane for the compound AB must reflect the newly introduced feature of dissociation of the compound at elevated temperatures. We assume that nowhere vapor and liquid have the same composition, and the compound S_{AB} is never in equilibrium with both vapor *and* liquid of composition AB. Hence the triple point of the non-dissociating case splits up here into two points:

1) The point M(T), at which the congruent (yet not strictly stoichiometric) melting point curve $S_{AB} = L$ is tangent to the solid-vapor-liquid coexistence line $S_{AB}VL$. Due to the insignificant slope of $S_{AB}VL$, M(T) is very close to, yet not identical with, the temperature maximum of $S_{AB}VL$. M(T) represents (for normal substances) the minimum temperature and pressure

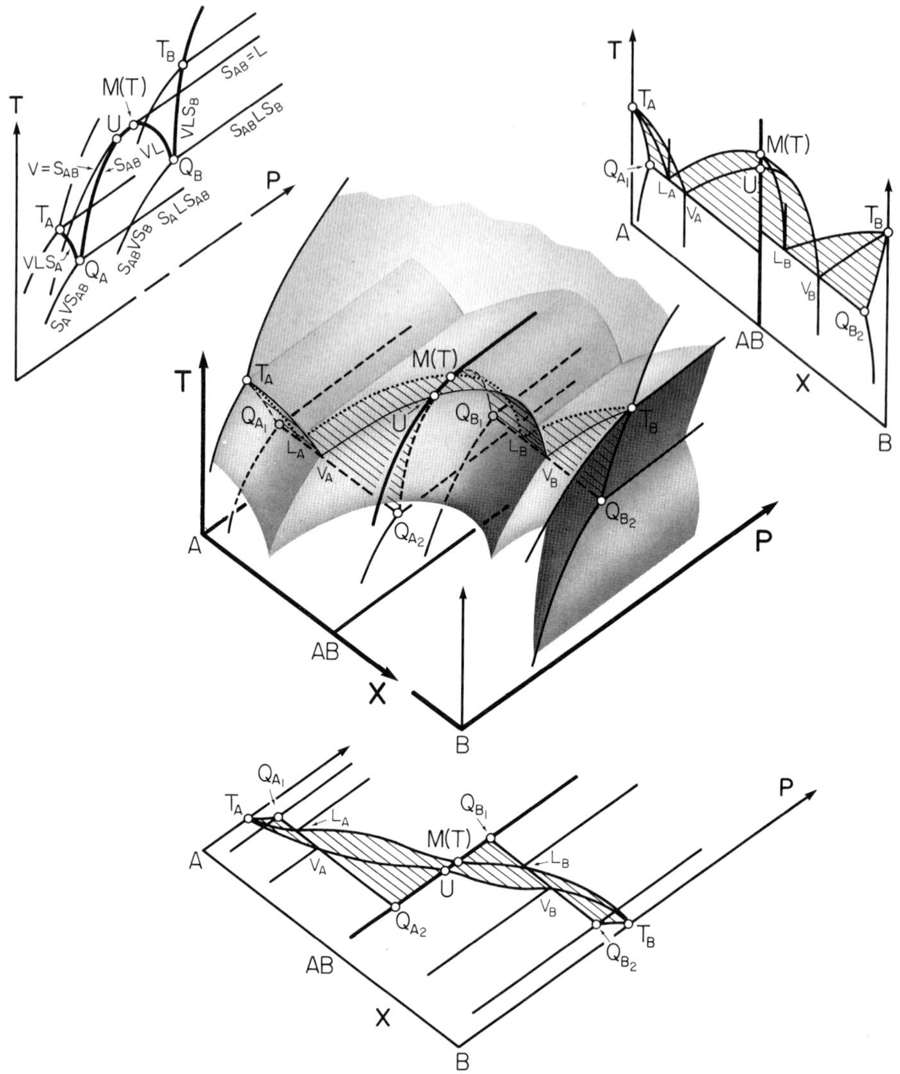

Fig.3.46. T-P-X phase diagram for a binary system in which a slightly dissociating compound forms which may be congruently molten and sublimed. Three-dimensional phase figure and corresponding T-P, T-X and P-X projections

combination at which S_{AB} is in equilibrium with liquid of the same composition only. Hence M(T) is often called the *minimum melting point*. At lower P-T combinations S_{AB} melts incongruently *and* gives off a vapor of yet a different composition.

2) The point U, at which the congruent sublimation curve $V = S_{AB}$ is tangent to the coexistence line $S_{AB}VL$. This point, just as M(T), does not coincide with the exact stoichiometric composition. U represents the maximum tem-

perature and pressure combination at which S_{AB} is in equilibrium with vapor of same composition only. Hence U is often called the *maximum sublimation point*. At higher temperatures (and pressures), i.e., between U and M(T), S_{AB} sublimes incongruently (vapor richer in B) and at the same time forms a liquid that is richer in A. Near M(T) much liquid is formed and little vapor, and near U it is just the other way around. In the T-X and P-X projections the compound AB is represented with heavier lines in order to indicate a stability range of finite width, which in higher resolution plots would require two lines. It is useful to visualize that, for instance, Q_{A_2} and Q_{B_1} are situated in two slightly different planes.

The three-phase strip in the center of the phase figure, which is suspended between the curves $L_AM(T)L_B$, V_AUV_B and $Q_{A_2}UM(T)Q_{B_1}$, is again ruled parallel to the X-axis as indicated in the T-X and P-X projections. The projection of this strip into the T-P plane is a single-curved line between Q_A and Q_B, the projections of the two quadruple axes. Note that $L_AM(T)L_B$ and V_AUV_B result from the intersection of liquid (-vapor) with liquid (-solid), and vapor (-liquid) with vapor (-solid) surfaces, respectively. The curve $Q_{A_2}UM(T)Q_B$, however originates from the intersecting AB "plane". For a non-dissociating compound U and M(T) would coincide and another steeply upward sloping curve L = V would originate in the resulting triple point.

The intersections and homogeneity ranges of the various equilibrium bodies that form the space figure are further illustrated in a sequence of isobaric sections, Fig.3.47. The dissociation behavior becomes particularly obvious in the isobaric sections through the maximum sublimation point and minimum melting point. At P_U one sees congruent – though not stoichiometric – solid and vapor coexisting with A-rich liquid. At $P_{M(T)}$ congruent solid and liquid are in equilibrium with a vapor of different composition, which in our case is B-rich. Note also that two separate lines are used to schematically indicate the finite width of the S_{AB} range. This helps also in visualizing that no structural change is observed during heating or cooling of S_{AB} on passing the temperature of a eutectic trough.

The space phase figure and its sections illustrate that the solid compound S_{AB} can be obtained from liquid and gaseous nutrients of a wide range of composition. Detailed criteria for the specific choice of a "working point", i.e., of the T-P-X combination during a crystal growth process, will be developed in later parts of this text. Yet we can at this point already deduce the following rules:

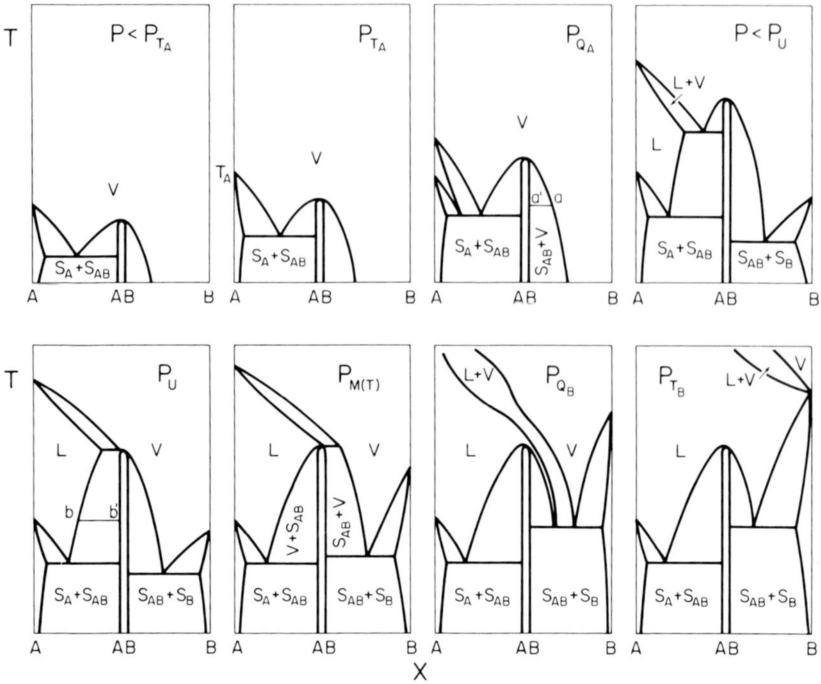

Fig.3.47. Isobaric sections through Fig.3.46 (schematic)

1) Although it may be possible to grow a compound from liquids and vapors of largely varying composition, the specific stoichiometry of the resulting S_{AB} depends on the specific composition of the nutrient phase. For instance, if S_{AB} is crystallized from congruent vapor it will be somewhat differently composed than B-rich S_{AB} obtained from B-rich vapor (see points a-a' in the P_{Q_A} isobaric section of Fig.3.47). Or, if in order to work at lower temperatures S_{AB} is solidified from A-rich melt (points b-b' in the P_U isobaric section) the solid compound is likely to contain more A than the congruently solidified S_{AB}.
2) In growth from melts or vapors one will try to avoid a working point close to U and M(T) where the intimate coupling between the three phases makes control more difficult. For example, if one would crystallize S_{AB} from congruent vapor close to U, slight fluctuations in temperature would cause fluctuating appearances of a liquid phase. Or if solidification would be conducted "at" M(T), fluctuations in the vapor composition would considerable modulate the liquid composition, solidification rate, etc.

For the choice of operating conditions in (crystal growth) processes that are governed by vapor-condensed phase equilibria, isothermal sections are also very useful. For instance, the consequences of compositional changes in the vapor phase from selective evaporation losses or from exterior adjustments of a partial pressure can more readily be deduced from P-X diagrams than from isobaric sections. Figure 3.48 represents a sequence of isothermal sections through the foregoing phase space figure with compound formation, Fig.3.46. It is instructive to follow through these diagrams how upon increase in temperature the liquid-solid "non-existence" ranges L + S_{AB} and S_{AB} + L "close in" on the S_{AB} stability range. At the temperature of M(T), L and congruent S_{AB} coexist at the pressure of M(T). At higher pressures, however, there remains a narrow wedge on both sides of S_{AB}, or, in other words, S_{AB} is in equilibrium with an L of slightly different composition. Another particularly noteworthy feature is the constriction in the L-V lens

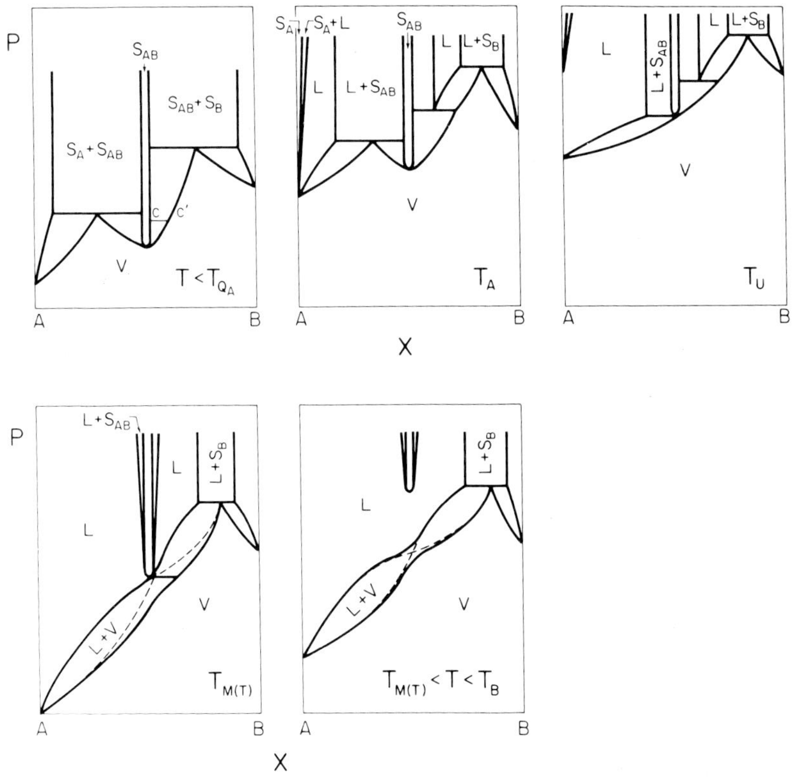

Fig.3.48. Isothermal sections through Fig.3.46 (schematic)

in the neighborhood of the dissociating S_{AB}. Even at high temperatures where there is no "direct contact" between S_{AB}-L and L-V coexistence lines, a pronounced constriction remains, indicating the existence of a dissociating compound. The dashed lines in the isothermal sections at M(T) and $M(T) < T < T_B$ indicate schematically the shape for the chaoticus in a corresponding system with triple point, i.e., with a "non-dissociating" compound.

The isothermal sections at lower temperatures reveal another feature of V-S_{AB} equilibria that is of considerable practical importance. If, for instance, S_{AB} of composition c is heated to $T < T_{Q_A}$, vapor of composition c' is given off. Under free sublimation conditions, that is, if the vapor is continuously removed, the compound loses more B than A. As a consequence, the solid composition will eventually be changed to that of the pressure minimum in the S_{AB} range from when on it sublimes congruently – but not necessarily stoichiometrically. The same applies to free sublimation from an initially A-rich S_{AB}. As discussed in connection with Fig.3.40, such a pressure minimum can be expected when both components have comparable volatility. Examples are the semiconducting IIB-VIA compounds, with the chalcogenides of Hg (which has the highest fugacity of the IIB elements) possessing the narrowest S_{AB} homogeneity range. Note that with increasing temperature the chaoticus maintains a pressure minimum up to T_U from where on the minimum gives way to the constriction in the L-V lens and congruent sublimation becomes impossible.

Let us conclude this section on three-dimensional V-L-S phase diagrams with a brief discussion of two special cases of the foregoing general system, which occur rather frequently in practice: the absence of a maximum sublimation and/or minimum melting point in a system with compound. Assume that the fugacity of component B is much larger than the fugacity of both A and S_{AB}. Such a situation will occur, for instance, in equilibria between an anhydrous salt (A), water (B) and their hydrate (S_{AB}). Other examples of more current interest are the III-V compounds where, as compared to the V-elements, the metallic components have negligible vapor pressure. Under no circumstances will there be coexistence of the compound and vapor of the same composition. The vapor will always be B-rich. Hence in Fig.3.40, V_A must be positioned to the right of the X_{AB}-T plane, V_B is moved closer toward B, and U must be missing. Figure 3.49 illustrates the resulting changes in schematic presentations of the T-P, T-X, and P-X projections of the space figure. Figure 3.50 depicts three isobaric sections through the space figure.

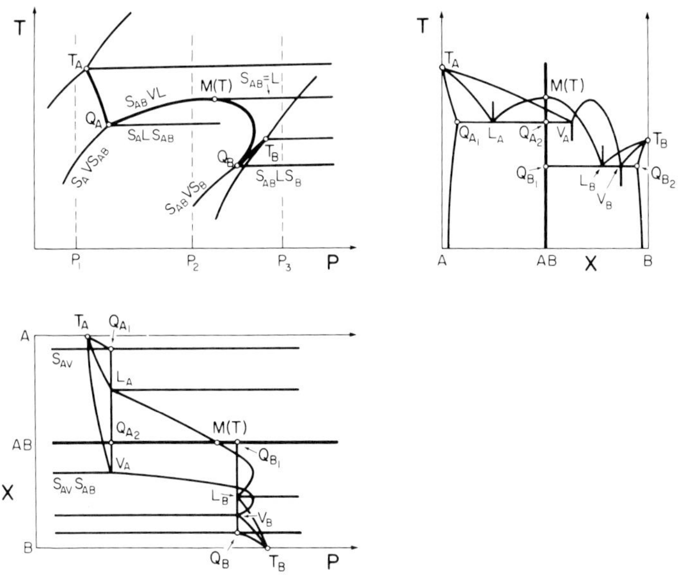

Fig.3.49. Projections of the T-P-X diagram of a binary in which a dissociating compound forms that melts congruently and sublimes incongruently (no maximum sublimation point)

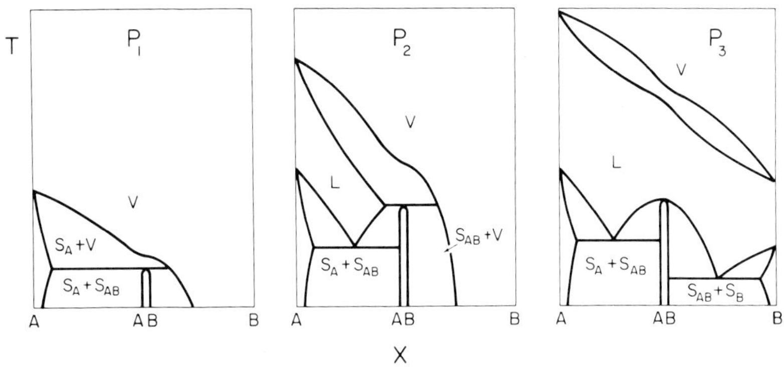

Fig.3.50. Isobaric sections through the T-P-X figure corresponding to Fig. 3.49 (schematic)

One sees that nowhere can the compound be sublimed without *decomposition*. At low pressure, P_1, on removal of the incongruent B-rich vapor the solid compound decomposes to S_A. At P_2, upon heating, S_{AB} "melts" incongruently to an A-rich liquid under B-rich vapor. Only at high enough pressures ("piston pressures", rather than inert gas pressures!), that is, at pressures above M(T),

S_{AB} can melt congruently. When a compound $S_{A_nB_m}$ sublimes incongruently, P-X diagrams lack a pressure minimum at $X_{A_nB_m}$. Yet pressure minima may even then occur at other composition coordinates if other volatile species are formed that contain the non-volatile component. An example, the siliconoxygen system with $SiO_2 \rightarrow SiO(g) + 1/2\ O_2$, will be discussed in Sect. 4.5.

The reader is urged to gain further insight into the T-P-X diagram on which Fig.3.49 is based by discussing some characteristic isothermal sections (Problem 3.15).

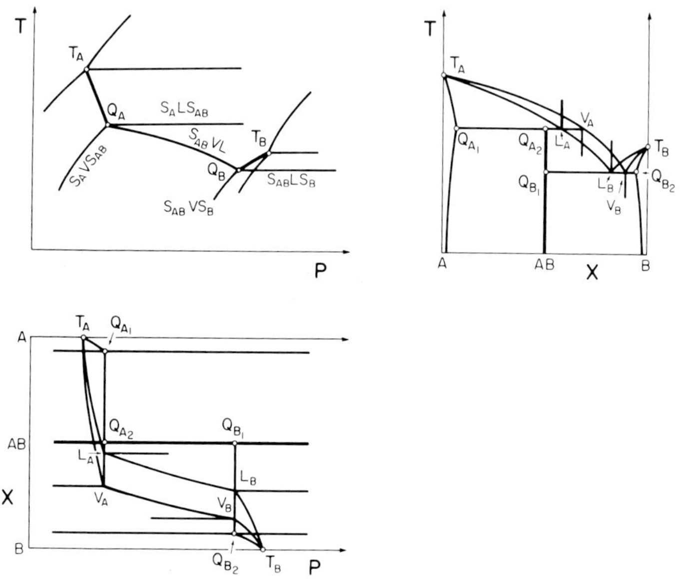

Fig.3.51. Projections of the T-P-X diagram of a binary system in which a compound forms that melts and sublimes incongruently

The basic difference between dissociation and decomposition should be emphasized. When a compound dissociates on heating, its composition is continuously shifted away from the strict (low temperature) stoichiometric ratio. However, it remains within its stability range, S_{AB}. At elevated temperatues the stability range will become increasingly assymmetric with respect to X_{AB} and will bulge toward the side of the component with the lower fugacity, as indicated in Fig.4.5 for CdS and ZnTe. Correspondingly, the composition of initially stoichiometric S_{AB} is also shifted to that side upon heating. Yet S_{AB} can be in equilibrium with congruent and hence also slightly off-stoichiometric liquid *or* vapor. Upon decomposition, however, solid S_{AB} is transformed into a different condensed phase (S_A or L, in the above case) and none of the coexisting phases are congruent.

With further increase in escaping tendency of one of the components from the compound, higher and higher pressure of the A and B species will be required at lower and lower temperatures to prevent decomposition of S_{AB} upon melting. The points L_A and L_B in Fig.3.46 will move more towards higher B concentrations. As a consequence the maximum in temperature on the three-phase strip of the compound is reduced and eventually disappears when L_A coincides with Q_{A_2}. At this point S_{AB} can still coexist with congruent L_{AB}. However, as L_A comes to the B-side of S_{AB}, congruent melting is no longer possible. T-X phase diagrams for incongruently melting compounds were discussed in the preceding section in connection with the peritectic reaction. Thus we present in Fig.3.51 only the projections of a possible space figure for a binary with a compound that can neither be congruently sublimed nor melted. A detailed discussion is left to the reader in Problem 3.16.

3.5 Ternary Systems

Since $X_1 + X_2 + X_3 = 1$ the composition of a three-component phase is defined upon specification of the concentration (mole fractions) of two constituents. Consequently, the state of a phase in a ternary system has four degrees of freedom, see (3.11). Hence, if we fix the pressure, i.e., for f = 3, three dimensions are required to represent an isobaric, ternary phase diagram.

In three-dimensional representations the z-axis is typically chosen as temperature. The simplest ternary phase diagram is then obtained for systems where all three components are isomorphic. Such a three-dimensional diagram with complete liquid solubility at high T and complete solid solubility at low T is shown schematically in Fig.3.52. Conventional two-dimensional ternary phase diagrams are either a complete projection of the three-dimensional figure down the T axis or projections of isothermal cuts. Figure 3.52 illustrates the intersection of an isothermal plane T_1 with the liquidus and solidus surfaces. The resulting liquidus and solidus lines (for T_1) are then projected onto the composition base, as also shown in Fig.3.53. The important difference to binary systems at constant pressure is that solidus and liquidus composition are not uniquely specified by the temperature. This simply

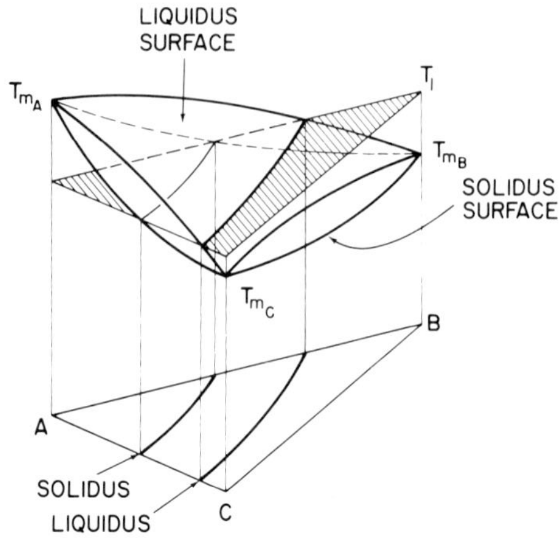

Fig.3.52. Three-dimensional phase diagram of a hypothetical isomorphous ternary system A-B-C with $T_{m_A} > T_1 > T_{m_C} > T_{m_B}$

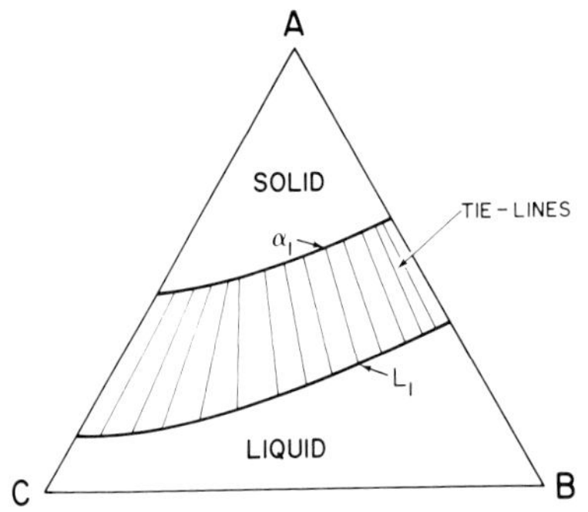

Fig.3.53. Isothermal section at T_1 of the spatial phase diagram in Fig.3.52

reflects the additional degree of freedom. Furthermore, the orientation of the tie-lines is not defined clearly either and has to be determined experimentally. Without specific knowledge of the system one can only predict, as indicated in Fig.3.53, that the direction of the tie-lines will gradually change from the direction of one bounding (binary) tie-line to that of the other. They do not need to go through the corner. L_1, for instance, is in equilibrium with the solid solution α_1.

Before we briefly discuss eutectics and miscibility gaps in ternary phase diagrams we will outline the conventions for equilateral composition triangles, as illustrated in Fig.3.54.

– Each corner represents 100% of the respective component; the opposite side = 0%.

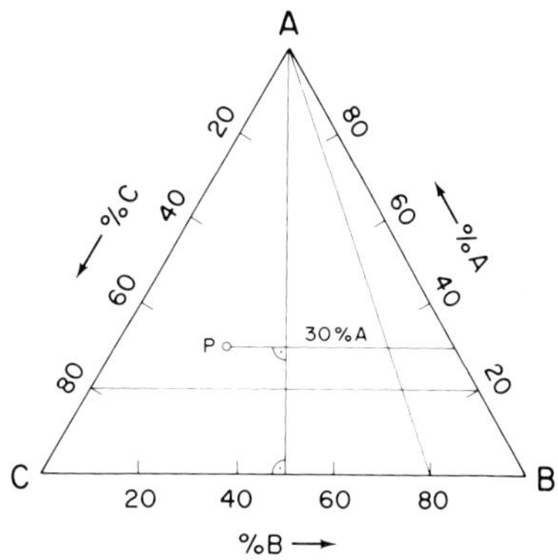

Fig.3.54. Representation of composition in a ternary system

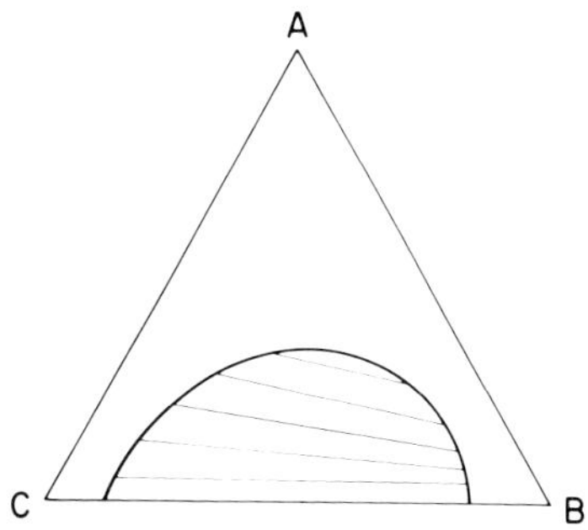

Fig.3.55. Three components, one pair partly miscible

– Points on lines parallel to one side represent compositions with constant concentration of the component in the opposite corner.

– Points on lines through a corner represent compositions with a fixed ratio of the components in the two other corners.

– To find, for instance, the percentage of component A in the alloy of composition P construct a normal from A onto the opposite side and determine the relative length of the projection.

A ternary system with miscibility gap is represented in Fig.3.55. Components B and C, which have a wide miscibility gap, become more and more miscible in each other with the addition of A. One example for such a case is the acetic acid-water-chloroform system. Another example is ethyl alcohol-water-salt systems. Although alcohol and water are miscible in all proportions, addition of salts with different solubility in the two components causes separation into an alcohol-rich and a water-rich solution. This effect of the addition of a third component can, in principle, be employed in separation processes. However, if high purities are required, one is reluctant in using an additive for driving a separation process since the added component itself may form an undersired contaminant.

Ternary systems will, of course, exhibit also invariant reactions. Fig. 3.56 depicts schematically a system where all three binary subsystems possess a eutectic transition, as for example, in reality for KF-LiF-CaF_2. The contour lines (light) are isotherms on the sloping liquidus surfaces. The

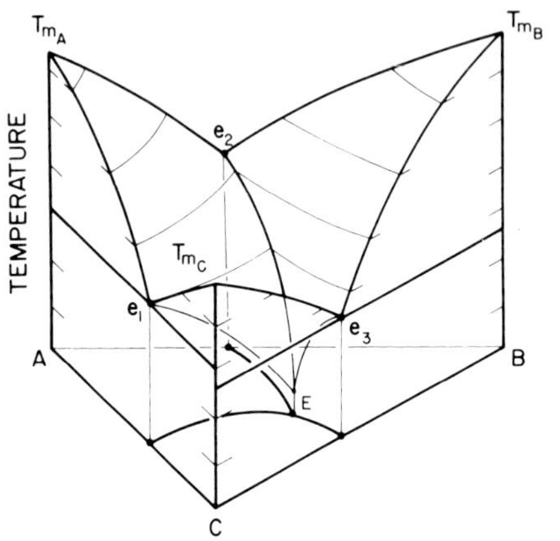

Fig.3.56. Three-dimensional ternary phase diagram for system with three eutectic binaries. Projection of eutectic troughs into triangle composition base

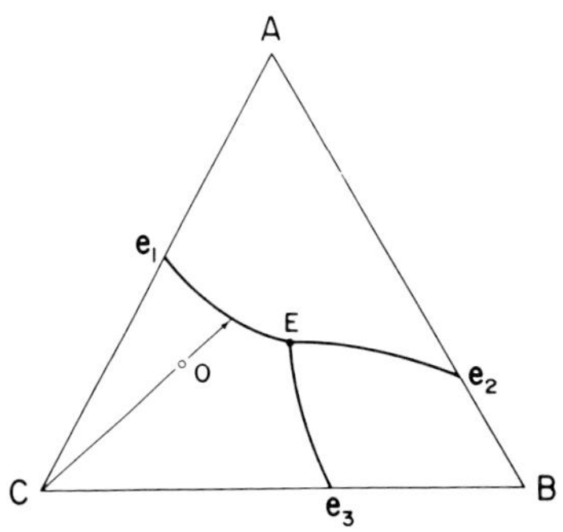

Fig.3.57. Path of liquid composition during freezing of ternary system

intersecting liquidus surfaces form the "eutectic troughs". Let us discuss the path that the composition follows upon lowering of the temperature in such a system. Starting with a liquid solution of composition O in Fig.3.57, as soon as the temperature reaches the liquidus surface in O, more or less pure C will solidify from the melt. This shifts the composition of the residual melt along an extension line from C through O until the binary eutectic trough is met. Note that the straight composition path through corner C comes from the fact that withdrawal of pure component C does not change the relative concentrations of A and B. At the eutectic trough, the melt is saturated with respect to C and A, which now start to freeze out upon further lowering of T. The system is thus brought along e_1-E until at E, at the ternary eutectic, the invariant reaction of the remaining liquid solution to A + B + C sets in.

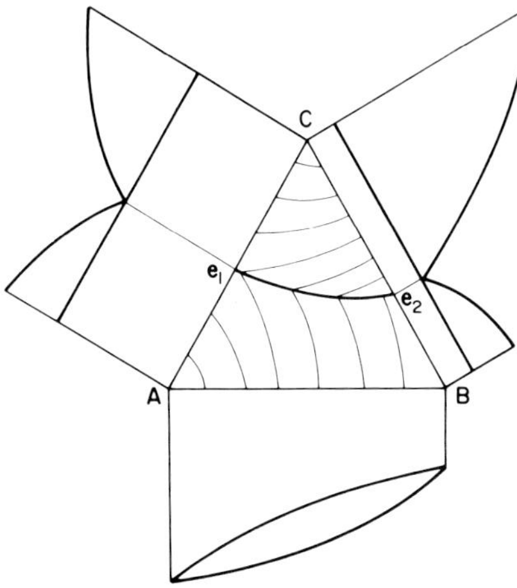

Fig.3.58. Ternary system composed of two binaries with simple eutectics and one binary of continuous solid solution type. After [3.21]

The system in Fig.3.58 has one binary subsystem with continuous solid solution formation, the other two are again simple eutectic systems. Liquid solutions (LS) with compositions in the area Ae_1e_2B are in equilibrium with solid solutions (SS) of A and B. Upon cooling, SS AB freezes out and the LS takes a curved path to the eutectic trough. There C begins to freeze out, too. The system moves toward e_2, where all A has been solidified so that the LS undergoes the binary invariant reaction.

Liquid on the surface Ce_1e_2 is in equilibrium with pure solid C. Cooling shifts the LS again along a straight line from corner C to the eutectic trough. The final melt composition in this case will also be e_2 – assuming that the trough does not go through a minimum.

Ternary phase diagrams are "responsible" for the formation of many of the most abundant minerals. Figure 3.59 for the system CaO-MgO-SiO_2 shows the complexity of actual ternary phase diagrams.

Detailed analysis of ternary eutectic, peritectic and other invariant transformation with emphasis on theory can be found in the monograph by REISMAN [3.29] and with extensive reference to experimental data in a review of HAASE and SCHÖNERT [3.30]. Calculations of ternary and quarternary III-V phase diagrams have been reviewed by STRINGFELLOW [3.31].

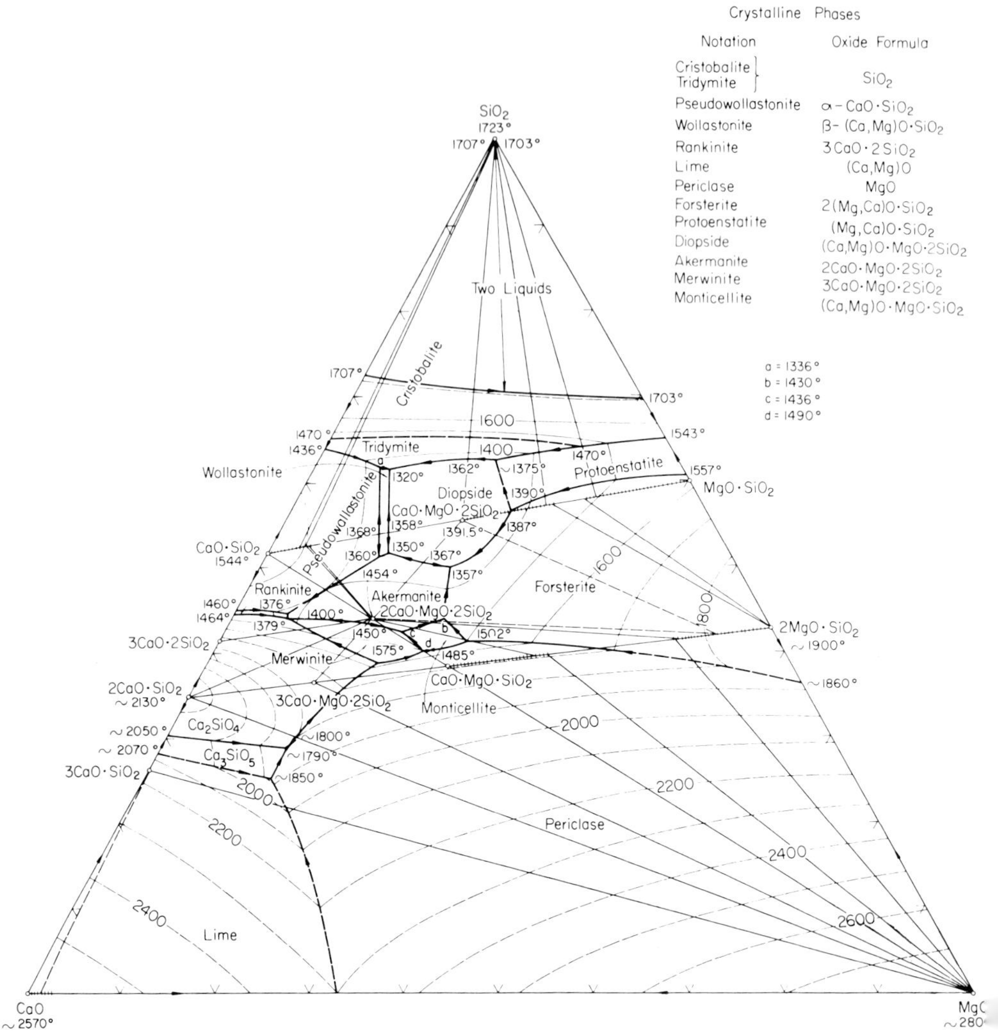

Fig.3.59. Phase diagram of the CaO-MgO-SiO_2 system. From [3.28] by permission of the American Ceramic Society

Obviously the complexity increases with the number of components in a system. Efficient discussions of quarternary and higher-order phase diagrams require considerable insight into topological schemes, conbinational methods, multidimensional vector spaces and matrix algebra. For an in depth discussion of phase equilibria in multicomponent systems see the monograph by PALATNIK and LANDAU [3.32].

3.6 Sources for Phase Diagrams and Solubility Data

A critical compilation of structural data, phase diagrams and solubility data for binary systems of elements up to 1969 has been given in

M. Hansen: *Constitution of Binary Alloys*, 2nd edition (McGraw-Hill, New York 1958),

R.P. Elliott: *Constitution of Binary Alloys*, 1st supplement, (McGraw-Hill, New York 1965) and

F.A. Shunk: *Constitution of Binary Alloys*, 2nd supplement (McGraw-Hill, New York 1969).

The systems are listed "in alphabetical order according to the chemical symbols of the component elements". Note that, for instance, the system Ti-O is treated under O-Ti since oxygen is the element that occurs earlier in this order. This "trilogy" is being supplemented and conveniently crossreferenced by

W.G. Moffat: *Binary Phase Diagrams Handbook* (General Electric Comp., Schenectady, NY 1976) Vol. 1 and 2, in a loose-leaf format which is updated twice annually.

Five thousand phase diagrams of two, three and four-component systems of compounds (mostly oxides, halides, nitrates, sulfates and sulfides) are compiled in

E.M. Levin, C.R. Robbins, H.F. McMurdie: *Phase Diagrams for Ceramists* (The American Ceramic Society, Columbus, Ohio 1964) with Supplement volumes in (1969) and (1975).

Phase diagrams, structural information and preparation aspects for many recently developed semiconductor and opto-electronic systems are given in

N.Kh. Abrikosov, V.F. Bankina, L.V. Poretskaya, L.E. Shelimova and E.V. Skudnova: *Semiconducting II-VI, IV-VI, and V-VI Compounds* (Plenum Press, New York 1969).

F.S. Gallaso: *Structure, Properties and Preparation of Pervoskite-Type Compounds* (Pergamon Press, Oxford 1969),

F.A. Kröger: *The Chemistry of Imperfect Crystals*; Vol. 1: Preparation, Purification, Crystal Growth and Phase Theory (North Holland, Amsterdam 1973).

A most recent review of structural and crystallochemical data for binary and ternary systems has been presented by O. Muller and R. Roy: *The Major Ternary Structural Families* and *The Major Binary Structural Families*, Crystal Chemistry of Non-Metallic Materials, Vols. 4 and 3 (Springer, Berlin, Heidelberg, New York 1974 and 1977).

An interactive computer system for rapid retrieval of information on phase diagrams and computation of phase diagrams from thermochemical properties of compounds, metal alloys, and dilute solutions of metals and gases in metals, has been developed by ManLabs, Inc., 21 Erie Street, Cambridge, Ma. 02139, USA.

In this context one should also mention the efforts of an international group of experimental and theoretical thermodynamicists towards establishment of a phase diagram data-bank and computer-based coupling of thermochemical and phase diagram data. Though the current efforts are concentrated on technologically important groups of metals (mostly refractory metals) investigations of ceramic, salt and other inorganic systems are underway. For references see the new journal *Calphad* (Pergamon Press) and also L. Kaufmann, H. Bernstein: *Computer Calculations of Phase Diagrams* (Academic Press, New York 1970).

Solubility data were compiled for:

– inorganic and metal-organic compounds in aqueous and organic solvents at low temperatures in

W.F. Linke: *Solubilities*, 4th ed. (American Chemical Society, Washington 1965) 2 volumes.

A. Seidell: *Solubilities*, 3rd ed. (Van Nostrand, New York 1940) 2 volumes.

H. Stephen, T. Stephen: *Solubilities of Inorganic and Organic Compounds*, Vol. 1: Binary Systems (McMillan, New York 1963) 2 parts.

Landolt-Börnstein Zahlenwerte und Funktionen, Vol. II/2b, ed. by K. Schäfer and E. Lax (Springer, Berlin, Göttingen, Heidelberg 1962)

– organic compounds at low temperatures in

J. Timmermans: *The Physico-Chemical Constants of Binary Systems in Concentrated Solutions* (Wiley, New York 1960) 3 volumes.

– gases in liquids and melts in

R. Battino, H.L. Clever: The Solubility of Gases in Liquids, Chem. Rev. 66, 395 (1966).

– oxides, oxide compounds, metals, alloys and semiconductors at elevated temperatures in

D. Elwell, H.J. Scheel: *Crystal Growth from High-Temperature Solutions* (Academic Press, London 1975) pp. 123-127.

Vapor-liquid equilibria:

A compilation of references on all vapor-liquid equilibrium data published from 1900 through 1972, systematically ordered by the chemical composition of the liquids (low temperature solvents and melts) has been put forward by

I. Wichterle, J. Linek, E. Hala: *Vapor-Liquid Equilibrium Data Bibliography* (Elsevier, Amsterdam 1973).

Problems

3.1 The vapor pressure of water at 0°C is 4.58 mm Hg. Use the Clausius-Clapeyron equation to estimate the boiling point of water. Treat the heat of vaporization as 595 cal/g, independent of temperature. Answer: the extrapolated "boiling point" is 96°C.

3.2 What contact area between a pair of ice skates and ice at -5°C must not be exceeded in order to melt the ice under an 80-kg man? The density of ice is 0.9168 g cm^{-3} and the heat of fusion of ice is 79.7 cal g^{-1}. What is implied by the necessary ΔP about the way ice skates are sharpened? Answer: Area $\leqslant$ 0.11 cm^2.

3.3 Consider the complex system that is formed by mixing NaCl, KBr, and H_2O. Suppose that we can isolate from this system also the constituents KCl, NaBr, $KBr \cdot H_2O$, $NaBr \cdot H_2O$ and $NaCl \cdot H_2O$. The possible distinct chemical reactions between these constituents are:

$$NaCl + KBr \rightarrow NaBr + KCl; \qquad NaCl + H_2O \rightarrow NaCl \cdot H_2O$$

$$KBr + H_2O \rightarrow KBr \cdot H_2O; \qquad NaBr + H_2O \rightarrow NaBr \cdot H_2O$$

What is the "number of components" C in this system? Answer: C = 3 for a closed system; C = 4 for an open system, i.e., if salts can be added separately. [3.8].

3.4 Estimate the vapor pressure of mercury under an external pressure of 1000 atm at 100°C. The density of liquid mercury at that temperature is 13.352 g cm^{-3} and the vapor pressure at 1 atm is 0.273 torr. Answer: P(1000 atm) = 0.455 torr. [3.8].

3.5 Derive the lever rule, (3.26), by simultaneously solving the mass balance equations for the number of moles and for the mole fractions in the liquid and solid phases.

3.6 The line X^s_{av} for an average composition of the solid solution in Fig.3.15 results from non-equilibrium solidification of a liquid solution that initially contains equal molefractions of Ge and Si. Discuss this function and its deviation from the solidus in detail. Assume a cooling rate that a) is slow enough to allow for complete compositional adjustment in the liquid, but b) is too rapid for any readjustment in the solid.

3.7 Discuss in detail a) the thermal changes (Fig.3.21b), and b) compositional changes (mass transfer), that occur in a system of overall composition X_m in Fig.3.21a upon continuous heating of the mixture of solid solutions, beginning at T_4.

3.8 Give a detailed account of the compositional changes that occur along the cooling paths 1 through 3 in Fig.3.26.

3.9 Draw schematically a set of free-energy vs. composition curves, g(X), that correspond to six characteristic temperatures (i.e., T_{m_A}, $T_{m_A} > T > T_p$, T_p, $T_p > T > T_{m_B}$, T_{m_B} and $T < T_{m_B}$) in the peritectic system represented in Fig.3.28.

3.10 Show schematically the T-P, P-X and T-X projections of a binary phase figure which differs from Fig.3.36 only in the solid-liquid equilibrium body that exhibits isobaric sections similar to Fig.3.17c, i.e., with common maxima in liquidus and solidus.

3.11 Draw schematically a sequence of isothermal sections through the phase figure shown in Fig.3.36 at the temperatures $T > T_B$, $T = T_B$, $T_B > T > T_A$, $T = T_A$ and $T < T_A$.

3.12 Prove that the vapor above ideal binary solutions is always richer in the more volatile constituent than the condensed phase. Answer:

$$Y_A/X_A = [1 + (D/p_A^0)X_B]^{-1} \quad \text{with} \quad D = p_B^0 - p_A^0 \quad .$$

3.13 Plot an isobaric section through the phase model shown in Fig.3.41 at a pressure between the triple point of A and P_2 (the pressure of the quadruple state).

3.14 Prove formally that a compound would have to have an infinite (negative) Gibbs free energy of formation in order to coexist with its striclty stoichiometric liquid and vapor.

3.15 Draw schematically isothermal sections at T_A, $T_{M(T)}$, T_{Q_A} and $T_{Q_A} > T > T_B$ through Fig.3.49.

3.16 Draw schematically isothermal sections at T_A, T_{Q_A}, T_B and T_{Q_B} through Fig.3.51.

4. Crystal Growth and Phase Diagrams

Given "all" thermochemical and thermophysical information on a specific material, one should be able to devise or select the optimum crystal growth method for it. The optimum method is defined as the sequence of preparative steps (preparation of starting material, growth and after-treatment of crystal) that leads to certain compositional (chemical) and structural properties of a real crystal which are desired for a specific experiment of application. "All information" goes far beyond the data needed to characterize the considered material itself under the envisioned range of preparative conditions. In addition, one needs a wealth of information to anticipate the interaction of the material with its environment during the preparation. Questions that must be answered include: How does the material chemically and physically interact with containers, "inert" atmospheres, vacua; what are the optical properties of the whole system, essential for estimates of radiative heat transfer; what are the consequences of fluctuations in environmental parameters like nutrient temperature and stresses on the growing solid, etc. The scheme presented in Fig.4.1 lists various parameters that are instrumental in selecting a specific crystal growth method. A good deal of these parameters, namely the ones shown in the left column, can be deduced from phase diagrams — *if* these are available. This will be illustrated in this chapter. The data on the right side have to be deduced from other sources. Beyond a mere listing of important system parameters, the scheme also illustrates their relation to growth parameters. Major portions of this text are concerned with the discussion and quantitative description of the relations that are qualitatively anticipated in Fig.4.1.

From this optimistic introduction one may gain the opinion that crystal growth processes can typically be designed in a rather ideal, academic manner to obtain optimum compositional and structural results. Well, typically they cannot. Firstly, only in a few cases will one be so fortunate as to deduce "all" the relevant information from the literature. Secondly, one has to

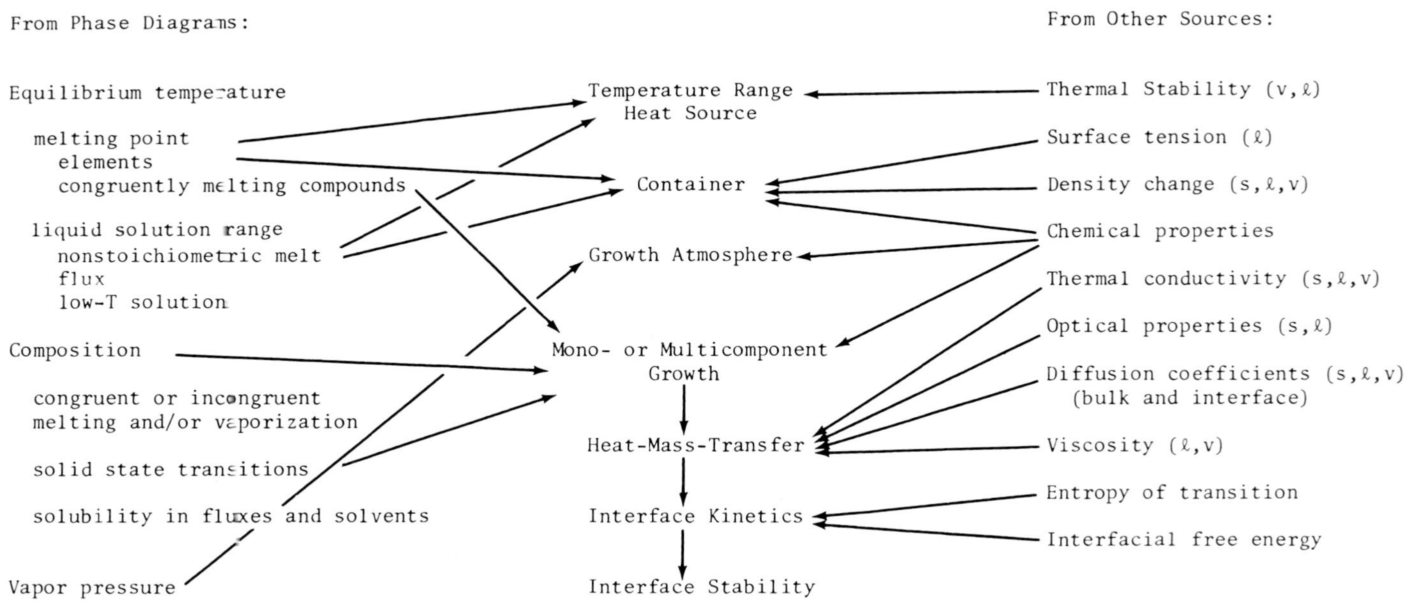

Fig.4.1. Material parameters important for the selection of a crystal growth method, and their relation to growth parameters (center column) (s: solid, ℓ: liquid, v: vapor)

realize that non-scientific parameters will influence and sometimes determine the selection of a method and its technical details. This will be briefly illustrated in the following section with the instructive, yet failing attempt of defining an "ideal system". Hence, in order to make Fig.4.1 more realistic one should add a third column of parameters that contains, for instance, available skill and tradition, available equipment and funds, position of the management or funding agency with respect to basic research, time scale and expectations with respect to marketability and last but not least *minimum* tolerable size and quality of the crystals needed for a specific research project or application. This last group of parameters is frequently the least defined one since it depends on such esoteric factors as the "professional pride" of the crystal grower involved and the "awareness" of the solid state investigator requesting the specific samples.

In Sect. 4.2 we will classify crystal growth processes according to the prevailing mode of mass transfer. In Sects. 4.3 and 4.4 we will demonstrate the importance of phase diagrams, vapor pressure and solubility data for crystal-growth process selection and conduction in a broad discussion of specific systems. Thereby, as an introduction to detailed treatments in later parts of the text we will draw attention to the mass-transfer problems that arise in the various types of crystal-growth situations. Section 4.5 concludes the chapter with a brief discussion of incompatibility between crystal growth materials (nutrients) and prospective containers, as it can be deduced from phase diagrams.

4.1 The Ideal System

The least problematic crystal-growth material consists of:

1) a pure element or congruently melting compound, with
2) low vapor pressure,
3) low melting point,
4) high thermal conductivity,
5) low heat of solidification, having
6) no solid state phase transitions between its melting point and room temperature.

Such a material can most easily be grown from its melt. No material transfer problems are encountered at the interface since the melt has the same composition as the solid. Growth rates are then typically heat transfer limited.

Growth can be performed under high vacuum, thus eliminating the growth atmosphere as potential contaminant. Due to the low temperature, little contamination will result from the container. After growth the crystal can be cooled down to room temperature without structural changes tha, in general, introduce high strain densities and proceed with low rates. The high thermal conductivity allows for high growth rates. Point 5 results in a non-faceted inte-face [1.1] and hence, minimum impurity segretation anisotropy (Sect. 6.5.2). Such ideal systems do exist in form of some metals and semimetals.

Some experienced readers will not agree on all the parameters of this "ideal" system. Workers with a preference for vapor growth may choose instead a material that

1) vaporizes congruently with a
2) high vapor pressure at a
3) modest temperature, and which has properties 4 and 6 as above.

They may argue that vapor growth allows for higher structural quality (lower dislocation density, etc.) and that the growth rates which one can obtain, in principle, are not that much lower than in melt growth. Purists may even point out that the morphological stability [1.1] of a vapor-solid interface is typically higher. Note, however, that both "schools" above (melt vs vapor) appreciate the advantage of a congruent fluid-solid phase transition. Hence, as a first step away from the ideal system we will look in Sects. 4.3 and 4.4 at the consequences that arise when the fluid deviates in composition (stoichiometry) from the crystal.

The above fictitious confrontation was chosen to demonstrate that the selection of a specific growth method depends as much on the structural and compositional qualities that one plans to achieve, as on available skill, experience, equipment, etc. For another, more humorous controversy between a "flux man" and a "Czochralski man", see [4.1]. Opinions as to *the* growth method will always differ. However, it appears to this author as more promising to base a selection initially solely on fundamentals regarding the desired quality of the "end-product" and *then* to modify the choice according to the given human and economical boundary conditions, rather than inversing the sequence.

4.2 Classification of Fluid to Solid Processes

For the classification of crystal growth processes we will make extensive use of the scheme suggested by LAUDISE [4.2] in which the underlying type of heterogeneous reaction

solid → solid

liquid → solid

vapor → solid

is used in combination with the distinction between mono- and multicomponent systems. In *monocomponent* processes, neglecting trace impurities and other deliberately added low level dopants, the nutrient consists only of the material that forms the desired crystal. In *polycomponent* processes an additional component or components have been added so as to lower the temperature at which the crystal constituents exist in liquid form (solution growth) or to increase the volatility of the material to be crystallized (vapor → solid). There is, of course, no distinct boundary between mono- and multicomponent growth. Transition cases are given by melts with high dopant concentrations and solutes with very high solubilities in certain solvents.

This scheme offers great advantages over the classical nomenclature where terms such as recrystallization, melt growth, flux growth, solution growth, vapor growth, etc., referred merely to the technical aspects of a growth method. With growing insight into the atomistic kinetics of the various processes, a classification scheme that emphasizes the analogy in transport and interfacial phenomena rather than in temperature range, solvents, heating technique, etc., became more appropriate. For instance, from the point of view of transport kinetics in the nutrient a distinction between "vapor growth" and "solution growth" appears artificial when one compares crystallization from gas mixtures with crystal growth from liquid solutions. In both cases the crystal building blocks are associated in the nutrient with other components from which they must separate at and on the interface in order to center the lattice. The processes may be carried out at very different temperatures. The characteristic parameters such as association (solvation, ligation) energies and diffusion coefficients may have drastically different numerical values. However, the molecular mechanisms can be expected to be so similar that a common treatment of the basic phenomena can be very fruitful.

Table 4.1 compares typical values for "kinetic parameters" that prevail in liquid and gaseous mono- and multicomponent systems. Anticipating later detailed discussion we see that typical growth rates in growth from multicomponent nutrients are lower than rates resulting from monocomponent nutrients. In liquid systems this is due to the material transport problem as-

Table 4.1. Comparison of growth methods

	NUTRIENT			
	Liquid		Vapor	
	mono- component	multi- component	mono- component	multi- component
Particle density $N[cm^{-3}]$	10^{23}	10^{22}	$10^{19} \frac{p[mm\ Hg]}{T[K]}$	
Diffusion coefficient $D[cm^2/s]$	10^{-3} - $5x10^{-6}$		10^{-2} - $5x10^{-2}$	
Viscosity $\mu[cp]$	1 - 10		10^{-2} - 5 x 10^{-2}	
Heat of transition ΔH[kcal/mole]	1 - 10	±10	50	±50
Thermal conductivity [cal/cm s deg]	10^{-3}		10^{-4} - 10^{-5}	
Solvation energy [kcal/mole]	-	10 - 100	-	?
Growth rate V[cm/day]	10 - 100	.02 - .2	2	0.2-1.0

sociated with low diffusion coefficients and high viscosity. In gaseous systems, in addition, heat transport can be rate limiting due to large heats of condensation involved and the typically low thermal conductivities of most materials that are commonly grown from vapors. But heat transfer and material transport in some gaseous systems has been enhanced by forced convection (carrier gas) to such an extent that the resulting growth rates exceed those of liquid multicomponent growth, where methods to enhance material transfer have not been used as effectively yet.

4.3 Liquid-Solid Equilibria

It is obviously somewhat arbitrary to divide a discussion of the use of phase diagrams in crystal growth into liquid-solid and vapor-condensed phase equilibria. As we will see, vapors do frequently play an important role even if growth occurs from a liquid. Hence, this separation refers here simply to the form of the nutrient, liquid or gaseous, that is, in direct contact with the growing solid.

This section has greatly benefited from a review by NIELSEN and MONCHAMP [4.3].

4.3.1 Congruent and incongruent solidification

If a compound melts congruently (Sect. 3.4.6) at a reasonable temperature and pressure and does not undergo solid-solid phase transitions on cooling one will likely choose a melt growth technique for the preparation of single crystals. The choice of the specific technique, however, will – among other parameters – depend on the shape of the liquidus in close proximity of the compound's stability range. Fig.4.2 depicts the two extreme types of liquidus curves which correspond to the enlarged center parts of Fig.3.45. Whereas type (b) shows a strong dependence of the equilibrium temperature on composition, type (a) requires only minor readjustment of the temperature even with significant changes in composition.

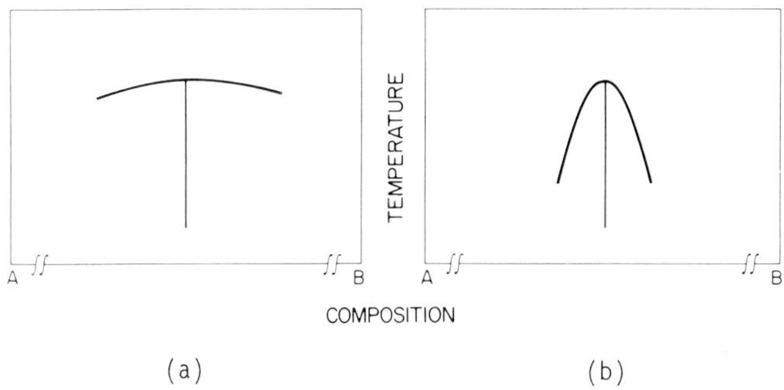

Fig.4.2. Enlarged liquidus curves in the vicinity of a solid compound in a binary system

In practice one will typically not be able to establish and/or maintain strictly congruent melt composition either due to slight weighing erros and/or preferential evaporation losses of one component. Depending on the specific shape of the liquidus, one has then to face the following consequences:

– Type (a): Even with considerable segregation, i.e., accumulation of an excess component in melt layers close to the interface no serious temperature programming problem arises. In addition, the interface will have higher shape stability [1.1] than an interface in a system of behavior (b) under otherwise identical conditions.

— Type (b): Small deviations of the melt from the composition of the growing solid will require major readjustment of the growth temperature. In connection with continuous preferential evaporation losses this may require elaborate temperature programming. The large liquidus slope reduces the interfacial shape stability (morphological stability) considerably [1.1]. In order to reduce these problems with compounds that exhibit a cusp-shaped liquidus one will select a growth technique which

1) allows for a reduction of the evaporation losses, and
2) determines the external shape of the crystal by the crucible rather than by the heat balance between the solid and liquid. Hence, the Bridgman-tech-

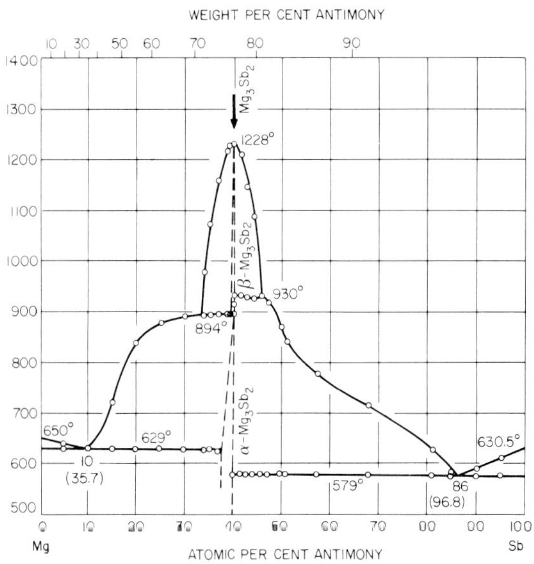

(a)

Fig.4.3. Phase diagrams of (a) Mg-Sb system and (b) Ag-Te system. From [4.42] by permission of McGraw-Hill Book Company

nique performed with a sealed crucible will often be more appropriate than the pulling-technique where a strong "pull-out" tendency exists with these materials (for technical details see the chapter on melt growth in [1.2]). Examples for this "inconvenient" behavior are the compounds Mg_3Sb_2 and Ag_2Te depicted in Figs.4.3a and b. Mg_3Sb_2 represents also an interesting example for stronger AB interaction as compared to the A-A and B-B bonds reflected in the considerably lower melting points of the pure components.

One of the most practically important questions for the growth of compound crystals concerns the (temperature dependent) width of their existence or homogeneity range. On the one hand, when the crystal property of interest is not very composition sensitive one may want to exploit the full existence range. On the other hand, with composition sensitive properties,

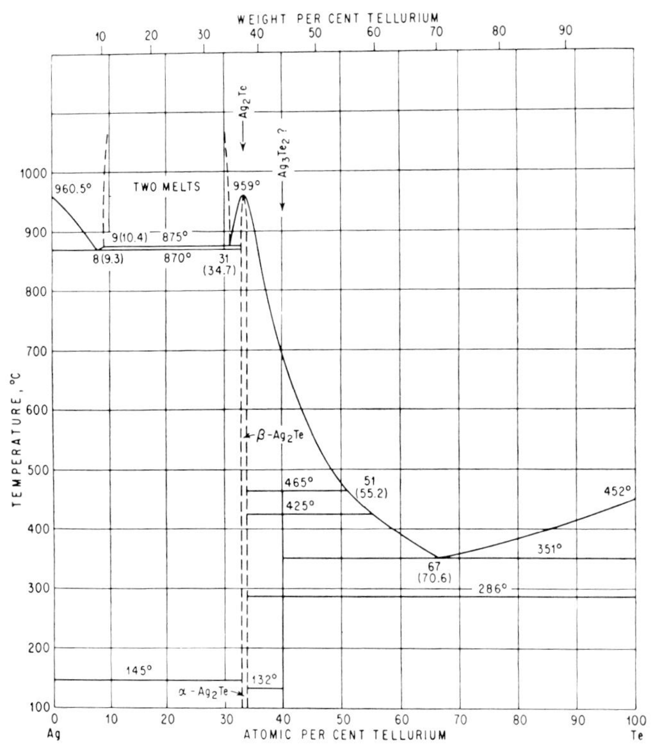

Fig.4.3. (b)

one must often know the consequences of very small deviations from strict stoichiometry, as, for instance, for the growth of electrically well characterized compound semiconductors.

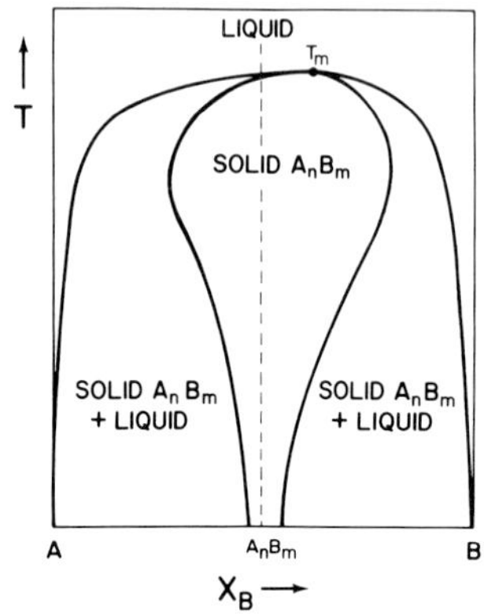

Fig.4.4. Schematic temperature-composition diagram for a compound-forming system A-B. Temperature dependence of the A_nB_m existence range

Compound stability ranges are typically not as wide as the ones in Fig. 4.3 and hence appear in most full-range phase diagrams as single vertical lines. However, it should be recalled that, in general, in solid solution and compound forming systems the intersulubility decreases with temperature as schematically indicated in Fig.4.4. The rather generally valid reasons for this behavior can be deduced from Fig.3.33. Hence, in order to avoid precipitation of the excess component upon cooling, one will often not be able to take advantage of the whole width of the existence range that is available at high temperature. Also, as anticipated in the discussion of Fig.3.46, the maximum of the liquidus, T_m, i.e., the congruently melting composition does not necessarily occur at the strict stoichiometric composition A_nB_m, where n and m are small integers. Furthermore, the homogeneity range can be rather asymmetric about A_nB_m (Fig.4.4). Hence, excess of one component in the melt may give more flexibility in the growth conditions than excess in the other component due to a less inward sloping solidus

These factors are, for instance, typical of many of the II-VI and III-V compounds, as depicted in Fig.4.5. Note the greatly enlarged scale of the compositon axis. The asymmetry of the existence range and the change in composition on cooling in these systems is governed by the prevailing vacancy formation mechanism. Corresponding to the high fugacities of these elements,

Zn, S and As vacancies, respectively, are the dominant intrinsic defects in ZnTe, CdS and GaAs. Thus, CdS, for instance, has always been found to be n-type or practically insulating. In such cases, where the vacancy concentration for one component dominates, even growth from strictly congruent melts results in a compositional shift as indicated by the thin lines within the stability fields of Fig.4.5a. The high vacancy concentrations that form at elevated temperatures cannot readily be quenched in. Quenching does occur only at lower temperatures where the diffusion rates become very low for

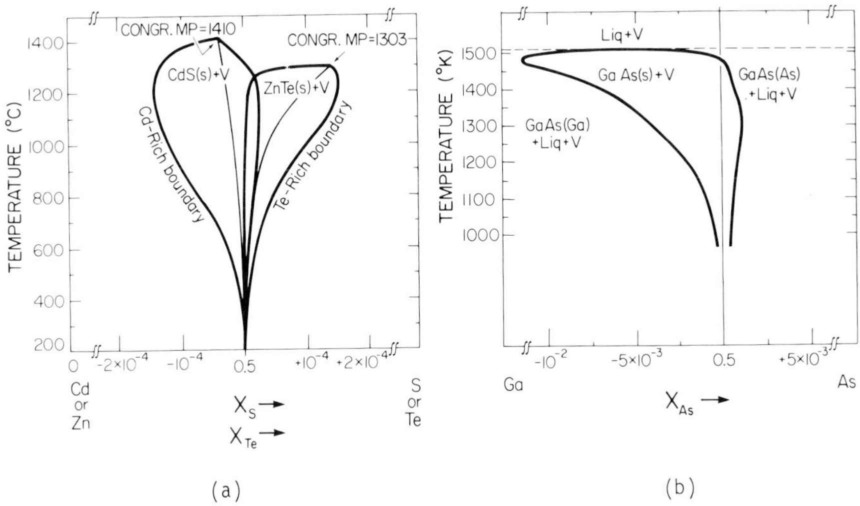

Fig.4.5. Existence fields for some semiconducting compounds. (a) CdS and ZnTe (tentative); after [4.4]. (b) GaAs (calculated); after [4.5]

these systems. So, during cooling in the upper temperature range, the vacancy concentrations will decrease to the appropriate equilibrium value. At the same time, referring to a CdS system, as the S vacancy concentration decreases, Cd must precipitate out in the crystal. Higher initial vacancy concentrations (from a less stoichiometric melt and vapor composition) merely result in the formation of larger amounts of precipitates. (See also Problem 4.1).

Homogeneity ranges may or may not contain the stoichiometric composition of the simple formula A_nB_m with n and m being small integers. In semiconducting compounds where A_nB_m lies outside the homogeneity range the minimal attainable carrier concentration is limited by the minimal attainable deviation from stoichiometry. Examples for such behavior, as shown in Fig.4.6, are the IV-VI compounds SnTe and GeTe. Note also that the polymorphic transitions occur at different temperatures on both sides of the homogeneity region. For instance, for GeTe + Ge the transition to the low temperature modification lies at 430° C, whereas on the tellurium side the transition temperature is about 365° C (dashed line; the full line at 380° C represents the GeTe + Te eutectic).

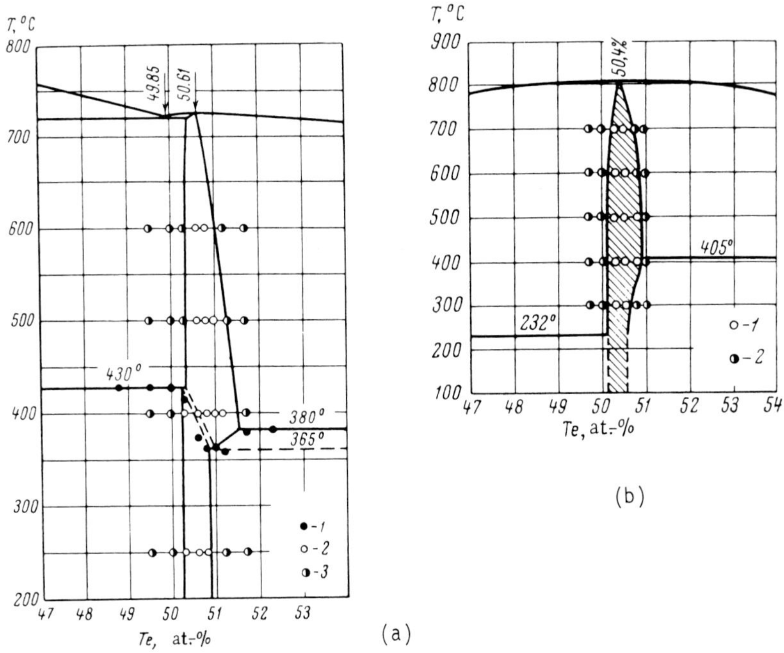

Fig.4.6. Homogeneity regions not containing the strict stoichiometric composition of compounds. (a) Phase diagram of the Ge-Te system in the region of the compound GeTe; 1) results of thermal analysis, 2) single-phase alloys, 3) two-phase alloys. (b) Sn-Te system in SnTe region; 1) single-phase state, 2) two-phase alloys. From [4.6] by permission of Plenum Press

A considerably broader existence range for a binary compound, with a larger deviation of the melting point maximum from the stoichiometric composition is shown in Fig.4.7a, the expanded center section of the phase diagram for the

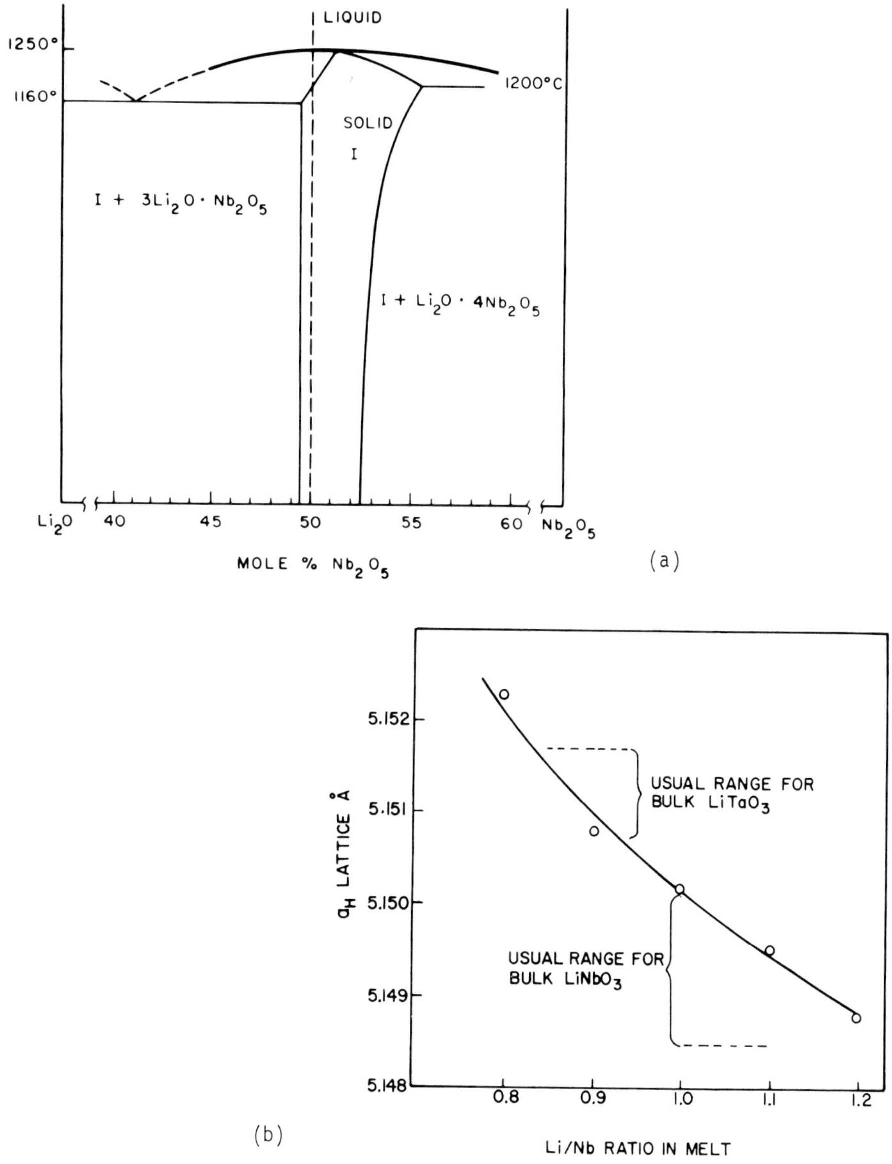

Fig.4.7. (a) Expanded center section of phase diagram for the Li_2O-Nb_2O_5 system. From [3.4] by permission of Academic Press.
(b) Lattice parameter versus Li/Nb ratio in melt. From [4.8] by permission of North Holland Publ. Co.

Li_2O-Nb_2O_5 system. One sees that for the growth of incongruently melting stoichiometric lithium niobate, $LiNbO_3$, one has no choice but employing an off-stoichiometric melt with 42% Nb_2O_5 [4.7]. For the growth of large crystals of (macroscopically) uniform composition one must then use either a very large melt volume or employ one of the nonconservative methods discussed in Sect. 6.4.3. Yet it has been demonstrated [4.9] that the birefringence fluctuations, which are typical for non-stoichiometrically grown $LiNbO_3$ and have in the past limited the useful phase matchable length of these crystals, can be minimized by growth from a congruent melt. The effect of unavoidable growth rate variations with respect to compositional fluctuations is then reduced and the slightly Li-deficient crystals are more homogeneous.

Another reason for the choice of a specific incongruent melt composition can also be illustrated with the Li_2O-Nb_2O_5 system. Fig.4.7b depicts the dependence of the a_H-lattice parameter of $LiNbO_3$ on the Li/Nb ratio of the melt from which it was grown. $LiNbO_3$ is very advantageously used for optical waveguide devices in the form of epitaxial layers on $LiTaO_3$. For successful epitaxial growth to take place a close lattice parameter match between substrate and epi-film is very important. One sees that an ideal match for these two materials can be obtained from a Li/Nb ratio of 0.9.

These examples have demonstrated that an understanding of the shifts in composition which occur on freezing is essential if some control over the crystal composition (and properties) is to be obtained – even if such changes are negligible on the scale of an ordinary $0 < X_B < 1$ phase diagram presentation. Hence, as in the above cases, crystal growers must frequently ascertain such "fine-structures" that go beyond the accuracy of typical phase diagram studies which in most cases were carried out for the understanding of less sensitive ceramic and metallurgical processes.

With increasing non-stoichiometry of the nutrient, the interfacial growth kinetics as well as the transport kinetics in the fluid close to the crystal becomes more complex. The larger the deviation, the higher is the concentration of the rejected excess component that impedes transfer of the deficient component towards the interface. Furthermore, the morphological stability of the interface is reduced [1.1]. Consequently, feasible growth rates in growth from off-stoichiometric fluids are significantly lower than in stoichiometric (congruent) systems (Table 4.1). In strongly non-stoichiometric melts, the transport conditions become comparable to those encountered in flux growth – or in solution growth in general.

Growth from *strongly* non-stoichiometric melts may be required for various reasons:

1) One component has an inconveniently *high fugacity* (vapor pressure) at the congruent melting temperature of the compound. Numerous III-V compound

semiconductors show this behavior. Examples are given in Fig.4.8. Gallium phosphide, for instance, has an equilibrium partial pressure of about 38 atm of phosphorus above its stoichiometric melt. Rather complex high pressure equipment [1.2] is necessary to reduce the phosphorus losses to a bearable amount. On the other hand, taking the problems of non-stoichiometric nutrient compositiion into account, one can grow GaP under normal pressure from a P-deficient melt where the fugacity of phosphorus is much lower (Figs.4.8 and 4.9). GaP crystals of cm-size have been obtained this way from melts with 5-10% concentration (for references see [4.3]).

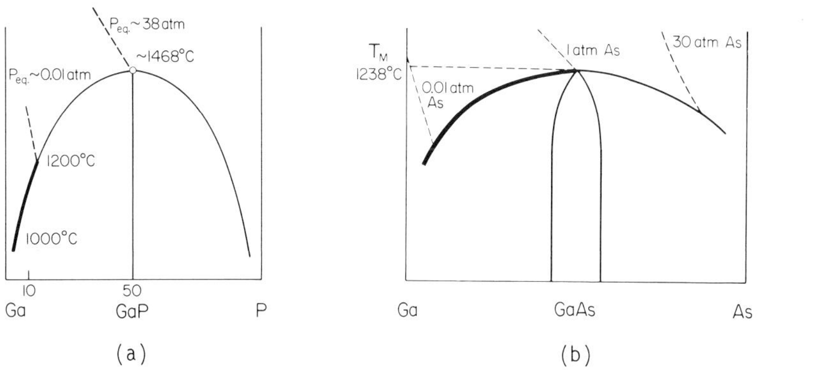

Fig.4.8. T-X projections of the phase figure for (a) the Ga-P system and (b) greatly expanded center portion of the Ga-As system. From [4.3] by permission of Academic Press

With gallium arsenide one encounters the same problem, although not as pronounced. Growth from Ga-rich melts due to the lower arsenic partial pressure (Fig.4.9) reduces the arsenic evaporation losses.

The vapor pressure functions of Fig.4.9 are a rather striking illustration of the validity of (3.9). One can deduce that the heat of vaporization for the V-component is independent of composition as long as III-V interactions dominate in the liquid solution. As soon as X approaches 0.5 (which can be seen by comparison with Fig.4.8), i.e., as V-V interactions become significant, L_V decreases and the fugacity increases with temperature more rapidly.

2) The *solid melts incongruently* – more than in the $LiNbO_3$ case. An example for this group is the solid electrolyte $RbAg_4I_5$ (specific conductivity of 0.26 $\Omega^{-1}cm^{-1}$ at RT). As the phase diagram of the RbI-AgI system in Fig. 4.10 reveals the compound can be grown from non-stoichiometric melts [4.11b]

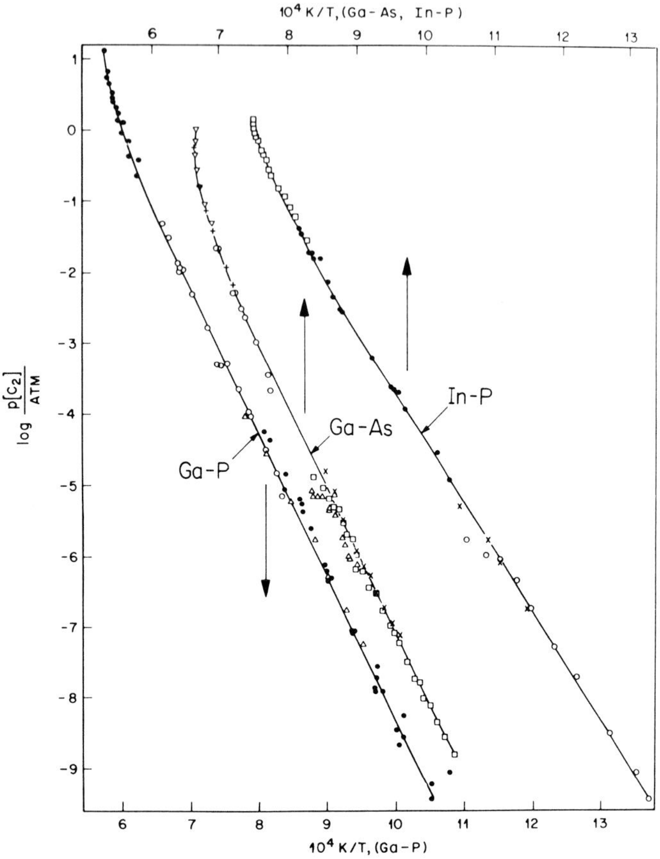

Fig.4.9. The Ga- and P-equilibrium partial pressures against reciprocal liquidus temperature for GaP, InP and GaAS. From [4.10] by permission of North Holland Publ. Co.

in the 70-76 mole percent AgI composition range, indicated with the heavy part of the liquidus between the AgI-rich peritectic and the eutectic (see Problem 4.2).

Probably one of the best investigated incongruently melting compounds is yttrium iron garnet (YIG), $Y_3Fe_5O_{12}$, an important ferromagnetic material. Whereas its isomorph yttrium aluminum garnet (YAG) can be grown from a congruent melt, the melt growth of YIG is considerably more intricate than the simplifying binary phase diagram of the iron oxide – $YFeO_3$ system in Fig.

4.11 suggests. The actual complexity results from the underlying reduction reaction $Fe_2O_3 \rightarrow 2FeO + 1/2\ O_2$. Hence an accurate account of the compositional changes upon solidification requires a ternary diagram that characterizes the Fe^{3+}/Fe^{2+} ratio as function of oxygen partial pressure. Fig.4.12a

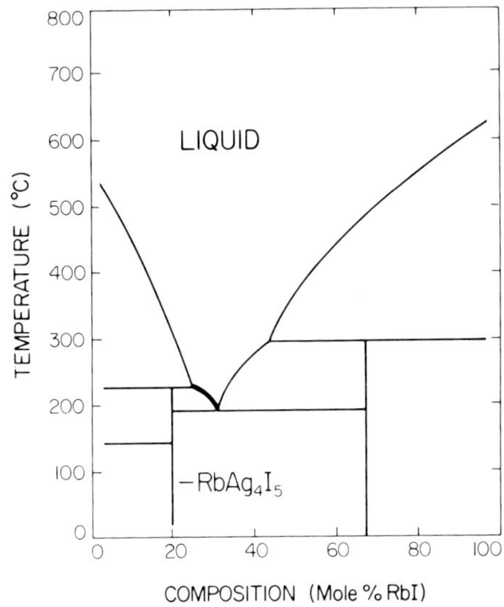

Fig.4.10. Phase diagram for the RbI-AgI system. After [4.11a]

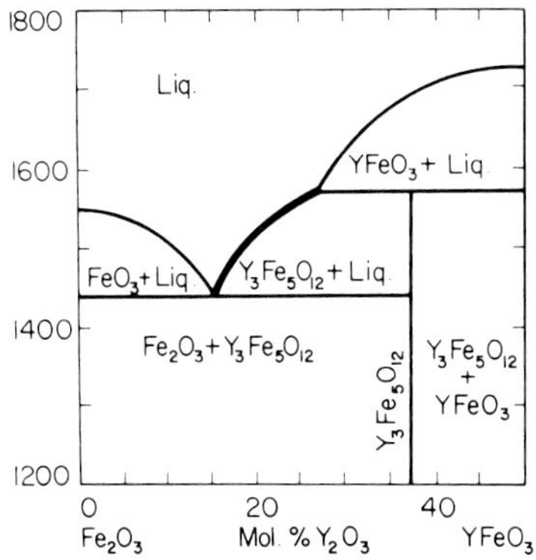

Fig.4.11. Partial phase diagram of the Fe_2O_3-$YFeO_3$ system. From [4.3] by permission of Academic Press

presents such a phase diagram in which three oxygen isobars have been plotted. $P_{O_2} \approx 2$ mm Hg corresponds to the oxygen partial pressure that exists in an equilibrium $CO_2 \rightarrow CO + 1/2\ O_2$ mixture at about 1500° C. $P_{O_2} = 159$ mm Hg = 0.21 atm

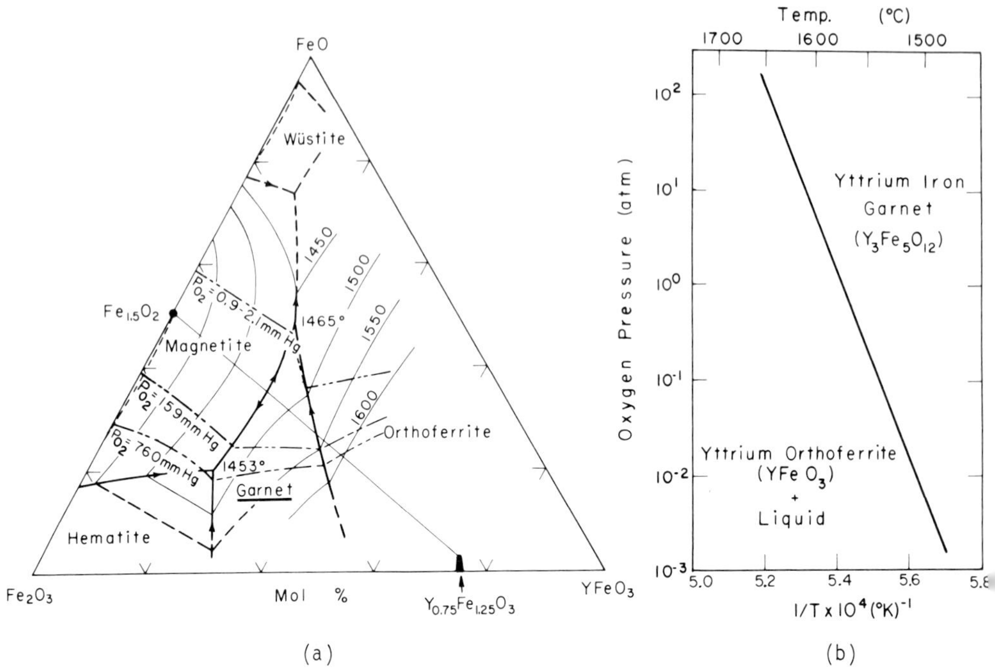

Fig.4.12. (a) $YFeO_3$-Fe_2O_3-FeO ternary phase diagram.
(b) Peritectic reaction temperature of YIG as a function of O_2 pressure. From [4.12] by permission of the American Ceramic Society

is the oxygen concentration of air. The third isobar corresponds to 1 atm of pure oxygen. As also reflected in Fig.4.12b, the higher the oxygen pressure above the melt, and hence the smaller the Fe^{2+}/Fe^{3+} ratio in the melt, the closer the isobar lies to the Fe_2O_3-$YFeO_3$ binary plane (see problem 4.3). Note that a P_{O_2} of about 100 atm would be required to make the deviations from the simpler binary negligible – simpler not only in the presentation but, more importantly, also in the resulting transport and interfacial growth kinetics. This complex situation is further illustrated in Figs.4.13 and 4.14. The heavy curve (A-B) in the three-dimensional presentation is the 0.2 atm oxygen isobar on the part of the liquidus surface that leads to garnet. A projection of this isobaric cut into the iron oxide – $YFeO_3$ plane is shown in Fig.4.14a. An isobar for lower oxygen pressure would lie even more remote from that plane as schematically indicated by the dashed line in the three-dimensional diagram (see Fig.4.13). A corresponding isobaric projection for the system "in CO_2" is given in Fig.4.14b. A comparison of these two quasi-

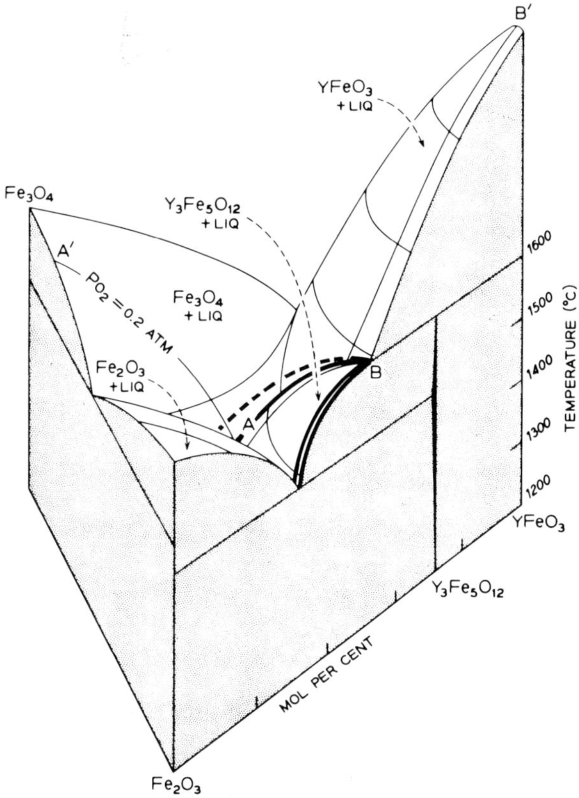

Fig.4.13. YIG phase diagram. From [4.2] by permission of Prentice Hall

binaries illustrates rather vividly the decrease in compositional changes required during solidification from the respective peritectic to YIG, as the oxygen pressure is increased. Because of this strongly incongruent melting behavior and the reduction problem YIG is typically not grown from melts. Flux growth at considerably lower temperatures as discussed in the following section is less problematic and is hence preferred.

3) The crystal undergoes a *solid-solid phase transition* on cooling. If the transition occurs at a high enough temperature – typically above two thirds of the melting temperature is kinetically still promising [Ref. 4.13, p. 331] – one may want to take advantage of the simplicity of congruent melt growth. If large crystals are needed then "growth from the solid" may introduce considerable strain, twinning, and mosaic structure. However if the transition temperature is above the lowest liquidus tem-

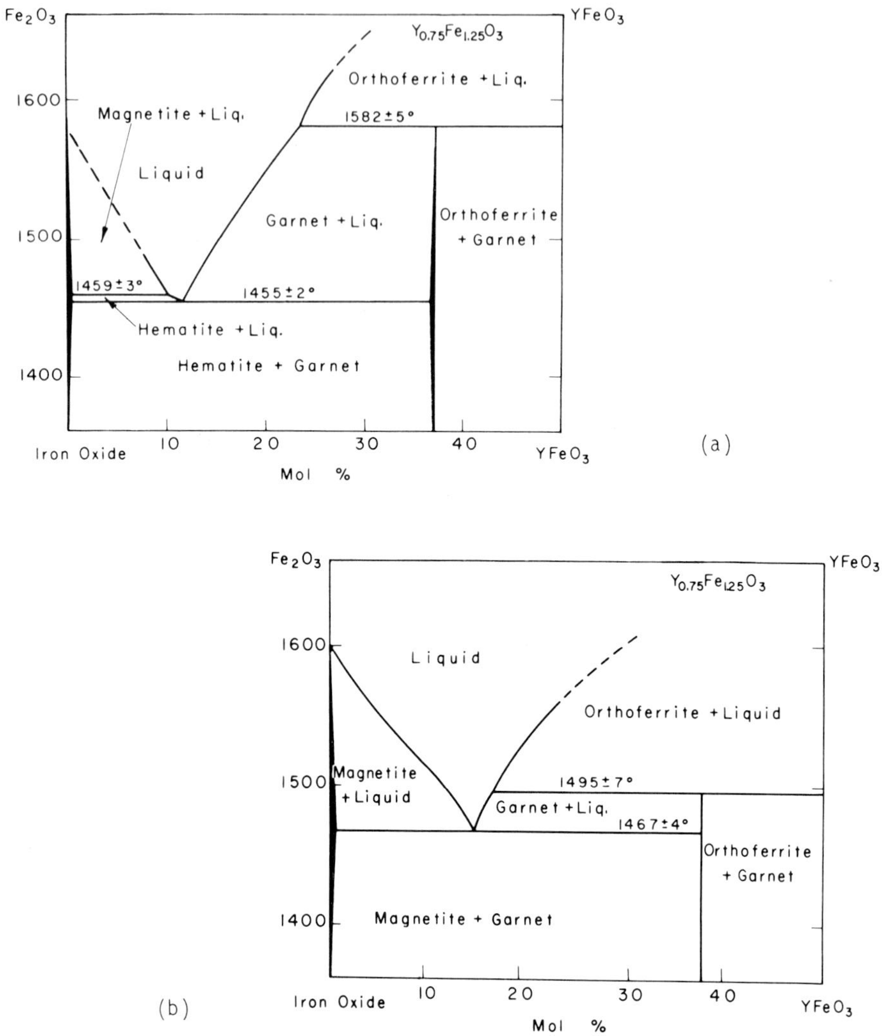

Fig.4.14. Isobaric cuts through the ternary YIG phase diagram. (a) for 1 atm oxygen pressure. (b) for CO_2, i.e., $\approx$ 2 mm Hg of O_2. From [4.12] by permission of the American Ceramic Society

perature of a non-stoichiometric melt in the system, then melt growth is still possible. For instance, *barium* titanate solidifies from its congruent melt in the hexagonal configuration (Fig.4.15). Upon further cooling, it goes at 1460°C through a destructive phase transition to the cubic modification. However, employing a 14% TiO_2-rich "melt" [4.14], one can obtain

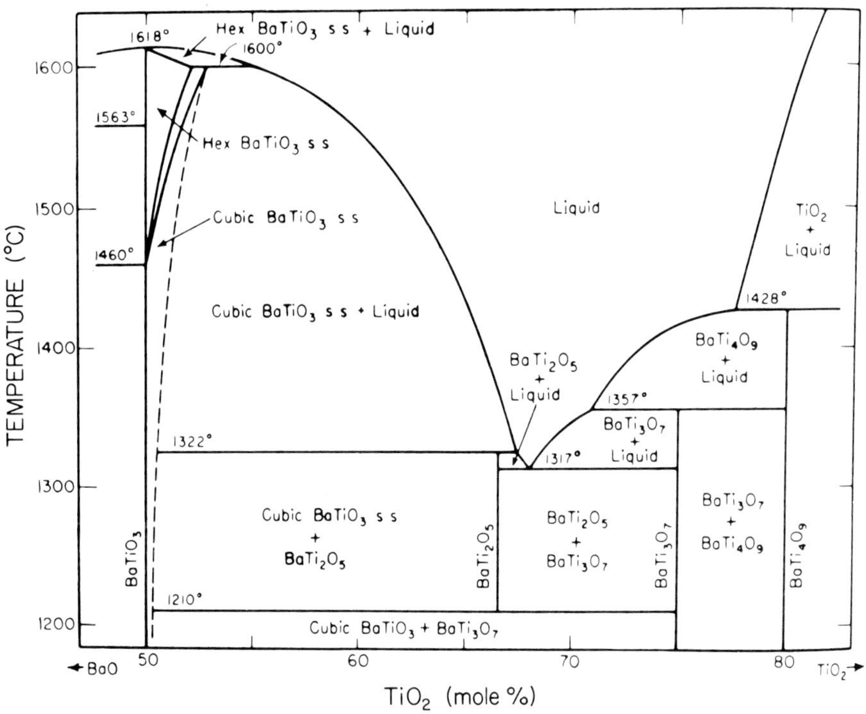

Fig.4.15. Section of phase diagram of BaO-TiO_2 system. From [4.14] by permission of Pergamon Press

directly large cubic $BaTiO_3$ crystals that are in many respects superior to the small "butterfly-twins" obtained earlier from KF fluxes.

Interestingly *strontium* titanate, as depicted in Fig.4.16, can be directly grown in its cubic configuration from the congruent melt. The transition from hexagonal to cubic structure depends in titanates sensitively on the size of the alkali earth ion. For instance cubic $BaTiO_3$ was grown from a congruent melt in which 3% of the Ba had been replaced by Sr [4.15].

Another interesting possibility to bypass a solid state transition is given in systems where the stability range of a high temperature phase can be narrowed or eliminated by application of pressure. Fig.4.17 represents the P-T phase diagram of CsCl [4.16]. One sees that at 660° C and 970 atm the high temperature face-centered cubic phase is in equilibrium with the "low temperature" simple cubic phase and their melt. This fact was employed to grow CsCl from the melt directly in the simple cubic configuration at a P-T combination slightly above the triple point [4.17].

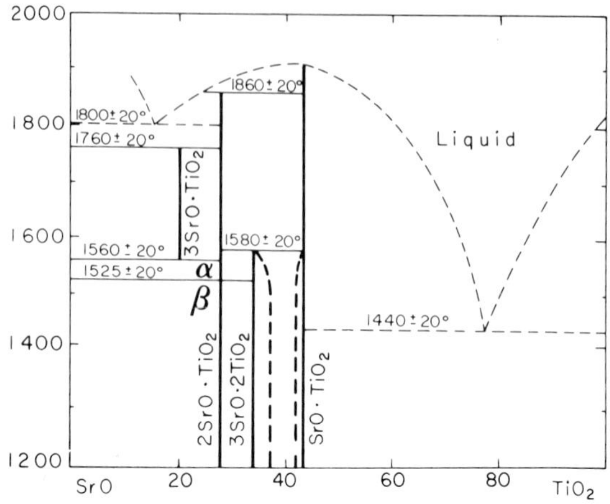

Fig.4.16. Phase diagram of the SrO-TiO_2 system. From [4.12] by permission of the American Ceramic Society

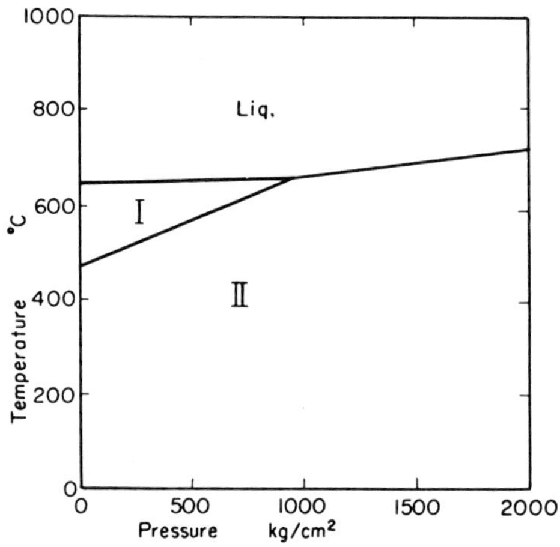

Fig.4.17. P-T phase diagram of CsCl. From [4.17] by permission of North Holland Publ. Co.

The semiconducting II-VI compounds are another example for materials where polymorphism makes it difficult to grow structurally satisfactory single crystals from the melt. They undergo upon cooling a transition from the hexagonal wurtzite structure to the cubic zinc blend structure. This causes twinning and stacking faults in melt-grown II-VI systems, most pronounced in ZnS, ZnSe and ZnTe. Unfortuantely, the solid phase transitions occur at temperatures below the lowest liquidus temperatures. Hence, one resorts for

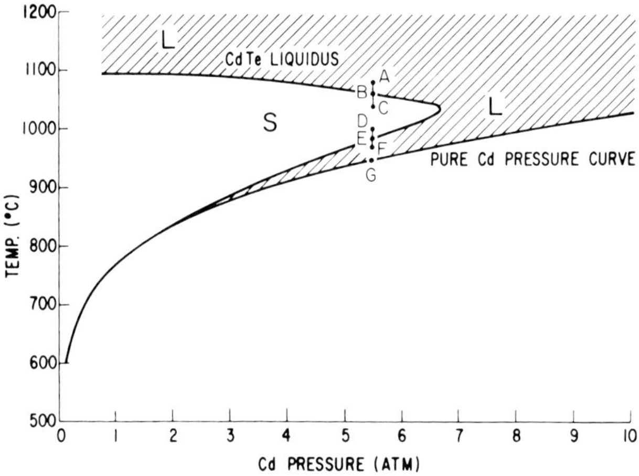

Fig.4.18. T-P projection of the "Cd-rich half" of the Cd-Te system. From [4.18] by permission of the American Institute of Physics

their growth typically to methods that allow for even lower temperatures. In particular, the high fugacities and practically congruent composition of most II-VI vapors makes them well suited for vapor growth.

A striking example for the usefulness of considering the whole phase space diagram for a solidification process was put forward by LORENZ [4.18]. Fig.4.18 displays part of the T-P projection of the Cd-Te phase space. Refer to Fig.3.46, and envision the above figure to be a projection of the right half (component A:Te, B:Cd), neglecting the lines that result from the (in this case very narrow) solid solution range S_B. Note that the projection of the three phase strip considerably undercuts the solid stability range S_{CdTe}, similar to the behavior in Fig.3.49. Note also that the Cd partial pressure rather than the total pressure is used as pressure coordinate. This is because in the experiment the vapor composition was adjusted by controlling the (dominating) Cd pressure by exterior means, as will be discussed in Sect. 4.4.1. As a consequence of this particular three-phase line shape, for every set of fixed component pressures there are two tie-lines in phase space on which solid, liquid and vapor of different composition are in equilibrium. Points on the upper "liquidus" curve (actually VLS_{CdTe} line) are associated with nearly stoichiometric melt composition. Points on the lower part of the curve are in equilibrium with liquid that is rich in Cd. This situation allows for the conduction of solidification from a melt in two basically different processes. The conventional way, corresponding to path A-B-C in Fig.4.18, consists of *extraction of heat* (lowering of T) from a nearly stoichiometric melt. The other process starts with Cd-rich melt, say in F, and requires *addition of heat* in order to achieve crystallization of CdTe. The heat supplied at point E, i.e., at constant temperature, corresponds to the heat of vaporization for the excess Cd from the melt minus the heat of solidification of CdTe. Further heating brings the solid to point D. As emphasized before, solids resulting from the two different liquids will also differ in their composition. In the CdTe case, "heating" will lead to a solid that is richer in Cd than the one obtained from conventional "freezing".

4.3.2 Crystallization from solutions

Both crystal growth from solutions and incongruent melts can be viewed as growth from multicomponent liquids. There is only a gradual but not a basic difference with respect to compositions. One may want to call solution growth a situation in which the excess component (solvent) forms a larger fraction of the crystal nutrient than in incongruent melts. But such a distinction is somewhat arbitrary. For instance, following this scheme, where should one "list" the growth of cesium dihydrogen arsenate (CDA) with a solubility of about 500 g CDA per 100 g of water at 50°C?

Similar to some applications of incongruent melt growth, one may call solution growth a situation where a *solvent has intentionally been added* in order to perform crystal growth at lower temperatures than the pure material *(solute)* would permit. As in growth from melts, solution growth may require greatly differing temperature and pressure conditions. It may be carried out in solvents near room temperature, or lack of adequate room temperature solvents may require dissolution in "melts" (flux growth) at elevated temperatures, or it may be necessary to extend by pressurization the liquid stability range of a solvent at high temperatures (hydrothermal growth).

The relatively high concentration of the solvent component aggravates the mass transfer problem as compared to incongruent melt growth. In addition, the polar nature of many solvents (which makes them good solvents) results in complexation (solvation) of the solute. Desolvation at the interface complicates the interfacial growth kinetics. Energetically there is then a competition between attachment of the solvent into the lattice and separation from the solute. This is well illustrated in some solubility versus temperature plots. For instance in Fig.4.19 one sees that at temperatures between 18 and 40°C, KF crystallizes in associated form with "2 crystal waters" (dihydrate). Above 40°C dehydration is energetically more favorable so that anhydrous KF forms from aqueous solution. In general the complex solvent-solute interactions require rather slow growth rates from solutions in order to avoid non-equilibrium solvent incorporation phenomena.

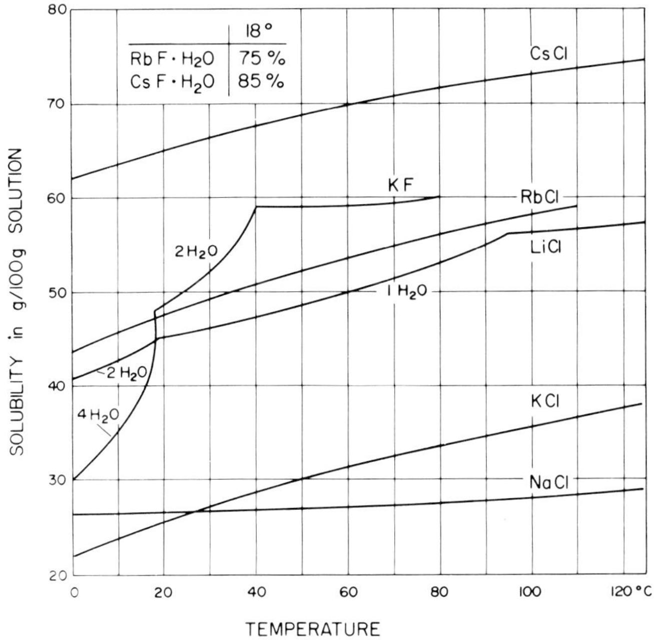

Fig.4.19. Solubility of alkali fluorides and chlorides in water as function of temperature. From [4.19] by permission of Marcel Dekker

Quartz, SiO_2, is a material with a solid-solid phase transition that excludes direct growth of the low temperature configuration from a non-stoichiometric melt. α-SiO_2 forms upon cooling at 573°C from β-SiO_2. The melting point is around 1600°C. High temperature solution growth (flux growth) cannot be applied here since most fluxes require temperatures above 600-800°C. Solution growth at ambient conditions cannot be applied either, since there are no solvents with any sizeable solubility for SiO_2 under these conditions. Consequently, the only way out is hydrothermal growth that has been very successfully applied to α-quartz. In hydrothermal growth, the solubility of water is increased by employing elevated temperatures and the correspondingly necessary high pressures. The upper bound for α-SiO_2 growth is typically chosen around 550°C. Solubility as well as the interfacial kinetics determine the lower bound.

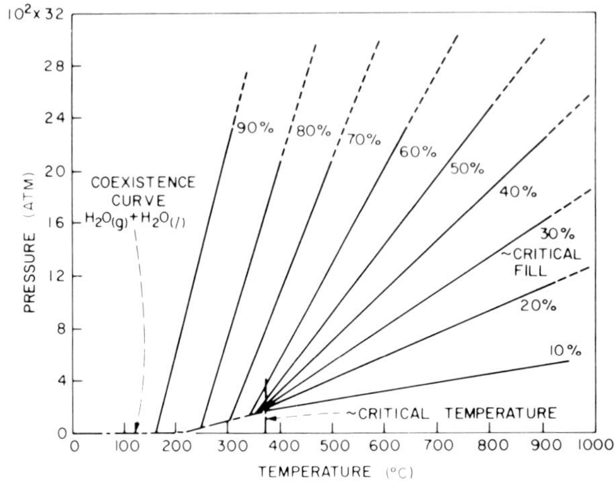

Fig.4.20. Pressure-temperature data for water for various "fills" of an autoclave. From [4.3] by permission of Academic Press

Figure 4.20 gives the temperature-pressure combinations that are required to fill a vessel uniformly with "one phase" of water as function of the fraction of the colume filled initially at ambient. Below the critical temperature of water, i.e., $T \leqslant 374°C$ the phase uniformly occupying the vessel is liquid water. Above that temperature the water is "gaseous" although its density and solvent power remain more or less unchanged. Fig.4.21 shows that the solubility of SiO_2 in H_2O even at high temperatures is rather low. (An interesting feature is the retrogarde solubility at "lower" pressures.) To increase the solubility further, "mineralizers" are added to the water. Their function is to form loose complexes with SiO_2 thereby shifting the dissolution equilibrium to higher values. Mineralizers that contain OH^- and CO_3^{2-} ions proved particularly useful.

Sapphire, Al_2O_3, is another material where hydrothermal growth without addition of a mineralizer would not be feasible due to negligible solubility in pure water. Fig.4.22 displays the solubility one can obtain at high P and

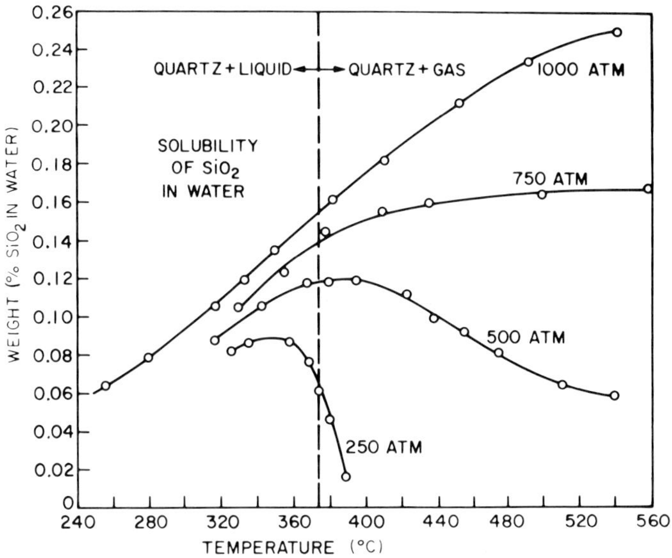

Fig.4.21. Solubility of SiO_2 in H_2O as function of T for various pressures. From [4.3] by permission of Academic Press

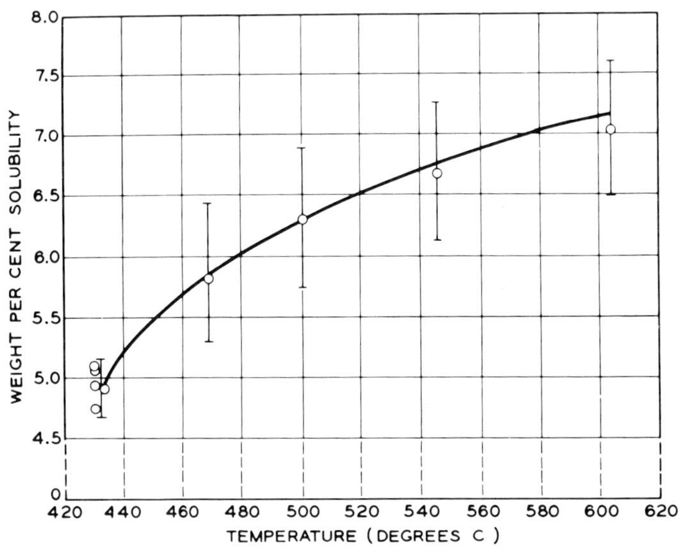

Fig.4.22. Solubility of Al_2O_3 in 3.4 m Na_2CO_3 at 1450 bars. From [4.2] by permission of Prentice-Hall

T upon addition of Na_2CO_3. Phase stability considerations give a low temperature bound in this case, as shown in Fig.4.23. One sees that below about

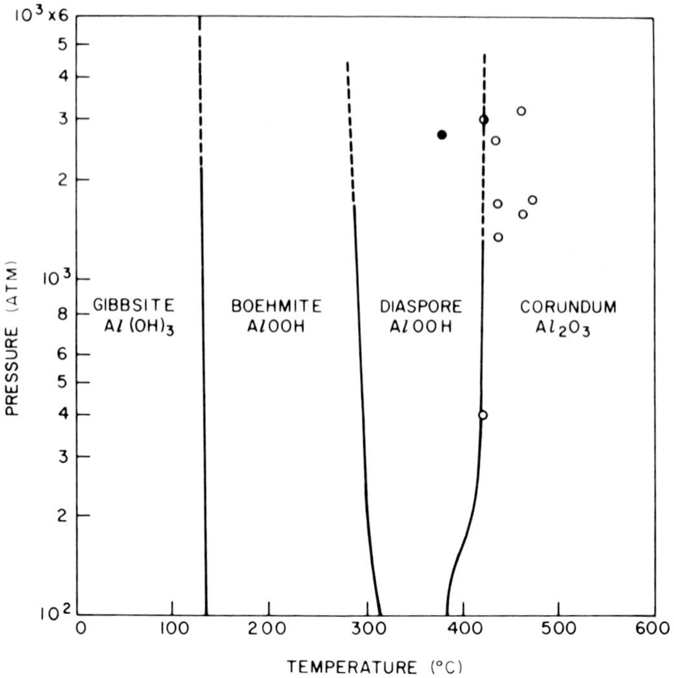

Fig.4.23. P-T diagram for the system H_2O-Al_2O_3. From [4.3] by permission of Academic Press

400°C the sapphire (α-corundum) phase is not stable anymore and diaspore is formed in the reaction $Al_2O_3 + H_2O \rightarrow 2AlOOH$.

For many of the refractory oxides and other materials as well it is more promising (or, in lack of hydrothermal equipment, necessary) to use a flux method in order to perform growth at temperatures way below the melting point of the pure material. For example, $BaTiO_3$ that was mentioned already in connection with growth from non-stoichiometric melts is successfully obtained from KF flux. The corresponding phase diagram is given in Fig.4.24.

The solution growth cases discussed so far have one important common feature: they saturate congruently in the mentioned solvent. This means that at saturation the constituents of the system to be grown are present in the solution in stoichiometric ratios. This appears to be trivial in the SiO_2 and Al_2O_3 cases, although it is not. For $BaTiO_3$, however, it means that the solubilities of BaO and TiO_2 in KF are the same, which is not obvious at all. Congruent saturation is required if the desired solid compound is the only solid phase to grow from the solution. This is illustrated with Fig.4.25. The

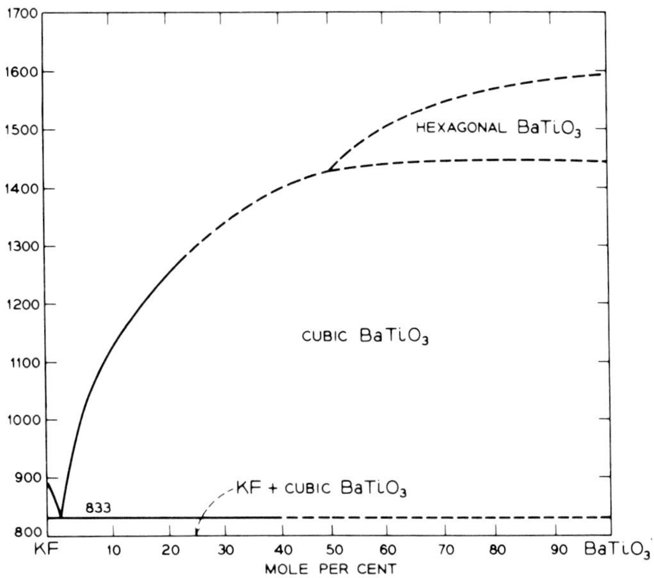

Fig.4.24. Solubility of $BaTiO_3$ in liquid KF. From [4.2] by permission of Prentice Hall

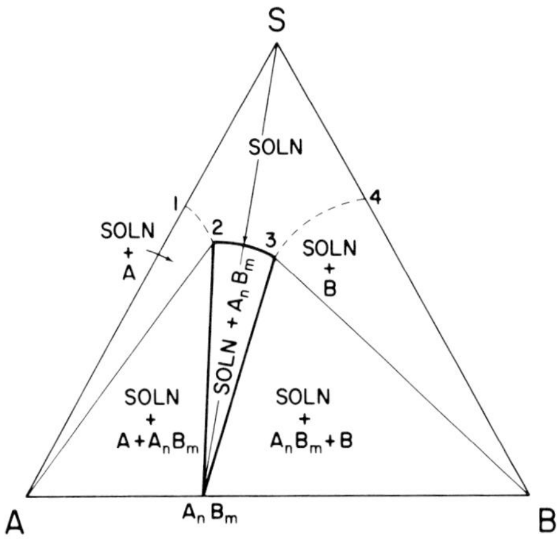

Fig.4.25. System A-B-solvent with A_nB_m congruently saturating in S. After [4.3]

straight line A_nB_m-S in this ternary phase diagram represents all possible combinations of A, B and solvent concentrations with A and B in a desired n:m stoichiometric ratio. If the solubility limit, line 2-3, intersects with

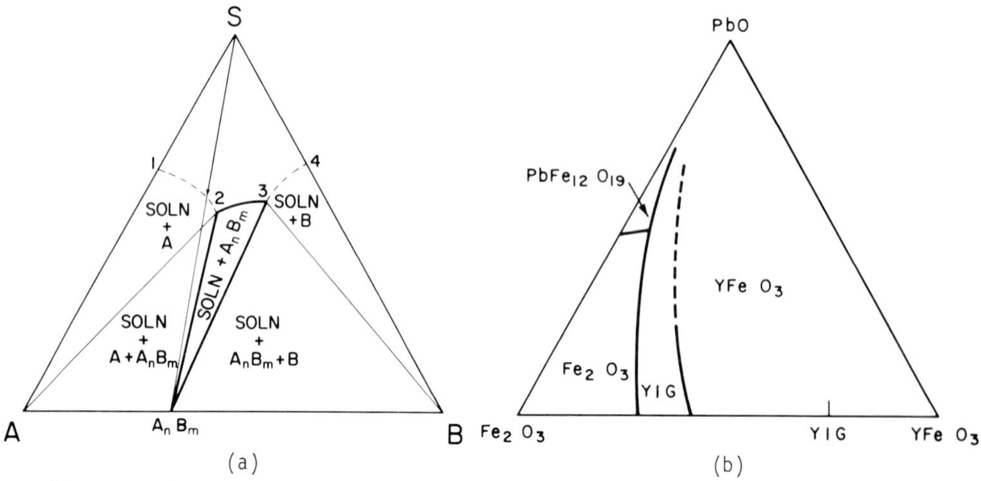

Fig.4.26. (a) System A-B-solvent with A_nB_m not congruently saturating in S. (b) Partial phase diagram of the system Fe_2O_3 - $YFeO_3$ - PbO. From [4.3] by permission of Academic Press

this stoichiometry line, solid A_nB_m can directly form from a solution that is saturated with respect to stoichiometric A_nB_m. A system in which the components A and B saturate incongruently is schematically represented in Fig. 4.26. The solubility and stoichiometry lines of the solution do not intersect. Attempts to crystallize stoichiometric A_nB_m from a stoichiometric solution would first lead to the formation of solid A until the solution composition is shifted to 2, where A_nB_m formation sets in. The undesirable formation of a second solid phase can be avoided if the solution is, from the beginning, saturated with A and B in some non-stoichiometric ratio between 2 and 3. In practice, however, it is often more promising to avoid the resulting complication of the interface kinetics and to search for a congruently saturating solvent.

An example for incongruent saturation is the afore-mentioned YIG (Fig. 4.26b). Y_2O_3 and Fe_2O_3 saturate in non-stoichiometric ratios in PbO used as a flux. In molten $BaO.6B_2O_3$, however, they saturate congruently [4.20]. Yet, in the selection of fluxes the advantages of congruent saturation may be overridden by other properties. It was found [4.21] that addition of B_2O_3 to the PbO flux lead to a remarkable stability with respect to garnet nucleation. Some complex B_2O_3-PbO-garnet melts can be subcooled by as much as 120°C without (spontaneous) nucleation. Fig.4.27 displays expanded parts of the pseudo-ternary phase diagrams for the Fe_2O_3 - Y_2O_3 - $PbO.nB_2O_3$ system for increasing

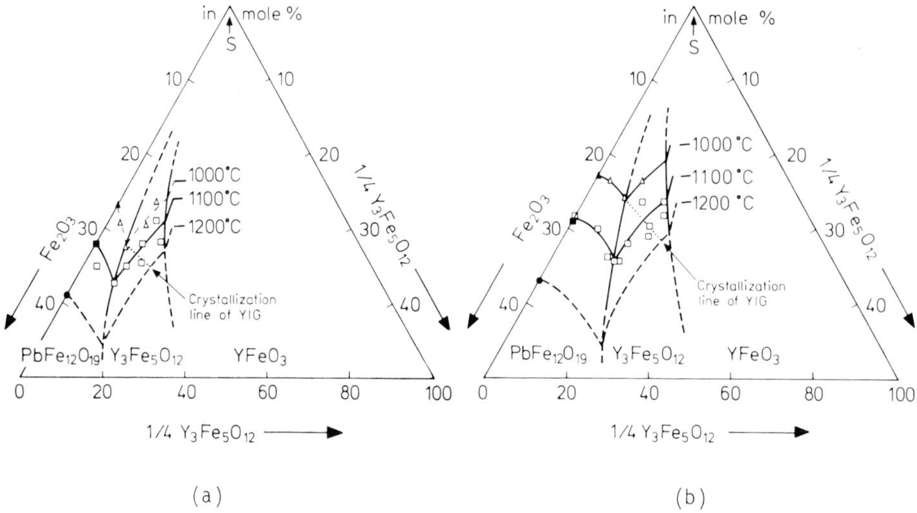

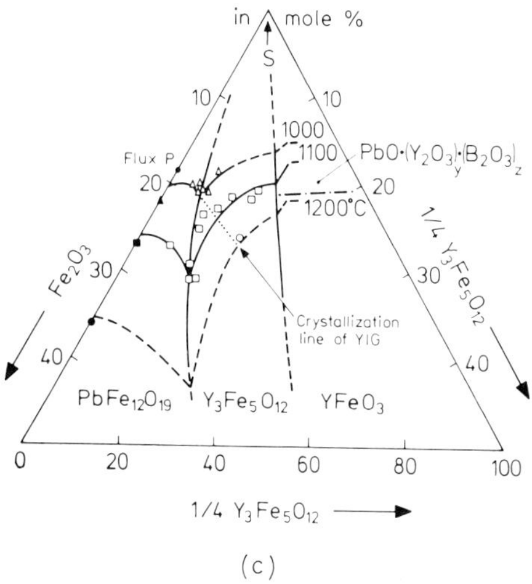

Fig.4.27. Pseudo-ternary phase diagrams of the system $PbO.nB_2O_3$-Fe_2O_3 - $1/4Y_3Fe_5O_{12}$ for $PbO.nB_2O_3$ = 50 mole %. (a) n = 0, (b) n = +.1, (c) n = 0.2. From [4.22] by permission of North Holland Publ. Co.

B_2O_3 additions. In order to enlarge the composition area of interest, 1/4 YIG ($\equiv Y_{3/4}\ Fe_{5/4}\ O_{12/4}$) rather than $YFeO_3$ used in the earlier figures is taken as the right corner component. In addition only the upper half of the diagram, that is, for solvent fractions > 50% is shown. It can be seen that the width of the YIG stability range increases with increasing B_2O_3 addition. Consequently, as can be deduced from the length of the YIG crystallization line between the $PbFe_{12}O_{19}$ and $YFeO_3$ phase boundaries, the maximum possible yield of YIG from a certain volume of flux increases significantly. Here a crystallization line represents the change in composition of the multicomponent liquid during the growth of YIG. The choice of a specific crystallization line is frequently dictated by kinetic considerations. In the present case the viscosity of the flux becomes too high below 1000°C. Hence a starting composition is chosen that, upon crystallization of YIG leads to an intersection of the liquidus of the residual flux with the $PbFe_{12}O_{19}$ - YIG phase boundary around 1000°C.

Lithium niobate, already discussed in connection with incongruent melt-growth, is also frequently grown from fluxes. The lower temperatures employed are, for instance, advantageous in the preparation of optical waveguide de-

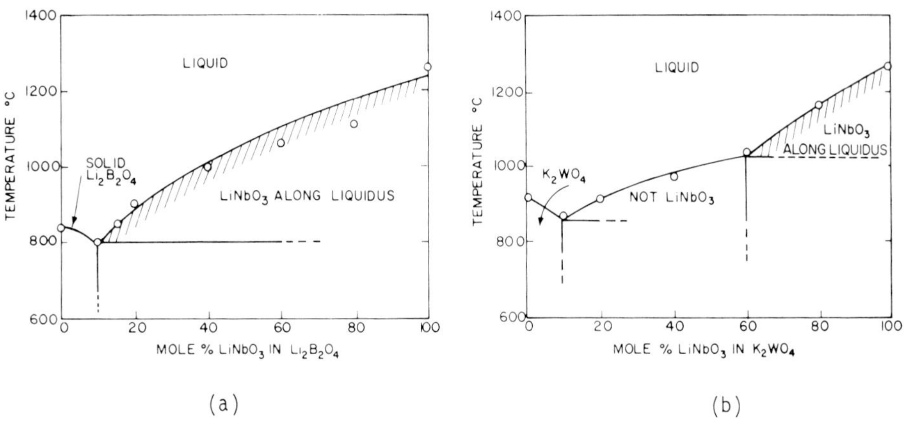

Fig.4.28. Partial phase diagrams of (a) $Li_2B_2O_4$ - $LiNbO_3$, (b) K_2WO_4 - $LiNbO_3$. From [4.8] by permission of North Holland Publ. Co.

vices, where the transitions between the $LiNbO_3$ epi-layer and a substrate are much sharper than in epi-growth at the high melt temperatures. Fig.4.28 shows the liquidus of $LiNbO_3$ with two solvents. $Li_2B_2O_4$ allows for growth at relatively low temperatures. However when some control of the lattice parameter of $LiNbO_3$ is desired, for instance for expitaxial growth onto $LiTaO_3$ (see also Fig.4.7b), one may want to choose a solvent such as K_2WO_4 that does not increase the Li/Nb ratio in the flux.

Melt growth of diamond requires a pressure of 140 kbar at 3700°C [4.23]. However, as mentioned in connection with the carbon P-T phase diagram, Fig. 3.7, diamond can be grown from carbon-saturated molten metal solvents at considerably lower and hence technologically more accessible P-T combinations. Fig.4.29 shows the Ni-C phase diagram at 54 kbar. As indicated by part A-B of the liquidus, the composition-temperature range (1455 to 1394°C) that can be used for diamond solution growth is narrow and the solubility rather low. Hence, in order to obtain large crystals in a cell of limited volume, much more carbon is required inside the cell than can be dissolved at one time in the solvent.

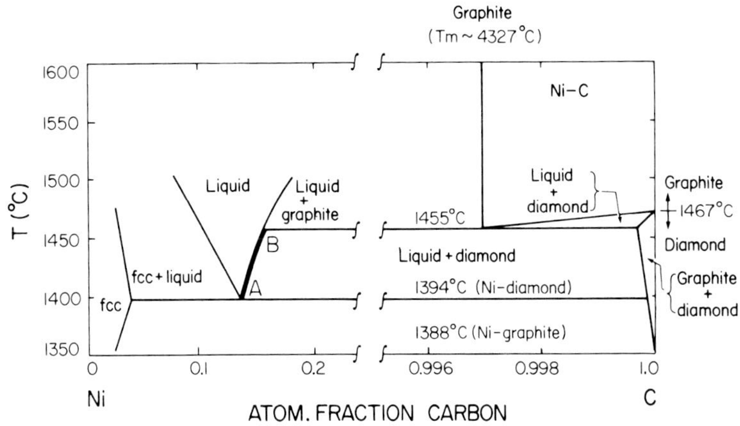

Fig.4.29. Nickel-carbon phase diagram at 54 kbar. From [4.24] by permission of Plenum Press

Another essential parameter that must be considered in high pressure solution growth is the pressure dependence of the melting temperature of prospective fluxes. A strong pressure dependence may impose severe technological problems on the realization of a specific process. Fig.4.30 shows some $T_m(P)$ functions for sodium halides. Note that, for instance, the melting point of NaCl at 65 kbar is 1550°C as compared to 800°C at atmospheric pressure. Another interesting possibility for the solution growth of oxides arises from the fact that low-melting oxides of a high valence state of the metal can be used as solvent for suboxides of the same metal. Fig.4.31 illustrates this for the V_2O_3 - V_2O_5 system. One sees that above 708°C the suboxide VO_2 can be crystallized from "V_2O_5". (The role of the oxygen concentration in the growth atmosphere of such complex oxide systems will be discussed in the following section, Sect. 4.4.2). This scheme has been used for the growth of VO_2 [4.27], as well as for CrO_2 (from CrO_3 at high pressure))[4.28].

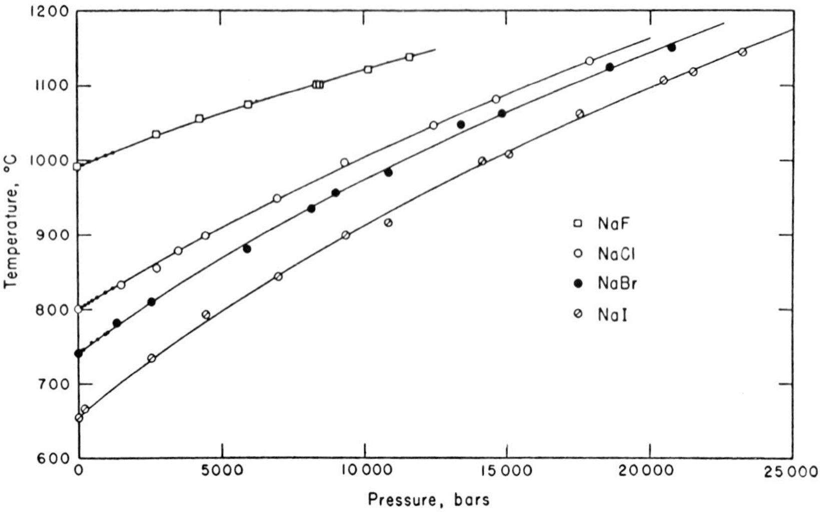

(a)

Fig.4.30. Melting point vs. pressure for sodium halides. (a) from [4.16] and (b) from [4.25] by permission of the American Institute of Physics

The supersaturation required to drive crystallization from a solution, can be brought about in various ways. One can:

1) lower the temperature (or raise it, in systems with retrograde solubility);
2) evaporate some solvent; or
3) add a component that lowers the solubility.

The last method is advantageous when a) the temperature dependence of the solubility is very weak or b) constant temperature throughout the growth process is desired. A typical example for a) is NaCl and H_2O where HCl or KCl can be used to precipitate NaCl from water. Curve AC in Fig.4.32 shows how the solubility for NaCl is decreased upon addition of KCl to the solution. This method is, of course, only feasible until the invariant composition is reached (point C). From a practical point of view, however, one must keep in mind that such a "common ion" precipitation technique, that is, the introduction of a third component into the solution, will always cause some contamination of the growing crystal with that additional component.

In this context it is also very instructive to consider the dehydration of one solute due to addition of another solute [4.30]. Fig.4.33 displays the

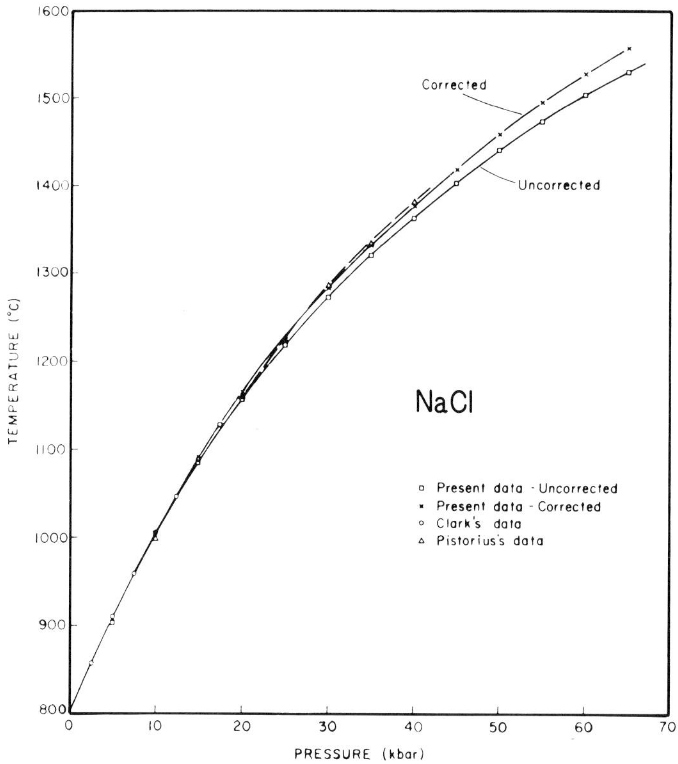

Fig.4.30 (b)

phase equilibria for the ternary system NaCl-Na_2SO_4-H_2O at 17.5°C and 25°C. Whereas NaCl does not combine with water at these temperatures, Na_2SO_4 forms under certain conditions a decahydrate, $Na_2SO_4 \cdot 10H_2O$. Points A and D represent the solubilities of NaCl and Na_2SO_4 in water. (Note the low temperature dependence for NaCl.) Point H represents the composition of the decahydrate. At the low temperature (Fig.4.33a) only NaCl and/or the decahydrate can precipitate from a saturated solution. However, at 25°C (Fig.4.33b) the solution can also be saturated with respect to anhydrous Na_2SO_4 along curve BC. If, for instance, NaCl is added to a solution which is in equilibrium with excess decahydrate, some more decahydrate will precipitate until point C is reached. There, additional NaCl can only dissolve by dehydrating some of the excess decahydrate. Only after all decahydrate in the system is dehydrated

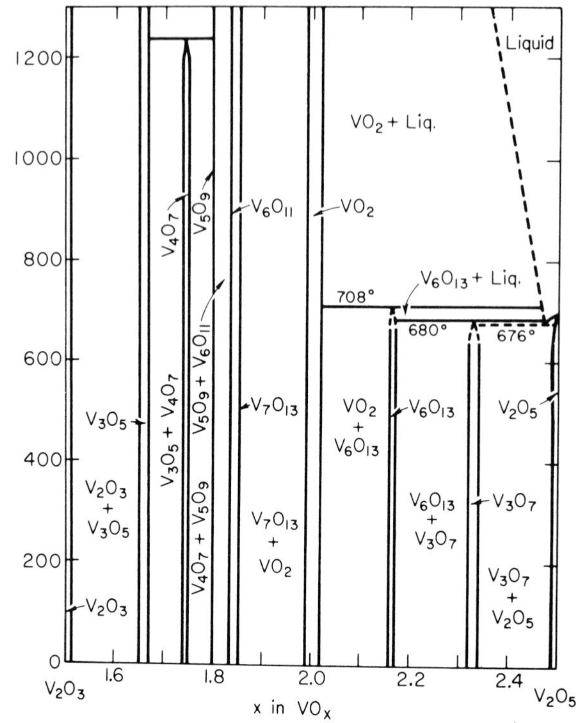

Fig.4.31. Phase diagram for the V_2O_3 - V_2O_5 system. From [4.26] by permission of the American Ceramic Society

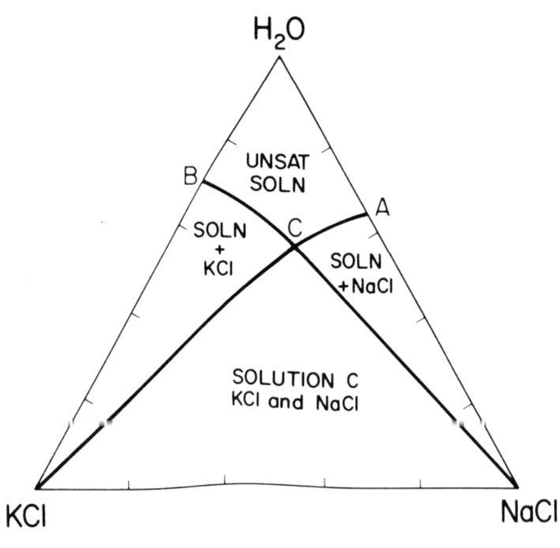

Fig.4.32. Phase diagram for the system NaCl-KCl-H_2O at RT. Concentrations in weight percent (approximate). After [4.29]

will further addition of NaCl shift the composition of the solution from C to B under concurrent precipitation of more anhydrous Na_2SO_4. At B, the

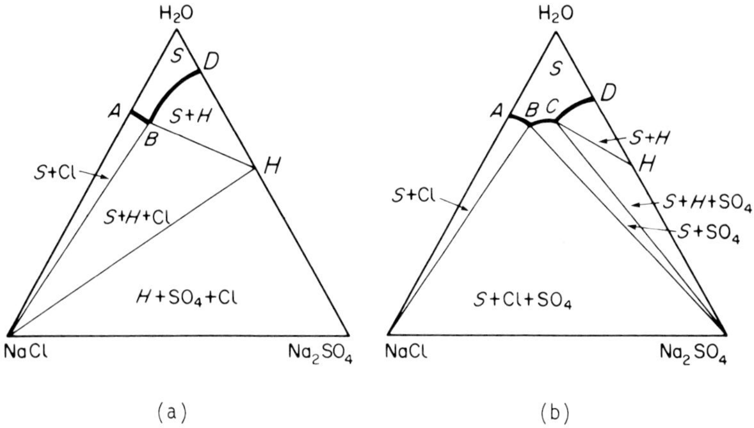

Fig.4.33. Isothermal phase diagrams for the system NaCl-Na_2SO_4-H_2O (a) at 17.5°C, (b) at 25°C. Concentrations in weight percent. Symbols: S = solution, H = hydrate $Na_2SO_4 \cdot 10H_2O$, SO_4 = Na_2SO_4, and Cl = NaCl. From [4.30] by permission of Butterworth Publishers

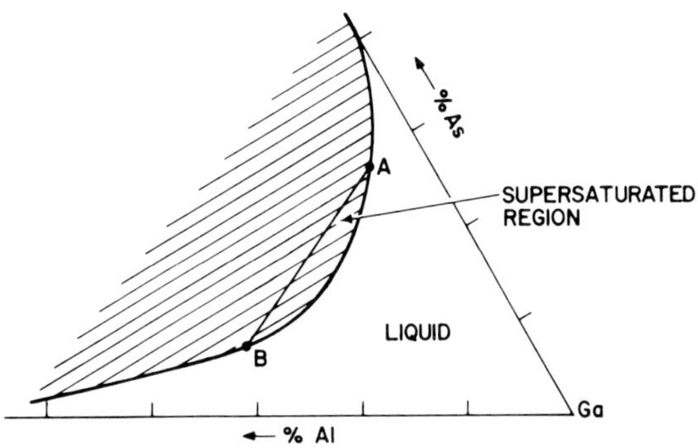

Fig.4.34. Schematic Ga-Al-As phase diagram at Ga-rich corner. After [4.31]

"drying-up point", the solution is also saturated with respect to NaCl. (See Problem 4.4).

Instead of adding a second solute to decrease the solubility for the first solute one can also obtain supersaturation by the (isothermal) mixing of solutions. This approach has been used for the growth of very thin $Ga_{1-x}Al_xAs$ layers from solutions of As in Ga and Al in Ga [4.31]. Fig.4.34, a schematic presentation of the Ga rich corner of the Ga-Al-As-phase diagram, illustrates the process. Points A and B on the liquidus line (for a certain temperature) represents two solutions ("melts") of different composition at solid-liquid equilibrium. Mixing of these melts brings the overall composition to a point somewhere on the line AB in the supersaturated region, the actual point depending on the volume ratio of the two melts. Solid $Ga_{1-x}Al_xAl$ will form until the residual solution composition resumes a point on the liquidus.

4.4 Vapor-Condensed Phase Equilibria

As we have seen, a vapor in equilibrium with a condensed phase contains the various components in the proportions of their fugacities rather than their relative concentrations in the condensed phase. Hence in free evaporation a compound loses predominantly the component that it contains in excess of the stoichiometry of the pressure minimum M in its existence range. This applies to all compounds whether their existence range has a congruently vaporizing composition or not (see Fig.4.35; and also Figs.3.46 and 3.48). With compounds that can congruently vaporize once the pressure minimum in the chaoticus is reached (which coincides with M, Fig.4.35a), vapor of the same composition as the solid will form until the solid compound has vanished. However, with intrinsically incongruently vaporizing compounds, i.e., with compounds that lack a chaoticus minimum in the P-X plot at a certain temperature, free evaporation at point D yields B-rich vapor until S_{AB} is completely decomposed into S_A (Fig.4.35b). Further free sublimation of S_A will eventually

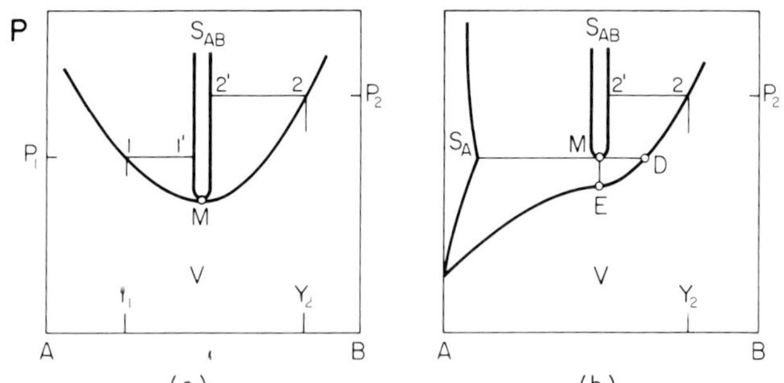

Fig.4.35. Free sublimation in a binary with compound. (a) congruently (b) incongruently vaporizing compound

pass through point E where the vapor contains both components in equal concentrations.

Details of vaporization mechanisms and consequences of various molecular types of vapor composition for the growth kinetics will be treated later [1.2]. However, for the benefit of the following discussion of vapor composition control and the application of vapor-solid equilibria, a brief preview on vaporization "reactions" appears useful. The selection and presentation of material in this section follows closely the corresponding chapter of the monograph by KRÖGER [4.32].

Vapors in equilibrium with a coexisting condensed phase of composition A_nB_m may *dominantly* consist of:

1) molecules A_nB_m or associates (polymers) thereof: *molecular vaporization*, e.g.,

$$PbTe(s) \rightarrow PbTe(g)$$

$$2NaCl(l) \rightarrow Na_2Cl_2(g)$$

2) molecules A_aB_b and B_c, where $a/b \neq n/m$: *partially dissociative vaporization*, e.g.,

$$SiO_2(s) \rightarrow SiO(g) + 1/2\ O_2(g)$$

$$CuCl_2(s) \rightarrow CuCl(g) + 1/2\ Cl_2(g)$$

3) molecules A_a and B_b, where $a \gtrless n$ and $b \gtrless m$: *dissociative vaporization*, e.g.,

$$NH_4Cl(s) \rightarrow NH_3(g) + HCl(g)$$

$$CdTe(s) \rightarrow Cd(g) + 1/2\ Te_2(g)$$

4) molecules of only one component, i.e., A_nB_m undergoes *decomposition*, e.g.,

$$AgCl(s) \rightarrow 1/2\ Cl_2(g) + Ag(s)$$

$$GaAs(s) \rightarrow As_2(g) + As_4(g) + Ga(l)$$

$$Na_2SO_4\cdot 10\ H_2O(s) \rightarrow 10\ H_2O(g) + Na_2SO_4(s)$$

$$CuSO_4\cdot 5\ H_2O(s) \rightarrow 2\ H_2O(g) + CuSO_4\cdot 3\ H_2O(s) \quad .$$

Note that in the decomposition case, type 4, non-gaseous products are formed during free evaporation. Let us further illustrate this with the complex behavior of the system copper sulfate-water. As can be seen from Fig.4.36, cop-

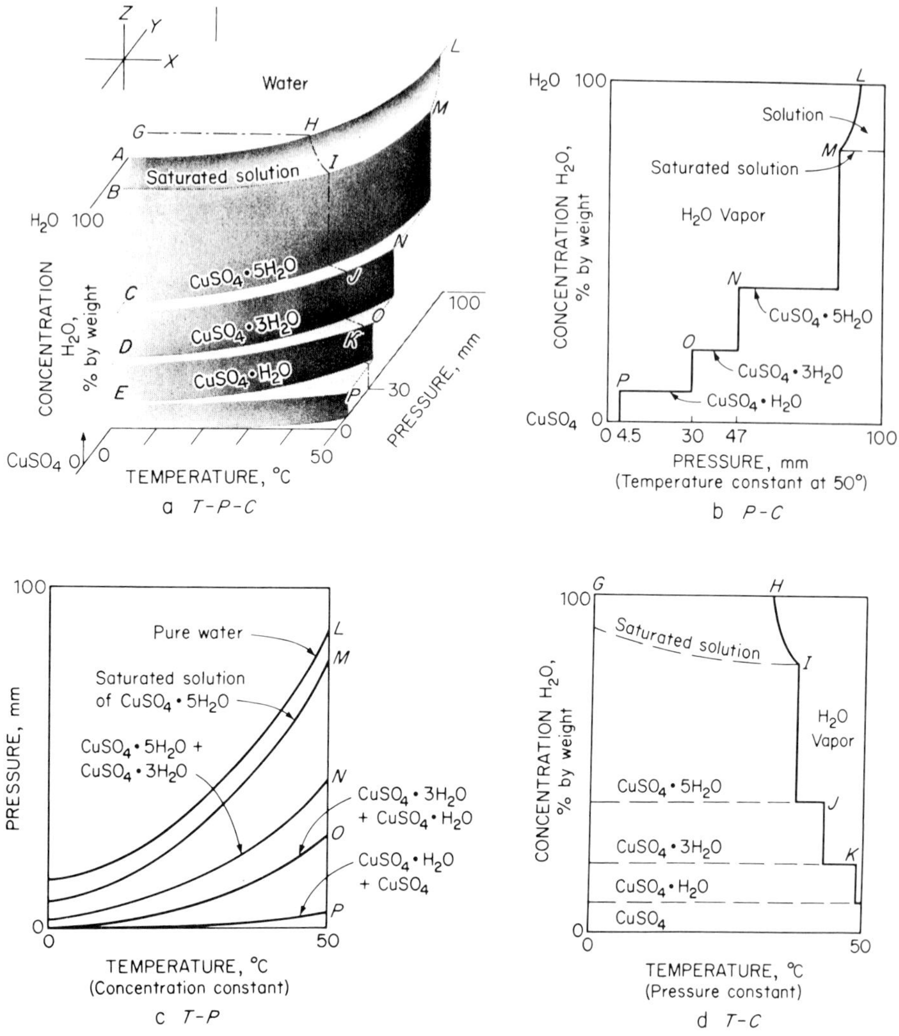

Fig.4.36. Phase diagrams for the system cupric sulphate-water (a) space figure, (b) isothermal section, (c) sections at various constant concentrations, (d) isobaric section. From [4.29] by permission of Prentice-Hall

per pentahydrate can exist in equilibrium with water vapor at any temperature as long as the pressure of water vapor is (a) above the decomposition pressure to the trihydrate and (b) below the vapor pressure of the saturated solution. If the partial pressure of water exceeds these limits, decomposition to trihydrate or formation of a saturated solution respectively, sets in.

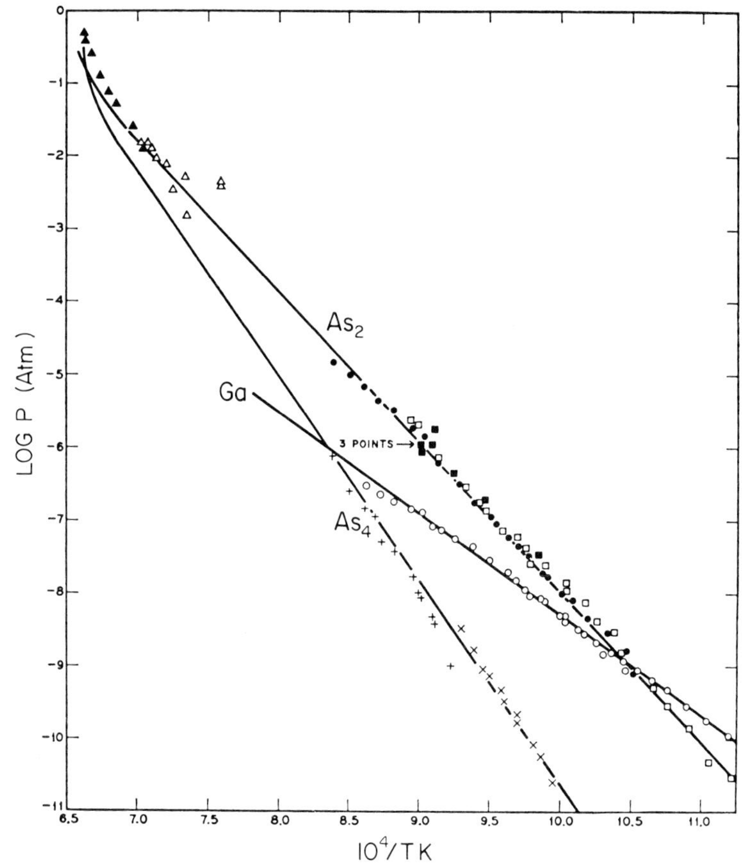

Fig.4.37. Equilibrium pressures of As_2, As_4 and Ga over GaAs. From [4.33] by permission of Pergamon Press

Though we have discussed independently various types of vaporization reactions, in reality vaporization may occur simultaneously through several mechanisms. Their relative weight will be governed by the energies of formation and dissociation of the various possible species in the system. Thus the temperature will determine which type of vaporization prevails in a system. With GaAs, for instance, at high temperatures the partial pressure of gallium is negligible as compared to p_{As_2} and p_{As_4}. Hence we have listed the system under decomposition (type 4 above). However, as shown in Fig.4.37, at $T < 1000$ K p_{Ga} approaches p_{As_2}. Hence, at lower temperatures one will rather list the system under dissociative (congruent) vaporization (type 3).

Vaporization from an incongruent phase can be envisioned, in essence, as a superposition of molecular (atomic) vaporization of the excess component and any of the above mechanisms that prevails in the considered (congruent) substance.

Note that the concentration ratio in a vapor rarely corresponds to the partial pressure ratio. In order to establish a certain concentration ratio in a vapor via (exterior) adjustment of the partial pressures one must know the molecular forms in which a binary system, for instance, A and B exist in the vapor under given P-T conditions. For example for the Ga-As system where $P = p_{Ga} + p_{As_2} + p_{As_4}$ the simple form of (3.40) must be modified to

$$\frac{Y_{Ga}}{Y_{As}} = \frac{p_{Ga}}{2p_{As_2} + 4p_{As_4}} \quad . \tag{4.1}$$

It is interesting to note here, subject to detailed discussions in later parts, that growth does not necessarily proceed via reversal of the dominant vaporization mechanism. Due to some kinetic barriers, growth may dominantly be based on the recombination of vapor species created by a "minority" vaporization process.

4.4.1 Vapor composition control

Materials preparation processes frequently require accurate control of the composition of a vapor phase. As a basis for the application of binary vapor-solid equilibria in the following section and more complex situations to be treated later we will review the principles of various methods for obtaining defined concentrations of a vapor species in a system.

1a) In a *uniformly heated* closed system (ampoule) one can predetermine the pressure of a vapor component, B by adding at ambient temperature more liquid or solid B than vaporizes at the temperature of the experiment. Given the heat of vaporization, the vapor pressure can then be calculated from $p_B \propto \exp(-\Delta H/RT)$. A disadvantage of this method is that T and p_B cannot be chosen independently. Yet, an advantage is that even when the material to be treated is either a source or sink for B, p_B will remain unchanged during the process due to the presence of the condensed B reservoir.

1b) If lower pressures are required than the vapor pressure of condensed B at the temperature of the process, one can simply add only that amount of B at which all B is vaporized and establishes via $p_B = nRT/V$ the desired pressure. However, this approach, unlike 1a), requires accurate knowledge of the experimental volume. Also unlike 1a), reactions that involve B will change p_B in the system.

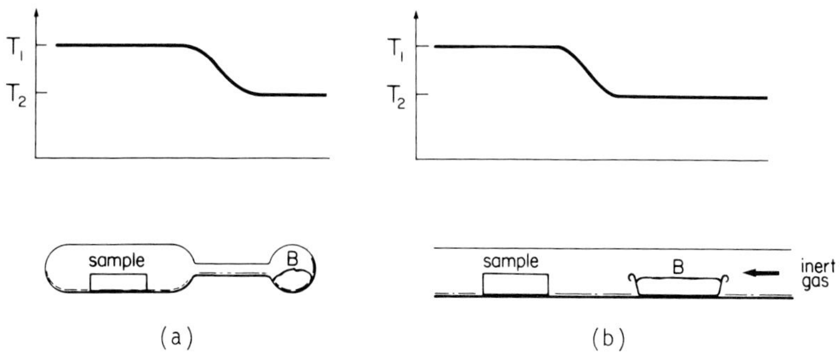

Fig.4.38. Vapor composition control by the two-temperature method in (a) closed and (b) open systems

2a) Greater flexibility for an independent choice of sample temperature and p_B is obtained in a *two-temperature* setup. In this method the material to be treated is heated to T_1 and a quantity of the volatile condensed B is contained in a different part of the system at T_2. Obviously this approach is limited by the requirement that $T_2 < T_1$. If $T_1 < T_2$, B would condense in the sample compartment, and T_1 would govern p_B. This method can be used in closed and open systems, as indicated in Fig.4.38. With open "flow-through" systems, one must allow for sufficient contact time between carrier gas and source B to ensure establishment of an equilibrium partial pressure p_B in the downstream part of the system.

2b) A very elegant, semi-closed version of the two-temperature method was introduced by VAN DOORN [4.34]. In this method the tube that contains the sample and component B is vertically oriented. As illustrated in Fig.4.39, the tube protrudes from the heat source that maintains the sample at T_1, out to ambient temperature where it is connected to a vacuum/gas-handling system. After filling with the material to be treated and a sufficient amount of solid or liquid B, the tube is evacuated and backfilled with an inert gas to pressure P. On heating the volatile B evaporates out of the hot zone and condenses as a liquid on the colder part of the wall. Droplets of sufficient size flow or drop back into the hotter zone where they re-evaporate. Similar to the flow in a diffusion pump this cycle will expell most of the inert gas from the sample part of the tube. More importantly, in steady state the partial pressure p_B throughout the lower part of the tube will be practically identical with the externally controlled P. This is because the locus of condensation for B in the temperature gradient must coincide with the boiling temperature of B under P, i.e., $p_B \approx P$. Advantages of this method are that neither the experimental volume nor the heat of vaporization must be known accurately, and the p_B can be "directly" measured with a standard manometer at room temperature, even if B is a corrosive vapor. Another advantage is the applicability of conventional sealing techniques (o-ring, etc.) to the room temperature end of the tube. A limitation is that the pressure to be established must not be too close to the triple point of B; otherwise significant condensation of solid B will occur in the temperature gradient.

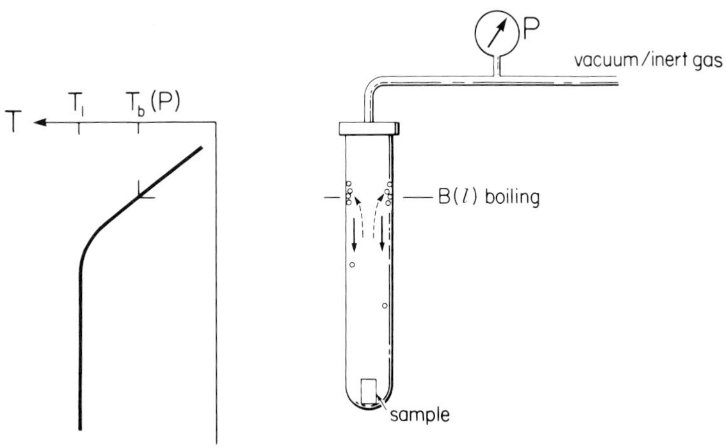

Fig.4.39. Van Doorn - method for heating materials in vapors of externally controlled pressure. Temperature profile and experimental scheme

3) The partial pressure of a component can be controlled by *buffer reactions*. Gaseous or condensed compounds which upon heating dissociate to the desired vapor and other components can be used for this purpose. By taking advantage of mass action one can establish a wide range of p_B at a given temperature. For instance, if a gas AB dissociates according to $AB(g) \rightarrow A(g) + B(g)$, in order to obtain a certain p_B one will fill the system with

undissociated AB *and* gas A in a ratio given by

$$p_B = k(T_1)\,\frac{p_{AB}}{p_A} \tag{4.2}$$

where $k(T_1)$ is the dissociation equilibrium constant at the temperature of the experiment. Besides this flexibility in the initial adjustment of B-concentrations the method has the advantage that the partial pressure of B is *buffered* due to the chemical equilibrium. If some if the initially formed B is used up in another process within the system, part of it will be replenished by further reaction of AB and A. In a closed system, with a finite supply of gaseous AB and B the newly formed p_{B_2} will, of course, be smaller than the initial p_{B_1}. Yet, as illustrated in Problem 4.5, the pressure variation Δp_B due to consumption of B will be significantly less than in an unbuffered system. In open systems Δp_B can be made negligible even with considerable amounts of B-consumption since the reactants AB and B can be supplied continuously (see chemical vapor deposition, [1.2]). Examples for this method are the use of the gas mixtures $H_2S + H_2$ and $H_2O + H_2$ or $CO_2 + CO$ for establishing defined sulfur and oxygen partial pressures, respectively (see also Sect. 4.4.2).

The buffering action in a closed system can be considerably improved by choosing a reaction in which one of the dissociation products, say A, exists at T_1 in condensed form; see Problem 4.5c. If both A and AB in the above reactions are (pure) condensed phases, p_B remains constant ("is fixed") in spite of competitive B-consuming reactions, as long as A and AB exist in the system. In terms of a thermochemical description, for the non-gaseous A and AB one replaces p_{AB} and p_A in (4.2) with the activities which in their pure states are unity [see discussion following (2.116)]; hence p_B becomes a function of T only. Expressed in terms of phase equilibria, one takes advantage of the fact that on a three-phase line, e.g., B(g) - AB(s) - A(s), the pressure is uniquely defined by the temperature. An example is the establishment of defined oxygen partial pressures by heating of mixtures of oxides of different valence states. For instance, varying ratios of TiO_2 and Ti_2O_3 result in the temperature range of 1000-1800°C in P_{O_2} ranging from $1\text{-}10^{-14}$ atm, which enabled PORTER and ROY [4.35] to prepare seven Magneli phases Ti_nO_{2n-1} (n = 4 through 10) with very narrow stability ranges. This complex system is one of the examples to be discussed in the following section on applications of vapor-solid equilibria in materials preparation.

4.4.2 Vapor-solid equilibria

Vapor-solid equilibria can be used to adjust the composition of a growing solid or to readjust the composition of a solid after growth. Solid composition control can readily be performed with most of the vapor control methods outlined in Sect. 4.4.1 if the vapor pressure of a component(s) of a compound S_{AB}, say p_B, predominates over the compound molecule pressure p_{AB}. This is the case with all incongruently vaporizing solids, and with (in their stoichiometric form) molecularly vaporizing solids of sufficiently large off-stoichiometric composition. In these cases it suffices to *fix* the partial pressure of one of the components. The other equilibrium component concentration(s) are then automatically established by sublimation from the solid compound. For instance one can stabilize the composition 2' of S_{AB} in Fig. 4.35 by fixing p_B to $p_B = Y_2 \cdot P_2$, where P_2 is the total pressure. [For this simple relation it is assumed that B exists in the vapor in atomic form only; otherwise modifications along the line of (4.1) must be made]. In the case of a congruently vaporizing compound (Fig.4.35a) S_{AB} of somewhat different composition 1' can be stabilized with vapor of composition Y_1. It has been pointed out that in unbuffered systems "attempts to regulate the vapor composition via the minority component are bound to be unsuccessful. This is because the majority component has to be furnished by evaporation of the solid. But this is, in general, accompanied by the formation of much more of the minority component than we wish to establish" [4.32] and the system behaves essentially as if no vapor control were present. This applies if one tried to establish Y_2 by method 1b), i.e., by sealing a certain amount of A together with S_{AB} into a uniformly heated tube. Yet if either p_A or p_B is truly fixed, for instance with a two-temperature method and a condensed reservoir (buffer), composition control through the minority component is possible. Then there is no principal difference between p_A- or p_B-control. Either A or B species, respectively must condense onto the reservoir material until equilibrium is established by vaporization from S_{AB}. A practical drawback with minority pressure control is that it requires larger sublimation losses from S_{AB} before an adequate majority concentration is reached in the vapor.

Figure 4.40 illustrates the partial pressure distributions in a closed two-temperature arrangement with S_{AB} at T_1 and p_B controlled via a reservoir at T_2. Part (a) depicts the case that S_{AB} sublimes predominantly in dissociated form, i.e., $p_{AB} << p_B$, and $p_A < p_B$. No significant gradient of the total pressure P will exist under these static conditions, unless the pressure

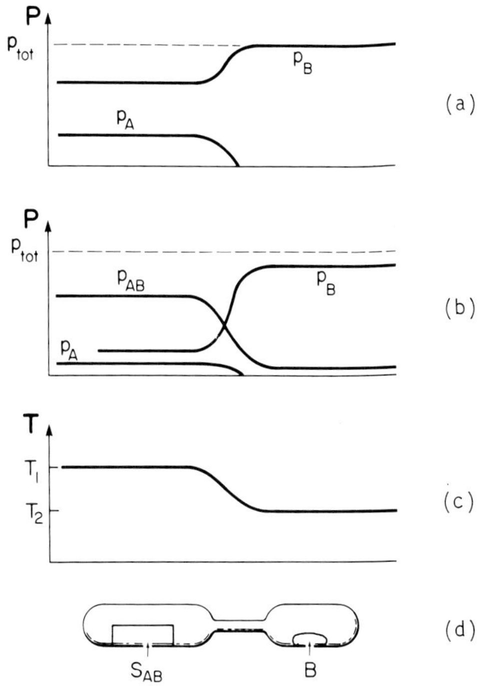

Fig.4.40. Partial pressure distributions in two-temperature setup. (a) for the case that $p_{AB} << p_B > p_A$. (b) $p_{AB} >> p_B > p_A$. (c) Temperature profile. (d) Closed system. After [4.32]

is so low that the mean free path is comparable to or longer than the tube dimensions (Knudsen gas) where $P_2/P_1 = (T_2/T_1)^{1/2}$. Hence, if no other gases are present in the part of the system that contains the B-source, $p_B(T_2)$ constitutes also the total pressure. In the hotter part, however, p_B will be reduced by the amount that corresponds to the equilibrium minority vapor pressure $p_A(T_1)$; hence, $p_A(T_1) + p_B(T_1) = p_B(T_2) = P$. The concentration gradients will drive mass transfer between the hotter and colder parts. Losses of A from the sample area can be minimized by choice of an appropriate geometry. A narrow tubular connection between the sample and B-source areas hinders convection (see Sect. 5.8.2) thus limiting mass transfer essentially to diffusion. Typically, the S_{AB} (of B-rich composition) precipitates out somewhere between T_1 and T_2 so that the purity of B in the source compartment and hence accurate control over the vapor composition can be maintained for long experimental periods.

In systems with predominantly molecular vaporization ($p_{AB} >> p_B \approx p_A$) a two-temperature arrangement gives much less control over composition and sublimation losses than a buffer reaction approach (method 3 of Sect. 4.4.1). In essence, B forms only a "non-contaminating" diffusion barrier against excessive AB losses. These loses can be further reduced by application of a high inert gas pressure.

In the following we will apply the above general considerations to specific materials. First we will familiarize ourselves with the various types of presentation used in the literature for vapor-solid (-liquid) equilibria of binary compound semiconductors. As a second example for a group of materials for which the vapor composition during preparation is important we will discuss the complex equilibria of some metal-oxide systems.

Most II-VI compounds are prepared from the vapor phase [4.6]. In order to properly interpret and control the vapor-solid equilibria one must know the molecular composition of the vapors that result from thermal dissociation of the specific compound. Zn, Cd and Hg are always monatomic in the vapor. Up to 1000°C practically all the Te vapor consists of diatomic molecules. The composition of the sulfur and selenium vapors is complex. However, in the temperature range in which the vapor pressures of zinc and cadmium sulfides and selenides have been measured, only the diatomic molecules S_2 and Se_2 are stable. The mercury sulfide and selenide vapors contain more complex polyatomic species of sulfur and selenium. The electrical conductivity of II-VI compounds, as with most compound semiconductors, is associated with departures from exact stoichiometric composition. Hence, their actual composition can readily be determined with high resolution by conductivity and Hall effect measurements.

Figure 4.41a depicts the T-X projection of the Cd-Te phase figure. Fig. 4.41b shows the partial pressure of Cd in equilibrium with the liquidus of the T-X projection, i.e., the three-phase line of the system. It is instructive to compare Fig.4.41b with the T-P projection of the more general case depicted in Fig.3.46. In comparing, however, note the following differences: 1) A log pressure versus 1/T plot is used here. This is advantageous since relationships for thermally activated processes (such as vaporization) tend to be linear in this type of representation, as long as the pre-exponential factor and the activation energy don't change significantly, see, e.g., (3.9). 2) The partial pressure of Cd is used as pressure coordinate rather than the total pressure. This simply reflects the conduction of the experi-

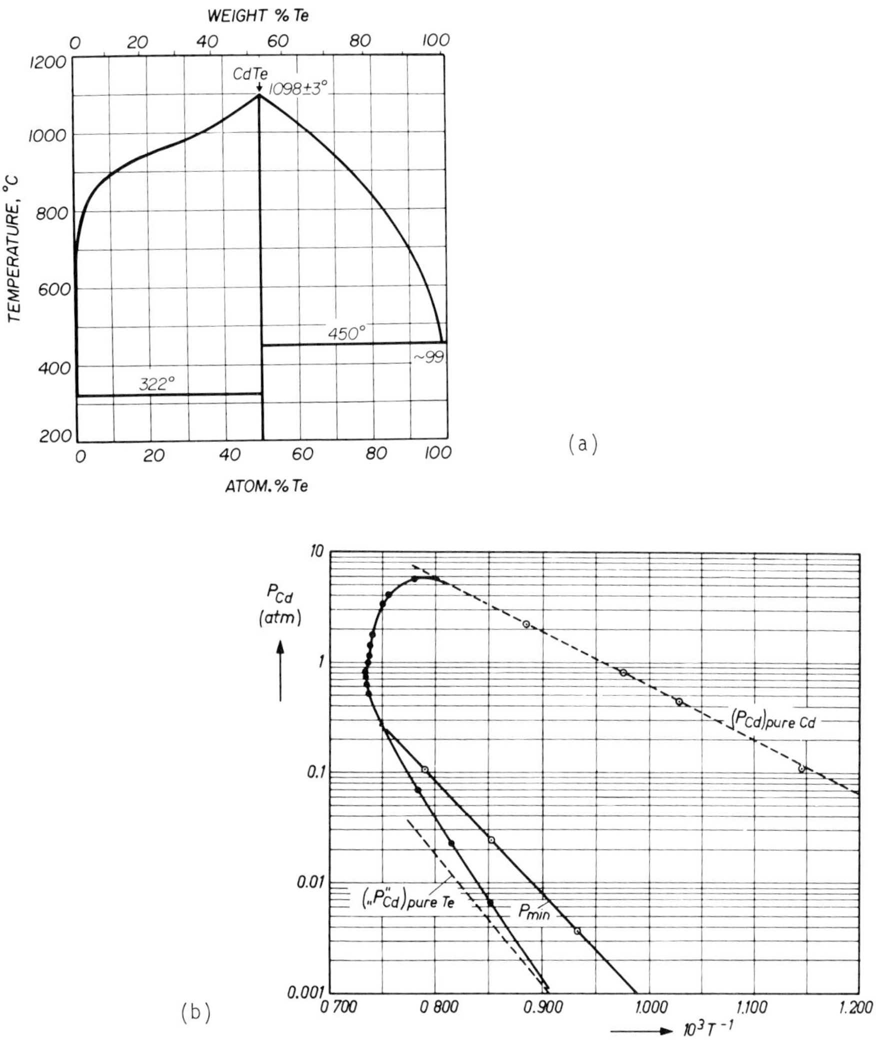

Fig.4.41. Equilibria in the Cd-Te system. (a) T-X projection, from [4.36] by permission of McGraw-Hill. (b) P_{Cd} versus 1/T, from [4.37] by permission of Philips Corp.

ment in which p_{Cd} in the closed ampoule used was controlled via a two-temperature setup (Fig.4.40). Consequently on the Cd-rich side (molten Cd with some CdTe dissolved in it) the three-phase line (Fig.4.41b) becomes practically identical with the pure Cd vapor pressure line which, in that range, also approaches the total pressure. With increasing Te concentration (in all phases) the shown p_{Cd} falls more and more below P_{total} (as will be more clearly illustrated in later figures). Similarly, on the Te-rich side (molten Te with some CdTe dissolved in it) where the composition-controlling Cd is the minority constituent, the three-phase line asymptotically approaches the Te liquid-vapor two-phase line (Te vapor pressure curve). 3) The solid solution ranges at the pure component ends are so narrow that they cannot be represented but by a vertical line at $X_{Te} = 0$ and 1. 4) Due to the lack of eutectics (actually there appears to be one at $X_{Te} = 0.99$) there are no quadruple points in the three phase line of Fig.4.41b.

If there is a congruently vaporizing composition within the stability range of a compound one finds that the total pressure of the vapor components in equilibrium with the condensed phase(s) of that composition is a minimum. Note the line marked P_{min} in Fig.4.41b which gives the (calculated) points where the total pressure above solid CdTe has minimum values at the corresponding temperatures (see Problem 4.6). Note also that this line intersects the three-phase line considerably below the maximum melting temperature, at about 1060°C. This would imply that congruently vaporizing/melting and stoichiometric CdTe differ in their composition by several percent. This, however is not confirmed by more recent phase equilibrium studies as a comparison with Fig.4.41a shows.

We have schematically indicated the minimal total pressure behavior in various phase diagrams before, in particular in Fig.4.35a. Now, taking CdTe as a specific example, we will discuss a "proof" of the P_{min} argument often found in the literature: Since the concentration of undissociated CdTe molecules in the vapor above solid CdTe is negligible as compared to the Cd and Te_2 concentrations [4.37] the total pressure is

$$P = p_{Cd} + p_{Te_2} \quad . \tag{4.3}$$

Furthermore, it is argued, p_{Cd} and p_{Te} are not independent but coupled by the equilibrium constant of the reaction

$$CdTe(s) \rightleftarrows Cd(g) + \frac{1}{2} Te_2(g) \quad , \tag{4.4}$$

which, assuming that solid CdTe has activity unity and that the fugacities of the gaseous components are equal to their vapor pressure, is

$$k_{CdTe} = p_{Cd} \cdot p_{Te_2}^{1/2} \quad . \tag{4.5}$$

With (4.5) one can rewrite (4.3) as

$$P = p_{Cd} + \frac{k^2_{CdTe}}{p^2_{Cd}} = \frac{k_{CdTe}}{p^{1/2}_{Te_2}} + p_{Te_2} \quad . \tag{4.6}$$

If P, at constant T, has a minimum with respect to composition then $dP/dp_{Cd} = 0 = dP/dp_{Te_2}$ which in turn, assuming constant p_{CdTe}, leads readily to $p_{Cd} = 2p_{Te_2}$; that is to say (it is argued) that the minimum lies at the congruently vaporizing composition. The critical reader realizes that this derivation is somewhat circular. First with (4.4) one states that vaporization is strictly congruent then with (4.5) that the congruent solid is strictly stoichiometric (i.e., activity unity) and then P is treated as if p_{Cd} or p_{Te_2} were variables – which they are not after (4.4) is stated in that form. Strictly speaking, if either partial pressure were altered with respect to the congruent ratio in the vapor, the right side of (4.4) would have to include vacancy and/or interstitial concentrations as product terms and k_{CdTe} would become composition dependent due to deviations of the CdTe activity from one. So, in essence, one obtains from this argument congruent vaporization at P_{min}, because one has assumed it with (4.4). Practically, the error introduced by assuming a constant k_{CdTe}, i.e., a composition independent Gibbs function of formation for the solid is very small, in particular for the narrow stability range of II-VI compounds. Yet for pedagogical reasons such potential pitfalls, which upon extension of the model to other materials can lead to cumbersome conclusions, must be pointed out.

Since deviations from strict stoichiometry determine via the excess carrier density the conductivity of the compound it has become customary to indicate within phase diagrams the carrier concentrations that result in the solid in equilibrium at certain P-T combinations. Fig.4.42, for instance, displays constant lines of equal carrier concentrations for PbS. These lines where determined by measuring the electrical properties of PbS crystals heated at various temperatures in an atmosphere with accurately defined partial pressure of sulfur. At low p_{S_2} in the vapor the solid phase shows n-type conductivity and contains an excess of lead. Higher p_{S_2} causes a decrease of this excess until stoichiometric PbS(n = p) is obtained. At even higher sulfur partial pressures excess sulphur and consequently p-type conduction is obtained in the solid.

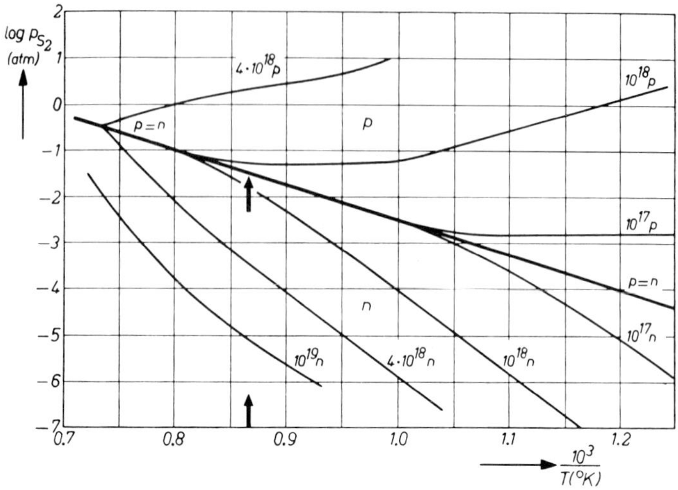

Fig.4.42. Contour lines of equal carrier concentrations for PbS in equilibrium at various temperatures with a vapor the composition of which was controlled through p_{S_2}. From [4.38] by permission of Akademische Verlagsgesellschaft

Another way of displaying that data is in form of plots for the dependence of the partial pressures of the components on the composition of the solid at a fixed temperature. Such plots for PbS at 900°C are given in Fig.4.43, which

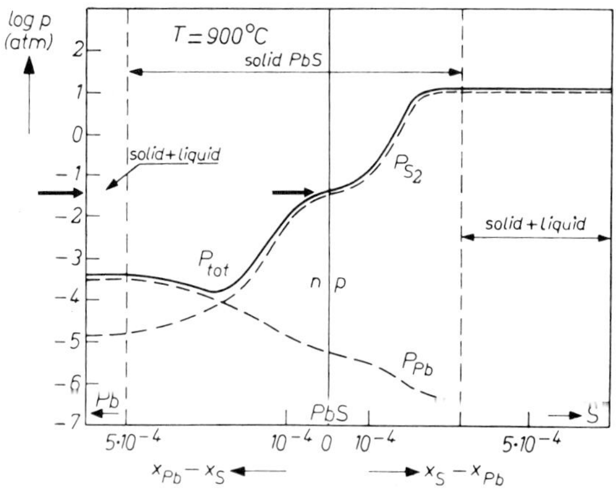

Fig.4.43. Partial pressures of lead and sulphur (p_{Pb} and p_{S_2}) versus composition of solid PbS at 900°C. From [4.38] by permission of Akademische Verlagsgesellschaft

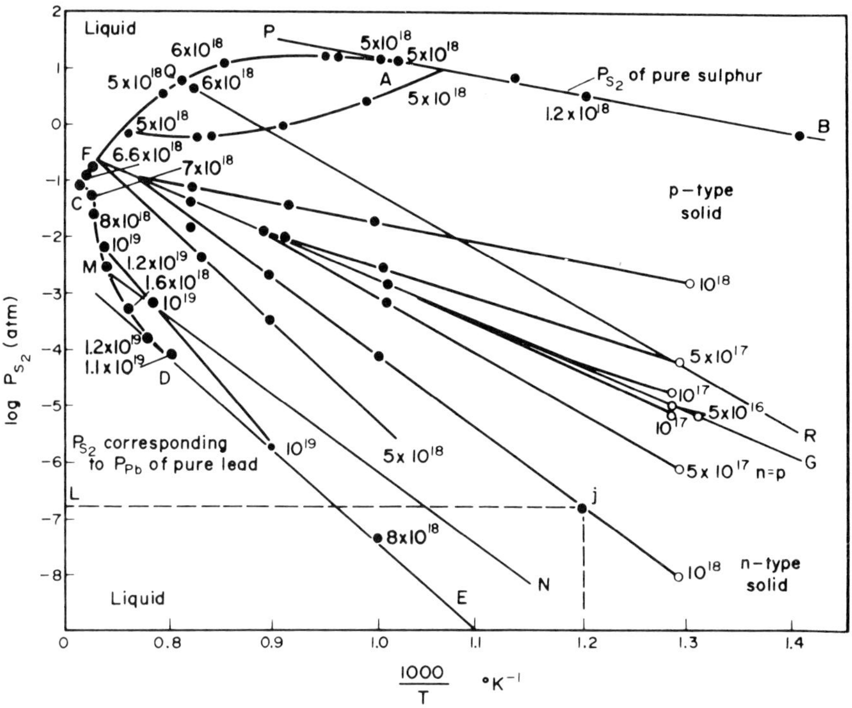

Fig.4.44. Log p_{S_2} versus 1/T and contour lines of equal carrier concentrations of the Pb-S system. From [4.39] by permission of Pergamon Press

illustrates very well the total pressure minimum within the narrow solid solution range. Note that P_{min} deviates by several 10^{-4} mole fractions from the stoichiometric composition. Though significant from a conductivity point of view, this difference is much too small to be displayed in a conventional full range phase diagram. It is instructive to relate Figs.4.42 and 43 by, for instance, locating the "stoichiometric point" (at 900°C) in both presentations (see arrows). The reader is also encouraged to discuss (again) the relations of partial pressures and total pressures on the lead- and sulfur-rich sides that are particularly well illustrated in that figure.

Figure 4.44 forms a combination of Fig.4.42 with the p_{S_2}-T projection of the three-phase blade of the Pb-S system. The lead-rich and sulphur-rich parts of the three phase line are smoothly adjusting to the lead and sulfur vapor pressure curves, respectively.

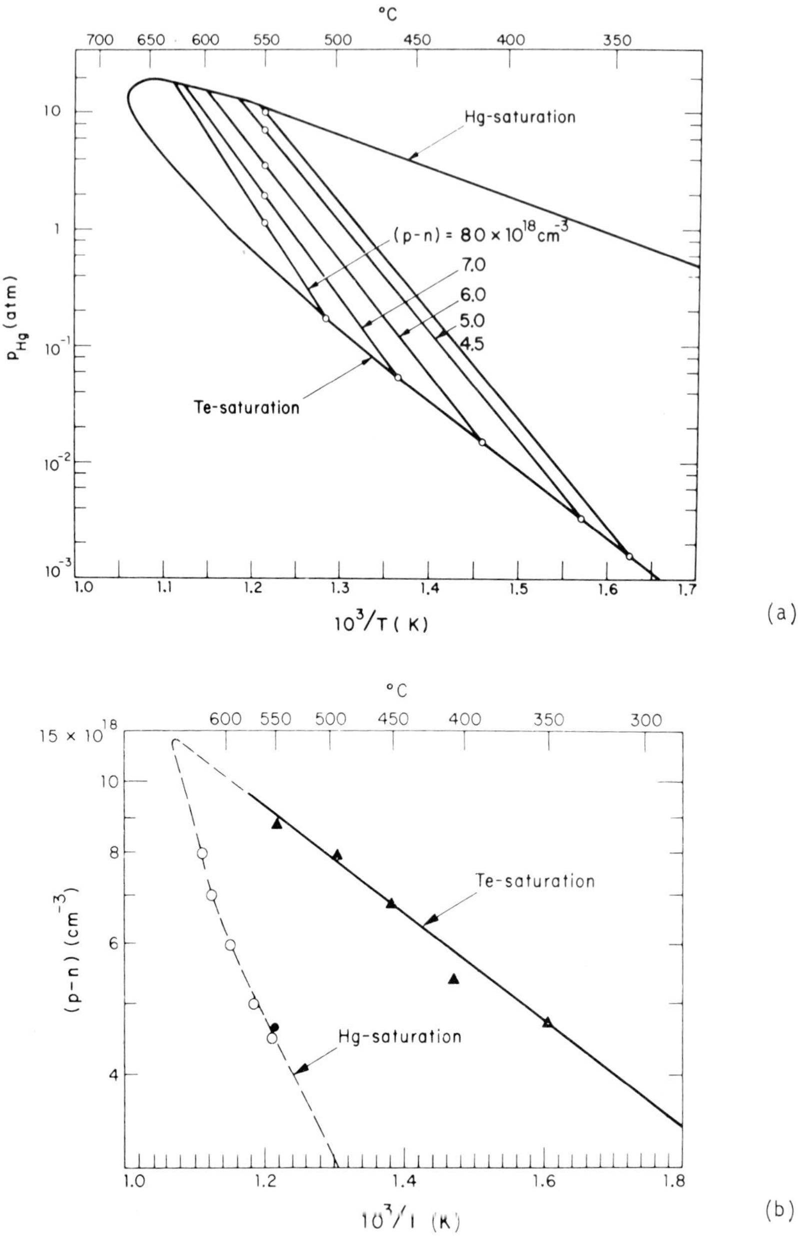

Fig.4.45. Solid-vapor equilibria for the Hg-Te system. (a) Log p_{Hg} versus 1/T with contours for equal acceptor concentration (b) Acceptor concentrations versus 1/T for Te-saturated HgTe (upper line) and Hg-saturated HgTe (lower line). From [4.41] by permission of Pergamon Press

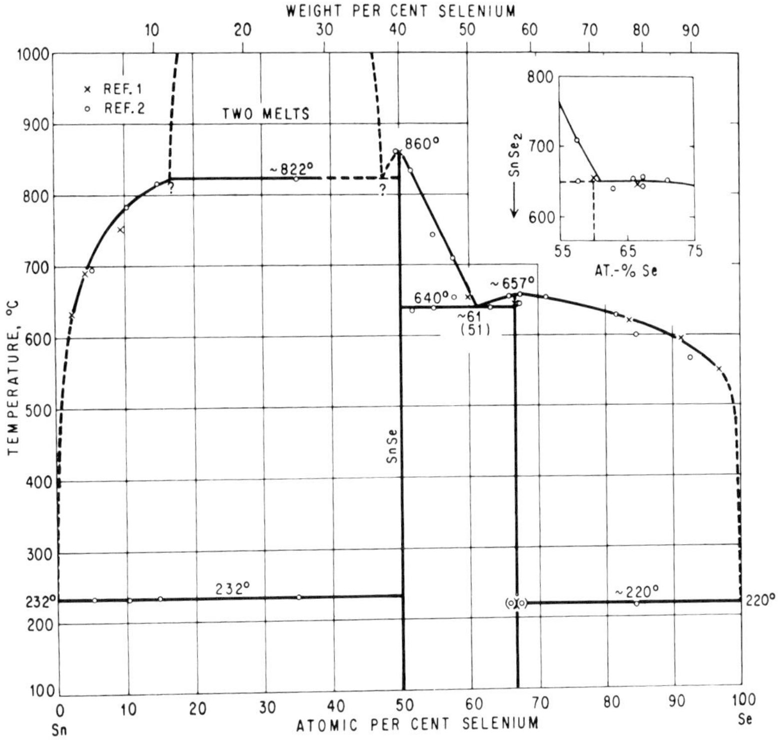

Fig.4.46. T-X projection of the Sn-Se system. From [4.42] by permission of the McGraw-Hill

As pointed out in Sect. 4.3.1 in connection with the solidification of GeTe and SnTe, the stability range of a compound may not contain the strict stoichiometric composition. A particularly well examined member of the II-VI family that shows this behavior is the Hg-Te system [4.40, 4.41].. HgTe, when grown from stoichiometric melts, is always found to be p-conducting due to Te-excess in the solid. But even after equilibration of solid HgTe with Hg vapor, the compound retains some Te-excess as displaced by Figs.4.45a and b. Note that in contrast to the phase diagrams shown earlier the partial pressure of the metallic component (Hg) is used in the P-T projection rather than that of the group VI component, which in this case has the lower fugacity.

As a last example for vapor-solid equilibria in compound semiconductors we show a system in which there are two compounds forming: the Sn-Se system. The T-X projection is given in Fig.4.46 and the corresponding P-T projection in Fig.4.47. Note that the total pressure is used as ordinate in the projection of the three-phase blade of Fig.4.47 since the vapor composition is very

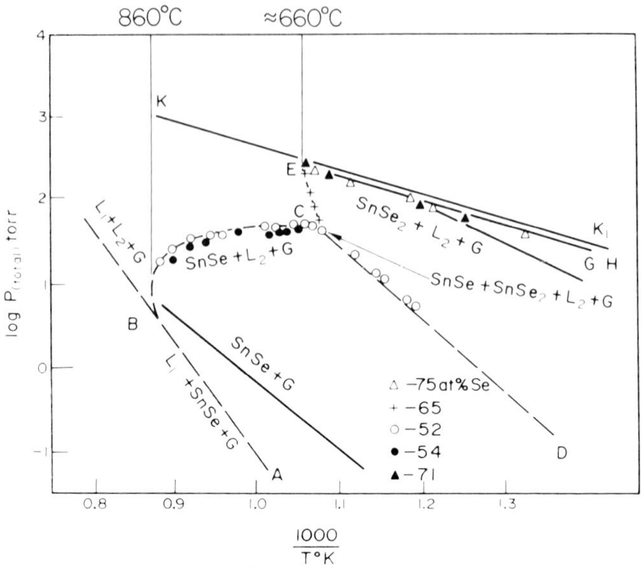

Fig.4.47. P-T projection of the Sn-Se system. From [4.39] by permission of Pergamon Press

complex. Mass spectrometer data indicate that the vapor over solid tin monoselenide is composed of SnSe, Se_2 and Sn_2Se_2 species [4.39]. Due to the presence of two solid compounds the system possesses a quadruple axis [SnSe + $SnSe_2$ + L($X_{Se} \approx 0.6$) + G] which in the P-T projection results in the characteristic cusp in the three-phase line (see also Fig.3.46).

For a concise comparison of phase diagrams of II-VI and III-V systems and their use for the choice of diffusion-doping conditions see the review by PANISH [4.43].

As a second group of examples for the importance of gas-solid equilibria in materials preparation we will now turn to the effect of oxygen pressure on phase relations in metal oxide systems, in particular 3d-metal oxides. Most of the material for this part was taken from reviews by REED [4.44], ROY and WHITE [4.45] and MUAN [4.46].

As we have seen in Fig.4.31, the phase relations in systems containing transition-metal oxides can be very complex. This is because the transition metals (from Sc to Zn) can exist in several oxidation (valence) states. The

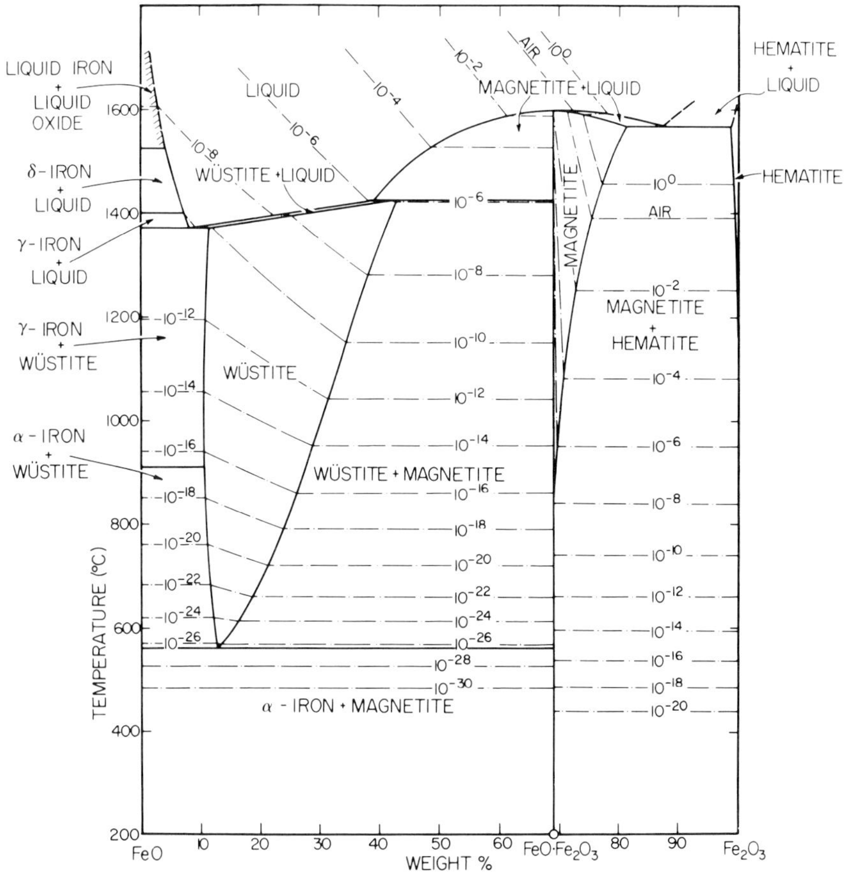

Fig.4.48. T-X projection of part of the Fe-O_2 phase figure. Dash-dot lines are equilibrium oxygen isobars. From [4.47] by permission of Addison-Wesley

width of the homogeneity (stability) range can vary largely even among oxides of the same metal. For instance, as depicted in Fig.4.48, at 800° C wüstite ("FeO") exists over a range of more than 10% in its oxygen content, whereas the stability range of hematite (Fe_2O_3) is too narrow to be resolved on the scale of that diagram. The fugacity of the undissociated oxides and their metal components are negligible as compared to their oxygen fugacity. Hence at a given temperature, the oxygen partial pressure in a system determines the composition of the condensed phase(s); see the oxygen isobars in Fig.4.48. Based upon the phase rule one can distinguish two different equilibrium cases:

1) If only one oxide is present, according to (3.11) f = 2 - 2 + 2 = 2, at a given temperature the oxygen pressure above an oxide can be varied as long

as the corresponding composition of the oxide is not shifted beyond its homogeneity range. From Fig.4.48 and data of other oxide systems one can deduce that in general the range over which the equilibrium oxygen pressure can be varied without jeopardizing the stability of a certain oxide is the wider the narrower the stability range is.

2) If two ("neighboring") oxides are to be present, from (3.11) f = 2 - 3 + 2 = 1 we see that a unique oxygen pressure is required at a given temperature. Hence the oxygen isobars within two-phase coexistence regions of T-X projections are strictly horizontal.

The oxygen pressure p_{O_2} necessary to form a specific *stoichiometric oxide from its metal*, M, can readily be calculated from the standard Gibbs free-energy of formation, ΔG_f^0, of the reaction

$$\frac{2}{m} M(s) + O_2(g) \rightleftarrows \frac{2}{m} MO_m(s) \quad . \qquad (4.7)$$

If it is justifiable to assume unit activity for the two solids in (4.7), then, as discussed in Sect. 2.12,

$$\Delta G_f^0 = -RT \ln(k) = RT \ln(p_{O_2}) \quad . \qquad (4.8)$$

Free energy of formation data for various 3d-metal oxides are given in Fig. 4.49. It follows from (4.8) that in such a plot of ΔG_f^0 versus T all points of a fixed p_{O_2} value form a straight line through the origin. This has been incicated for $p_{O_2} = 10^{-10}$ atm, for instance, with the dashed line in Fig. 4.49. Hence, if one presents ΔG_f-data based upon the "peculiar" form of (4.7) and calculates the corresponding nomographic scale, one can read the p_{O_2} that is in equilibrium with the metal and the desired oxide at each temperature directly on the chart.

Correspondingly, the pressure of oxygen required for the coexistence (mechanical mixture) of two oxides of the same metal at a given temperature can be calculated from the change in standard Gibbs function ΔG_r^0 of the reaction in which an oxide, MO_n, is formed from the next lower oxide, MO_m,

$$\frac{2}{n - m} MO_m + O_2 \rightarrow \frac{2}{n - m} MO_n \quad . \qquad (4.9)$$

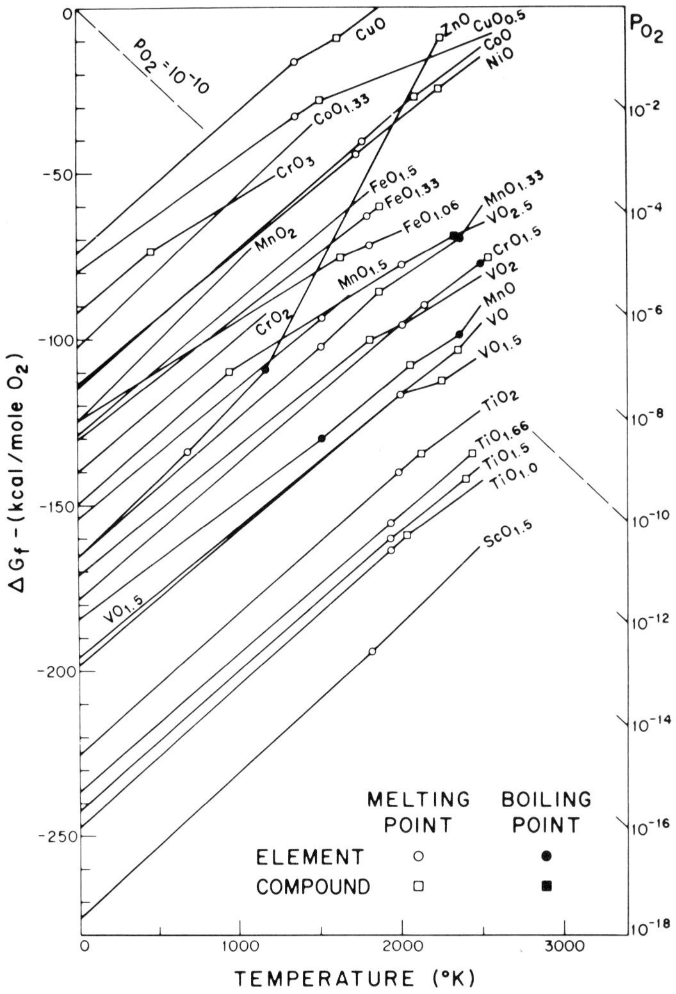

Fig.4.49. Free energy of formation of 3d-metal oxides per mole of oxygen as function of temperature. From [4.44] by permission of North-Holland Publ. Co.

Assuming again that the activity of both solids is one, i.e., that both oxides are strictly stoichiometric, one obtains for the equilibrium oxygen pressure

$$\ln(p_{O_2}) = \frac{\Delta G_r^0}{RT} \quad . \tag{4.10}$$

Figure 4.50 displays calculated p_{O_2} values for pairs of oxides for 3d-metal oxides. The horizontal lines representing the respective coexistence pres-

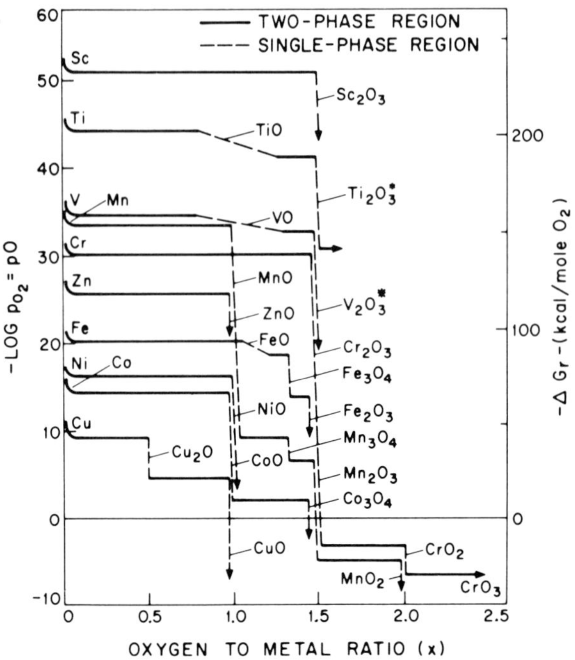

Fig.4.50. Variations of oxygen pressure with overall composition for 3d-oxides, calculated for 1000 K. From [4.44] by permission of North-Holland Publ. Co.

sures terminate at the phase boundaries of the neighboring oxides. One sees that progressively higher p_{O_2} values are required for oxide pairs of higher valence state (see Problem 4.7). The horizontal lines in Fig.4.50 are joined diagonally with dashed lines to indicate *schematically* the variation of the equilibrium oxygen pressure with the stoichiometry over the stability range of each oxide. Though we know from our earlier discussion of phase figures that p_{O_2} (which in these systems is practically identical with P) must be continuous across the homogeneity (stability) range of a compound, for its specific functional shape one needs more experimental data than just the ΔG_f^0 of the stoichiometric compounds from which the coexistence lines were calculated. For that matter p_{O_2} could even have a minimum at the strict stoichiometric composition. Yet, as an estimate in Problem 4.8 shows (and as is chemically intuitively obvious), the p_{O_2} in equilibrium with an oxide and its

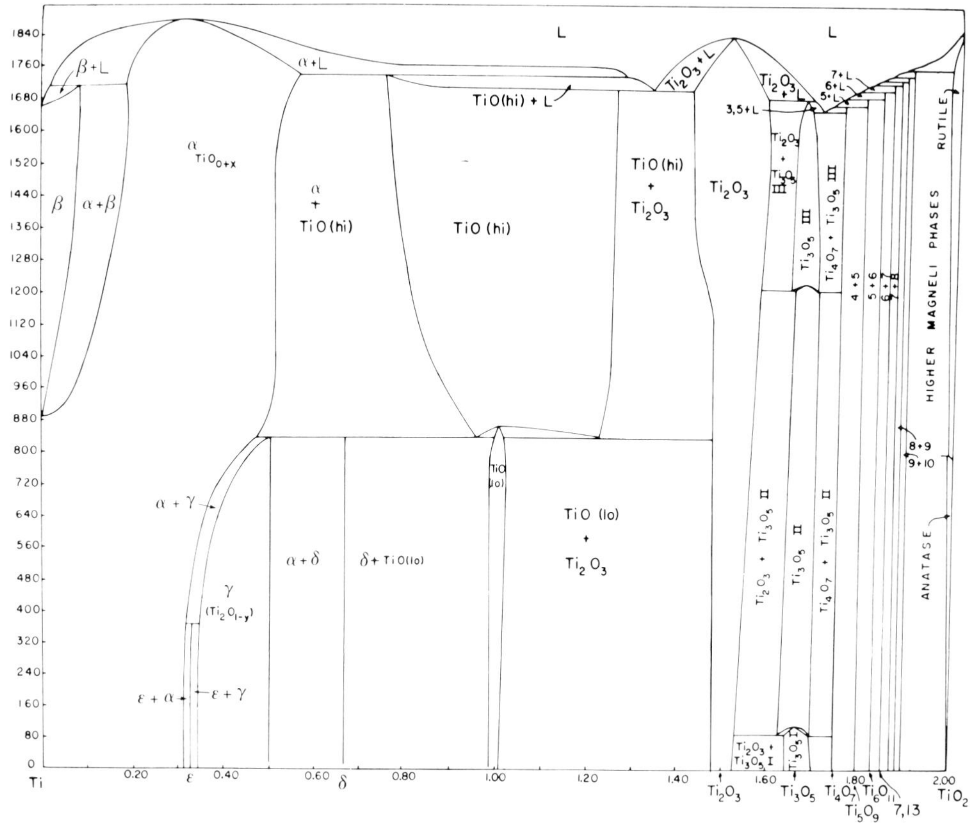

Fig.4.51. Condensed T-X projection for the Ti-O system. From [4.45] by permission of North-Holland Publ. Co.

metal is always considerably lower than any possible p_{O_2} in equilibrium with the oxide without the metal present (Fig.4.50).

Probably the most complex but also most carefully investigated nonferrous system, the Ti-O system [4.45], is represented in Fig.4.51. In the region between Ti_3O_5 and the higher oxide, TiO_2 (rutile, anatase) one notes a particular high density of different phases. These "shear phases", originally identified by MAGNELI et al. [4.48], demonstrate particularly well the precision that is required in the control of the oxygen partial pressure during the preparation of certain oxides. As can be deduced from Fig.4.52, stability of p_{O_2} to within a factor of two or three *at very low absolute values* is necessary if a specific Magneli phase is to be obtained. In addition to the rather precise p_{O_2} control required for certain systems, the range of oxygen pressures in equilibrium with 3d-metal oxides, as shown in Fig.4.50, is enormous, namely 10^{-50} to 10^{5} atm. The practical question arises here, of course, how can p_{O_2} be controlled, in particular, in the lower part of this range.

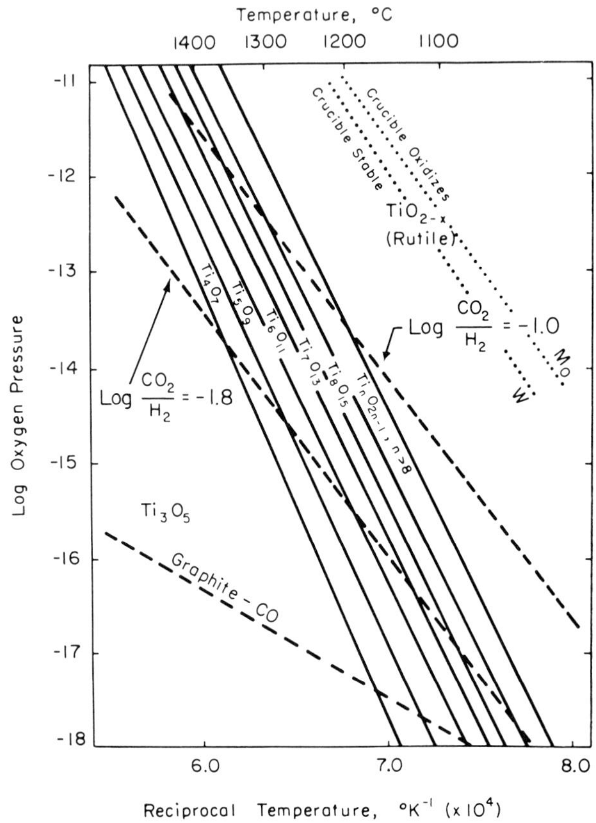

Fig.4.52. Stability relations of titanium oxides. Solid lines are phase boundaries separating regions of stability of various Magneli phases. For dashed and dotted lines see text. From [4.45] by permission of North-Holland Publ. Co.

With *direct control*, i.e., by supplying oxygen gas through external means, one can cover a range from $p_{O_2} = 10^{5}$ atm (compressors) down to 10^{-6} atm or so (dilution in inert gas). One may be tempted to try establishing more reducing conditions by heating an oxide in hydrogen. But even highly purified hydrogen (diffusion through palladium) contains typically 0.1-1 ppm O_2. Thus, for instance, efforts to obtain Ti_2O_3 through hydrogen reduction of Ti_3O_5 have failed [4.45].

The commonly used means of adjusting lower oxygen pressures over a wide range is through *gas buffers*, in particular of CO_2 - CO, H_2O - H_2 and CO_2 - H_2 mixtures. Figures 4.53a-d display the p_{O_2} of these mixtures as a function of mixing ratio and temperature. True buffer capability (Sect. 4.4.1 and Problem 4.5) is, of course, lost for mixing ratios very different from unity. Hence, limiting ourselves to, say, ratios between 10^{-1} and 10, we see that at one atmosphere total pressure one can readily cover p_{O_2} values between 10^{-4} - 10^{-19} atm with these systems. A disadvantage of this method, as pointed out earlier, is that the sample temperature and the temperature which determines the chemical equilibrium in the gas, and hence p_{O_2}, cannot be chosen independently. Thus cooling a furnace while maintaining a constant-ratio gas mixture will cause the oxygen partial pressure to vary grossly. Furthermore, with CO_2 - CO and CO - H_2 mixtures one must be aware that under CO_2-deficient conditions O_2 and also carbon may be formed (see the dash-cross equilibrium curves in Fig.4.53a and c and the lowest dashed line in Fig.4.52). Use of hydrogen, on the other hand, may result in hydrogen incorporation or even hydroxyl formation in some oxides. CO_2 - H_2 mixtures have, for instance, been used to establish the stringent p_{O_2} conditions during the preparation of Magneli phases [4.45]; see the upper two dashed lines in Fig.4.52.

In real-life experiments conditions are often far from what we would expect from externally controllable gas mixture considerations. Many of the materials used as high-temperature furnace components contain significant amounts of oxidizing agents, in particular water, which upon heating upset the externally adjusted balance for p_{O_2}. Hence, the use of *condensed phase buffers* with their intrinsically greater buffering capacity is often preferable. Also, with condensed phase buffers, as long as one of the phases involved is not used up in the buffering reaction, one can truly fix p_{O_2} in a system. The additional advantage of independent adjustability of the buffer temperature and hence of p_{O_2} has also been pointed out before (Sect. 4.4.1). As the discussion of Fig.4.49 showed, extremely low oxygen pressures can be buffered through equilibria between oxides and their metals (often in vapor

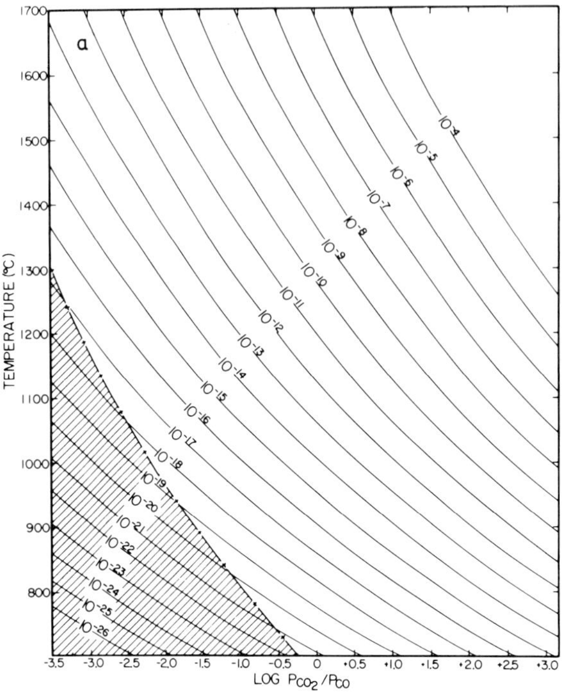
a
TEMPERATURE (°C)
LOG P_{CO_2}/P_{CO}
10^{-4}
10^{-5}
10^{-6}
10^{-7}
10^{-8}
10^{-9}
10^{-10}
10^{-11}
10^{-12}
10^{-13}
10^{-14}
10^{-15}
10^{-16}
10^{-17}
10^{-18}
10^{-19}
10^{-20}
10^{-21}
10^{-22}
10^{-23}
10^{-24}
10^{-25}
10^{-26}

(a)

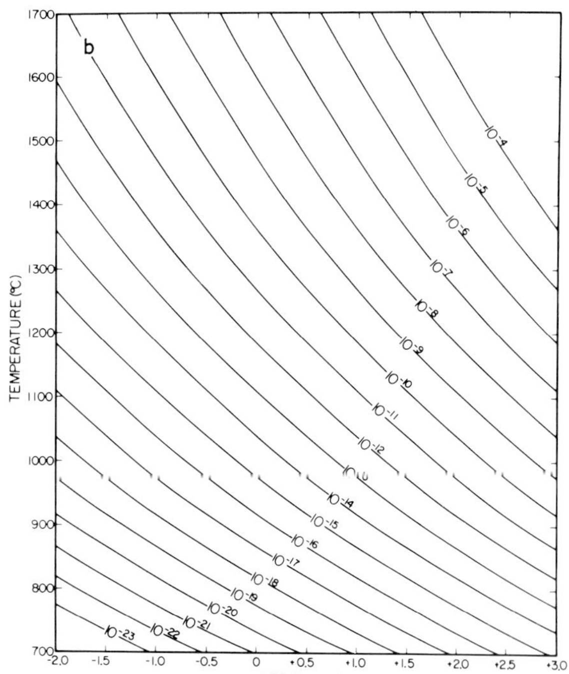
b
TEMPERATURE (°C)
LOG P_{H_2O}/P_{H_2}
10^{-4}
10^{-5}
10^{-6}
10^{-7}
10^{-8}
10^{-9}
10^{-10}
10^{-11}
10^{-12}
10^{-14}
10^{-15}
10^{-16}
10^{-17}
10^{-18}
10^{-19}
10^{-20}
10^{-21}
10^{-22}
10^{-23}

(b)

Fig.4.53a,b.

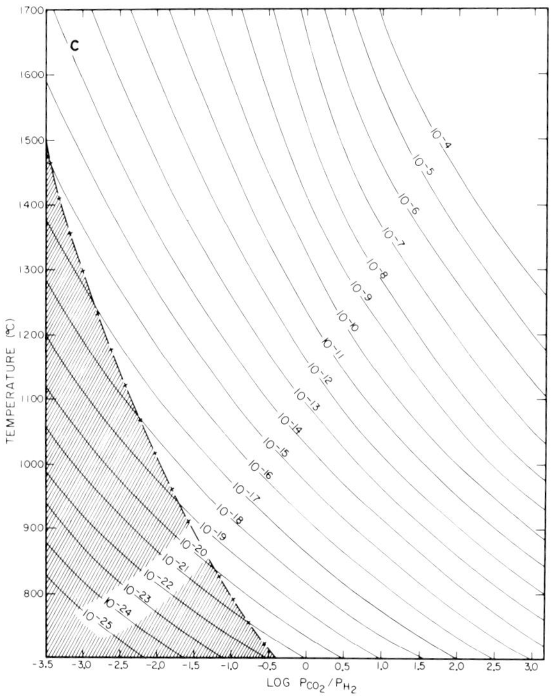

(c)

Fig.4.53a-c. Oxygen pressures (atm) as function of temperature and mixing ratio at a total pressure of 1 atm for (a) CO_2/CO, (b) H_2O/H_2 and (c) CO_2/H_2. The dash-cross curves in (a) and (c) represent equilibrium between the gas phase and carbon and the shaded areas below these curves represent metastable conditions of the gas phase. From [4.47] by permission of Addison Wesley

form). An extremely wide range of p_{O_2} can be covered by using the temperature dependent equilibria between oxide mixtures, as shown, for instance, in connection with Fig.4.50.

In this context one should point out the limitations of calculations along the line of (4.9) and (4.10) for predictions of p_{O_2} resulting from oxide mixture equilibria. Though very useful for *rough estimates*, the resulting values can often be expected to considerably deviate from reality. This is due to the fact that at elevated temperatures solid state diffusion will cause shifts in the higher oxide and lower oxide toward oxygen-deficient and oxygen-rich compositions, respectively, within their stability ranges. The calculation, however, is (supposedly) based upon the ΔG_f^0's for stoichiometric oxides. Thus calculated values will tend to be somewhat high.

This and the possibility of uncontrolled shifts of p_{O_2} due to contaminants makes the measurement of the oxygen partial pressure during demanding oxide preparations almost mandatory. Oxygen partial pressures down to approximately 10^{-25} atm can be measured with solid state electrodes in which one utilizes the strong dependence of the electrical conductivity of some oxides (typically zirconia) on stoichiometry and hence on the ambient p_{O_2} [4.49].

4.5 Container Materials

Phase diagrams and thermochemical phase stability considerations can provide information for the choice of, or more obviously, for the incompatibility between container materials and specific fluids. This will be illustrated with a few selected examples.

All phase diagrams for oxide systems shown before (Figs.4.11-16) indicate that combinations of these oxides lead to compounds that melt at lower temperatures than the pure components. As inspection of other refractory oxide phase diagrams shows, this is a general trend. Fig.4.54 represents another example. TiO_2, if molten in an Al_2O_3 crucible, will react with the container material and form β-$Al_2O_3.TiO_2$. Hence, contamination of the melt and destruction of the crucible are inevitable.

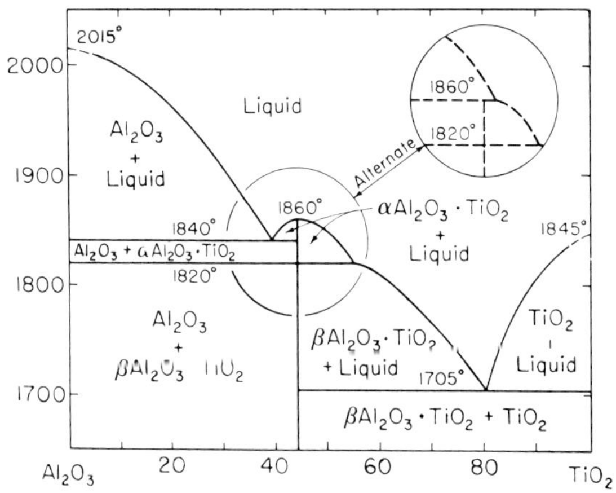

Fig.4.54. Al_2O_3-TiO_2 phase diagram. From [4.50] by permission of American Ceramic Society

Refractory metal crucibles can be used when the oxygen partial pressure is so low that the metal is the stable phase. For instance, for tungsten and molybdenum phase equiliberia studies have shown [4.51, 4.52] that the lowest oxides in these systems are WO_2 and MoO_2. Univariant reaction curves for the reactions

$$W + O_2 \rightleftarrows WO_2 \quad \text{and} \quad Mo \rightleftarrows O_2 \text{ x } MoO_2$$

calculated according to (4.7) are plotted in Fig.4.52 (dotted lines) and give the upper limit in oxygen fugacity of a melt at which these crucibles can be used.

Excessive loss of silica from glass and refractories in high temperature setups forms another example for the importance of the appropriate oxygen partial pressure in a system [4.53]. Volatilization of SiO_2 can result from the reaction

$$2SiO_2(s) \rightleftarrows 2SiO(g) + O_2(g) \quad .$$

The equilibrium constant of this reaction, $k = p^2_{SiO}p_{O_2}/a^2_{SiO_2}$ has been estimated to be 10^{-25} at 1320°C. Hence under reducing conditions (H_2 or CO atmosphere, etc.) of, say, $p_{O_2} = 10^{-18}$atm, assuming unit activity for silica, the SiO partial pressure is 3×10^{-4}atm (0.23 torr)!

Another example for melt-container incompatibility is given in Fig.4.55. Although bismuth melts about 1500°C below platinum, molten Bi dissolves Pt, forming α-Bi_2Pt. Similar effects occur in the Pb-Pt system and other combinations of refractory and heavy metals even when the T is considerably below the melting point of the high melting component.

Another situation of incompatibility predictable from phase diagrams can occur in heat treatment processes of solids. If one would, for example, try to anneal some KBr and KI samples which are not in mechanical contact but are in the same compartment so that vapors can be exchanged, one may run into problems even at temperatures considerably below the melting point of the lower melting material, here of KI. As Fig.4.56 shows, at temperatures as low as 580°C, surface melting will set in as the vapor of KBr and KI reaches the KI and KBr sample, respectively. Given enough "annealing" time, this can lead to the complete destruction of large samples.

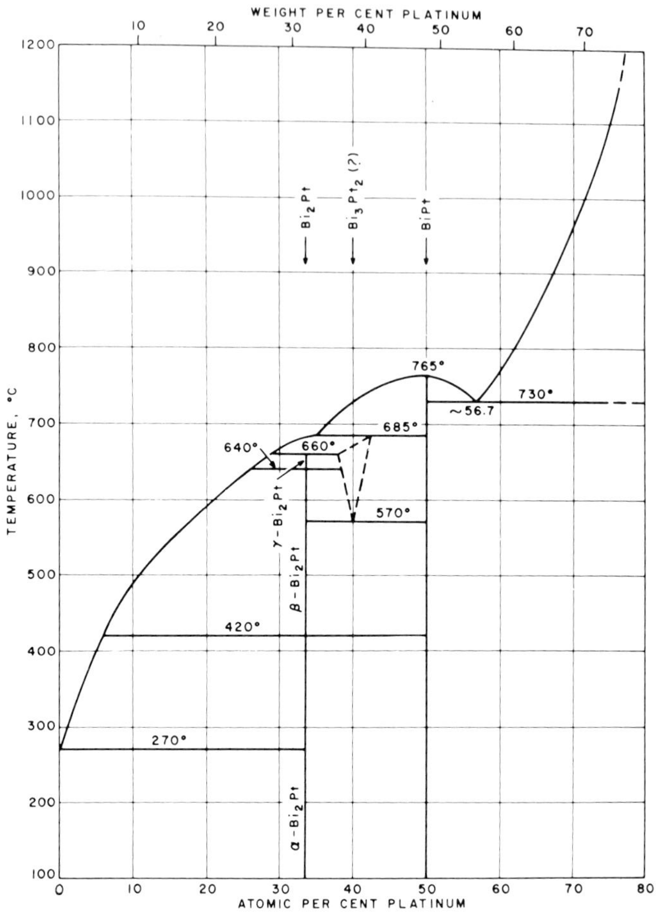

Fig.4.55. Bi-Pt phase diagram. From [4.54] by permission of McGraw-Hill

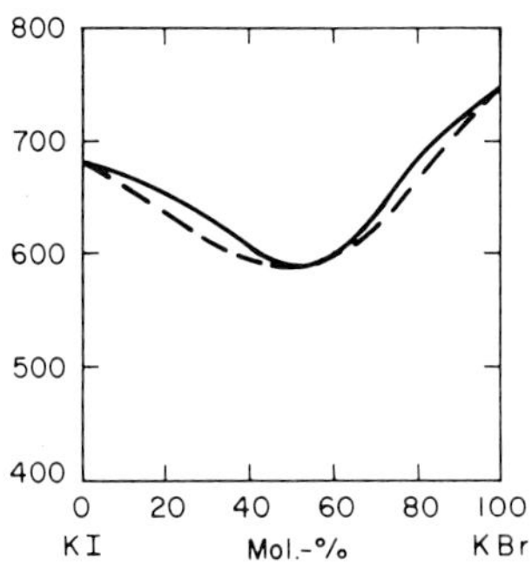

Fig.4.56. KBr-KI phase diagram. From [4.50] by permission of the American Ceramic Society

Problems

4.1 Derive a quantitative relation between the Gibbs potentials for the formation of Cd and S vacancies in CdS at 1400° C, based upon the data in Fig.4.5. Hint: Consider the vacancy concentration as resulting from a reaction equilibrium with occupied lattice sites and "ejected" atoms, for instance, $\boxed{S} \rightarrow \square + 1/2\ S_2$.

4.2 Suppose one needs to grow a crystal of $RbAg_4I_5$ weighing 20 grams. What is the minimum initial charge containing 75 mole percent of AgI? Refer to Fig.4.10. Answer: ≈40 grams.

4.3 Confirm Fig.4.12b (3 points) from Fig.4.12a. Resketch the parts of the latter figure that are required for this discussion.

4.4 One hundred grams of an aqueous solution of Na_2SO_4 is in equilibrium with 50 grams of $Na_2SO_4 \cdot 10H_2O$ at 25° C. How much NaCl must one add to the solution in order to completely dehydrate the excess decahydrate. Base the estimate on Fig.4.33. Note that the concentrations are given in weight percents, i.e., in weight of *anhydrous* solute per 100 parts by weight of solvent. For instance the solubility of Na_2SO_4 at 25° C is 21.6% w/w as represented by point D. Answer: About 23 grams.

4.5 A gaseous compound AB dissociates at elevated temperatures according to $AB(v) \rightarrow A(v) + B(v)$ (1).

(a) To what pressure of undissociated AB, p_{AB}, must one fill an ampoule at $T_1 = 300$ K to obtain $p_B = 0.316$ atm at $T_2 = 900$ K where the equilibrium constant for the dissociation reaction $k(T_2) = 0.1$?

(b) Assume that at T_2 in the sealed ampoule another reaction takes place that converts into non-volatile species an amount of B that, in an unbuffered system, would correspond to $\Delta p_B^0 = -0.1$ atm. Calculate the actual Δp_B in the buffered system.

(c) If A in reaction (1) were a non-volatile condensed phase, what would the required $p_{AB}(T_1)$ in (a) and the resulting Δp_B in (b) be?

(d) If both AB and A were condensed phases, what would $k(T_2)$ have to be and how would Δp_B in (b) change?

4.6 DeNOBLE [4.37] has computed the equilibrium constant for dissociation during sublimination of congruent CdTe as a function of temperature as

$T[°C]$	$k_{CdTe}[atm^{3/2}]$	$p_{Cd}[atm]$	$p_{Te_2}[atm]$
700	4.78×10^{-6}		
800	1.62×10^{-4}		
900	2.76×10^{-3}		
1000	3.16×10^{-2}		

Calculate the corresponding minimum partial pressures of Cd and Te_2 and compare the result with the p_{min} curve in the p_{Cd}-T diagram of Fig. 4.41b.

4.7 From the ΔG_f data in Fig.4.49, confirm the variation of p_{O_2} with overall composition for the various mixtures of oxides of manganese plotted in Fig.4.50.

4.8 Calculate p_{O_2} for the equilibrium $Mn + 1/2\ O_2 \rightleftarrows MnO$ at 1000 K and compare it with the value obtained nomographically from Fig.4.49.

5. Mass Transport and Heat Transfer

Macroscopic mass and heat transport play a central role in crystal growth processes. Before "molecules proceed from a position in the fluid medium, just outside the crystal-fluid interface, to their (relatively) permanent resting place in the crystal surface" [5.1], i.e., before molecules are "attached" to the crystal, they must typically be transported in the fluid over macroscopic distances towards the interface. This transport can be either fast or slow compared to the attachment kinetics. Then the rate with which a crystal grows is limited by interfacial kinetics or by the macroscopic transport, respectively. Interfacial growth mechanisms and their kinetics are the subject of later chapters. In this chapter we concern ourselves with the macroscopic transport part of crystal growth processes. Emphasis is put on the fundamentals of the various transport modes, their distinction and interplay. The general concepts obtained will then be applied to more specialized configurations and geometries in the later chapters on segregation, morphological stability, and growth from melts, vapors and solutions.

From the discussion of multicomponent phase equilibria in Chapters 3 and 4 we know that in most systems of interest the composition of a crystal is different from the composition of the nutrient from which it grows. Due to this selective incorporation of components the interface of a growing crystal represents not only a sink for matter but also a source of both heat and matter. Both latent heat that is released during the interfacial attachment process *and* matter that is not incorporated into the solid, must be transported away from the progressing interface. Thereby heat transport proceeds through the fluid and solid. The rejected species (by definition) are transported through the fluid only.

The rate with which "crystal building blocks" are incorporated into the solid is governed by the difference of their chemical potential in the two phases in contact. The chemical potentials, however, depend on the concentration of all species present *and* on the local temperature, pressure, etc. Thus the mass and heat transport problems are coupled. Hence, in order to determine, for instance, the time-dependent position and shape of an interface one must *simultaneously* consider the conservation of mass, momentum and energy for the system in question. To convey to the reader a full appreciation of the complexity of the general problem and in order to supply physical criteria for the structuring of this chapter we will introduce here the most general forms of the three governing equations [Ref. 5.2, Chap. 18], which are commonly referred to as the equations of change. Some of the vector operations used in these equations will be briefly reviewed in later sections with emphasis on their physical implications. However, at this point the reader is urged to "brush-up" on the fundamentals of vector calculus. For a very concise review see, for instance, [Ref. 5.2, Appendix A].

For a non-isothermal multicomponent fluid the consideration of mass conservation results in a *continuity equation* or *diffusion equation* for each species, i,

$$\frac{\partial}{\partial t}\rho_i = -\nabla\cdot(\rho_i\underline{v} + \underline{j}_i) = -\nabla\cdot\underline{n}_i \qquad (5.1)$$

rate of gain in mass density	from:	convective flux	diffusive flux

where ρ_i is the mass density of component i, $\underline{j}_i$ represents the diffusive flux with respect to the mass average velocity $\underline{v}$, defined in (5.7), and $\underline{n}_i$ is the total (local) mass flux of component i with respect to a fixed coordinate system. Hence (5.1) describes the rate of gain in mass density of component i at a fixed point (infinitesimal volume element) in space as the negative "divergence" of the sum of a convective flux (flux of component i due to bulk motion of the fluid) and diffusive flux (which is entrained in that bulk flow). The divergence, as we will discuss below, is a measure for the difference of in-flux and out-flux for a volume element. So, simply speaking, (5.1) states that the rate of mass accumulation in a volume element must equal the difference of in-flux and out-flux, i.e. matter is conserved.

Considering the momentum conservation in a multicomponent system one obtains from Newton's second law the *Navier-Stokes equation*

$$\frac{\partial}{\partial t}\rho\underline{v} = -[\nabla\cdot\{\rho\underline{v}\underline{v} + \underline{\underline{\Pi}}\}] + \sum_i \rho_i\underline{g}_i \qquad (5.2)$$

rate of gain in momentum	from:	convection	viscous transfer + pressure forces	external forces

where ρ is the total mass density, $\underline{vv}$ is a dyadic product (second-order tensor), $\underline{\underline{\Pi}}$ is the pressure tensor, $\underline{g}_i$ is the sum of the external forces per unit mass on component i. This equation of motion, (5.2), states that the momentum of the mass in a stationary fluid (volume) element changes because of the convective flow through the element and due to the acceleration (or deceleration) from internal pressure forces and viscous (shear) forces (both contained in the $\nabla\cdot\underline{\underline{\Pi}}$-term) as well as from the external forces acting on the mass. It is very useful to distinguish here between *surface forces* and *field forces*. Surface forces (also "contact forces") in the classical model of matter, in order to be propagated, require direct contact between matter elements. Examples for surface forces are the force exerted by a (moving) piston on a gas or the frictional force exerted by a rotating crystal on a melt. Field forces (also "body forces") such as the gravitational force or an electric field, in contrast, may act even through empty space. This distinction will allow us below to readily define various different (dominant) mass transport modes in fluids.

Considering energy conservation in a non-isothermal, multicomponent moving fluid the first law of thermodynamics yields the *energy transport equation*

$$\frac{\partial}{\partial t}\rho(u+\frac{1}{2}v^2) = -\nabla\cdot\{\rho(u+\frac{1}{2}v^2)\underline{v} + \underline{q} + [\underline{\underline{\Pi}}\cdot\underline{v}]\} + \sum_i(\underline{n}_i\cdot\underline{g}_i), \tag{5.3}$$

rate of gain in internal and kinetic energy	from:	convection	heat conduction	compression and viscous dissipation	external forces

where u is the internal energy per unit mass, and $\underline{q}$ the conduction heat flux. In (5.3), the rate of gain in energy in a volume element that is fixed in space (i.e. stationary) is seen as sum of the rates of energy input from convection, conduction, compression and viscous (frictional) dissipation, and from external forces.

As complex as (5.1-3) are, they still do not supply a complete description of the thermo-fluid dynamic state of a system. For a complete description we also need the thermal equation of state $P = P(\rho_i,T)$ and the caloric equation of state $u = u(\rho_i,T)$. In addition, we need expressions for the fluxes $\underline{j}_i$, $\underline{\underline{\Pi}}$ and $\underline{q}$ in terms of the gradients and transport coefficients (diffusivity, viscosity, thermal conductivity, etc.) which may themselves be functions of pressure, temperature and composition. Furthermore, one may have to account for chemical reactions and radiative heat transfer.

We see that the general formulation of coupled mass and heat (energy) transport is unwieldy. One is faced with a system of nonlinear differential equations. Hence solutions cannot be built up by superposition. Thus the motivation for simplification is great. Fortunately, crystal growth processes are often conducted in a manner that allows for considerable "shortcuts" in the formal description of the considered transport problem. In particular when low flow velocities are encountered, drastic simplifications can be made in (5.2) and (5.3). Furthermore under steady-state conditions the time-dependent terms can be dropped. When the thermal properties depend only weakly on composition or when only small temperature differences exist in a system one may approximate reality rather well by independently solving for the temperature and concentration distributions; or even by treating the mass transport part as isothermal. On solid-fluid interfaces, where due to viscous interactions the tangential velocity of the fluid relative to the solid becomes zero, one may even uncouple the diffusive (mass and heat) flux from the rest. Such simplifications, however, should only be the result of scrutiny rather than custom. As our discussion will show, great confusion can be noted in the crystal growth literature due to inconsiderate neglect of "small" terms in transport discussions.

In the following we will first introduce a rigorous framework of definitions for fluxes, in particular for diffusive (component) fluxes and convective (bulk, average) fluxes (Sect. 5.1). Then in Sect. 5.2 we will treat mass transport by *diffusion* only, i.e., we assume [see (5.2)] that the surface forces acting on the fluid are negligible and that the interaction of (always present) external forces (gravity) with density gradients (due, for instance, to heat transfer) does not lead to bulk motion. Section 5.3 is concerned with *forced convective flow*. Ignoring diffusive transport, i.e., in essence treating a "monocomponent" fluid, we assume that the surface forces outweigh the external forces by far. The combination of diffusive and convective transport is then dealt with in Sect. 5.4. The case that the surface forces are insignificant as compared to external forces, which may lead to the dominance of *free convection* or *surface tension-driven convection* and combinations thereof, is discussed in Sect. 5.8. Non-radiative heat transfer and its limited analogy to mass transfer, and radiative heat transfer are the subject of Sects. 5.5 and 5.6, respectively. A compilation of references to Stefan problems (heat and/or mass transfer with moving boundary, i.e., time-dependent solutions) is given in Sect. 5.7.

5.1 Fluxes and Reference Frames

The description of mass transfer is typically more intricate than that of heat conduction. In mass transfer, molecules of different mass and sometimes significantly different volumes must be exchanged between different parts of a system. Hence, in general, the mass fluxes continuously change, for instance, the position of the center of mass of the "matrix" through which diffusion occurs. In contrast, in heat conduction only momentum transfer is required. The molecules, in first order and neglecting self-diffusion, remain at their time-averaged coordinates. Hence the "substance" being transported does not change the "position" of the matrix through which the transport proceeds. This, as we will see below, puts severe limitations on "analogies" between solutions to mass and heat transport problems, though the underlying transport equations, such as Fick's first law and Fourier's law, may in some circumstances be *mathematically* similar.

Let us now define an unambiguous framework for the description of diffusive and convective mass transport in multicomponent fluids. This topic has been treated in various elegant presentations in the literature, in particular by BIRD et al. [5.2], KIRKWOOD et al. [5.3], and GEANKOPLIS [5.4]. However, for historical reasons that are discussed in a later part, these concepts have found only slow acceptance in crystal growth transport discussions.

Now, what is actually meant with (mass) diffusion? To clearly formulate this we refer to a volume element small compared to any system dimension but large compared to atomic dimensions (the usual continuum approximation). In such a volume element the local statistically averaged velocity of one species is

$$\underline{v}_i = \frac{1}{n_i} \sum_{j=1}^{n_i} \underline{v}_{ij} \quad , \tag{5.4}$$

where n_i is the number of molecules of species i in the sample volume and the $\underline{v}_{ij}$'s are the molecular velocities. The $\underline{v}_i$'s, however, for different species may be different. It is this relative motion of unlike molecules that we call diffusion. We shall neglect here self-diffusion and hereafter the term diffusion shall mean multicomponent or binary diffusion.

One of the most useful ways to mathematically describe this relative motion centers on the concept of an average flow velocity. A frame of observation is chosen, and the local average velocity $\underline{v}_i$ of each species i is measured. Then a weighted average flow velocity is constructed according to

$$\underline{v}_a = \sum_i a_i \underline{v}_i \tag{5.5}$$

where $\sum_i a_i = 1$. The weights a_i may be, for example, concentration fractions (see below). The diffusive motion of species i is now described relative to the average flow velotiy. This flow velocity, in effect, defines a frame of reference, different, in general, from the initial frame of observation, in which the (diffusive) motion of each species is gauged. Thus the diffusion velocity of i with respect to v_a is defined as

$$\underline{v}_{ia} = \underline{v}_i - \underline{v}_a \quad . \tag{5.6}$$

Many different average velocites, i.e., many different ways to choose the weights a_i, have been used in the literature. Each one allows for a simple analysis of certain special experimental configurations. We will discuss three approaches that have proved most useful. The first is to let $a_i = \rho_i/\rho = W_i$ where ρ, ρ_i and W_i are the total local mass density, the local mass density of i and the mass fraction of i, respectively. This choice of a_i defines the *mass average* or *barycentric velocity*

$$\underline{v} = \frac{1}{\rho} \sum_i \rho_i \underline{v}_i = \sum_i W_i \underline{v}_i \quad . \tag{5.7}$$

Alternatively, one could let $a_i = C_i/C = X_i$, where C_i is the local molar concentration of i, C is the total concentration, and X_i is the mole fraction of i. The velocity thus defined is called the *molar (or molal) average velocity*

$$\underline{v}^{*} = \sum_i X_i \underline{v}_i \tag{5.8}$$

and is denoted with a superscript *. Still another formulation for the local mean velocity in a multicomponent fluid is obtained by letting $a_i = C_i \overline{V}_i$ where $\overline{V}_i$ is the partial molar volume of component i defined in (2.39). We use here the symbol $\overline{V}_i$ rather than v_i to avoid confusion with the species average velocity $\underline{v}_i$. With this choice of a_i one defines the *volume average velocity* (superscript □)

$$\underline{v}^{\square} = \sum_i C_i \overline{V}_i \underline{v}_i \tag{5.9}$$

which will allow for a particularly simple formulation of Fick's first law (5.31). The products $C_i\overline{V}_i$ in (5.9), unfortunately, have frequently been called "volume fractions F_i". This designation, suggesting that the term stands for the fractional volume occupied by component i, in general, grossly oversimplifies matters. As pointed out earlier, in real solutions the $\overline{V}_i$'s are concentration dependent and may even become negative at certain concentrations. Only with ideal solutions, where $\overline{V}_i$ = const, i.e., independent of the composition of a system, the descriptive "volume fraction" becomes meaningful.

It should be pointed out that the thus defined average velocities may have quite different magnitudes. Even in uniform systems $\underline{v} \neq \underline{v}^* \neq \underline{v}^{\square}$ unless by chance all $W_i = X_i$, and there is no way, in general, to convert one to the other, e.g., $\underline{v}$ to $\underline{v}^*$, unless the X_i's are known.

Various other average velocities (solvent-centered, Kirkendall, etc.) have been used in the literature. Detailed accounts of these "reference frames" and of interrelations of the resulting different flux formulation are presented in [5.2-5] and in particular comprehensive form by WILCOX [5.6].

Emplying any of the above-defined average velocities, we can now formulate the *total flux of a component* relative to the frame of observation (typically the stationary coordinates or laboratory frame), in which $\underline{v}_i$, (5.4), is determined as sum of a *bulk flow* velocity and an *entrained diffusive* velocity. For instance, the local mass flux of i, $\underline{n}_i$, i.e., the product of the local mass density of i, ρ_i and $\underline{v}_i$, can, in general, be written as

$$\underline{n}_i = \rho_i\underline{v}_i = \rho_i\underline{v}_a + \rho_i(\underline{v}_i - \underline{v}_a) \quad ,$$

and using (5.6) as

$$\underline{n}_i = \rho_i\underline{v}_a + \rho_i\underline{v}_{ia} = \rho_i\underline{v}_a + \underline{j}_{ia} \quad , \tag{5.10}$$

i.e. as the sum of a bulk mass flux ("convective flux") $\rho_i\underline{v}_a$ and an entrained diffusive mass flux $\underline{j}_{ia}$.

Choosing now a specific average (reference) velocity, e.g., the mass average velocity $\underline{v}$ defined in (5.7), the total mass flux of i as the sum of the convective flux and the *diffusive flux with respect to* $\underline{v}$ is

$$\underline{n}_i = \rho_i\underline{v}_i = \rho_i\underline{v} + \rho_i(\underline{v}_i - \underline{v}) = \rho_i\underline{v} + \underline{j}_i \quad . \tag{5.11}$$

Similarly for the total molar flux if i, using the molar average velocity defined in (5.8), one obtains

$$\underline{N}_i = C_i\underline{v}_i = C_i\underline{v}^* + C_i(\underline{v}_i - \underline{v}^*) = C_i\underline{v}^* + \underline{J}_i^* \tag{5.12}$$

where $\underline{J}_i^*$ is the molar *diffusive flux of i with respect to* $\underline{v}^*$, i.e., with respect to the molar bulk flow.

Correspondingly one can, for instance, write the total mass flux of i, using the volume average velocity defined in (5.9) as

$$\underline{n}_i = \rho_i\underline{v}_i = \rho_i\underline{v}^{\square} + \rho_i(\underline{v}_i - \underline{v}^{\square}) = \rho_i\underline{v}^{\square} + \underline{j}_i^{\square} \tag{5.13}$$

where $\underline{j}_i^{\square}$ is the *diffusive mass flux of i with respect* to the local bulk velocity with which the "volume-center" of the fluid flows. Note, for instance by comparing (5.11) and (5.13), that the same total flux of a component is divided into different relative amounts of difusive and convective contributions depending on the reference frame used for the description of the diffusive motion.

Similarly one could express the molar diffusive flux of i with respect to $\underline{v}$ or $\underline{v}^{\square}$, or the diffusive mass flux of i with respect to $\underline{v}^*$, giving a total of six possible combinations: three each in terms of diffusing mass and moles, respectively, with the average velocities (reference frames) defined in (5.7)-(5.9).

In addition to expressing fluxes of components in terms of the transported mass density ρ_i and mole number density C_i as above, one finds in the literature component fluxes expressed in terms of the transported volume $C_i\underline{V}_i$. We will not use this concept in later chapters, yet we will employ it for some of the following proofs. Hence we must familiarize ourselves with a few more definitions. The concept of volume flow is "dangerous unless handled with care". For instance, as one can see from the continuity equation (5.1), in steady-state one-dimensional flows the total mass fluxes $\rho_i\underline{v}_i$ are constant throughout and even molar fluxes $C_i\underline{v}_i$ remain unchanged unless the fluid undergoes chemical reactions while it flows. Of course, there is no general conservation principle for volume. Since the partial molar volume $\overline{V}_i$ of a specific component depends on the concentration of the other components in the fluid, a volume flux $C_i\overline{V}_i\underline{v}_i$ will, in general, change in space even if the specific C_i remains constant throughout. Below we will have to discuss some of the consequences of this basic difference to mass and mole fluxes. In order to clearly mark this difference we will use the "odd" letters "I" and "O" for volume fluxes. For instance, we will express the total volume flux (volumetric flux) associated with a component i, as

$$\begin{aligned} \underline{I}_i = C_i\overline{V}_i\underline{v}_i &= C_i\overline{V}_i\underline{v}^{\square} + C_i\overline{V}_i(\underline{v}_i - \underline{v}^{\square}) \\ &= C_i\overline{V}_i\underline{v}^{\square} + \underline{O}_i^{\square} \\ &= C_i\overline{V}_i\underline{v}^{\square} + \overline{V}_i\underline{J}_i^{\square} \end{aligned} \tag{5.14}$$

where $\underline{O}_i^{\square}$ is the *volumetric diffusion flux of i with respect to* $\underline{v}^{\square}$.

Now we have a maze of reference frames and transport fluxes. However, the situation is just the same as with the numerous thermodynamic potentials (Sect. 2.9). Which specific frame of reference is to be preferred depends solely on which one yields the most concise description of a given experimental situation. This will become clearer in later sections. Some of the particular significance of the particular reference frames, however, can be anticipated from the relations

$$\sum_i \underline{j}_i = 0 \quad , \quad \sum_i \underline{J}_i^* = 0 \quad \text{and} \quad \sum_i \underline{O}_i^{\square} = 0 \quad , \tag{5.15}$$

the derivation of which is left to the reader in Problem 5.1. These relations show also that in a system with n components only n-1 of the diffusion fluxes are independent.

5.2 Diffusion

5.2.1 Driving potentials

Having defined the diffusive flux we will now proceed to relating it quantitatively to its causes, i.e., its driving forces. *In principle,* one should be able to derive flux-cause relationships from molecular kinetic models based upon quantum mechanical methods, i.e., from first principles. Indeed, for some rather simple systems such as fluids with spherically symmetric interaction potentials, (i.e., monatomic gases and liquids), at low densities (so that three-body collisions can be neglected), one has developed good semiquantitative molecular theories of the basic mass, heat, and momentum transport properties [5.7]. These models are semiquantitative to the extent that the governing intermolecular force laws, due to mathematical difficulties, have still to be described in terms of simple empirical functions, the form of which is suggested by quantum mechanics. These functions contain several adjustable constants which can be deduced from experimental data of the bulk behavior by means of statistical mechanical relations. For less idealized systems, however, one is left with *phenomenological relations* between transport fluxes and driving forces. For conditions not too far from equilibrium, i.e., not too steep gradients, where local equilibrium is maintained in spite of the irreversible nature of the flow process (see Sect. 2.10), one finds *linear relations*.

For diffusive mass transfer, for instance, experiments show that the flux of a component in the absence of significant temperature and pressure gradients and external fields is approximately proportional to its concentration gradient (Fick's first law). Analogous relations are found between heat flux and temperature gradient (Fourier's law) and for momentum transfer, where proportionality between viscous stresses and deformation velocities is encountered (Newton's law). Another example for a linear transport relation is Ohm's law. But not all experimental transport relationships are simply a proportionality between a flux and one force. When two different types of transport processes take place simultaneously, cross phenomena will occur. For instance, interferences between mass diffusion and heat conduction give rise to thermal diffusion, i.e., the maintenance of a concentration gradient as a result of a temperature gradient.

Although it cannot contribute to an atomistic understanding, *irreversible thermodynamics* [5.5, 8, 9] can yield important information about these complex cross phenomena. Once the functional dependence between the *simple* fluxes and their *proper* forces is adopted, one can predict the form of the cross coupling phenomena and assign upper bounds to their magnitudes in terms of the simple phenomena. For mass diffusion in a system with non-uniform temperature, pressure and concentration, in the presence of body forces (external fields) one obtains for the molar flux of component i in a n-component system [5.5]

$$\underline{J}_i^r \equiv C_i(\underline{v}_i - \underline{v}^r) = L_{i0}^r \nabla \ln(T) - \sum_{j=1}^{n} L_{ij}^r \nabla[\mu_j]_T \qquad (5.16)$$

where the chemical potentials μ_j are taken as though the system were isothermal and, hence, are denoted by square brackets and subscripts T. The superscript r is used to emphasize that the *phenomenological coefficients* (transport coefficients) L_{ij} are different in different reference frames. Eq.(5.16) separates the diffusive flux into a *thermal diffusion* contribution (the cross coupling term, $L_{io}\nabla \ln(T)$, between pure heat and mass flux) and the *isothermal diffusion* part, which is governed by the gradients in the chemical potentials. Considering various contributions that can be significant in crystal growth mass transfer one can write for the μ_j's in (5.16)

$$\mu_i = \mu_j(C_1, \ldots C_n)_{T,P} + \mu_j(P) + p_j F + z_j \mathcal{F}\phi + \ldots \quad . \tag{5.17}$$

The first term represents the dependence of the chemical potential on the composition of the system at constant temperature and pressure. The second term reflects the pressure dependence. The third and fourth term with F, the electric field and ϕ its potential are examples for the influence of an external field that acts differently on the different components; p_j, z_j stand for the polarization and ionic valence of the j-th species and $\mathcal{F}$ for Faradays constant. In combining (5.16) and (5.17) one can regard the diffusive mass flux in the form

$$\underline{J}_j = \underline{J}_j^{(C)} + \underline{J}_j^{(F)} + \underline{J}_j^{(P)} + \underline{J}_j^{(T)} \tag{5.18}$$

as the sum of terms describing *concentration* (ordinary) diffusion $J_j^{(C)}$, *pressure* diffusion $J_j^{(P)}$, *forced* diffusion $J_j^{(F)}$ (all proceeding isothermally) *and* the before mentioned *thermal* diffusion $J_j^{(T)}$. The physical significance of these fluxes will be discussed separately in later sections.

Previously, in connection with (2.89), we have stated that two phases, which then were assumed each to be uniform, cannot be in equilibrium unless the chemical potential for each component i is equal in both phases. Now, (5.16) yields considerably more insight into the full meaning of that statement and allows for inclusion of non-uniform systems into the mass transfer equilibrium condition. We see that the isothermal diffusion flux in (5.16) consists of a sum of terms. Note that only the coefficient L_{ii} determines the (in most cases dominant) part of the diffusion flux of i that arises from "its own" chemical potential gradient $\nabla\mu_i$. [Its own is used with quotation marks since even $\mu_i = \mu_i(C_1, \ldots C_n)$]. In addition, the codiffusion coefficients L_{ij} determine those contributions to J_i arising from the chemical potential gradients of the other components. Even if $\nabla\mu_i$ were zero, the i-th isothermal mass flux would not necessarily vanish. True mass transfer equilibrium for component i requires simultaneous vanishing of all $\nabla\mu_j$'s. The origin of this co-diffusion lies in the forces acting between the molecules of different components. If the average intermolecular force between components i and j is attractive, L_{ij} is positive and diffusion of component j induces a diffusion current of i in the same direction. With dominant repulsion between i and j, L_{ij} is negative, and counterdiffusion currents of i and j result. Only at very low concentrations will it be reasonably accurate to neglect the off-diagonal ($i \neq j$) terms and to regard isothermal diffusion of component i as being driven only by its own chemical potential gradient.

5.2.2 Concentration or ordinary diffusion

In most cases the dominant mode of diffusion results from the concentration dependent term of the chemical potential in (5.17). Hence for a system in which the effects on mass transfer of temperature, pressure and external fields can be neglected, the diffusive flux (5.16) reduces to

$$\underline{J}_i^r = - \sum_{j=1}^{n} L_{ij}^r \nabla[\mu_j]_{T,P} \quad \text{with} \quad j = 1,2,\ldots n \quad . \tag{5.19}$$

In order to replace the practically inconvenient chemical potential gradients we introduce, following [5.6], general composition variables Y_j^r in the form

$$\nabla\mu_i = \sum_{\substack{j=1 \\ j\neq i}}^{n} \left(\frac{\partial\mu_i}{\partial Y_j^r}\right)_{\substack{Y_k,P,T \\ k\neq i,j}} \nabla Y_j^r \tag{5.20}$$

in order to transform (5.19), see [5.5], to

$$\underline{J}_i^r = - \sum_{j=1}^{n-1} \alpha_{ij}^r D_{ij}^r \nabla Y_j^r \tag{5.21}$$

the multicomponent generalization of *Fick's first law*. Here the *multicomponent diffusion coefficients or diffusivities* D^r_{ik} are defined by

$$D^r_{ik} = \frac{1}{\alpha^r_{ik}} \sum_{j=1}^{n} L^r_{ij} \frac{\partial \mu_i}{\partial Y^r_k} \quad . \tag{5.22}$$

The summation to n-1 rather than n in (5.21) reflects the fact that only n-1 of the component diffusive fluxes are independent as already reflected in relations (5.15). The advantages of introducing a reference-system dependent ("correction") factor α^r_{ik} will later become obvious. Among other things with α^r_{ik} one can ensure, independent of the specific choice of the composition coordinate, that D^r_{ik} always has the dimensions area/time.

Equation (5.21) relates the local molar diffusion flux of component i, with respect to a *yet to be specified average velocity*, to the concentration gradients of n-1 components in the system. Since concentration gradients are much easier to measure than chemical potential gradients, (5.21) has obvious practical advantage over (5.19) and typically the diffusion coefficients D_{ik} rather than the phenomenological coefficients L_{ij} are determined experimentally.

In the following we will formally specify Fick's first law for *binary* diffusion for various composition coordinates in the three reference frames defined earlier. For treatments of the complex forms of multicomponent diffusion refer to [5.2, 5.6, 5.7, 5.40].

The definition of the diffusion coefficients in (5.22) must yet be completed by the choice of an average velocity (frame of reference) with respect to which the flux of matter is to be measured *and* of the specific composition coordinates used. We will first consider diffusion with respect to the volume average velocity $v^\square$. Choosing C_i for Y_j, setting $\alpha^\square_{ij} = 1$, and dropping the vector symbols, Fick's first law for a binary system can be written in form of two equations

$$\begin{aligned} J^\square_A &= -D^\square_{AB} \nabla C_A \quad \text{and} \\ J^\square_B &= -D^\square_{BA} \nabla C_B \end{aligned} \tag{5.23}$$

where, at this point, we distinguish between the coefficients for diffusion of A "through B" and of B through A. Now with $O^\square_i = \overline{V}_i J^\square_i$, (5.14), and

$\sum_i O_i^{\square} = 0$, (5.15), one can rewrite (5.23) as

$$\overline{V}_A D_{AB}^{\square} \nabla C_A + \overline{V}_B D_{BA}^{\square} \nabla C_B = 0 \quad . \qquad (5.24)$$

Since (see Appendix C)

$$\sum \overline{V}_i C_i = 1 \quad , \qquad (5.25)$$

it follows that

$$\sum \nabla[\overline{V}_i C_i] = 0 \quad . \qquad (5.26)$$

However, as also shown in Appendix C, in general

$$\sum C_i \nabla \overline{V}_i = 0 \quad , \qquad (5.27)$$

hence (5.27) can be rewritten as

$$\sum \overline{V}_i \nabla C_i = 0 \quad , \qquad (5.28)$$

which for a binary system is

$$\overline{V}_A \nabla C_A + \overline{V}_B \nabla C_B = 0 \quad . \qquad (5.29)$$

Consequently, if $\overline{V}_A$ and $\overline{V}_B$ are non-zero, which holds for most practical situations, comparison of (5.24) and (5.29) yields

$$D_{AB}^{\square} = D_{BA}^{\square} = \mathcal{D}_{AB} \quad . \qquad (5.30)$$

Thus, $\mathcal{D}_{AB}$ is a *mutual* binary diffusion coefficient. Equation (5.30) is sometimes considered as a unique feature of the volume-centered description of diffusive fluxes. However this equality of the coefficients for the interdiffusion of two components applies to mass- and mole-centered binary fluxes as well.

The choice of the "volume-centered frame" as the first one in which we specified Fick's first law is not accidental. As we will show later, for quasi-ideal systems (i.e., for $\overline{V}_A$ and $\overline{V}_B$ constant) diffusive fluxes with respect to $v^{\square}$ are particularly easy to measure. Most of the earlier diffusion studies were performed with gases, where typically volume changes upon mixing as well as the concentration dependence of $\mathcal{D}_{AB}$, are insignificant. Hence, diffusion coefficients are customarily tabulated as *volumetric diffusion coefficients* $\mathcal{D}_{AB}$, even for non-gaseous systems. This is also why above we set $\alpha^{\square}_{AB} = 1$. For diffusion fluxes with rexpect to avarage velocities other than $v^{\square}$ and composition coordinates other than C_i, yet *based upon the same* $\mathcal{D}_{AB}$, Table 5.1 lists various equivalent flux-, α- and Y-combinations for Fick's first law in the form

$$(\text{diffusive flux})^r_A = -\alpha^r_{AB}\mathcal{D}_{AB}\nabla Y^r_A \quad . \tag{5.31}$$

Table 5.1. Appropriate parameters for frequently used forms of Fick's first law (5.31). After [5.6]

Flux	α^r_{AB}	Y^r_A	Restrictions
J^X_A	1	C_A	
J^X_A	$C^2\overline{V}_B$	X_A	$\overline{V}_A$ and $\overline{V}_B$ constant
J^X_A	$\dfrac{\rho^2\overline{V}_B}{M_AM_B}$	W_A	$\overline{V}_A$ and $\overline{V}_B$ constant
j_A	ρ	W_A	
j_A	1	$\rho_A = C_AM_A$	constant ρ
j_A	$\dfrac{C^2M_AM_B}{\rho}$	X_A	
$J^\star_A$	C	X_A	
$J^\star_A$	1	$C_A = CX_A$	constant C
$J^\star_A$	$\dfrac{\rho^2}{C\,M_AM_B}$	W_A	

Other combinations can readily be formed based upon the relations

$$J_A = j_A/M_A = \frac{M_B}{M} J_A^{\star} \quad , \qquad j_A^{\star} = \frac{M}{M_B} j_A = J_A^{\star} M_A \qquad (5.32\text{-}35)$$

where $M = X_A M_A + X_B M_B$.

This compilation is a combination of tabulations by BIRD et al. [5.2] and WILCOX [5.6]. A derivation of $J_A^{\star}(X_A)$ from $J_A^{X}\ (C_A)$ is given in Appendix C. Derivations of other flux connections can be found, e.g., in [5.7].

Now, having established the formal framework for the description of diffusive fluxes we will concern ourselves with the question: Under what experimental conditions does one directly observe diffusive fluxes and when must one correct for bulk flow set up by diffusion? Though this section will not bring any conceptually new material it should emphasize some of the above considerations *and* based upon realistic examples clarify some misconceptions found in the (crystal growth) literature.

The most simple condition for the deduction of diffusion coefficients from measured fluxes arises if an experiment is set up so that no bulk flow occurs with respect to stationary coordinates – say with respect to the coordinates of a diffusion cell. Then, since

$$N_A = J_A^r + C_A v^r \qquad (5.36)$$

and since there is zero bulk flow, i.e., $v^r = 0$,

$$N_A = J_A^r \quad . \qquad (5.37)$$

In applying (5.36) and (5.37) one sees from Table 5.1 that for one of the most direct ways of determining $\mathcal{D}_{AB}$ one wants to devise an experiment in which $v^{\square} = 0$ and thus $N_A = J_A^{\square}$. Then measurements of $C_A(t)$ and dC_A/dx, by optical means, for example, can be directly evaluated for $\mathcal{D}_{AB}$ without knowing the absolute values of C, ρ, or $\overline{V}_A$ etc. Now, for a binary system, from (5.9) $v^{\square} = C_A \overline{V}_A v_A + C_B \overline{V}_B v_B$, which with (5.14) is $v^{\square} = I_A + I_B$. So in order to have $v^{\square} = 0$ the sum of the volumetric fluxes must be zero.

An experimental scheme in which this condition can be obeyed is depicted in Fig.5.1. Components A and B are initially separated by an impermeable thin wall. After removal of this partition interdiffusion sets in as indicated by the arrows for the fluxes N_A and N_B. Consider a stationary differential volume element of unit cross-sectional area.

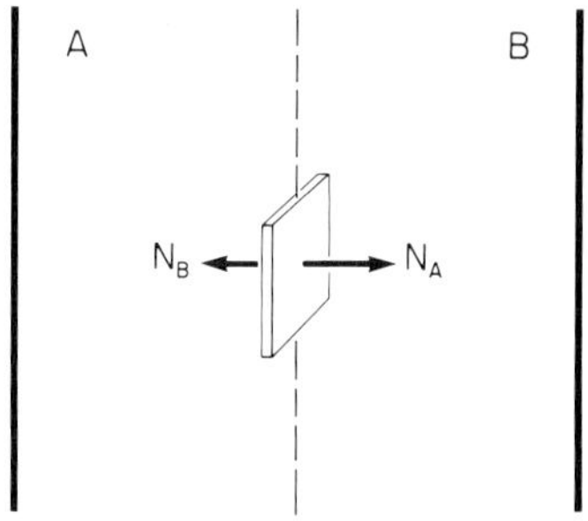

Fig.5.1. Binary counterdiffusion; see text

The continuity equation (5.1) with $\rho_i = C_i/M_i$ requires that

$$\frac{\partial C_A}{\partial t} + \frac{\partial}{\partial x}(C_A v_A) = 0 \quad \text{and} \quad \frac{\partial C_B}{\partial t} + \frac{\partial}{\partial x}(C_B v_B) = 0 \quad . \tag{5.38}$$

For systems in which the partial molar volumes do not change upon mixing, a condition which is well approximated, for instance, in various aqueous solutions, one can further simplify (5.38). By multiplying the component equations with the constants $\overline{V}_A$ and $\overline{V}_B$, respectively, and then adding them, using (5.14), one obtains

$$\frac{\partial(C_A\overline{V}_A + C_B\overline{V}_B)}{\partial t} = \frac{\partial}{\partial x}(I_A + I_B) = 0 \quad .$$

Now, since according to (5.25) $\sum_i C_i\overline{V}_i = 1$, it follows that $(I_A + I_B)$ is constant throughout the system. However, this condition must also hold on the walls where both I_A and I_B vanish. Thus we can conclude that $I_A + I_B = 0$. Consequently, $v^{\square} = 0$ and the fluxes N_A and N_B measured with respect to the diffusion cell coordinates are mere (volumetric) diffusive fluxes. Note that with ideally behaving gases, where $\overline{V}_A = \overline{V}_B$, this case respresents *eqimolar counterdiffusion*, i.e., the total molar flux with respect to stationary coordinates, ΣN_i, is zero, Then besides $v^{\square}$, v^* is also zero, and the volume-average velocity (volume-centered frame) loses its unique advantage. Note that in the above derivation we have tacitly assumed that neither of the $\overline{V}_i$'s is zero, a condition which may be violated in certain ("pathological") systems.

For the above experimental arrangement the condition that no volume change occurs on mixing is equivalent to the statement that $v^{\square} = 0$. Unfortunately it is frequently assumed (see Problem 5.2) that constant total volume V is in turn a sufficient condition for the vanishing of volume- or molar-average velocities and, hence, for the direct observation of mere diffusive fluxes. Let us disprove this supposition in the following with the discussion of an experimental scheme resembling an idealized crystal growth situation.

Assume two parallel solid walls of a component A separated by a fluid consisting of A and an inert component B (solvent), as depicted in Fig.5.2.

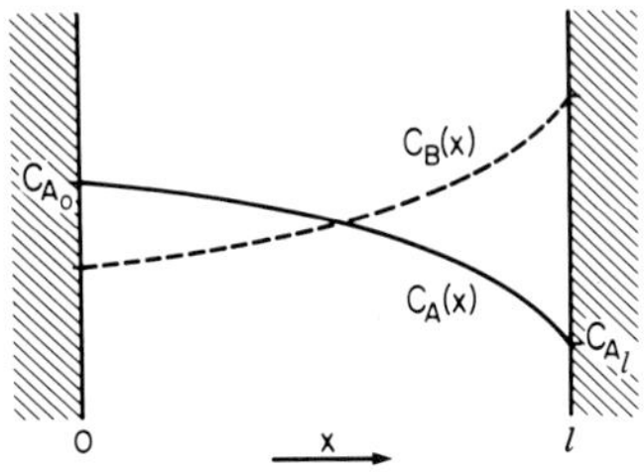

Fig.5.2. Diffusion of A through stagnant B (unimolar unidirectional diffusion); steady-state concentration profiles

B is assumed stagnant, i.e., $N_B = n_B = 0$ throughout. The boundary concentrations C_{Ao} and C_{Al} are kept constant at different values with $C_{Ao} > C_{Al}$. Hence there will be a flux of A, $N_A > 0$. From (5.9) and (5.7) we obtain for this system

$$v^{\square} = \overline{V}_A N_A + \overline{V}_B \cancelto{0}{N_B} > 0 \quad \text{or} \quad v = \frac{1}{\rho}(n_A + \cancelto{0}{n_B}) > 0 \tag{5.39}$$

inspite of constant V during transport, and consequently

$$N_A = J_A^{\square} + C_A v^{\square} = J_A^{\square} + C_A \overline{V}_A N_A \quad \text{or} \quad n_A = j_A + \rho_A v = j_A + W_A n_A \quad . \tag{5.40}$$

Hence the molar flux of A with respect to the container is not solely diffusive. For instance, for a deduction of $\mathcal{D}_{AB}$ from N_A one must correct for the convective, or better, *advective* flow that is set up by the diffusing A. So, even in the absence of any buoyancy-driven or forced convection in a crystal growth nutrient the molar or mass flux, respectively, of solute towards the growing crystal *is not*, as frequently claimed,

$$J_A^{\square} = -\mathcal{D}_{AB}\frac{dC_A}{dx} \quad \text{or} \quad j_A = -\rho\mathcal{D}_{AB}\frac{dW_A}{dx} \tag{5.41}$$

but rather

$$N_A = -\frac{\mathcal{D}_{AB}}{1 - C_A\overline{V}_A}\,\frac{dC_A}{dx} \qquad \text{or} \qquad n_A = -\frac{\rho\mathcal{D}_{AB}}{1 - W_A}\,\frac{dW_A}{dx} \tag{5.42}$$

as follows from (5.39) and (5.40). For the molar flux expressed in terms of ∇X_A see Problem 5.3.

From (5.42) one sees that in any situation where the crystal building components (solute) contribute a significant fraction of the nutrient the simple description of (5.41) becomes intolerably erroneous. As pointed out elsewhere [5.10], failure to observe this has led to fluid dynamically unjustified claims in the literature. If, however, the solute forms only a small fraction of the nutrient, the error introduced by using (5.41) becomes small as compared to typical uncertainties in $\mathcal{D}_{AB}$. Hence (5.41) is well suited for the description of interfacial mass transport, for instance, in low solubility solution growth and certain types of vapor-deposition processes such as Si-epitaxy where the silicon carrying species often form only a few percent of the total pressure. This will be further elaborated upon in Sects. (5.4) and (6.4).

Figure 5.2 schematically shows *steady state concentration distributions* for A and B as they could have been calculated from the relations derived in Problem 5.4. Note that for this "unimolar" diffusion situation the concentration profiles are curved, and hence the slopes dC_i/dx are not constant. Consequently, if $\mathcal{D}_{AB}$ = constant, the diffusive contribution to, say, the actual flux N_A must increase with decreasing distance from the sink at the wall. Yet, since there are no sources or sinks for A within the fluid, continuity requires that in steady state N_A = const. throughout. Thus the convective part $C_A v^{\square}$ in (5.40) must decrease as the diffusive J_A increases between bounding source and sink. This illustrates the *local character* of most formulations given above. It should be emphasized, though, that with a constant N_A also $v^{\square}$ is constant throughout the above system as long as $\overline{V}_A$ is constant, see (5.39).

These facts were already clearly recognized by STEFAN in 1882 [5.11] in his study of diffusive water evaporation through a stagnant column of air (Problem 5.5). Hence, in particular in the European literature $v^{\square}$ and v^* for the configuration of Fig.5.2 are frequently called *Stefan's velocity*, or even "Stefan's wind" (see, e.g., [5.12]). The latter term very vividly describes the "sweeping" flow of A towards the sink which leads to an accumulation of B at that interface. Correspondingly A is depleted in that area, as is reflected by the profiles in Fig.5.2. These phenomena play an important role in discussions of mass transfer and morphological stability in later sections.

The emphasis on fluids in the above example may give the impression that in the formulation of diffusion in solids the choice of an appropriate average velocity is an "unnecessary luxury". This is largely true. Since in most situations the concentration of diffusing species is low and the host is (in first order) at rest, typically $N_A = J_A$, as shown in (5.39) and (5.40). However, with comparable concentrations of "diffusent" and matrix the above refinement of the description beyond Fick's simple formulation is required. This was confirmed by Kirkendall and coworkers in various experiments beginning in 1942; it is now commonly known as the Kirkendall effect [5.13]. Macroscopic markers, originally imbedded in the interface between a metal and an alloy, were found to move towards the alloy when the temperature of the diffusion couple is raised to the point where significant diffusion sets in.

5.2.3 Diffusion coefficients

Diffusion coefficients for solutes in fluids are customarily determined far from saturation. As has been pointed out before [5.14], extrapolations to a crystal growth situation are open to question since in saturated or oversaturated fluids different structural properties may prevail.

Concentration diffusion coefficients *in gases* are typically several orders of magnitude higher than in liquids and solids and range from $10 - 10^{-2} cm^2/s$. Table 5.2 presents a compilation of experimental data. High values are found for systems containing a gas with low molecular weight (hydrogen, helium). A total pressures P of a few atmospheres and below, $\mathcal{D}_{AB}$ is found to be inversely proportional to P; it increases with increasing temperature and depends only very weakly on the relative concentrations for a given gas pair (Table 5.3). In fact, for most systems experimental data are not sufficiently accurate to examine the concentration dependence.

For an estimate of $\mathcal{D}_{AB}$ at pressure and temperature conditions that differ from the P_0 and T_0 of tabulated values, the following semi-empirical relation may be used:

Table 5.2. Mutual diffusion coefficients D_{AB} for various inorganic gas pairs. From [5.15] by permission of Pergamon Press

Diffusing gas	Medium into which the gas diffuses	P [mm Hg]	T [°C]	Diffusion coefficient D[cm^2/s]
Ammonia	Air	760	0	0.198
Argon	Helium	751	15	0.703
Bromine	Argon	760	0	0.0861
Bromine	Carbon dioxide	760	0	0.0863
Bromine	Hydrogen	760	0	0.563
Bromine	Hydrogen chloride	760	0	0.0869
Bromine	Methane	760	0	0.156
Carbon dioxide	Hydrogen	749	18	0.606
Carbon monoxide	Hydrogen	760	0	0.622
Carbon monoxide	Oxygen	760	0	0.185
Hydrogen	Ammonia	760	25	0.849
Hydrogen	Helium	760	25	1.64
Hydrogen	Hydrogen chloride	760	21	0.795
Hydrogen	Nitrogen	755	12.5	0.739
Hydrogen	Nitrogen	760	15	0.743
Hydrogen	Nitrous oxide	760	0	0.535
Hydrogen	Oxygen	753	14	0.778
Hydrogen	Sulphur dioxide	760	0	0.483
Iodine	Air	760	20	0.081
Iodine	Nitrogen	10	19.4	5.70
Mercury	Hydrogen	760	0	0.53
Mercury	Nitrogen	3	19.4	32.3
Nitrogen	Carbon dioxide	760	15	0.158
Nitrogen	Carbon monoxide	760	15	0.211
Nitrous oxide	Carbon dioxide	760	15	0.107
Oxygen	Ammonia	760	20	0.253
Sodium	Nitrogen	760	15	20.4
Water	Air	760	0	0.219
Water	Carbon dioxide	760	0	0.138
Water	Hydrogen	760	0	0.747

$$D = D_0 \frac{P_0}{P} \left(\frac{T}{T_0} \right)^b \qquad (5.43)$$

where the exponent b is found to be

b = 1.82 for non-polar gas pair

b = 2.33 for water vapor with a non-polar gas .

Table 5.3. Examples for variation of D_{AB} with relative composition of gas pair. From [5.15] by permission of Pergamon Press

Mole ratio	D_{AB}	Mole ratio	D_{AB}
$H_2 - CO_2 = 3/1$	0.594	He - Ar = 2.65/1	0.678
1/1	0.605	1.66/1	0.696
1/3	0.633	1/1	0.706
		0.311/1	0.731

This relation holds very satisfactorily up to around 1000°C and several atmospheres.

All these experimental facts can be understood from rigorous kinetic gas theory. Employing the appropriate potentials for the intermolecular forces (Lennard-Jones "6-12" potential for non-polar gases and Stockmayer potential for polar gases) between atoms and simple (spherical) molecules, quantum statistical methods yield data for diffusivities, viscosities and thermal conductivities of gases that are in very good agreement with experiment [5.7]. However, in the case of interactions between pairs of molecules of complicated structure, where the specific electronic configuration must be accounted for, it is impossible to carry out exact calculations. Then one typically makes use of the reverse procedure. Provided some experimental values, e.g., if thermal conductivity or viscosity for the complicated system are known, given the rigorous theoretical relations for simple molecule systems, one can calculate the other properties (here diffusivity) in terms of a few empirical parameters. The accuracy of the data thus obtained is comparable or sometimes even better than that of experimental values. For instance, it is difficult to accurately determine diffusivities in high-temperature vapors. Then one will rather resort to easier viscosity determinations and taking advantage of their theoretical interrelation, estimate the diffusivities from these data.

The methods of calculations are discussed in detail by HIRSCHFELDER et al. [5.7]. For the case of two non-polar gases A and B these authors give the approximate relationship

$$D_{AB} = \frac{0.00263}{P\sigma_{AB}^2\Omega^{(1,1)*}} \sqrt{\frac{T^3(M_A + M_B)}{2M_AM_B}} \tag{5.44}$$

Table 5.4. Functions for estimating the properties of non-polar gases. From [5.15] by permission of Pergamon Press

T^*	$\Omega^{(1,1)*}$	$\Omega^{(2,2)*}$	T^*	$\Omega^{(1,1)*}$	$\Omega^{(2,2)*}$
0.30	2.662	2.785	2.50	0.9996	1.093
0.35	2.476	2.628	2.60	0.9878	1.081
0.40	2.318	2.492	2.70	0.9770	1.069
0.45	2.184	2.368	2.80	0.9672	1.058
			2.90	0.9576	1.048
0.50	2.066	2.257	3.00	0.9490	1.039
0.55	1.966	2.156	3.10	0.9406	1.030
0.60	1.877	2.065	3.20	0.9328	1.022
0.65	1.798	1.982	3.30	0.9256	1.014
0.70	1.729	1.908	3.40	0.9186	1.007
0.75	1.667	1.841	3.50	0.9120	0.9999
0.80	1.612	1.780	3.60	0.9058	0.9932
0.85	1.562	1.725	3.70	0.8998	0.9870
0.90	1.517	1.675	3.80	0.8942	0.9811
0.95	1.476	1.629	3.90	0.8888	0.9755
1.00	1.439	1.587	4.00	0.8836	0.9700
1.05	1.406	1.549	4.10	0.8788	0.9649
1.10	1.375	1.514	4.20	0.8740	0.9600
1.15	1.346	1.482	4.30	0.8694	0.9553
1.20	1.320	1.452	4.40	0.8652	0.9507
1.25	1.296	1.424	4.50	0.8610	0.9464
1.30	1.273	1.399	4.60	0.8568	0.9422
1.35	1.253	1.375	4.70	0.8530	0.9382
1.40	1.233	1.353	4.80	0.8492	0.9343
1.45	1.215	1.333	4.90	0.8456	0.9305
1.50	1.198	1.314	5.0	0.8422	0.9269
1.55	1.182	1.296	6.0	0.8124	0.8963
1.60	1.167	1.279	7.0	0.7896	0.8727
1.65	1.153	1.264	8.0	0.7712	0.8538
1.70	1.140	1.248	9.0	0.7556	0.8379
1.75	1.128	1.234	10.0	0.7424	0.8242
1.80	1.116	1.221	20.0	0.6640	0.7432
1.85	1.105	1.209	30.0	0.6232	0.7005
1.90	1.094	1.197	40.0	0.5960	0.6718
1.95	1.084	1.186	50.0	0.5756	0.6504
2.00	1.075	1.175	60.0	0.5596	0.6335
2.10	1.057	1.156	70.0	0.5464	0.6194
2.20	1.041	1.138	80.0	0.5352	0.6076
2.30	1.026	1.122	90.0	0.5256	0.5973
2.40	1.012	1.107	100.0	0.5170	0.5882
			200.0	0.4644	0.5320
			300.0	0.4360	0.5016
			400.0	0.4170	0.4811

Table 5.5. Force constants in the Lennard-Jones equation. From [5.15] by permission of Pergamon Press

Substance	Constants computed from viscosity data		Constants computed from compressibility factors	
	ε/k[K]	σ[Å]	ε/k[K]	σ[Å]
H_2	10.22*	2.576*	6.03	2.63
H_2	33.3*	2.918*	29.2	2.87
	38.0†	2.915†	–	–
D_2	39.3	2.948	31.1	2.87
Ne	27.5†	2.858†	34.9	2.78
A	116†	3.465†	122	3.40
Kr	190	3.61	171	3.60
Xe	229	4.055	221	4.100
Air	97.0*	3.617*	99.2	3.522
	84.0†	3.689†	102	3.62
N_2	91.5*	3.681*	95.05	3.698
	79.8†	3.749†	–	–
O_2	113*	3.433*	117.5	3.58
	88.0†	3.541†	–	–
CO	110*	3.590*	100.2	3.763
	88.0†	3.706†	–	–
CO_2	190*	3.996*	189	4.486
	213†	3.897†	–	–
NO	119*	3.470*	131	3.17
N_2O	220*	3.879*	189	4.59
CH_4	137*	3.882*	148.2	3.817
CCl_4	327	5.881	–	–
SO_2	252*	4.290*	–	–
F_2	112	3.653	–	–
Cl_2	357	4.115	–	–
Br_2	520	4.268	–	–
I_2	550	4.982	–	–
HCl	360	3.305	–	–
HI	324	4.123	–	–
AsH_3	281	4.06	–	–
$SnCl_4$	1550	4.540	–	–
Hg	851	2.898	–	–
C_2H_2	185	4.221	–	–
C_2H_4	205	4.232	199.2	4.523
C_2H_6	230	4.418	243	3.954
C_3H_8	254	5.061	242	5.637
n-C_4H_{10}	410	4.997	297	4.971

Table 5.5 (Continued)

Substance	Constants computed from viscosity data		Constants computed from compressibility factors	
	ε/k[K]	σ[Å]	ε/k[K]	σ[Å]
$i\text{-}C_4H_{10}$	313	5.341	—	—
$n\text{-}C_5H_{12}$	345	5.769	—	—
$n\text{-}C_6H_{14}$	413	5.909	—	—
$n\text{-}C_7H_{16}$	—	—	282	8.88
$n\text{-}C_8H_{18}$	320	7.451	—	—
$n\text{-}C_9H_{20}$	240	8.448	—	—
Cyclohexane	324	6.093		
C_6H_6	440	5.270	—	—
CH_3OH	507	3.585	—	—
C_2H_5OH	391	4.455	—	—
CH_3Cl	855	3.375	—	—
CH_2Cl_2	406	4.759	—	—
$CHCl_3$	327	5.430	—	—
C_2N_2	339	4.38	—	—
COS	335	4.13	—	—
CS_2	488	4.438	—	—

Note: The values marked with an asterisk and a dagger have been calculated on the basis of measurements performed in the temperature range 80-300 K and 300-1000 K, respectively.

where M_A and M_B are the molecular weight for gases A and B, σ_{AB} [Å] is a "force constant" (5.45), $\Omega^{(1,1)*}$ (T^*) is a function the value of which is taken from Table 5.4 for $T^* = kT/\varepsilon_{AB}$ with ε_{AB} a "force constant" calculated from (5.46).

The defining equations for these force constants are

$$\sigma_{AB} = \frac{\sigma_A + \sigma_B}{2} \quad \text{and} \quad \varepsilon_{AB} = \sqrt{\varepsilon_A \varepsilon_B} \;, \tag{5.45,46}$$

where the individual σ's and ε's can be taken from Table 5.5. For a review of other theoretical work and semi-empirical relations, such as the Gilliland equation used for estimating D_{AB}, see BRETSZNAJDER [5.15].

Of course, at pressures above a few atmospheres, where the basic assumptions for the kinetic treatment (two-body collision only; molecular dimensions are small compared to mean distance between molecules) do not hold anymore; (5.44) gives unsatisfactory results and $P \cdot D_{AB} \neq$ const. Table 5.6 illustrates the pressure dependence as well as a particularly strong concentration dependence of the diffusivity.

Table 5.6. The coefficient of diffusion of nitrogen into a mixture of nitrogen and carbon dioxide under elevated pressures ($D_{N_2} \times 10^3$, cm^2/s). From [5.15] by permission of Pergamon Press

Temperature and pressure	Mole fraction of nitrogen in the mixture			
	0.25	0.30	0.45	0.60
Temperature 25°C				
6.0 atm	22.2	22.7	23.0	23.5
13.8	8.45	9.20	9.7	10.4
24.0	4.55	5.20	5.7	6.35
47.0	1.83	2.30	2.65	3.13
58.6	1.07	1.45	1.80	2.20

Diffusion coefficients for selected *liquids and solids* and their dependence on temperature are given in Fig.5.3. One sees that solid-state diffusion is on the average four orders of magnitude slower than diffusion in liquids and, as a comparison with Table 5.2 shows, seven orders of magnitude slower than in gases. We will take advantage of this large difference in later sections. For instance, in discussing segregation we will assume that on the time-scale of typical crystal growth processes the diffusive flux in solids is negligible ("frozen-in") as compared to the diffusion in nutrient fluids. Note also the exponential T-dependence of D_{AB} in condensed phases.

Figure 5.4 presents an extensive compilation of diffusion coefficients in refractory materials.

The exponential temperature dependence of diffusion coefficients in condensed phases can be understood from simple models. In order to move from one position to another, a molecule must possess a certain *activation energy* ΔQ to overcome an energy barrier which arises from having "to push its way between surrounding atoms". In solids ΔQ is typically much larger than the average thermal energy per atom, kT. Hence, for a successful diffusion jump

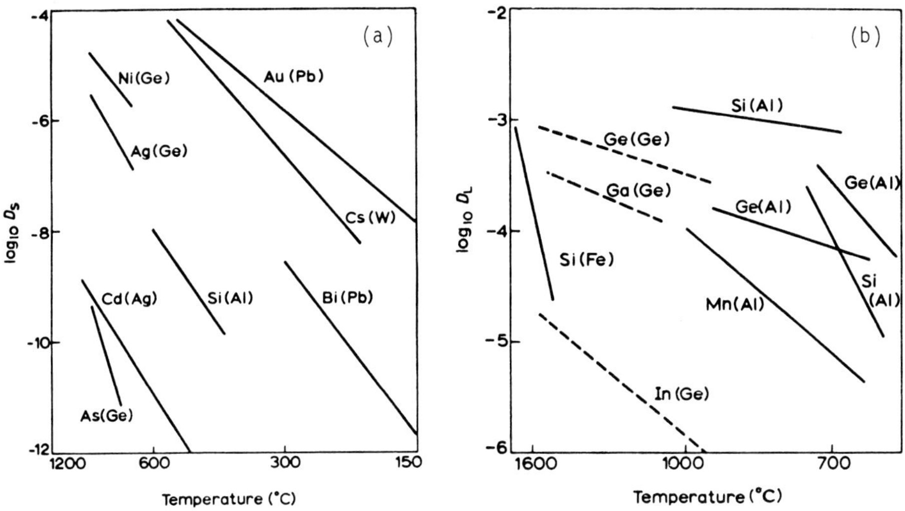

Fig.5.3. Diffusion coefficients in (a) solids and (b) liquids. Log D_{AB} versus 1/T, in cm^2/s. Symbol in parenthesis is the solvent. From [5.16] by permission of North Holland Publ. Co.

an atom must acquire ΔQ by collisions with its neighbors. This occurs with the Boltzmann probability $\exp(-\Delta Q/kT)$. Thus the diffusion coefficient can be formally expressed as

$$D = D_0 \exp(-\Delta Q/kT) \tag{5.47}$$

where the pre-exponential factor D_0 as well as ΔQ are material-specific constants over a wide temperature range. Representative values for D_0 are 0.1-1 and 10^{-3} - 10^{-4} cm^2/s for liquids and solids, respectively. Correspondingly ΔQ ranges from 1-10 and 10-80 kcal/mole as compared to about 2 kcal/mole for RT at 1000 K.

In contrast to gases, $\mathcal{D}_{AB}$ for condensed phases are found to depend strongly on concentration.

For estimating diffusivities in liquids no generally satisfactory method exists. For a recent review of the various approaches for polar and non-polar systems see [5.15]. It has been pointed out [5.16] that 90% of the available data falls within a factor of two of the value deduced from

$$\mathcal{D}_{AB} = 5 \times 10^{-9} \frac{T}{\eta r} \, [cm^2 \, s^{-1}] \tag{5.48}$$

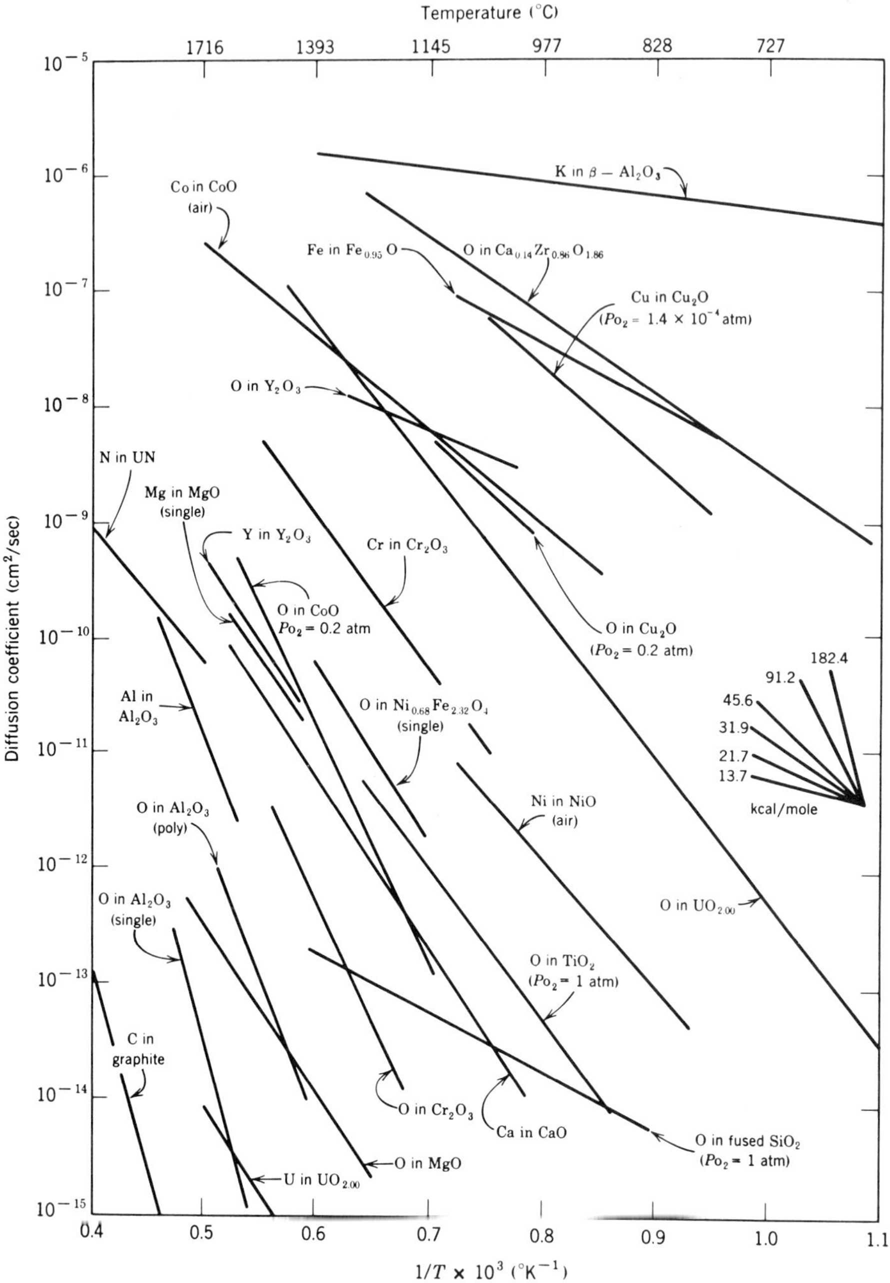

Fig.5.4. Diffusion coefficients in some refractories. The activation energy ΔQ can be estimated from the slope and the insert. Example: for O in $Ca_{0.14}Zr_{0.86}O_{1.86}$, $\Delta Q \approx 29$ kcal/mole. From [5.17] by permission of Wiley and Sons

where η is the viscosity in poise (see Sect. 5.3.3) and r is the radius of the diffusing particle in Å.

In electrolyte solutions the ions tend to diffuse separately. Yet they are coupled by the requirement of approximate net electrical neutrality of the solution. Hence it is only meaningful to define a D_{AB} for an ionized molecule where no other ions but the "proper" anion and cation are present. Values for such aqueous electrolyte solutions are given in Table 5.7. In the presence of other ions, these values will strongly depend on the relative concentrations. It has been found, for instance, that the diffusivity of H^+ varies by a factor of two in various mixtures of HCl and NaCl in water. For more details see [5.14].

In molten salts one can define D_{AB} for diffusants containing an ion in common with the solvent. Some values for such systems are presented in Table 5.8. Note also that concentration diffusion in electrolytes will, in general, lead to potential gradients due to differences in the fluxes of the components.

Table 5.7. Diffusion coefficients for some aqueous solutions of ionic salts. From [5.14] by permission of Marcel Dekker

Diffusant	C[moles/liter]	T[°C]	D_c[cm^2/s]
$CuSO_4$	0.005	20	0.58×10^{-5}
$MgSO_4$	0.2	10	0.39×10^{-5}
$Na_2C_4H_4O_6$	0.125	4	0.50×10^{-5}
LiCl	→ 0	25	1.4×10^{-5}
NaCl	→ 0	25	1.6×10^{-5}
KCl	→ 0	25	2.0×10^{-5}
KNO_3	→ 0	25	1.9×10^{-5}
$AgNO_3$	→ 0	25	1.8×10^{-5}
Na_2SO_4	→ 0	25	1.2×10^{-5}
$LaCl_3$	→ 0	25	1.3×10^{-5}
KBr	→ 0	25	2.0×10^{-5}
$(NH_4)_2SO_4$	→ 0	25	1.5×10^{-5}
$Pb(NO_3)_2$	0.2	12	0.82×10^{-5}

Table 5.8. Selected value for diffusion coefficients of (and in) molten salts. From [5.14] by permission of Marcel Dekker

Solvent melt	Diffusing species	C [moles/liter]	T[°C]	D_c [cm^2/s]
$LiNO_3$	Li^+	–	350	2.9×10^{-5}
	NO_3^-	–	350	1.2×10^{-5}
Na_2CO_3	Na^+	–	895	19×10^{-5}
	CO_3^-	–	895	28×10^{-5}
$PbCl_2$	Pb^{2+}	–	530	1.1×10^{-5}
	Cl^-	–	530	1.5×10^{-5}
$NaNO_3$	Ag^+	1.9×10^{-5}	400	3.2×10^{-5}
KBr	Ag^+	?	780	4.9×10^{-5}

5.2.4 Thermal diffusion

Thermal diffusion consists in the relative motion of the components of a mixture or solution due to a temperature gradient within the solution. Thermal diffusion is observed in solid, liquid and gaseous multicomponent systems. It was first discovered in aqueous solutions by Ludwig in 1856. Soret, around 1880, began systematic investigations. Hence, thermal diffusion in liquids is frequently referred to as the Soret effect.

Even in an initially uniform system, diffusion can occur if an imposed temperature gradient causes chemical potential gradients for some of the components. Mass flow and concentration gradients will develop to lower the Gibbs function of the system. Eventually a steady state will be attained in which the separating effect of thermal diffusion is balanced by the remixing effect of concentration diffusion. Of course, this distinction of "two mechanisms" is mostly made in the interest of manageability of the formal description of the complex real situation. After all it is the temperature, concentration, pressure, etc. dependent *chemical potential* that governs mass transfer. Yet one bases the description on a sum of linear laws as indicated in (5.16). For instance, in a system with a temperature gradient, concentration diffusion is treated as if it were occuring isothermally, choosing D_{AB} at some mean temperature. The consequences of the temperature dependence of the chemical potentials are then added on in form of the thermal diffusion flux term. Here another mean temperature is judiciously chosen for a given situation such that the nonlinear overall temperature dependence is well approximated.

As with concentration diffusion, thermal diffusion is theoretically best understood for gases [5.7, 18]. Thermal diffusion cannot be simply explained in terms of hard-sphere kinetic gas theory. The predicted transport coefficients are very sensitive to the specific interaction potential assumed. In fact, Maxwell's original kinetic gas theory, based upon repulsive forces that vary with r^{-5} (where r is the mutual distance between unlike colliding molecules) did not predict thermal diffusion because the corresponding coefficients cancel under this particular power law. In most binary systems (at elevated temperatures) one observes that the heavier molecules are concentrated at the colder end and the lighter ones at the hotter end of the temperature gradient. The rigorous theory (Enskog, Chapmann, 1910-20) based upon a Lennard-Jones potential predicts inversions of the separation at lower temperatures which has been experimentally confirmed. At higher temperatures the most important contribution to the separation process comes from collisions between molecules with high relative velocities. On the average, molecules coming from higher temperatures have higher thermal velocities than those coming from lower temperatures. The thermal velocity at a given temperature decreases with increasing molecular weight. Hence, collisions with highest relative velocity occur between lighter molecules coming from higher temperatures and heavier molecules coming from lower T. The resulting average momentum transfer for lighter molecules is towards higher T and for heavier ones towards lower T. The separation effect increases with the mass ratio between the molecules.

The lack of uniformity in the nomenclature used for formal descriptions of thermal diffusion in the literature is, to say the least, bewildering. Terms such as "Soret coefficient" or "thermal diffusion factors" are used for several completely differently defined entities. Hence, publications on thermal diffusion which lack clear definitions or references to definitions for the terms used should not be expected to communicate much information. In the following formal discussion we will point out some of the most frequently found pitfalls.

For the description of diffusive fluxes in the presence of a temperature gradient it is disadvantageous to express concentrations in terms of volume related units. A change in temperature causes already a density change, typically via bulk flow to colder parts, without diffusive fluxes *(relative* motion of components) being established yet. Hence in non-isothermal systems, i.e., for simultaneous occurence of concentration and thermal diffusion one preferrably uses relative concentrations W_i and X_i. Following the conventions presented earlier, we write for the mass flux of component A that occurs in a binary system with non-uniform temperature and concentration field [see (5.31)] in the absence of pressure gradients and external fields

$$j_A = j_A^{(C)} + j_A^{(T)} = - \frac{c^2 M_A M_B}{\rho} \mathcal{D}_{AB} \nabla X_A - D_A^T \nabla \ln(T) \quad . \tag{5.49}$$

Within the above sign convention, when D_A^T is positive, component A moves to the cold region. This, as mentioned above is the case for the heavier component in most gas mixtures and also for the solute in most aqueous solutions. Comparison with (5.16) shows that the *thermal diffusion coefficient* $D_A^T = -D_B^T$, so defined is identical with the transport coefficient L_{io} for A [5.2, 5.7]. Yet another thermal diffusion coefficient is frequently used which is defined by [5.6]

$$D_i^{T'} = \frac{1}{\alpha_i^r T} L_{io} \tag{5.50}$$

where the reference-system dependent α_i^r is different from the α_{ik}^r used earlier. Within this convention, since $\nabla \ln(T) = \nabla T/T$ the thermal diffusive flux contribution in (5.49) becomes

$$j_i^{r^T} = \alpha_i^r D_i^{T'} \nabla T \quad . \tag{5.41}$$

In experiments one determines the result of "simultaneously occurring" concentration and thermal diffusion. Hence it is formally more advantageous to set D_A^T in relation to $\mathcal{D}_{AB}$ even in the "mere" thermal part of the diffusive flux. Consequently one rewrites (5.49) as

$$j_A = - \frac{c^2 M_A M_B}{\rho} \mathcal{D}_{AB} \left[\nabla X_A + \frac{\rho}{c^2 M_A M_B} \frac{D_A^T}{\mathcal{D}_{AB}} \nabla \ln(T) \right] \tag{5.52}$$

which, defining a *thermal diffusion ratio*

$$k_T(X_i) = \frac{\rho}{C^2 M_A M_B} \frac{D_A^T}{\mathcal{D}_{AB}} \quad , \tag{5.53}$$

is then

$$j_A = -\frac{C^2 M_A M_B}{\rho} \mathcal{D}_{AB}(\nabla X_A + k_T \nabla \ln(T)) \quad . \tag{5.54}$$

Note that based upon $D_i^{T'}$ of (5.50) the $\nabla \ln(T)$ in (5.52) and (5.54) becomes ∇T. The ratio of thermal diffusion coefficient to concentration diffusion coefficient

$$\sigma = \frac{D_A^T}{\mathcal{D}_{AB}} \tag{5.55}$$

is sometimes called the *Soret coefficient*, or also thermal diffusion ratio [5.18]. Beware: There are at least two more different definitions for "Soret coefficients" in use. For a comparison of the conventions used in the recent literature see [Ref. 5.13, pp. 348-9].

Another frequently used quantity is the *thermal diffusion factor* α which BIRD et al. [5.2] defined for the case that the concentration diffusion flux is expressed in terms of ∇X_i, i.e., based upon $k_T(X_i)$ as

$$\alpha = \frac{k_T}{X_A X_B} \quad . \tag{5.56}$$

For gases α is almost independent of concentration. (Beware: Various authors call $\alpha' = \sigma T$, where σ is the same as in (5.55), the "thermal diffusion factor").

Equivalent to (5.54) one obtains in terms of a weight fraction gradient [5.19]

$$j_A = -\mathcal{D}_{AB}\rho[\nabla W_A + k_T^r \nabla \ln(T)] \tag{5.57}$$

where now

$$k_T^r = k_T(W_i) = \frac{W_A W_B}{\rho} \frac{D_A^T}{\mathcal{D}_{AB}} \tag{5.58}$$

Correspondingly, for molar flux with respect to the mole-average velocity

$$J_A^* = -\mathcal{D}_{AB}C[\nabla X_A + k_T^* \nabla \ln(T)] \quad , \tag{5.59}$$

with

$$k_T^*(X_i) = \frac{X_A X_B}{C} \frac{D_A^T}{\mathcal{D}_{AB}} \quad . \tag{5.60}$$

A good deal of the experimental results is presented in terms of "k_T, the thermal diffusion ratio". Of course, we realize now that for a meaningful comparison of data one must know whether $k_T(W_i)$, $k_T(X_i)$ or $k_T^*(X_i)$, etc. is being referred to.

In steady state $j^{(C)}$ and $j^{(T)}$ balance and the net diffusive flux is zero. Hence, from (5.54)

$$j_A^{(C)} + j_A^{(T)} = -\frac{C^2}{\rho} M_A M_B \mathcal{D}_{AB}\left(\frac{dX_A}{dx} + \frac{k_T}{T}\frac{dT}{dx}\right) = 0 \quad , \tag{5.61}$$

and consequently

$$dX_A = -\frac{k_T}{T}\, dT \quad . \tag{5.62}$$

For the integration of (5.62) it is customary to ignore the concentration dependence of k_T and to take its value at some mean temperature T_m. For the typical experimental arrangement of two interconnected fluid-filled bulbs that are at T_1 and T_2, respectively, one then obtains

$$X_{A_1} - X_{A_2} = k_T \ln\left(\frac{T_2}{T_1}\right) \quad . \tag{5.63}$$

The difference $X_{A_1} - X_{A_2}$, called the *separation* S, is often stated in experimental papers. The ratio $S/\ln(T_2/T_1)$ then gives a mean value of k_T for the range of temperature and concentration considered. Coordination of the mean k_T to

$$T_m = \frac{T_1 T_2}{T_2 - T_1} \ln\left(\frac{T_2}{T_1}\right) \tag{5.64}$$

is recommended [5.2], which more satisfactorily than the arithmetic means approximates the complex temperature dependence of k_T.

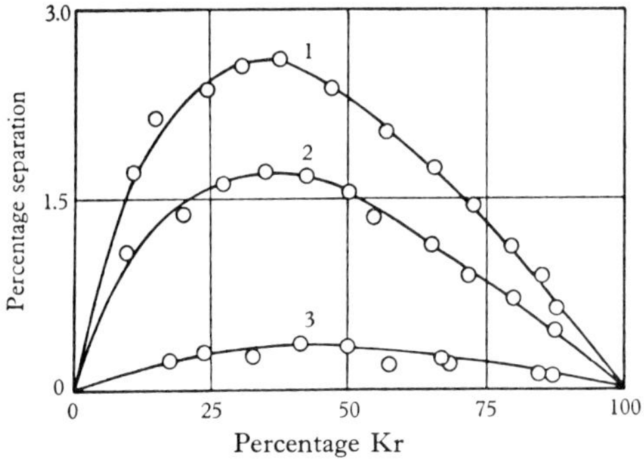

Fig.5.5. Separation as a function of composition for mixtures of krypton with (1) helium, (2) neon, (3) argon; $T_1 = 15°C$, $T_2 = 100°C$. From [5.18] by permission of Cambridge University Press

An alternative procedure of evaluating experimental results that is theoretically preferable since it does not invoke the assumption that k_T is concentration independent is as follows [5.18]: In terms of the thermal diffusion factor defined in (5.56), the steady-state equation (5.62) can be written as

$$dX_A = -\alpha X_A X_B d\ln(T) \quad . \qquad (5.65)$$

Now, we assume that α is independent of temperature and concentration. With regards to concentration this is much more realistic than in the case of k_T. Integration, making use of $X_B = (1 - X_A)$, yields

$$\frac{X_A}{X_B} = AT^{-\alpha} \qquad (5.66)$$

where A is a constant. Hence for a two-temperature arrangement

$$\left(\frac{T_2}{T_1}\right)^{\alpha} = \frac{X_{A_1}/X_{B_1}}{X_{A_2}/X_{B_2}} = q \quad \text{and} \quad \alpha = \ln(q)/\ln\left(\frac{T_2}{T_1}\right) \qquad (5.67)$$

with q the *separation factor*.

The experimental data presented in Fig.5.5 and Table 5.9 show that separation due to thermal diffusion in stagnant systems is typically rather small. However considerable separation enhancement can be obtained in combination

Table 5.9. Experimental thermal diffusion ratios, (5.55), for liquids and low density gases. From [5.2] by permission of Wiley and Sons

Liquids				Gases			
Components A — B	T[K]	x_A	k_T	Components A — B	T[K]	x_A	k_T
$C_2H_2Cl_4$—n-C_6H_{14}	298	0.5	1.08	Ne—He	330	0.20	0.0531
$C_2H_4Br_2$—$C_2H_4Cl_2$	298	0.5	0.225			0.60	0.1004
$C_2H_2Cl_4$—CCl_4	298	0.5	0.060	N_2—H_2	264	0.294	0.0548
CBr_4—CCl_4	298	0.09	0.129			0.775	0.0663
CCl_4—CH_3OH	313	0.5	1.23	D_2—H_2	327	0.10	0.0145
CH_3OH—H_2O	313	0.5	-0.137			0.50	0.0432
cyclo-C_6H_{12}—C_6H_6	313	0.5	0.100			0.90	0.0166

with convection. In special free convection columns (Clusius-Dickel columns) thermal diffusion is employed on a large scale for the separation of gases and liquids which are chemically difficult to separate, such as isotope mixtures. For an earlier, detailed review see [5.155].

In later sections we will see that in many crystal growth problems considerable temperature gradients are sustained in fluids. Hence one might suspect that thermal diffusion will significantly contribute to various mass transfer phenomena in crystal growth. Most recent results which will be discussed in the appropriate sections confirm this expectation.

Thermal diffusion coefficients D_i^T and thermal diffusion ratios k_T for gases may be estimated from rigorous kinetic theory. For the intricate computational procedure the reader is referred to [5.7].

In electrolyte solutions the presence of a temperature gradient gives rise, in general, to an electric potential gradient. There have been extensive studies of the thermoelectric powers of such thermocells. The earlier literature is reviewed in [5.19]. One can envision that this mechanism may be of considerable influence on mass transfer near solid-liquid interfaces in crystal growth from ionic melts and solutions.

5.2.5 Pressure diffusion

Pressure gradients can result in *pressure diffusion* through $\mu_i(P)$ in (5.17). The tendency for a solution or mixture to separate under a pressure gradient is very small. Hence, the effects of the small pressure gradients that result from viscous forces in a crystal growth fluid from velocity gradients (Sect. 5.3.2) and from the gravitational field of the earth can typically be neglected. However, in centrifugal fields, where [5.5]

$$\mu_i(P) = \frac{1}{2}(M_i - \rho\overline{V}_i)\omega^2 r^2 \tag{5.68}$$

with $\overline{V}_i$ the partial molar volume as defined in (2.39), ω the angular frequency and r the distance from the rotation axis, pressure diffusion can be the dominant mode of mass transfer.

For a binary system in steady state, i.e., when sedimentation equilibrium is reached, one obtains [5.2] a Boltzmann distribution of the components in the form

$$\left(\frac{X_A}{X_A(r=0)}\right)^{\overline{V}_B}\left(\frac{X_B(r=0)}{X_B}\right)^{\overline{V}_A} = \exp\left[(\overline{V}_A M_B - \overline{V}_B M_A)\frac{g_\omega r}{RT}\right] \tag{5.69}$$

where $g_\omega = \omega^2 r$ is the centrifugal acceleration. Hence, in most systems the heavier molecules will accumulate at larger radii. Yet (5.69) shows that for pathological partial molar volumes this is not necessarily the case.

Pressure diffusion has recently been applied to growth of crystals from solutions [5.20].

5.2.6 Forced diffusion (electromigration)

Although there are some indications that magnetic fields may influence diffusive mass transfer, we will here only deal with diffusion forced by electric fields. As indicated in (5.17), an electric field may change the chemical potential of a component j by either polarizing it and/or exerting a force on it equal to the product of the ionic charge of j and the local electric field strength. Hence, each species may be under the influence of a different force and relative motion of the components, i.e., diffusion, may set in. If ionized species are present, electromigration due to field-charge interactions will typically dominate over the polarization term which we will omit in the following.

It is instructive to compare the steady-state concentration profile obtained in electromigration with that resulting from concentration diffusion through a

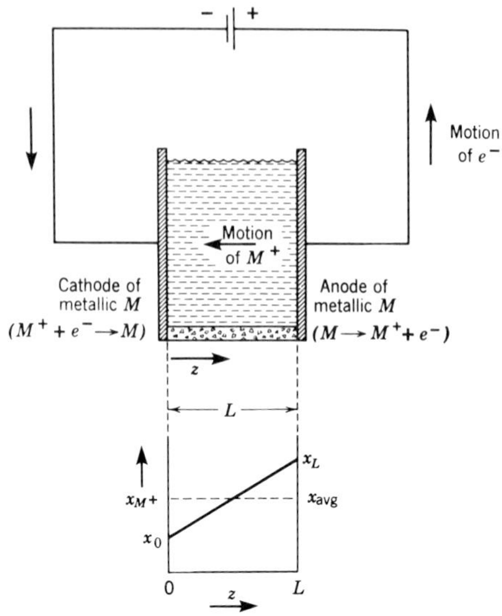

Fig.5.6. Electrocrystallization. Experimental scheme and steady-state concentration profile. From [5.2] by permission of John Wiley and Sons

stagnant film. Following [5.2], we base the discussion on the scheme of Fig. 5.6 which corresponds to a certain type of electrocrystallization experiment [1.2]. A salt M^+X^- is dissolved in water. A constant DC current passes between the anode (source) of metallic M and cathode of M (growing crystal). The molar flux of M^+ is proportional to the current in the cell. The electrochemical conditions are chosen such that no electrolysis sets in; hence, the fluxes of X^- and water are zero. Furthermore, electrical neutrality requires that the local molar concentrations of M^+ and X^- are approximately equal. We may write for the diffusive flux of either species

$$J_i^* = -C\mathcal{D}_{iw}\left(\frac{dX_i}{dz} + \frac{X_i z_i F}{kT}\right) \tag{5.70}$$

where z_i is the ionic charge, F is the local electric field and $\mathcal{D}_{iw}$ the diffusion coefficient of i in water. $J^*_{M^+}$ times $z_i F$ (F being the Faraday constant 9.65 x 10^4 coulombs/gram equivalent) is the current density in the cell. Consequently, with $z_{M^+} = z_{X^-} = z$

$$J^*_{M^+} = -C\mathcal{D}_{M^+W}\left(\frac{dX_{M^+}}{dz} + \frac{X_M zF}{kT}\right)$$

$$0 = -C\mathcal{D}_{X^-W}\ \frac{dX_{X^-}}{dz} - \frac{X_{X^-}zF}{kT} \qquad (5.71)$$

which, since $X_{M^+} \approx X_{X^-}$, can be summed to give

$$J_{M^+} = -2C\mathcal{D}_{M^+W}\ \frac{dX_{M^+}}{dz}\quad . \qquad (5.72)$$

For *dilute* solutions $C\mathcal{D}_{M^+W}$ is approximately constant. Hence, integration of (5.72) yields

$$\frac{X_{M^+} - X_0}{z} = -\ \frac{J_{M^+}}{2C\ \mathcal{D}_{MX(W)}} \qquad (5.73)$$

where X_0 is the mole fraction of M^+ at the cathode (Fig.5.6). Thus at the low concentrations assumed for the above derivation the concentration gradient is constant throughout. Note that the exponential concentration profiles that one obtains in concentration diffusion through a stagnant film also degenerates to straight lines as $C_A \to 0$ (Problem 5.4).

5.3 Forced Convective Flow

5.3.1 Equations of change

In this section we will concern ourselves with fluid flow systems in which the surface forces clearly dominate the external forces, as discussed in the introduction of Chap. 5. Ignoring diffusive mass transfer, that is treating the fluid as a monocomponent system, we will deal with convective mass flow in the bulk of fluids as well as near solid-fluid interfaces. Such interfaces, however, are at this point considered as momentum *and* species barriers. Reaction surfaces, that is, surfaces with mass transfer across them (the central part of all crystal growth processes), will be introduced in Sect. 5.4.

Fluid flow velocities associated with crystal growth processes are, in general, very small as compared to the speed of sound in the respective fluid. Hence density changes that may result from *inertial forces* are almost always negligible. For instance, a gas in a stagnation flow against a wall (Fig.5.8) is compressed so little at the low flow velocities considered, that the gas density close to the wall is practically identical to that in the unperturbed approaching flow. Thus, as a given mass of fluid is followed in the flow, its volume and hence its density remains essentially unchanged. In such a situation the fluid is, somewhat misleadingly, said to be *"incompressible"*, meaning that it is not being compressed. Frequently this condition is considered as identical with the assumption of constant (uniform) density throughout. Yet ρ = const. is only a sufficient criterion for incompressible flow; it is not necessary. In order to be incompressible it suffices that the density in any control-volume that is followed in the flow remains unchanged. This, however, as illustrated in Fig.5.7, is also the case in a density-stratified flow and does not require uniform density. Stratification may have solutal and/or thermal origin. Of course, in order to be stratified in the above sense, there must be little mass transfer normal to the flow direction. In other words, the diffusive mass flux must be negligible as compared to the respective convective flux. This sets a lower limit for the flow velocity of stratified fluids to be dealt with in a "ρ = const." model.

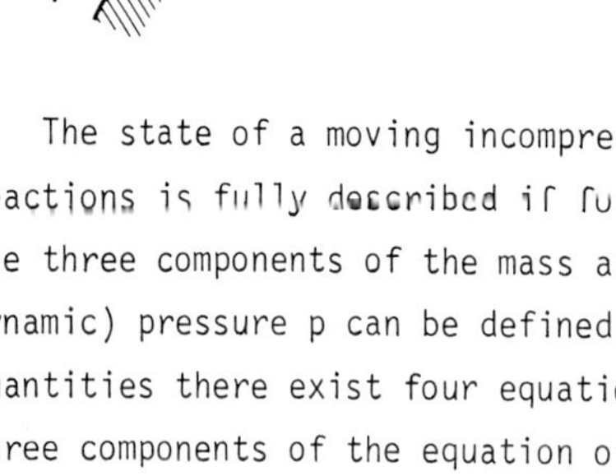

Fig.5.7. Flow of a stratified fluid

The state of a moving incompressible fluid in the absence of chemical reactions is fully described if for each point in space and for each instant the three components of the mass average velocity $\underline{v}$ (5.7) and the (thermodynamic) pressure p can be defined. For the determination of these four quantities there exist four equations: the continuity equation (5.1) and the three components of the equation of motion (5.2).

For fluids with "constant density", the *continuity equation* (5.1) reduces to

$$\nabla\cdot\underline{v} \equiv \mathrm{div}(\underline{v}) = \frac{\partial v_x}{\partial x} + \frac{\partial v_y}{\partial y} + \frac{\partial v_z}{\partial z} = 0 \tag{5.74}$$

and expresses the fact that for a stationary control volume the masses entering and leaving per unit time must balance. (Note that besides the "del" or "nabla" operator ∇ also the older notation is used throughout this text, in which $\mathrm{grad}(s) \equiv \nabla s$, $\mathrm{div}(\underline{v}) \equiv \nabla\cdot\underline{v}$ and $\mathrm{curl}(\underline{v})$ or $\mathrm{rot}(\underline{v}) \equiv \nabla x \underline{v}$.)

The *equation of motion* (5.2) can also be considerably simplified for incompressible fluids and reads for a unit of fluid volume

$$\rho\,\frac{\partial \underline{v}}{\partial t} = -\rho(\underline{v}\cdot\nabla)\underline{v} - \nabla p + \eta\nabla^2\underline{v} + \underline{f} \quad . \tag{5.75}$$

The three components of (5.75) are known as the *simplified Navier-Stokes* equations. Particularly revealing derivations are found in [5.21, 22]. Here ρ, η and p are the density, viscosity (both considered constant), and local total pressure, respectively. Note that μ is also frequently used for viscosity in the literature. In this text μ is used for the chemical potential only. The vector $\underline{f}$ represents the sum of all external or body forces (gravity, electric field, etc.) that act on a control volume of the fluid.

All aforementioned formulations of the equations of change refer to a *stationary* control volume. However, it is frequently desirable to rather refer to a fixed amount of fluid as it is seen by an observer *moving with the fluid*. In order to formally perform this transformation from the "lab frame" to the moving frame one must collect the local contribution (i.e., the Eulerian time differential $\partial/\partial t$ of the parameter considered) and the convective contributions on the left side of the applicable equation [5.4, 22]. The resulting Lagrangian *total* or *substantial time derivative*, e.g., of the velocity vector $\underline{v}$ reads

$$\frac{D\underline{v}}{Dt} = \frac{\partial \underline{v}}{\partial t} + (\underline{v}\cdot\nabla)\underline{v} = \frac{\partial \underline{v}}{\partial t} + (\underline{v}\ \mathrm{grad})\underline{v}$$

and (5.76)

$$\frac{D\underline{v}}{Dt} = \frac{\partial \underline{v}}{\partial t} + v_x\frac{\partial \underline{v}}{\partial x} + v_y\frac{\partial \underline{v}}{\partial y} + v_z\frac{\partial \underline{v}}{\partial z}$$

in Cartesian coordinates. The first term on the right side of (5.76) represents the *local* contribution to the acceleration (in non-steady flow) and the other terms are the *convective* contribution (due to translation). Employing (5.76) we obtain from (5.75) the most compact formulation of the Navier-Stokes equation for fluids with ρ = const.

$$\rho \frac{D\underline{v}}{Dt} = -\nabla p + \eta\nabla^2\underline{v} + \underline{f} \quad . \tag{5.77}$$

The right side of (5.77) comprises the sum of the forces acting on the moving mass in a unit volume. The negative of the pressure gradient, $\nabla p \equiv \mathrm{grad}(p)$, is that force which acts on the fluid element when the pressure changes from point to point. The term $\eta\nabla^2\underline{v}$ (where ∇^2 denotes the Laplace operator, e.g., $\partial^2/\partial x^2 + \partial^2/\partial y^2 + \partial^2/\partial z^2$) accounts for the effect of the viscous forces. Hence the viscous nature of a fluid, its *internal friction*, comes only to bear when one region of the fluid moves relative to another. Then momentum is transferred from the faster to the slower layers. The body forces $\underline{f}$ are regarded as given external forces. The surface force terms (with respect to the surface of the control volume), grad p and $\eta\nabla^2\underline{v}$, however, depend on the rate at which the fluid is strained by the velocity field present in it. The functional form of this dependence has to be determined empirically. For the derivation of (5.75) it has been assumed that the normal and shearing stresses are linear functions of the rate of deformation, which restricts the Navier-Stokes equations to "Newtonian" fluids.

Note that (5.75) does not contain the temperature as a variable. This does not necessarily imply that the flow must be isothermal but rather that density (in flow direction) and viscosity must not be significantly changed by the prevailing temperature field. This model of a *constant-property fluid* is a good approximation for liquids only as long as the temperature differences involved are moderate since most liquids exhibit a strong temperature dependence of viscosity. With gases, on the other hand, where η is typically proportional to the square root of T only (Sect. 5.3.3), limitations of the tolerable temperature differences with respect to the constant property treatment result rather from the strong temperature dependence of density.

As indicated in the introduction to Chap. 5, if the temperature dependence of the transport properties ρ, η and p must be accounted for, a complete description of the fluid requires three more equations in addition to (5.74) (where the mass has to balance with the density change) and a considerably more complex form of (5.75) [5.2, 21]. First, one needs an *equation of state* which relates pressure, density and temperature. Secondly, applying the First Law of Thermodynamics (2.13), one can draw up a balance between heat and mechanical energy (kinetic and potential energy) for the control volume of the fluid. This *energy equation* furnishes a differential equation for the temperature distribution (Sect. 5.5) if, thirdly, the *caloric equation of state* is known, e.g., $C_p = C_p(T)$. Furthermore, this set of equations must be supplemented with an empirical viscosity law $\eta = \eta(T)$, unless the prevailing conditions allow for assuming constant properties.

5.3.2 Flow near solid-fluid interfaces

In order to solve the set of equations (5.74) and (5.75) for a specific flow configuration, boundary conditions must be specified. For the contact area between fluids and solids the boundary condition takes on a particularly simple form. Numerous experimental studies have established that the flow of Newtonian liquids, past the surface of a solid body wetted by them, ceases immediately adjacent to the surface. The liquid adheres to the solid surface. The thickness of this no-slip layer is only a few molecular distances. An analogous phenomenon is observed in not too rarefied gases, i.e., when their free mean path is much smaller than a characteristic dimension of the solid. Hence the boundary condition for (stationary) solid-fluid interfaces without mass flux across can be written as

$$\underline{v} = 0 \quad . \tag{5.78}$$

Note that in multicomponent fluids (5.78) can be obeyed even when the $\underline{v}_i$'s $\neq 0$ as will be discussed in connection with vapor transport in closed ampoules [1.2]. As a consequence of this no-slip condition and of viscosity, the flow of the fluid must become increasingly retarded as it approaches the solid. The velocity must undergo a continuous change from its value in the bulk of the fluid to stagnation. As solutions of the Navier-Stokes equations will show, for fluids of not too high viscosity this region of considerable velocity change is very narrow in its extent from the solid. Hence it is commonly termed *boundary layer* or shear layer. Beyond this boundary layer the viscosity of a fluid has little bearing. The inertial forces dominate over the viscous forces, and the fluid can be treated as inviscid or ideal.

In general the nonlinearity of the Navier-Stokes equations (5.75) presents considerable mathematical difficulties. Exact solutions have been found in but a few special cases. In the following we will present two configurations which, although idealized, resemble some actual arrangements in crystal growth: stagnation in flow normal to a plane and flow near a rotating disc. After that we will discuss the parallel flow along a flat plate, also frequently encountered in crystal growth systems. Based upon this latter configuration we will then outline the essential steps in simplifying the Navier-Stokes equations to the Prandtl boundary layer equations.

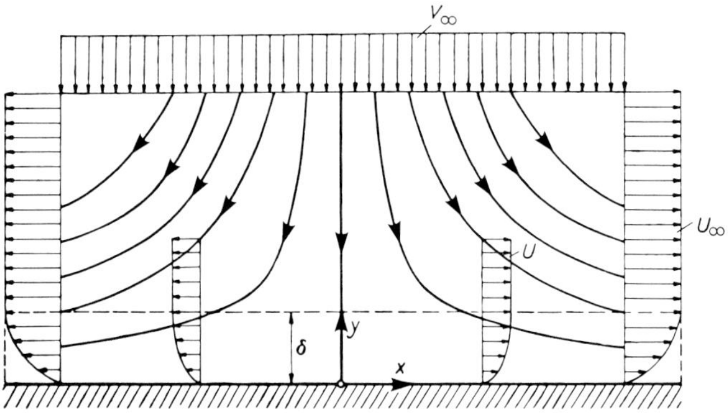

Fig.5.8. Stagnation in plane flow. From [5.21] by permission of McGraw-Hill Book Company

Stagnation flow, as depicted in two dimensions in Fig.5.8 occurs when a fluid stream impinges on a wall at right angles to it and flows away radially in all directions. This flow configuration is approximately encountered, for instance, in non-stirred solution growth when the solutal gradient above an interface drives convection towards the interface. At the stagnation point (x = 0, y = 0), by definition, fluid flow ceases completely. Far away from the stagnation point *and* the plane, the flow is only deflected, the *free stream velocity* being maintained, i.e., $|\underline{V}_\infty| = |\underline{U}_\infty|$ as schematically indicated in Fig.5.8. If there were no viscous interaction with the plane, the flow [then termed potential flow; see (5.88)] in the neighborhood of the stagnation point would be given by U = ax and V = -ay where a is a constant. In our case with friction, however, the velocity component parallel to the wall (which we now denote with u to distinguish it from the U of the "inviscid" flow pattern) will be progressively retarded as one approaches the wall. On the plane u = 0 (and, of course, v = 0). Far away from the plane, i.e., at "$y = \infty$" the inviscid flow remains unperturbed, that is to say that $u = U(x)$. Most significantly, the solution of this problem (e.g., [5.21]) shows that the *viscous* or *velocity boundary layer* δ_0, defined here as the distance from the plane at which the velocity component u parallel to the wall becomes 0.99 U, is independent of x, namely

$$\delta_0 = 2.4\left(\frac{\nu}{a}\right)^{1/2} \tag{5.79}$$

where ν is the kinematic viscosity, $\nu = \eta/\rho$. (It is useful to memorize that ν has the same dimension as the diffusion coefficient D, e.g., cm^2/s). Figure 5.9 shows the functional dependence of u/U on y. It also allows for a comparison of the solutions to the two-dimensional and axially symmetrical problems, which demonstrates that a two-dimensional treatment can well approximate the solution of the more complex three-dimensional configuration.

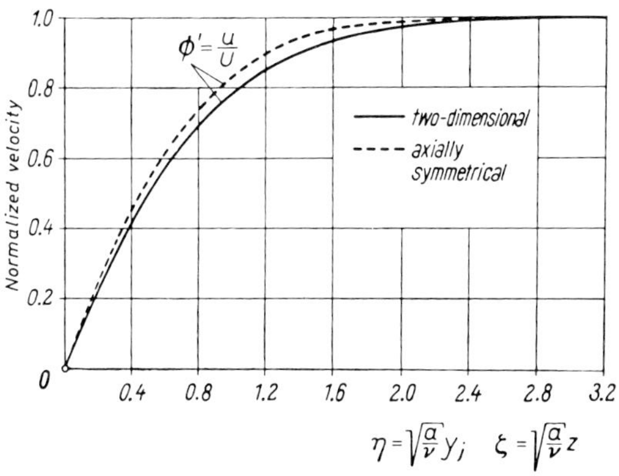

Fig.5.9. Velocity distribution of plane and rotationally symmetrical flow at a stagnant point. From [5.21] by permission of McGraw-Hill

Flow near a rotating disk is a further example of an exactly solvable configuration that is encountered, for instance, in pulling of a crystal from its melt. The layer near the rotating disk is carried by it through friction and is thrown outward due to the centrifugal forces acting upon it. This outward flow, as shown in Fig.5.10, is balanced by a flow in axial direction towards the disk to be in turn carried and ejected centrifugally. This problem has been solved by COCHRAN [5.23]. Again the analysis reveals the fact that the thickness of the layer carried by the disk, i.e., the boundary layer, is independent of the distance from the rotation axis, and the axial velocity component attains 99% of its limiting bulk value at

$$\delta_0 \simeq 4\left(\frac{\nu}{\omega}\right)^{1/2} \tag{5.80}$$

where ω is the angular velocity of the disk. Although this solution is based upon an infinite extent of the disk and fluid, it has been shown that the solution holds well even for finite systems as long as the radius of the disk is very much larger than δ_0.

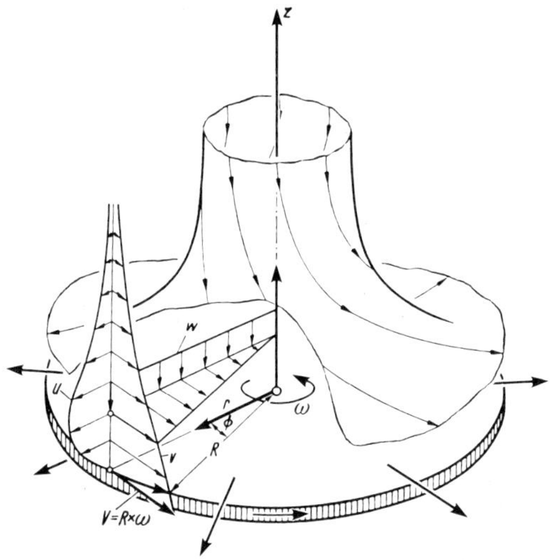

Fig.5.10. Flow near a disk rotating in a fluid at rest. From [5.21] by permission of McGraw-Hill

It is seen from these and other exact solutions of the Navier-Stokes equations that the boundary layer thickness is proportional to the square root of the kinematic viscosity

$$\delta_0 \sim \sqrt{\nu} \quad . \tag{5.81}$$

Evaluation of (5.79) and (5.80) for typical values of the material parameters and velocities involved (see Sect. 5.3.3) reveals that the solution for both the rotating disk and the flow with stagnation are of boundary layer type. This is to say that the effect of the internal friction, in particular in liquids, extends only over a narrow layer in the fluid at the solid unless the free-stream velocity V or the angular velocity ω are "very small" and/or the kinematic viscosity of the fluid is "very large". (See Problem 5.7).

For a quantitative general criterion as regards the significance of viscous effects in a given flow configuration, one advantageously introduces the dimensionless ratio

$$\frac{\text{inertial force}}{\text{viscous force}} = \frac{\rho u \, \partial u/\partial x}{\eta \partial^2 u/\partial x^2} \quad , \tag{5.82}$$

which can be transformed as follows. The velocity u at any point in the velocity field is proportional to the free-stream velocity V. Based upon a characteristic linear dimension L of the flow problem (e.g., size of a body in the flow or diameter of a conducting tube) $\partial u/\partial x$ is proportional to V/L and $\partial u/\partial x^2$ is proportional to V/L^2. Hence the dimensionless ratio, (5.82) can be expressed as

$$\frac{\rho V^2/L}{\eta V/L^2} = \frac{\rho VL}{\eta} = \frac{VL}{\nu} = N_{Re} \quad , \tag{5.83}$$

which is known as the *Reynolds number*. Going back to (5.79) and (5.80), we see that the larger the N_{Re} of a given flow problem, i.e., the more the inertial forces dominate over the viscous forces, the narrower is the fluid region adjacent to a fluid-solid interface in which one must account for internal friction.

Consequently, for sufficiently *large Reynolds numbers* ($N_{Re} >> 1$) one may subdivide the entire zone of flow into two regions: a region of inviscid flow and a viscous boundary layer region.

1) For the region of *inviscid* flow the Navier-Stokes equations (5.75) or (5.77) can be considerably simplified by omitting the viscosity term.* The resulting form

$$\frac{D\underline{v}}{Dt} = \frac{\partial \underline{v}}{\partial t} + (\underline{v}\cdot\nabla)\underline{v} = -\frac{\nabla p}{\rho} + \underline{f}\,\frac{1}{\rho} \quad , \tag{5.84}$$

---

* Although very large values of N_{Re} are considered *mathematically* for the limiting procedure, *physically* one assumes $N_{Re} \leq 10^4$, i.e., laminar flow. $N_{Re} > 10^4$ lead typically to turbulence.

known as *Euler's equation,* is of first order and hence can often be integrated in general form. Note, however, that the reduced order of (5.84) excludes simultaneous satisfaction of the boundary conditions $v_{normal} = v_{tangential} = 0$.

In the absence of external forces and for steady flow (5.84) for a linear problem becomes

$$u \frac{du}{dx} = - \frac{1}{\rho} \frac{dp}{dx} \tag{5.85}$$

which may be integrated to yield

$$p + \frac{1}{2} \rho u^2 = \text{const.} \quad , \tag{5.86}$$

known as *Bernoulli's equation.* Thus under steady "non-viscous" flow conditions the sum of the inner pressure and the dynamic head are constant. Note that here again "constant" refers to a control volume as it moves along in the flow. It is not necessary that the fluid is uniform, i.e., that the constant has the same value throughout the fluid. It may change as one determines it along a path that has a component normal to the flow direction.

If in a flow a fluid can be treated as incompressible and frictionless, it is said to behave ideally. If the flow of such an ideal fluid is *non-rotational,* i.e., if its *vorticity* defined by

$$\underline{\omega} = \text{curl}(\underline{v}) = \nabla \times \underline{v} \tag{5.87}$$

is zero, the velocity field $\underline{v}$ can be represented as the gradient of a (scalar) potential field ϕ

$$\underline{v} = \text{grad}(\phi) \quad . \tag{5.88}$$

Consequently, from the continuity equation it follows that the potential must satisfy the Laplace equation

$$\nabla^2 \phi = 0 \quad . \tag{5.89}$$

Note that a closed-path fluid flow may have zero vorticity provided that the fluid elements do not revolve about their centers of gravity as they move. Such irrotational flows of ideal fluids are also called *potential flows.*

2) In the viscous boundary layer region, the thickness of which is small compared to its length along the solid body, viscosity plays an important role. In this layer the velocity gradients normal to the interface exceed by far the gradients that prevail parallel to it. This allows again for a simplification of the Navier-Stokes equations (5.75), which was introduced by PRANDTL in 1904. For two-dimensional steady-state flow onto and along a plane (assumed to be oriented in x-direction) and negligible external forces, the resulting *boundary layer equation* has the form [5.21, 24]

$$u \frac{\partial u}{\partial x} + v \frac{\partial u}{\partial y} = -\frac{1}{\rho}\frac{dp}{dx} + \nu \frac{\partial^2 u}{\partial y^2} \tag{5.90}$$

where u and v are the x- and y-components, respectively, of $\underline{v}$.

The *continuity equation*, (5.74), reduces correspondingly to

$$\frac{\partial u}{\partial x} + \frac{\partial v}{\partial y} = 0 \quad . \tag{5.91}$$

The pressure gradient in (5.90) is written as a total differential with respect to x since the pressure gradient normal to the surface is $\partial p/\partial y \approx (\delta_0/x)\partial p/\partial x$ and hence can be neglected. Thus the pressure variation within the boundary layer in the x-direction is determined by the change in pressure outside. Note that the loss of the term $\partial^2 u/\partial x^2$ now makes the boundary layer equation parabolic, whereas the Navier-Stokes equations are elliptic. Physical consequences of this mathematical change will be pointed out later.

In fluid flow that is characterized by very *low Reynolds numbers* ($N_{Re} << 1$), that is, in very slow flows and/or flows with very large viscosity, the viscous forces dominate the intertial forces. In this situation the effects of internal friction must be considered throughout the fluid. A separation into boundary layers at walls and inviscid flow in the bulk is no longer physically meaningful. The dominance of the viscous forces allows for neglect of the inertial term. For this *creeping motion* without body forces the Navier-Stokes equation for incompressible flow assumes the form

$$\nabla p = \eta \nabla^2 \underline{v} \quad , \tag{5.92}$$

known as the *Stokes equation*. The continuity equation, of course, remains unchanged; $\nabla \cdot \underline{v} = 0$, (5.74). This system of equations must be supplemented with the same boundary conditions as the full Navier-Stokes equations, for instance at walls $v_n = v_t = 0$.

From (5.92) one can obtain an important characteristic of creeping motion. Forming the divergence of both sides and realizing that the operations div and ∇^2 may be performed in reversed order, with (5.74) one gets

$$\mathrm{div}[\mathrm{grad}(p)] = \nabla\cdot\nabla p = \nabla^2 p = 0 \quad . \tag{5.93}$$

Hence, the pressure field in creeping motion satisfies the potential equation, i.e., the pressure is a potential function.

The equation for *two-dimensional* creeping motion becomes particularly simple in form if one introduces a *stream function* ψ via $u = \partial\psi/\partial y$ and $v = -\partial\psi/\partial x$. For this special case the stream function must satisfy the equation (e.g., [5.21])

$$\nabla^4\psi = 0 \quad . \tag{5.94}$$

Consequently the stream function of planar creeping motion is a bipotential (biharmonic) function. Due to its mathematical simplicity (5.94) is often used for obtaining velocity fields, although one must keep in mind that its physical condition ($N_{Re} \ll 1$, two-dimensionality, no external body-forces) are not too frequently met in reality.

As another example of fluid flow in a configuration relevant to crystal growth let us briefly discuss viscous flow parallel, i.e., with zero incidence to a solid plane with leading edge, as depicted in Fig.5.11. This configuration is well approximated in certain vapor growth and solution growth systems. The boundary conditions are

$$y = 0 : u = v = 0$$

$$y \to \infty : u \to U_\infty$$

where the second condition matches the flow inside the boundary layer to the outer free-flow velocity U_∞. An exact solution of the boundary layer equation shows that the boundary layer width

$$\delta_0 \approx 5\left(\frac{\nu x}{U_\infty}\right)^{1/2} \tag{5.95}$$

increases with the square root of the distance x from the leading edge. The corresponding velocity distribution across the boundary layer is shown in

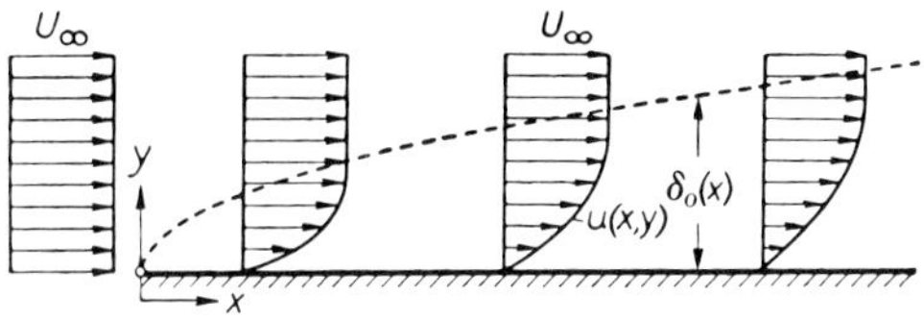

Fig.5.11. Development of a boundary layer on a flat plate in parallel flow at zero incidence. From [5.21] by permission of McGraw-Hill

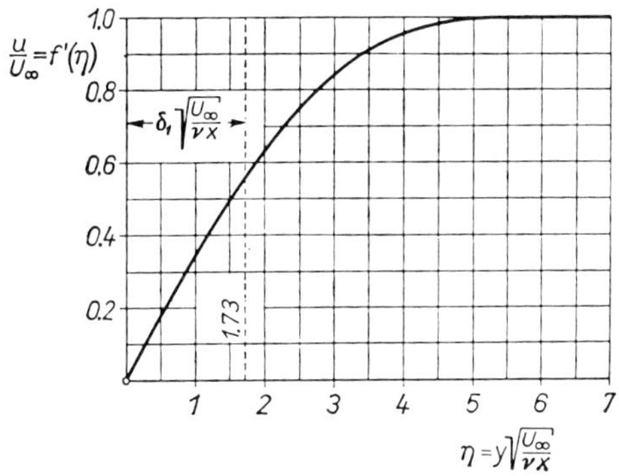

Fig.5.12. Velocity distribution in the boundary layer along a flat plate, after Blasius. From [5.21] by permission of McGraw-Hill

Fig.5.12. This solution, of course, does not apply to short distances from the leading edge, i.e., for $x \simeq \delta$, since the boundary layer assumptions are not valid if N_{Re} is not $\gg 1$. On the other hand, although the solution is based upon an infinite extent of a plane, (5.95) holds well for finite and even slightly curved bodies as long as $x \gg \delta$.

As a last example for a configuration that occurs also frequently in crystal growth we will discuss the steady flow of a viscous fluid in a conduit of constant cross section; this is referred to as a *Poiseuille flow*. We assume that the free-flow velocity U_0 at the entrance section of the pipe is uniform over the whole cross-sectional area. As schematically indicated in Fig.5.13, a boundary layer develops in a similar manner as on a flat plate at zero incidence. Eventually the boundary layer will reach the center of the pipe from where the velocity profile will become independent of x. The distance over which the asymptotic transition from the uniform free-flow velocity profile at the entry section to the fully developed Poiseuille flow is accomplished is called the *inlet length* or *entry length*, l. For pipes of a circular cross

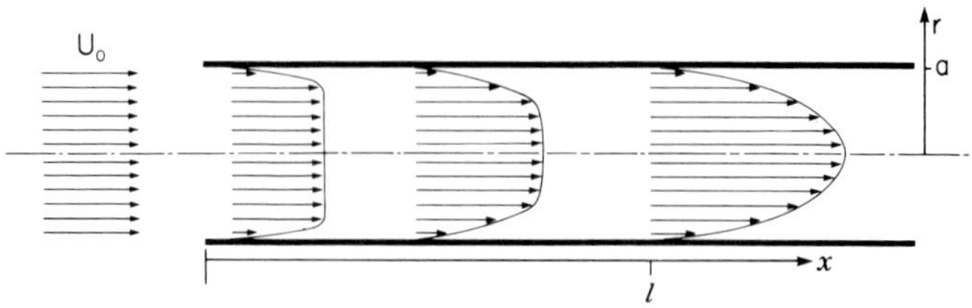

Fig.5.13. Definition sketch for the entrance length in Poiseuille flow

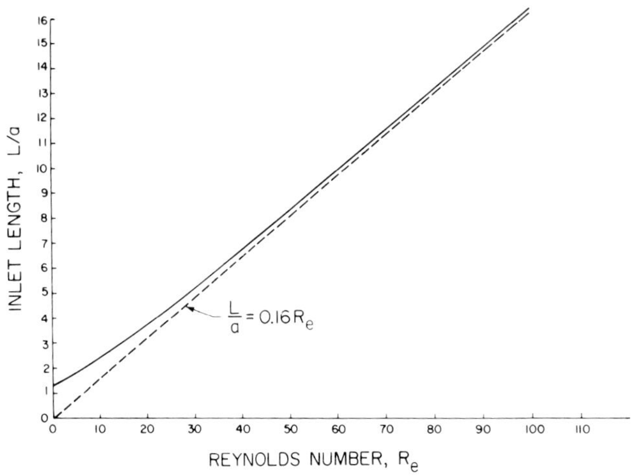

Fig.5.14. Relative change of inlet length l/a vs. Reynolds number for circular cylindrical tube. From [5.25] by permission of Pergamon Press

section of radius a, for large N_{Re} one obtains [5.25]

$$l \approx 0.16\, aN_{Re} \tag{5.96}$$

where the Reynolds number is based upon the radius a. However, the proportionality factor (0.16 at large N_{Re}) is a continuous function of N_{Re} and, as shown in Fig.5.14, increases with decreasing Reynolds number. When $N_{Re} \to 0$, the inlet length approaches $l = 1.3$ a. For a derivation of the fully developed Poiseuille velocity distribution one realizes that the transverse velocity components v and w will be zero, while u will be a function of y and z only. Also, since there is no motion or forces in these directions, the pressure will be a function of x only. Under these conditions the Navier-Stokes equation (5.75) reduces to

$$\frac{dp}{dx} = \eta \left(\frac{\partial^2 u}{\partial y^2} + \frac{\partial^2 u}{\partial z^2} \right) \tag{5.97}$$

which is *formally* equivalent to the creeping motion equation (5.92) yet is the result of quite different physical conditions. For a circular cross section, after transformation into cylindrical coordinates and assuming that the pressure gradient is independent of the radial coordinate r, integration of (5.97) yields

$$u(r) = - \frac{1}{4\eta} \frac{dp}{dx} (a^2 - r^2) \quad , \tag{5.98}$$

a parabolic velocity profile. If one neglects the entry and exit flow transition, the volume flux Q can be easily evaluated from (5.98) since the volume of a paraboloid of revolution is equal 1/2 x base area x height. Hence,

$$Q = \frac{\pi}{2} a^2 u_m = \frac{\pi a^4}{8\eta L} (p_1 - p_2) \tag{5.99}$$

where u_m is the maximum velocity at $r = 0$, L the total length of the pipe and $\Delta p = p_1 - p_2$, the pressure difference between entry and exit section. Equation (5.99) is known as the *Hagen-Poiseuille equation*.

For further emphasis of the powers of dimensionless characteristic numbers, such as N_{Re}, it is very instructive to rewrite the Navier-Stokes equations (5.75) in terms of dimensionless parameters by referring to characteristic magnitudes of the physical terms. Based upon the free-stream velocity V, a characteristic length L and a typical pressure p_0, one introduces the dimensionless ratios

velocity $\underline{W} = \underline{v}/V$,

lengths $X = x/L$, $Y = y/L$, $Z = z/L$ and

pressure $P = p/p_0$.

In substituting these ratios into the steady-state form of (5.75) one obtains the dimensionless Navier-Stokes equation

$$\rho \frac{V^2}{L} (\underline{W} \text{ grad})\underline{W} = - \frac{p_0}{L} \text{grad}(P) + \frac{\eta V}{L^2} \nabla^2 \underline{W}$$

or $\qquad$ (5.100)

$$(\underline{W} \text{ grad})\underline{W} = - \frac{p_0}{\rho V^2} \text{grad}(P) + \frac{1}{N_{Re}} \nabla^2 \underline{W} \quad .$$

Equation (5.100) can now conveniently be used for deductions of steady-state solutions for flow problems from existing solutions of geometrically similar configurations. "*Geometrical similarity*" is here understood as interchangeability of boundaries and streamlines after a mere change in the scale of length, for instance in the form $L_1 = nL_2$. Fluid motions under comparison are *dynamically similar* only if the solutions of (5.100) are identical. This requires that for both motions the respective dimensionless Navier-Stokes equations differ only by a factor common to all terms. Since in "incompressible" flows pressure changes are assumed to cause only insignificant density changes, the first term on the right side of (5.100) is not important for dynamic similarity. However, the second term contains all the parameters that govern the dynamics of the flow. Hence, dynamic similarity is assured if for two geometrically similar systems (1) and (2), $N_{Re}(1) = N_{Re}(2)$. This similarity principle, discovered by Reynolds, is the very basis for modeling in fluid dynamics. (See Problem 5.8).

Similarity considerations are not only useful for the comparison of different flow configurations. Solutions can be similar in themselves. In fact, we have tacitly made use of the similarity principle for the presentation of the velocity distribution in the stagnation flow problem in Fig.5.9. There, as is also schematically indicated in Fig.5.8, the velocity profile u(y) within the boundary layer is a *similarity solution* with $(a/\nu)^{1/2}$ as scaling factor. This means that the profile of the velocity distribution does not change with x; only the magnitude of each individual velocity vector at each individual y grows with $(a/\nu)^{1/2}x$.

It should be pointed out that a prerequisite for the characterization of a flow with *one* dimensionless number is the parabolic nature of the governing equation (5.90), i.e., the lack of an *intrinsic* length scale in the problem. Had there been a second interface at some plane y = h not too far from the first one, the geometric length scale h would be impressed upon the problem, and the conditions for similarity solutions only based upon N_{Re} would no longer exist. It should also be stressed that N_{Re} as a criterion is only valid in the absence of gravitational effects (i.e., balance between gravitational and buoyancy for each fluid element) since (5.100) does not account for external body forces. If external forces must be taken into consideration, different similarity principles must be deduced in which other dimensionless numbers are the key parameter. Some of these *controlling numbers* will be made use of in later sections.

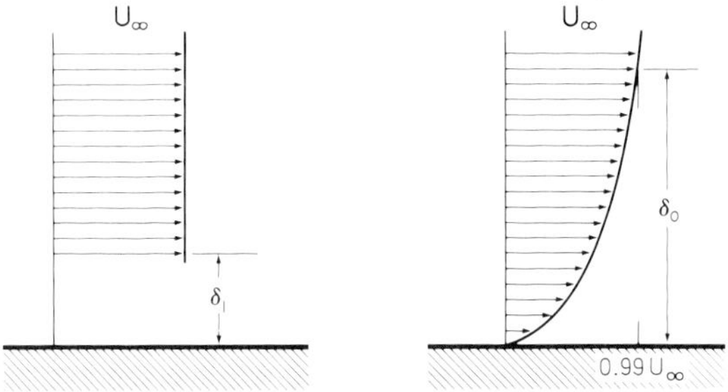

(a) (b)

Fig.5.15. Definition sketch for (a) displacement thickness δ_1 and comparison with (b) hydrodynamic boundary layer thickness δ_0. After [5.22]

The notion of a boundary layer in which the flow velocity changes from its free-flow magnitude or, more precisely, from 99% of it, to the velocity of the solid surface is fully adequate for the upcoming discussion of mass transfer in crystal growth. It should be kept in mind, however, that besides this laminar or *viscous hydrodynamic boundary layer* of width δ_0 other boundary layer concepts are advantageously used in fluid dynamics. These are the *displacement thickness* δ_1 and *momentum thickness* δ_2, also frequently called (mass) flow defect and momentum defect layer, respectively. The (fictitious) displacement thickness is used to *simulate* the same total flux by replacing the actually slowed-down flow near a solid wall by a somewhat wider region of inviscid flow and a zone of width δ_1 in which no flow occurs, as illustrated in Fig.5.15. In order to replace the reduced flux within δ_0 by extending the region of free-flow velocity U_∞ closer to the impeding wall,

$$U_\infty \delta_1 = \int_0^\infty (U_\infty - u)dy \quad , \tag{5.101}$$

and hence

$$\delta_1 = \int_0^\infty \left(1 - \frac{u}{U_\infty}\right) dy \quad . \tag{5.102}$$

Similarly one characterizes the momentum loss that a flow encounters, compared to the case of a fictitious inviscid flow extending all the way to the wall, through the momentum defect layer thickness δ_2. The mass flow which would exist if U_∞ would prevail throughout δ_2 would be $\rho U_\infty \delta_2$. Hence, the momentum defect due to this layer is $\rho U_\infty^2 \delta_2$, which is equal to the momentum defect in the actual boundary layer, i.e.

$$\rho U_\infty^2 \delta_2 = \rho \int_0^\infty u(U_\infty - u) dy \quad ,$$

and hence

$$\delta_2 = \int_0^\infty \frac{u}{U_\infty} \left(1 - \frac{u}{U_\infty}\right) dy \quad . \tag{5.103}$$

Usually $\delta_0 > \delta_1 > \delta_2$.

In conclusion, it should be pointed out that the width of "the (hydrodynamic) boundary layer" is not always understood as the distance from the solid at which the tangential velocity attains 0.99 of the value of the free-stream velocity U_∞. For instance for *boundary layer approximations*, one will be more concerned about the distance at which the assumptions that lead to (5.90) are not obeyed anymore. At 1% deviation from U_∞ the velocity gradients normal to the surface are typically still comparable to the parallel gradients, yet (5.90) requires a clear dominance. Hence for boundary layer approximations a better choice of the "cutoff" distance is a coordinate at which, e.g., $u = U_\infty(1 - 1/e)$ — as is used by many authors.

5.3.3 Viscosities

Most fluids, excluding colloids and high polymers, obey *Newton's law*. That is to say that the frictional force between two neighboring "layers" in a flowing fluid is proportional to the relative velocity of the layers and their area. The shearing stress, or shear force per unit area is proportional to the velocity gradient perpendicular to the direction of flow. Hence, for a one-dimensional flow Newton's law is then

$$\tau_{yx} = -\eta \frac{dv_x}{dy} \tag{5.104}$$

where τ_{yx} is the shear stress in the x-direction due to a gradient of the flow velocity v_x in y direction. The material dependent proportionality constant η is called *coefficient of internal friction, dynamic coefficient of viscosity, dynamic viscosity* or simply *viscosity*. τ_{yx} can also be interpreted as the viscous flux of x-momentum in y-direction. This, in the sense of the linear transport laws, gives the velocity gradient the meaning of a driving force for momentum transport from regions of high flow velocities to regions of low velocities. As we have seen before, sometimes it is useful to use instead of η the *kinematic coefficient of viscosity* or, in short, *kinematic viscosity* ν defined by

$$\nu = \eta/\rho \quad . \tag{5.105}$$

From these introductory remarks and the defining (5.104) one deduces that η has the dimensions

$$\frac{\text{force/area}}{\text{velocity/length}} = \frac{\text{momentum/(area x time)}}{\text{velocity/length}} = \frac{\text{mass}}{\text{length x time}} \quad .$$

In c.g.s. units the dimension of viscosity, i.e., 1[g/(cm s)] is called 1 *poise* [p]. Most frequently viscosities are stated in centipoise (cp). Note that 1[g/cm s] is equivalent to 1[dyne s/cm^2] and 10^{-1}[Newton s/m^2], which are also sometimes used for η. Correspondingly ν has the same dimensions as the diffusivity, i.e., 1[cm^2/s] which is called 1 *stoke* [s].

In analogy to the diffusivities of gases discussed in Sect. 5.3.3, viscosities for gases of simple molecules can be predicted from rigorous kinetic gas theory in good agreement with experimental values. The theoretical models for calculations and estimates of η for pure gases, gas mixtures and liquids were reviewed by SVEHLA [5.26], HIRSCHFELDER et al. [5.7], REID and SHERWOOD [5.27], BRETSZNAJDER [5.15] and in particular comprehensive form (including an outline of the experimental techniques) by TOULOUKIAN et al. [5.28]. For estimating the viscosity of non-polar gases one can employ the approximate relationship [5.7]

$$\eta \times 10^7 = 226.93 \frac{\sqrt{MT}}{\sigma^2 \Omega^{(2,2)*}} \quad \text{[poise]} \tag{5.106}$$

where M is the molecular weight, T the absolute temperature, σ the (Lennard-Jones) force constant, compiled for various gases in Table 5.5, and $\Omega^{(2,2)*}$ is a function (collision integral) the value of which is taken from Table 5.4 for $T^* = kT/\varepsilon$, where a compilation of k/ε - values is given in Table 5.5.

Table 5.10. Viscosities of water and air at 1 atm pressure. From [5.2] by permission of John Wiley and Sons

	Water (liq.)		Air	
Temperature $T[°C]$	Viscosity μ[cp]	Kinematic Viscosity $\nu[cm^2 s^{-1}]$	Viscosity μ[cp]	Kinematic Viscosity $\nu[cm^2 s^{-1}]$
0	1.787	1.787×10^{-2}	0.01716	0.133
20	1.0019	1.004×10^{-2}	0.01813	0.130
40	0.6530	0.658×10^{-2}	0.01908	0.169
60	0.4665	0.474×10^{-2}	0.01999	0.189
80	0.3548	0.365×10^{-2}	0.02087	0.209
100	0.2821	0.294×10^{-2}	0.02173	0.230

For estimating viscosities in liquids and solids no generally satisfactory model exists. A detailed review of the various approaches has been presented in [5.15]. Table 5.10 illustrates the large difference in viscosity between liquids and gases. Note that η of water is 1 cp around room temperature. As expected from (5.106) for all gases (and gas mixtures), one sees that the viscosity of air increases with temperature. According to the kinetic theory, the viscosity of a gas should not depend on pressure. This is experimentally verified at moderate pressures. The upper limit of the pressure range for which η is constant for a given temperature varies from a few to several tens of atmospheres, depending on the gas. Beyond that limit, viscosities of gases increase with gas pressure.

Table 5.11. Liquid metal viscosities in the representation $\eta = Ce^{E/RT}$. From [5.30] by permission of Marcel Dekker

Metal	C[cP]	E[kcal mole^{-1}]	Melting point [K]
Ag	0.4532	5.3	1234
Al	0.1492	3.95	933
Au	1.1320	3.799	1336
Bi	0.4458	1.541	544
Ca	0.0651	6.5	1123
Cd	0.3001	2.6	594
Co	0.2550	10.6	1768
Cs	0.1022	1.15	302
Cu	0.3009	7.3	1357
Fe	0.3699	9.9	1809
Ga	0.4359	0.955	303
Hg	0.5565	0.60	234
In	0.3020	1.59	430
K	0.1340	1.2	337
Li	0.1456	1.33	454
Mg	0.0245	7.3	923
Na	0.1525	1.25	371
Ni	0.1663	12.0	1725
Pb	0.4636	2.061	601
Pu	1.0879	3.1	913
Rb	0.0940	1.23	312
Sb	0.0812	5.25	904
Sn	0.5382	1.3	505
Tl	0.2983	2.50	577
U	0.4848	7.262	1405
Zn	0.4131	3.030	693

Unlike the viscosity of gases, that of most liquids decreases with increase of temperature at constant pressure, see e.g., the water data in Table 5.10. This difference in temperature dependence is well understood from a molecular point of view [5.8]. Here we can only note that the difference stems from the fact that in gases the free mean path is typically large as compared to the molecule diameter, whereas in liquids collisions occur after very short travel distances. Following a suggestion by Reynolds one often presents viscosity data for liquids in an Arrhenius relation of the form

$$\eta = C \exp(E_\eta/RT) \tag{5.107}$$

where C, and E_η (the "energy of activation for viscous flow") are rather independent of temperature. Many liquids, including molten metals and salts, have been found to obey this simple relation which also finds some justification in theoretical treatments of liquid viscosity. Table 5.11 presents viscosity data for various liquid metals in that form.

Compilations of viscosity data for systems of current interest in crystal growth are rather scarce. An extensive compilation of estimated viscosities (and thermal conductivities) of gases at high temperatures was put forward by SVEHLA [5.26], TOULOUKIAN et al. [5.28] give a critical review of experimental values for gaseous and a mostly low-temperature liquid mono- and multicomponent systems. Viscosities of a large selection of molten salts can be obtained from KLEMM [5.29]. The liquid metal data collected by BEYER and RING [5.30] are presented in Table 5.11. SAMSONOV [5.31] gave a tabulation of dynamic and kinematic viscosities of many elements in liquid form. ELWELL and SCHEEL [5.32] have compiled viscosity data for a large number of halides and oxides that are frequently used as solvents (fluxes) in high-temperature solution growth, see Fig.5.16.

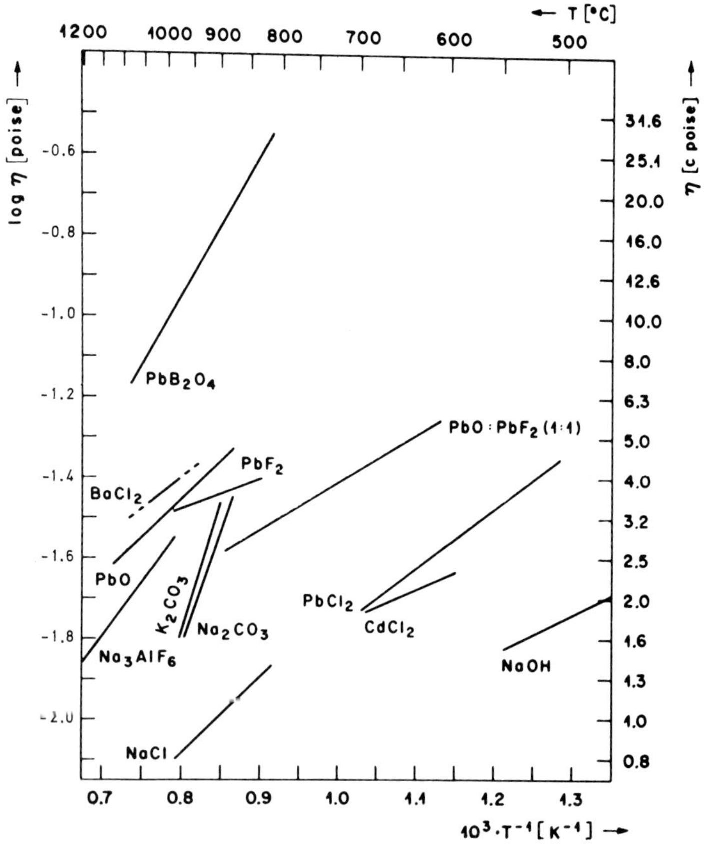

(a)

Fig.5.16. Viscosity of high-temperature solvents: (a) low-viscosity solvents, (b) high-viscosity solvents. From [5.32] by permission of Academic Press

Table 5.12 illustrates the wide range of viscosity encountered in crystal growth processes. One sees that the η-values for gases are in general 50-100 times smaller than those of simple, non-polymerizing liquids. The gaseous viscosities increase with temperature, as expected. Note that η-values for hydrogen lie within those of the other gases, in contrast to its diffusivity values (Sect. 5.2.3). The powers of rigorous kinetic theory are well illustrated by the comparison of the estimated and experimentally obtained values for water vapor — which, due to the polar nature of H_2O, must be considered a "worst case". Zinc is representative for most liquid metal η-data, with a few centipoise at the respective melting points (see also Table 5.11). Germanium and silicon are somewhat lower. Molten salts range from 1-10 cp with the majority again at a few cp. Many oxide melts

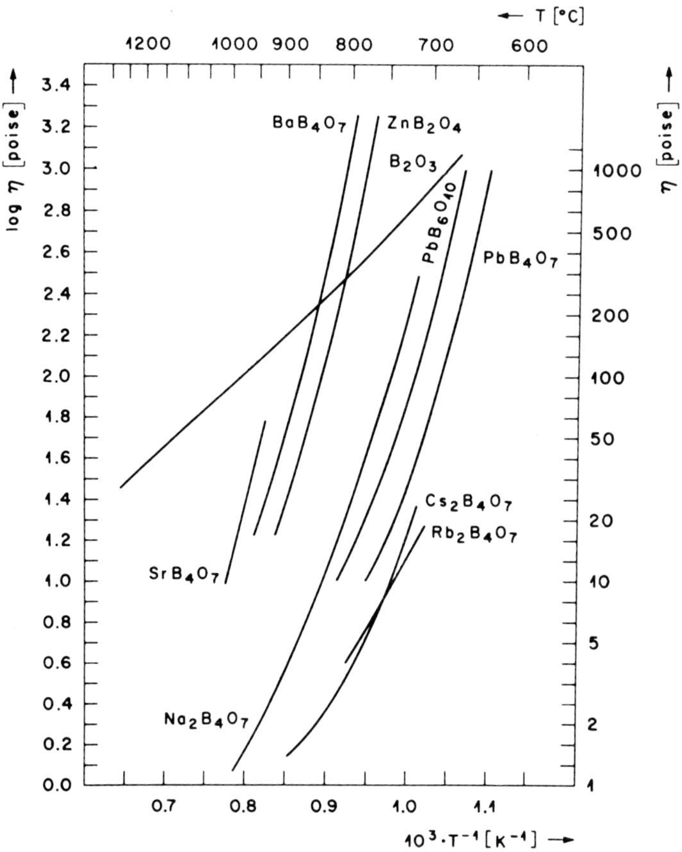

Fig.5.16. (b)

Table 5.12. Viscosity data for crystal growth fluids

Fluid	T_m[°C]	T[°C]	Viscosity η	ν[cm^2/s]	References for η
H_2	-259	700	0.020cp	7.4	[5.28]
		1000	0.024	11.6	
		1300	0.028	16.7	
O_2	-218.4	700	0.047	1.1	[5.28]
		1000	0.059	1.8	
I_2 (vapor)	113.5	20	0.013cp		[5.26]
		500	0.035		
H_2O	0.0	700	0.037cp		[5.28]
		700	0.035		[5.26]
$SiCl_4$	- 70	700	0.028cp	0.12	[5.26]
		1000	0.034	0.17	
		1300	0.039	0.27	
AsH_3	-116.3	1000	0.053cp		[5.26]
		1300	0.060		
Zn	420	450	3.17 cp	4.4×10^{-3}	[5.31]
		500	2.78		
Ga	30	52	1.89 cp	3.2×10^{-3}	[5.31]
		500	0.81		
Ge	937	950	0.70 cp	1.3×10^{-3}	[5.31]
		1000	0.65		
Si	1410	1430	0.62 cp	2.7×10^{-3}	[5.31]
NaCl	801	801	1.556cp	7.3×10^{-3}	[5.29]
		900	1.502		
$GaBr_2$	167	167	6.92 cp	1.9×10^{-2}	[5.29]
		181	5.72		
$K_2Cr_2O_7$	398	400	13.2 cp	5×10^{-2}	[5.29]
		450	9.7		
$MgAl_2O_4$	2135	2290	0.115 p	3.3×10^{-2}	[5.33]
UO_2	2500	2755	0.46 p	4.3×10^{-2}	[5.33]
Al_2O_3	2045	2070	1.38(0.55)p	0.35	[5.33] and
		2260	0.65(0.29)p		[5.39]
S	119	123	11 cp	5.7×10^{-2}	[5.31] and
		149	7 cp		[5.34]
		160	15 cp		
		270	180 cp		
		≃ 300	≃ 20 cp		
Se	220	220	30 p	6.3	[5.35]
Te	452	452	30 cp	4.8×10^{-2}	[5.35]
$K_2Si_2O_5$	1030	1050	2×10^3 p	833	[5.36]
		750	10^6 p		
B_2O_3	≃ 450	600	1.6×10^3 p	660	[5.16]
		750	500 p		[5.32]
		1000	100 p		

possess considerably higher viscosities which, as briefly indicated in Problem 5.7 and emphasized in later chapters, has significant consequences for interfacial mass transfer. The viscosity values for the chalcogens lie clearly above those for the other elemental liquids. In particular, sulphur displays probably the most complex temperature dependence of η of an inorganic liquid. Note the drastic *increase* in viscosity with T followed by a just as rapid decrease. This extraordinary behavior results from the complex polymerization phenomena observed in molten chalcogens. For instance, for liquid selenium one has estimated that it consists to a large fraction of long polymeric chains containing up to 10^5 atoms and it is these chains which appear to govern the crystal growth kinetics during solidification [5.35]. Glass-forming liquids such as the silicates, which can be drastically supercooled without undergoing spontaneous nucleation, exhibit an exceptionally wide viscosity range as well as high η-values. Boron trioxide, widely used as a high-temperature solvent (flux) and as a melt encapsulant to reduce evaporation losses of volatile components (e.g., the V-elements of III-V compounds) shows a rather high viscosity even considerably above its melting temperature. B_2O_3 is also presented in Fig.5.16 together with the viscosity-temperature data of a large number of other frequently used fluxes.

Table 5.13. Viscosities of binary fluids, see text

System	Tm[°C]	T[°C]	Viscosity	Mole Ratio	Reference
$H_2(-O_2)$		500	0.0126 cp	X_{O_2} = 0.00	[5.28]
			0.0224	0.25	
			0.0269	0.50	
			0.0291	0.75	
			0.0302	1.0	
Sn(-Pb)	505	504.2	2.75 cp	X_{Pb} = 0.000	[5.28]
		509.7	2.11	0.000	
	≃ 500	503.2	8.64	0.025	
	≃ 500	513.2	2.05	0.025	
	≃ 465	503.7	2.47	0.300	

Viscosities of mixtures differ often considerably from values that one would predict from a simple additive model. The system H_2-O_2 characterized in Table 5.13 is representative for the behavior of many gas mixtures. One sees that upon addition of a component with higher viscosity the overall η-value rises initially more rapidly and then levels off.

Even very small additions of a second component can result in large viscosity changes as, for instance, illustrated by the Sn-Pb system in Table 5.13. Pure tin exhibits the typical weak temperature dependence even upon slight supercooling. However, upon addition of only 2.5% of Pb the melt increases its viscosity fourfold furing cooling by 10 degrees even without supercooling. On further addition of lead this anomaly seems to disappear.

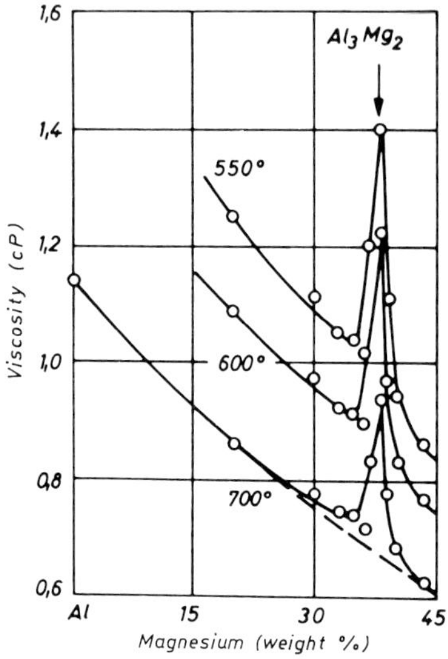

Fig.5.17. Viscosity isotherms for liquid aluminum-magnesium alloys. T_m of Al_3Mg_2 is 451° C. From [5.37] by permission of Dr. Riederer-Verlag

Another possibility for the composition dependence of viscosity is illustrated in Fig.5.17. As the composition of a binary melt approaches that of a solid compound the viscosity may increase significantly. In the case of the metallic Al-Mg system depicted one sees an increase in η by about 40% even 100° C above the solidification temperature of the intermetallic compound Al_3Mg_2. In contrast to this finding, the viscosity of eutectic-forming liquid alloys does not deviate from its monotonic composition dependence at the eutectic composition [5.38].

5.4 Convective-Diffusive Mass Transport

Let us now combine the concepts of mere diffusive and convective mass transfer dealt with in the foregoing sections. Thereby we will concentrate on convective-diffusive transport adjacent to and across a fluid-solid reaction surface, i.e., a growing or dissolving (melting) crystal-nutrient interface.

In addition to the *advective* flux induced by diffusion (Sect. 5.2.2) we have now to account for *convective* bulk flow. Yet we will not be concerned with the origin of that bulk flow. It may, for instance, result from surface forces, i.e., forced convection (Sect. 5.3). The bulk flow may also be due to field forces, e.g., from the interaction of gravity with unstable density gradients as discussed in Sect. 5.8 on free convection. Here we are "only" interested in the consequences of convection in the nutrient for mass transport at and across interfaces.

As crystal growers we are actually concerned with two (coupled) aspects of the nutrient-to-crystal transport:

(a) with the mass flux across an interface, which we will call the *interfacial flux* and which determines the crystal growth rate; and

(b) with the *concentration profile* of growth species in the nutrient adjacent to the crystal, which is an essential parameter in morphological stability discussions [1.1].

We will find that solutions to the first problem are, in general, easier to obtain than for the second problem. This can be readily understood. We know that directly at the interface the no-slip condition (5.78) demands that the tangential component of the mass average velocity of the fluid vanishes, regardless of the fluid flow conditions in the bulk nutrient. This implies, referring to a binary nutrient, for instance, and an interface which *remains* planar during growth (Fig.5.18), that the interfacial component fluxes $\underline{n}_A(0)$ and $\underline{n}_B(0)$ are strictly normal to the interface. Furthermore, we know that these interfacial component fluxes must be balanced by the "flux of solid from the interface into the crystal" to maintain continuity. (With this phrasing we have tacitly chosen the *stationary interface frame* as the frame of observation.) As soon as we leave the interface, however, out in the fluid the flow conditions are, in general, much more difficult to describe. Neither the no-slip condition applies, nor are, in general, the component fluxes $\underline{n}_A$ and $\underline{n}_B$ normal to the interface; also they are not balanced by the "solid growth flux". In order to specify the component fluxes and/or the corresponding concentration distribution, complete knowledge of the prevailing fluid dynamics is required; in other words, simultaneous solutions to the coupled continuity (diffusion) and momentum transfer equations must be obtained. Only in the absence of free or forced convection, again assuming a planar growing interface, one can resort to a one-dimensional description in the nutrient (Fig.5.18b) since in steady state the $\underline{n}_A$

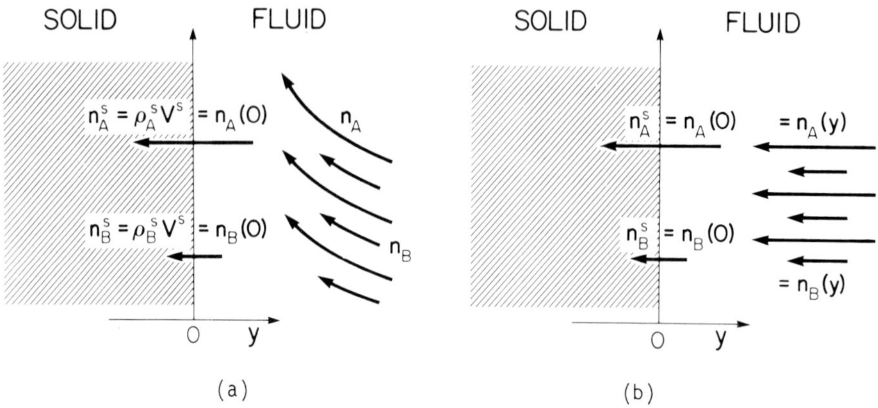

Fig.5.18. Component fluxes near planar crystal-nutrient interface in stationary interface frame. (a) convective fluid motion, (b) no fluid motion with respect to laboratory frame. Schematic

and $\underline{n}_B$ are constant throughout and the bulk flow term $\rho_i\underline{v}$ in (5.11) results from the diffusion-driven advection only, as discussed in Sect. 5.2.2.

Before we set up the various advantageous forms of the (convective)-diffusion equation, it is instructive to compare two basic situations (extreme cases) for interfacial component fluxes encountered in crystal growth. Our "master equation" for mass fluxes (Sect. 5.1)

$$\underline{n}_i = \rho_i\underline{v} + \rho_i(\underline{v}_i - \underline{v}) = \rho_i\underline{v} + \underline{j}_i \quad , \tag{5.11}$$

transformed with (5.7) for a binary system to

$$\underline{n}_i = W_i(\underline{n}_A + \underline{n}_B) + \underline{j}_i \tag{5.108}$$

(where i = A,B) lends itself particularly well to the distinction of the two cases.

First we consider the case in which the bulk flux $\rho_i v$ dominates in the total interfacial flux (5.11) of a component, irrespective of the convective conditions in the bulk nutrient. This situation arises in the solidification from an impure melt, or in close-to-congruent vapor growth. Here the interfacial mass average velocity is determined almost exclusively by the host. For instance, denoting the impurity with A and the melt with B, and using the superscript f for the fluid phase, one obtains

$$v = \frac{1}{\rho^f}\left(\rho_A^f v_A^f + \rho_B^f v_B^f\right) \approx \frac{1}{\rho_B^f}\left(\rho_B^f v_B^f\right) = v_B^f \tag{5.109}$$

to the extent that $\rho^f = \rho_A^f + \rho_B^f \approx \rho_B^f$. Consequently, from (5.11) above with $j_A \approx \rho_A(v_A - v_B) \to 0$

$$n_A^f(0) \approx \rho_A^f(0)v_B^f \quad . \qquad (5.110)$$

Of course, in this formulation it is tacitly assumed that A is *not segregated* at the interface, i.e., that the relative concentration of A and B does not change upon solidification. From the discussion of phase equilibria in Chaps. 3 and 4, however, we know that this is an exception rather than the rule. Hence in Chap. 6 on segregation we will expand (5.110) by a segretation coefficient (here taken as unity) to account for the compositional change during phase transitions.

An example for the other extreme, namely for dominant diffusive interfacial transfer, is the growth of a crystal from a solvent in which the solubility of the crystal-forming component (solute) is very low (or correspondingly, growth from a vapor component the partial pressure of which is small as compared to the total pressure). Here, since the solvent (intert gas, etc.) is (we hope!) completely rejected at the interface, i.e., $n_B^f(0) = 0$,

$$n_A^f(0) = W_A^f(0)n_A^f(0) + j_A(0) \approx j_A(0) \qquad (5.111)$$

to the extent that $W_A^f(0) \to 0$.

In general, however, when the solute and solvent concentrations in a nutrient are not greatly different, both diffusive and bulk flow must be accounted for. Assumed that the concentration gradient *and* the concentration of the crystal-building component (solute) directly at the interface are known, the interfacial growth flux can be calculated. Assuming quantitative solvent rejection one obtains in analogy to (5.42)

$$\underline{n}_A = \frac{-\rho^f(0)\mathcal{D}_{AB}}{1-W_A^f(0)}\left(\frac{\partial W_A^f}{\partial y}\right)_{y=0} \quad \text{or} \quad \underline{N}_A = \frac{-\mathcal{D}_{AB}}{1-C_A^f(0)\overline{V}_A^f}\left(\frac{\partial C_A^f}{\partial y}\right)_{y=0} \qquad (5.112)$$

where $W_A^f(0)$ and $C_A^f(0)$ are the weight fraction and molar concentration respectively, of the solute at the interface, i.e., at $y = 0$. Though these relations appear rather simple, one must be aware of the fact that they "contain the whole fluid dynamics" that prevails in the bulk nutrient adjacent to the interface. The actual interfacial concentration gradient can either be obtained experimentally (see, e.g., [5.62]), or, as we will discuss below for some simple configurations, if the nutrient bulk flow is fully understood, the gradient can be predicted theoretically. If in addition to complete interfacial rejection of B, $n_B = 0$ throughout some volume of nutrient adjacent to the interface (e.g., in the absence of free or forced convection), then (5.112) yields also the steady-state concentration profile in that region. This case was discussed before in connection with (5.42) and Problem 5.4.

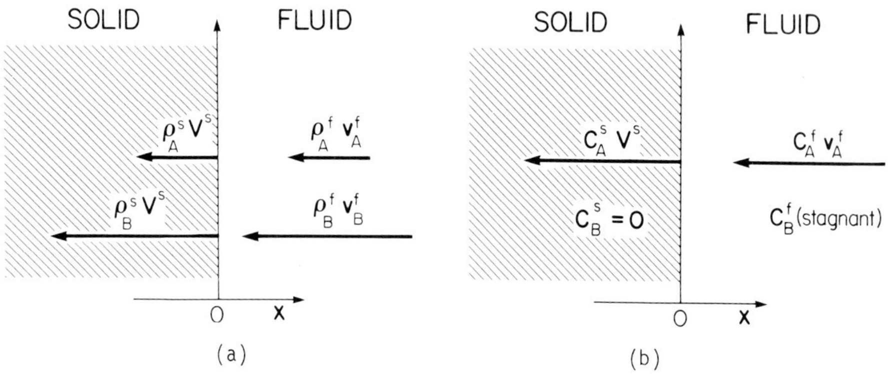

Fig.5.19. Fluxes *immediately* at interface. Stationary interface frame. (a) Impurity A in melt B. (b) Solute A in solvent B

For all of these cases, whether some fluid components are rejected or segregated at the interface or not, continuity at the interface leads to a very simple and useful boundary condition with respect to the *linear growth rate* V^S. Remember that in the stationary interface frame growth is considered as flux of nutrient components into the interface balanced by flux of solid from the interface into the crystal. Hence, assuming that diffusion in the solid can be neglected, all components (once crystallized) "leave the interface for the crystal" with V^S as depicted for a binary system in Fig.5.19, i.e., $v_A^S = v_B^S = V^S$.

Since the fluid in-fluxes must balance the solid out-fluxes,

$$\rho_A^S V^S = \rho_A^f(0) v_A^f(0) \quad , \qquad \rho_B^S V^S = \rho_B^f(0) v_B^f(0)$$

and (5.113)

$$(\rho_A^S + \rho_B^S) V^S = (\rho_A^f v_A^f + \rho_B^f v_B^f)_{y=0} \quad ,$$

which with (5.7) can be rewritten as

$$v(0) = \frac{\rho^S}{\rho^f(0)} V^S \qquad (5.114)$$

where ρ^s and $\rho^f(0)$ are the *total* mass density of the solid and of the fluid (taken at $y = 0$), respectively. Note that (5.113) holds under all circumstances, even when some components are completely rejected at the interface (solution growth, etc.) as depicted in Fig.5.19b. Looking back, (5.113) seems almost trivial since it simply expresses mass conservation irrespective of the complexity of the two phases. Yet this useful relation is sometimes misunderstood in the literature.

Let us now derive the convective-diffusive equation (continuity equation) in its various forms. Consider an isothermal, isobaric binary fluid consisting of a solute A and a solvent B. Diffusion, "entrained" in the overall motion of the fluid, is assumed to be solely due to concentration gradients. Continuity requires that the rate with which matter, say component A, accumulates (or depletes) in a stationary volume element dV in the fluid,

$$\int \frac{\partial \rho_A}{\partial t}\, dV \quad , \tag{5.115}$$

is balanced by the net mass flux of A entering and leaving the volume through its surface S,

$$-\oint \underline{n}_A\, dS \quad , \tag{5.116}$$

where the negative sign results from choosing a direction outward from the surface as the positive direction for the normal vector. Employing Gauss' divergence theorem one can convert the surface integral (5.116) to a volume integral in the form

$$\oint \underline{n}_A\, dS = \int \mathrm{div}(\underline{n}_A)\, dV \quad . \tag{5.117}$$

Thus from (5.115) and (5.117) one obtains the *continuity equation*

$$\frac{\partial \rho_A}{\partial t} = -\mathrm{div}(\underline{n}_A) \quad . \tag{5.118}$$

By choosing, for instance, the mass average velocity as a reference frame for the division of $\underline{n}_A$ into a convective and diffusive part [see (5.11)] we obtain from (5.118) the continuity equation as presented before

$$\frac{\partial}{\partial t}\, \rho_i = -\nabla\cdot(\rho_i \underline{v} + \underline{j}_i) \quad . \tag{5.1}$$

Which particular form of Fick's first law (Table 5.1) one will choose for $\underline{j}_i$ in (5.1) will depend on the specific experimental conditions. For instance, with systems of *uniform total mass density* one may choose $\rho\nabla W_i$, which for component A results in

$$\frac{\partial}{\partial t}\rho_A = \mathcal{D}_{AB}\,\mathrm{div}[\rho\ \mathrm{grad}(W_A)] - \mathrm{div}(\rho_A\underline{v}) \quad . \tag{5.119}$$

Since div $\rho_A\underline{v} = (\underline{v}\ \mathrm{grad})\rho_A + \rho_A\ \mathrm{div}(\underline{v})$, with ρ = const. and hence $\mathrm{div}(\underline{v}) = 0$, using $\rho_A/\rho = W_A$, one gets then from (5.119) the convenient form

$$\frac{\partial}{\partial t}W_A + (\underline{v}\cdot\mathrm{grad})W_A = \mathcal{D}_{AB}\,\nabla^2 W_A \quad . \tag{5.120}$$

Note the similarity of this *(convective-)diffusion* equation to the Navier-Stokes equation, (5.75). As we have discussed before, the diffusion (continuity) equation represents the conservation of matter, and the Navier-Stokes equation ensures balance of momentum in a moving fluid. The formal correspondence can be further emphasized by introducing a mass source term Q into (5.120) as a counterpart of the volume force term f in (5.75). It should be understood, however, that Q is a volume source term. Sources and sinks on interfaces must be accounted for by appropriate boundary conditions.

Equation (5.120) is often derived in such a "generous" manner with respect to its diffusive flux contribution that the fact is overlooked, that ρ = const. is a precondition for its validity, (e.g., [5.24]). Hence we want to stress that (5.120) will only yield reasonably accurate results for fluids with a strongly dominating component or for the rare case that all components have approximately the same density. Thus, with regard to crystal growth, its validity is limited more or less to melts (with impurities), and vapors/solutions with low concentrations of the crystal building components in the carrier gas/solvent, respectively.

For mass transfer into the interface we obtain, combining (5.120) and (5.114), for steady state

$$\underline{\mathcal{D}}_{AB}\left(\frac{d^2W_A}{dy^2}\right)_{y=0} + \frac{\rho^s}{\rho^f(0)}\,V^s\left(\frac{dW_A}{dy}\right)_{y=0} \approx 0 \quad . \tag{5.121}$$

For fluids that undergo pronounced total density variations along the mass transfer path, one must base the description of convective diffusion on other criteria than ρ = const. For instance for isothermal, isobaric gas mixtures the *total molar concentration is uniform throughout*. Instead of (5.118) one will then take advantage of

$$\frac{\partial}{\partial t} C_A = -\mathrm{div}(\underline{N}_A) \quad , \tag{5.122}$$

the conservation of moles. Analogous to the foregoing derivation, assuming a concentration independent $\mathcal{D}_{AB}$, one obtains from (5.122)

$$\frac{\partial}{\partial t} C_A = \mathcal{D}_{AB}\, \mathrm{div}[C\, \mathrm{grad}(X_A)] - \mathrm{div}(C_A \underline{v}^*) \quad , \tag{5.123}$$

where $\underline{v}^*$ is the mole-average velocity vector. Since from (5.8) and (5.12) $\mathrm{div}\, \underline{v}^* = \mathrm{div}(\Sigma \underline{N}_i)/C$, and realizing that $\Sigma \underline{N}_i = 0$ for fluids with constant C, (5.122) can be rewritten to a final "molar form" of the diffusion equation

$$\frac{\partial}{\partial t} X_A + (\underline{v}^* \cdot \mathrm{grad}) X_A = \mathcal{D}_{AB} \nabla^2 X_A \quad . \tag{5.124}$$

For the interfacial transfer in relation to the growth rate V^s one obtains from (5.124), see Problem 5.10,

$$\mathcal{D}_{AB} \left(\frac{d^2 X_A}{dy^2} \right)_{y=0} + \frac{C_A^s}{C^f(0)}\, V^s \left(\frac{dX_A}{dy} \right)_{y=0} = 0 \quad . \tag{5.125}$$

Correspondingly, for nutrients in which the total molar concentration *and the partial molar volumes* of all components remain unchanged as the flow and the entrained diffusive mass transfer proceed,

$$\frac{\partial}{\partial t} C_A + (\underline{v}^{\square} \cdot \mathrm{grad}) C_A = \mathcal{D}_{AB} \nabla^2 C_A \quad , \tag{5.126}$$

the derivation of which is left to the reader in Problem 5.9. For the interfacial form one obtains in this case, see Problem 5.10,

$$\mathcal{D}_{AB} \left(\frac{d^2 C_A}{dy^2} \right)_{y=0} + C_A^s \overline{V}_A^f V^s \left(\frac{dC_A}{dy} \right)_{y=0} = 0 \quad . \tag{5.127}$$

Of course, (5.120), (5.124) and (5.126) apply to any of the components in a fluid nutrient as long as the diffusion can be dealt with independently in terms of a binary diffusion coefficient. Note that, as $\underline{v}$, $\underline{v}^*$ and $\underline{v}^\square$ become negligible, all three equations reduce to *Fick's second law*

$$\frac{\partial C_A}{\partial t} = \mathcal{D}_{AB}\nabla^2 C_A \quad . \tag{5.128}$$

In the general case the velocity field and the concentration field mutually interact. The concentration distribution depends on the velocity distribution and vice versa. Hence, in general, to obtain the appropriate concentration distribution in a flowing fluid, one must *simultaneously* solve the diffusion equation with the equations of momentum (Navier-Stokes equation). This, in most circumstances, presents a formidable mathematical task and can often only be performed numerically. However, in the special case of constant total mass density [see (5.120)] mutual interaction ceases and the velocity field no longer depends on the concentration fields (of the individual components) — although the dependence of the concentration field(s) on the velocity field still persists. Hence, in this special case, one may first solve the Navier-Stokes equations and then use the obtained velocity distribution as input function for solving the diffusion (continuity) equation.

In order to facilitate a characterization of the different types of concentration fields that result from different orders of magnitudes in the controlling system parameters, let us reduce the diffusion equations to a dimensionless form. We will treat (5.120) as representative for all three equations. The dimensionless form of (5.124) and (5.126) can then simply be obtained by replacing W_A with X_A, respectively. Based upon the free-stream velocity V and a characteristic length one defines the dimensionless ratios for velocity and length as in (5.100). A meaningful choice for the characteristic length is, for instance, the distance over which the major change in concentration takes place. Substitution of the dimensionless ratios into (5.120) yields

$$\frac{\partial W_A}{\partial t} + (\underline{W}\ \mathrm{grad})W_A = \frac{\mathcal{D}_{AB}}{VL}\nabla^2 W_A \quad . \tag{5.129}$$

One sees that the relative weight of the convective and diffusive flux contribution is characterized by the dimensionless ratio

$$N_{Pe} = \frac{VL}{\mathcal{D}_{AB}} \tag{5.130}$$

which is known as the *Peclet number*. The N_{Pe} plays an analogous role in mass transfer to the Reynolds number $N_{Re} = VL/\nu$, (5.83), in the momentum transfer of a fluid. When $N_{Pe} \ll 1$, which may occur at very small flow velocities and/or in regions of small dimensions and which is facilitated by high diffusivities (gases), the diffusive contribution to the mass flux will override the convective flux throughout the system. However, when $N_{Pe} \gg 1$ and concentration gradients in the bulk fluid are not too high, the concentration field will be of the boundary layer type. That is to say that in the bulk fluid convective transfer will dominate (see Problem 5.11). Only in the vicinity of a reaction surface, due to the reduction of the flow velocity, will diffusion prevail and will the concentration change more or less rapidly. This condition is, of course, favored by high flow velocities and in fluids with low diffusivities (liquids).

It appears intuitively advantageous to base a correlation of the velocity and concentration distribution at an interface on the ratio of the velocity boundary layer width δ_0 [(5.79), etc.] and the analogously defined *concentration layer width* δ. From the formal similarity of the Navier-Stokes and convective-diffusion equations one can expect that δ is governed by N_{Pe} in a similar manner as the Reynolds number is decisive for the extent of the velocity boundary layer. Hence, subject to a rigorous discussion for each specific system, we can anticipate that in general

$$\frac{\delta_0}{\delta} = C\left(\frac{N_{Pe}}{N_{Re}}\right)^n \quad . \tag{5.131}$$

For the constant C and exponent n one might expect, from the foregoing discussion of momentum transfer, values around unity and 1/2, respectively. Yet even in this general form, (5.131) allows already for a very valuable identification of basic trends and behavior patterns. The dimensionless ratio of Peclet number and Reynolds number, known as the *Schmidt number* (or diffusive Prandtl number)

$$N_{Sc} = \frac{N_{Pe}}{N_{Re}} = \frac{VL}{\mathcal{D}_{AB}} : \frac{VL}{\nu} = \frac{\nu}{\mathcal{D}_{AB}} \quad , \tag{5.132}$$

is simply the ratio of the kinematic viscosity and the diffusivity and thus depends only on the properties of the fluid and not on the specific geometry (boundary conditions) and free-stream velocity. Inserting typical values for

ν (Sect. 5.3.3) and D_{AB} (Sect. 5.2.3) into (5.132) we obtain a characteristic difference between the Schmidt numbers for gases and liquids. Whereas N_{Sc}'s for gases range between 0.1–1, with the majority around unity, liquids possess Schmidt numbers of 10 and higher. Thus, as schematically indicated in Fig.5.20a, in gases the concentration and velocity profiles extend about the same distance into the bulk independent of the free-stream velocity. In liquids, however, δ_0 will exceed δ considerably (Fig.5.20b). Thus, in the region of major concentration change at the interface the flow velocity is reduced to relatively small values.

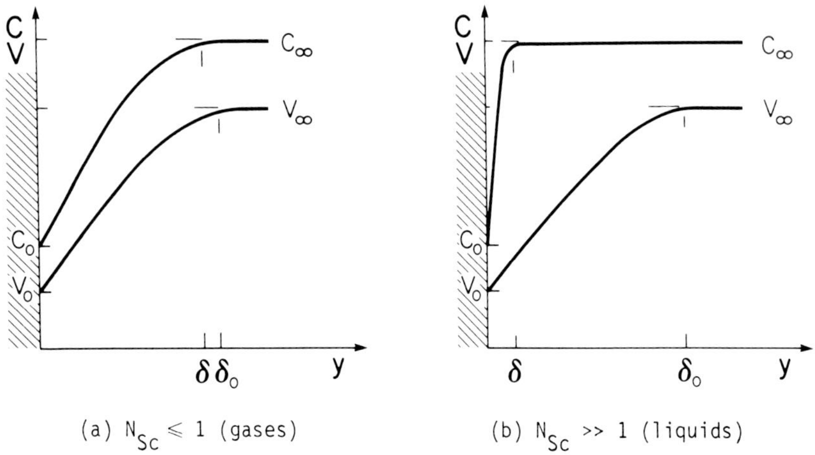

Fig.5.20. Schematic comparison between concentration and velocity distribution at solid-fluid interfaces for fluids of small and large Schmidt number

With regards to mass transfer in the bulk of liquids we see from (5.132), i.e., $N_{Pe} = N_{Sc} \cdot N_{Re}$, that for Reynolds numbers as low as a few 10^{-3} the convective flux is still overriding diffusive transfer (see Problem 5.11), mostly due to the low diffusivities in liquids. Thus, with the N_{Re}'s usually met in practice the N_{Pe} of liquids is very high. Consequently, in analogy to momentum transfer in fluids with $N_{Re} \gg 1$, a liquid system (i.e., all fluids with $N_{Pe} \gg 1$) may be divided with respect to mass transfer into two regions:

1) The *bulk liquid* in which diffusion is comparably negligible. As a consequence the convective-diffusion equation can be reduced to

$$\frac{\partial W_A}{\partial t} + (\underline{v}\ \mathrm{grad})W_A = 0 \quad . \tag{5.133}$$

A physically meaningful solution of (5.133) is given by W_A = const., i.e., the liquid is *completely mixed*.

2) A *boundary layer region* of rapidly changing concentration in the immediate vicinity of a reaction surface (crystal-liquid interface). Here diffusion must be taken into account, just as in fluids with $N_{Re} >> 1$ where the effect of viscosity comes to bear only in the velocity boundary layer. Note, however, that the width of "the" concentration boundary layer, for a given velocity, is proportional to the third root of the diffusivity of the considered component [see, e.g., (5.139)]. Hence in multicomponent fluids several concentration boundary layers of significantly different widths can exist simultaneously.

Since the concentration gradients normal to the interface (in y-direction) exceed by far the gradients parallel to it, the steady-state convective-diffusion equation (5.120) can be reduced to the *concentration boundary layer equation*

$$u \frac{\partial W_A}{\partial x} + v \frac{\partial W_A}{\partial y} = \mathcal{D}_{AB} \frac{\partial^2 W_A}{\partial y^2} \qquad (5.134)$$

where u and v are the x- and y-components of the velocity distribution which satisfies the continuity equation (5.91). An elegant similarity treatment of mass transfer through laminar boundary layers and a review of solutions to the "b-equation" can be found in a sequence of papers by SPALDING and coworkers [5.41].

There are numerous solutions of the convective-diffusion equation and the formally identical, simplified energy equation (see Sect. 5.5) in the mass and heat transfer literature. Various configurations pertaining to crystal growth are reviewed in [5.6, 14, 42]. We will restrict ourselves here to two of the geometries which we have also discussed in the momentum transfer section: the rotating disk and flow parallel to a flat plate. Other configurations will be treated in later sections in connection with specific growth methods.

Mass and heat transfer about a rotating disk in contact with a large body of a quiescent, constant-property fluid of any N_{Sc} or Prandtl number (Sect. 5.5) was treated by SPARROW and GREGG [5.43, 44]. From their analysis one can deduce the concentration boundary layer width, i.e., $y(0.99\ C_\infty)$, as

$$\delta \simeq 4\left(\frac{\nu}{\omega}\right)^{1/2} f(N_{Sc}) \quad , \qquad (5.135)$$

independent of the location on the disk. Discrete values for $f(N_{Sc})$ are presented in Table 5.14. With (5.80) we can rewrite (5.135) as

$$\delta \approx \delta_0 f(N_{Sc}) \quad , \tag{5.136}$$

which for gases with a typical $N_{Sc} = 0.7$ gives $\delta \approx 2\delta_0$ and for a liquid with $N_{Sc} = 100$, $\delta \approx 0.16\delta_0$, thus nicely confirming the intuitively obtained (5.131). Note that Table 5.14 also contains a column for the dimensionless concentration gradient at the interface

$$\left(\frac{\nu}{\omega}\right)^{1/2} \frac{\partial}{\partial y}\left(\frac{C - C_\infty}{C_0 - C_\infty}\right)_{y=0} = N_{Sh} \tag{5.137}$$

to be used later.

Table 5.14. $f(N_{Sc})$ in (5.136) and N_{Sh} (5.137) as function of Schmidt number. Deduced from [5.43]

N_{Sc}	$f(N_{Sc})$	N_{Sh}
$\to 0$	$\to \infty$	$1.13\ (N_{Sc})^{-1}$
0.01	131	115
0.1	15	13.0
0.7	2.1	3.0
1	1.5	2.52
10	0.4	0.88
100	0.16	0.37
$\to \infty$		$1.61\ (N_{Sc})^{-1/3}$

For $N_{Sc} \gg 1$, i.e., for liquids the steady-state concentration profile can be written in the form [5.24]

$$\frac{C - C_0}{C_\infty - C_0} = \frac{\int_0^Y e^{-u^3} du}{\int_0^\infty e^{-u^3} du} \tag{5.138}$$

where C_0 and C_∞ are the concentrations on the surface and at the outer edge of the diffusion boundary layer (bulk concentration), respectively. The variable $u = \omega^{1/2} y/(5.88\ D_{AB}\nu^{1/2})^{1/3}$, and $Y = y/\delta$. Eq.(5.138) shows that for small values of y, the concentration increases more rapidly than simply exponentially (Fig.5.21). The forced convection introduced by the rotation steepens the concentration profile beyond the exponential dependence that results merely from the diffusion of A through stagnant B considered in Problem 5.4c, i.e., the forced flow increases the interfacial flux substantially.

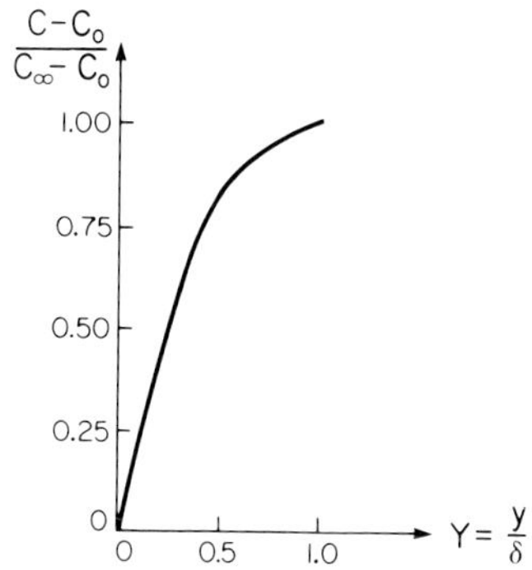

Fig.5.21. Concentration profile in the vicinity of a rotating disk. $N_{Sc} \gg 1$. After [5.24]

Mass and heat transfer in parallel flow past a semi-infinite flat reaction surface and heated plate, respectively, has been treated by several authors; for references see [5.21, 24]. From solutions of the boundary layer equations for momentum transfer (velocity) and concentration for incompressible laminar flow and constant properties, one can deduce that

$$\delta \simeq 3\left(\frac{\mathcal{D}_{AB}}{\nu}\right)^{1/3}\left(\frac{\nu x}{U_\infty}\right)^{1/2} \tag{5.139}$$

which with (5.95) transforms to

$$\delta \simeq (N_{Sc})^{-1/3}\delta_0 \quad . \tag{5.140}$$

Hence, the diffusion boundary layer along the flat plate increases with the square root of the distance from the leading edge. Consequently, as schematically indicated in Fig.5.22, the mass flux onto the plate decreases with $x^{-1/2}$ from the leading edge, a fact that has severe consequences in crystal growth from vapors and solutions. Concentration profiles for various N_{Sc} are plotted in Fig.5.23. This figure was adapted from the heat transfer literature and originally presented the temperature distribution in the vicinity of

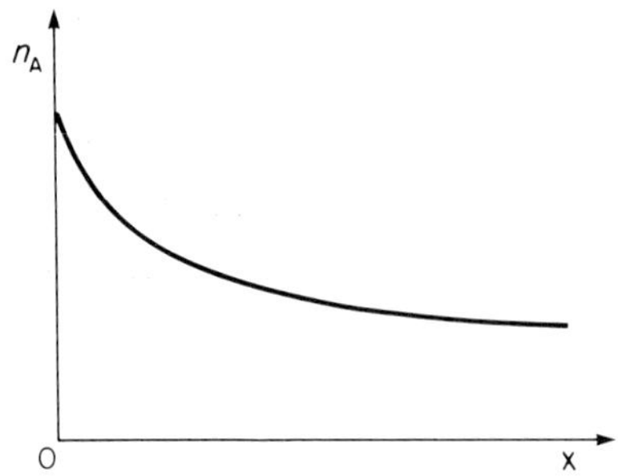

Fig.5.22. Mass flux onto a planar reaction surface as a function of distance from the upstream edge of the plane (schematic)

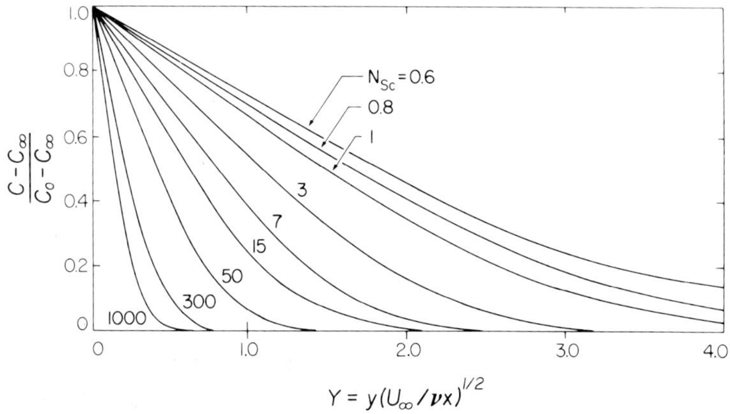

Fig.5.23. Concentration distribution on a flat reaction surface at zero flow incidence, plotted for various Schmidt numbers N_{Sc}. Adapted from [5.21]

a heated plate.* Here, where the Prandtl numbers (Sect. 5.5) are replaced by the Schmidt numbers of same values, the figure presents the concentration distributions during dissolution of a crystal or rejection of a fluid component at the growing interface. The concurrent velocity distribution can be deduced from the function with $N_{Sc} = 1$. One sees that, for example, for fluids such as aqueous solutions with $N_{Sc} = 10^3$ the concentration boundary layer extends only over one tenth of the velocity boundary layer, as was anticipated in Fig.5.20b.

It has become common practice, in particular in the chemical engineering literature, to present and correlate theoretical as well as experimental results for interfacial mass transfer in terms of a dimensionless concentration gradient [5.61]

---

* For limitations of this analogy see Sect. 5.5.3.

$$N_{Sh} \equiv -L\left[\frac{\partial}{\partial y}\left(\frac{W_A - W_{A\infty}}{W_{A0} - W_{A\infty}}\right)\right]_{y=0} = f\left(N_{Re}, \frac{v_0}{V}, N_{Sc}\right) \tag{5.141}$$

known as the *Sherwood number*. Here L is a length that is characteristic for the system investigated. W_{A0} and $W_{A\infty}$ are the weight fractions of the transferred component immediately at the interface and in the (completely mixed) bulk fluid (outer edge of boundary layer), respectively. For instance, Sherwood numbers for the rotating disk problem, i.e., dimensionless concentration gradients based upon $L = (\nu/\omega)^{1/2}$, are tabulated in Table 5.14. Also, for flow parallel to flat reaction surfaces the N_{Sh}'s based upon $L = (U_\infty/\nu_x)^{1/2}$ can be deduced from Fig.5.23 by extending the tangents onto the various curves in the origin.

The actual concentration gradient at y=0 can readily be obtained from (5.141) and is

$$\left(\frac{\partial W_A}{\partial y}\right)_{y=0} = \frac{W_{A\infty} - W_{A0}}{L} N_{Sh} \quad . \tag{5.142}$$

Hence with the Sherwood number, (5.112) and (5.114), we can express the growth rate as

$$V^s = -\frac{\rho^f(0)}{\rho^s} \frac{\mathcal{D}_{AB}}{1 - W_{A0}} \frac{W_{A\infty} - W_{A0}}{L} N_{Sh} \quad . \tag{5.143}$$

The reader should be aware that there has been a considerable amount of confusion in the crystal growth literature with respect to interfacial mass transport and boundary layer concepts. The modern notion of N_{Sh} as a dimensionless concentration gradient at a phase boundary has yet to be generally recognized. Historically, N_{Sh} resulted from the *stagnant film model* and is entangled with the terms *film width* and *mass transfer coefficient*. We will not employ these parameters in this text, yet they must be understood for a proper interpretation of older results.

In the stagnant-film concept, or better "effective film for mass transfer" [5.46] concept, one simulates the complex fluid dynamic conditions at a phase boundary with a stagnant film the width d of which is chosen such that the model mass-flux normal to the interface (envisioned to solely result from diffusion and advection) is the same as in the real case with diffusive-convective flow. (In the literature the symbol δ is widely used for the film width. To avoid confusion with the true concentration boundary layer width, however, we will not follow this custom). Within this stagnant-film picture one can describe mass transfer at a phase boundary based upon (5.40), (5.42) and the solutions obtained in Problem 5.4. Note that even in this simplifying model the advective term normal to the interface causes an exponential concentration profile, as schematically shown in Fig.5.24. Note further that only for $v = 0$ [see (5.40)] will there be a linear interfacial concentration gradient. In reality this takes very special provisions. Zero v or $v^{\square}$ may be approximated in a crystal growth system when the solute concentration is very low [see (5.108) and (5.112)] or when $\underline{n}_A$ and $\underline{n}_B$ are opposite and comparable, as it may occur in processes involving chemical reactions at the interface.

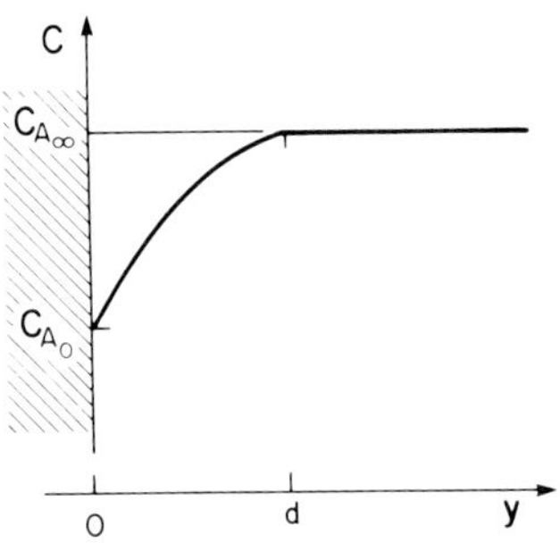

Fig.5.24. The stagnant-film model for mass transfer at a phase boundary

Closely related to the stagnant film concept is another approach in which the actual interfacial transport conditions are simply lumped into a fictitious resistance to mass transfer that is placed at the interface. The (well mixed) bulk is treated as if it had zero resistance. Then, in the spirit of a linear transport law (e.g. Ohm's law) one defines transport coefficients as the ratio of the flux (normal to the interface) to the driving concentration difference. This can be done in various ways depending on the concentration parameters used [5.63]. For instance for the *total* molar flux one may write

$$N_A = -K_c(C_{A\infty} - C_{A0}) \tag{5.144}$$

with K_c[cm/s] the mass transfer coefficient with respect to molar concentrations. It must be kept in mind, that this approach in defining mass transfer coefficients is not generally accepted. Numerous authors define K's with respect to the *diffusive* term in the total flux, i.e., as in [5.61] as

$$J_A = -K(C_{A\infty} - C_{A0}) \quad . \tag{5.145}$$

This approach has the "advantage" that the resulting mass transfer coefficient can readily be related to N_{Sh}. The potential fallacy of this choice, however, is that it lends itself to the confusion of J_A with N_A, i.e., interfacial mass transfer may inadvertently be treated as if there were no net mass flux ($v = 0$) normal to the interface. This case, however, is, at least in crystal growth, the exception rather than the rule, as emphasized before.

In the foregoing we have sketched the stagnant film model based on the generally valid flux formulation, i.e., in terms of the total flux, as sum of the diffusive and convective contribution. Many presentations in the literature, in particular in older works, ignore the convective part and define the steady state concentration profile in the film ($0 \leqslant y \leqslant \delta^*$) as

$$\frac{C_A - C_{A0}}{C_{A\infty} - C_{A0}} = \frac{y}{\delta^*} \quad . \tag{5.146}$$

thus *assuming* a linear concentration profile right at the start. The asterisk indicates that the effective film width δ^* refers to zero interfacial flow. From the above discussion we know, of course, that this approach can only be expected to give reasonably accurate results at low values for transfer rates and/or concentrations. For a detailed discussion of the necessary corrections to δ^* for significant interfacial flux see [5.64].

The reason for the slow acceptance of the appropriate flux formulation in interfacial mass transport is twofold. First, as we will discuss in Sect. 5.5.3, it has become customary to (often rather uncritically) deduce mass fluxes from solutions to similar heat transfer problems. (Note that, ignoring radiation, heat flow across an interface is truly diffusive – the *advective* term doesn't exist there.) The second reason is mostly traditional. In 1904 (the same year in which Prandtl published his boundary-layer simplification of the Navier-Stokes equations!) in an often referred to paper, NERNST and BRUNNER [5.45] investigated dissolution and growth rates for salts in aqueous solutions. They postulated a stagnant ("unstirred") layer of solution that "adheres to the solid". The width δ of that layer was thought to be governed by the forced flow conditions in the solution, with more vigorous stirring resulting in a narrower δ. Nernst and Brunner assumed (ignoring

Stefan's work) that the solute flux through their fictitious layer can be described with Fick's first law in the form $N_A = D_{AB}(C_\infty - C_0)/\delta$. Furthermore, they constructed the width δ by extending the tangent on the actual concentration profile at the interface, as indicated by the dashed line in Fig.5.25. It is true that this construction yields (by definition) the correct interfacial concentration gradient. The Nernst and Brunner supposition, however, that consequently Fick's first law would yield *in general* the correct interfacial flux was erroneous as a comparison with (5.112) shows. Of

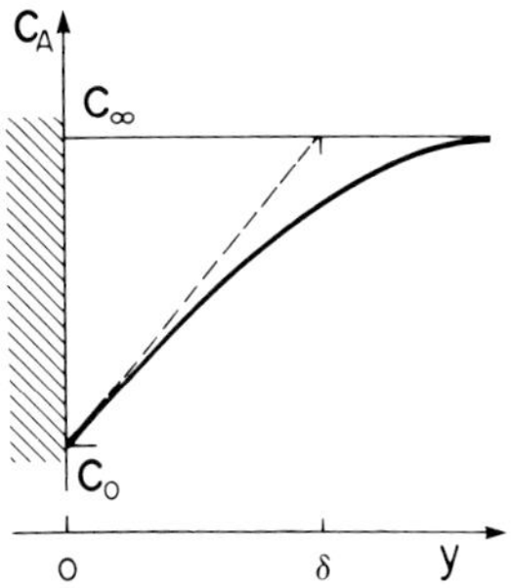

Fig.5.25. Actual solute concentration profile near crystal-nutrient interface. (δ: "unstirred film width", as introduced by NERNST and BRUNNER [5.45])

course, an "unstirred layer" model lacks physical justification. As we have discussed in Sect. 5.3.2, near solid-fluid interfaces, due to viscous shear forces the tangential velocity of a fluid relative to the solid *approaches* zero. True "no-slip" conditions, however, are only found in a layer of a few atoms width. However, unfortunately the fictitious nature of the (erroneous) Nernst-Brunner model and of the stagnant-film model are often forgotten. Even in a most recent crystal growth text one finds "a layer of solution which is considered stagnant because of adhesion to the crystal surface while the remainder of the solution is flowing past surface."

5.5 Non-Radiative Heat Transfer

The transfer of thermal energy within materials and experimental setups is as important for the understanding of crystal growth as the (coupled) transport of matter. Specific temperature ranges and profiles are chosen, or in other words, specific heat transfer configurations must be engineered in order to 1) control interphase mass transport rates through the temperature dependence of the chemical potentials; 2) dissipate or supply latent heat that is generated or consumed at interfaces; and 3) control transport properties such as diffusivities, viscosities, emissivities, densities. This separate listing of "reasons" is not to imply that these parameters can be controlled independently; they cannot. The list is to indicate the complexity of the heat transport topic in materials preparation. In the following we will deal with the various modes of heat transfer emphasizing the differences and analogies to the formal description of mass transport.

5.5.1 Basic equations for non-radiative heat transfer

As an elementary amount of fluid moves along its path, it may exchange thermal energy by conduction and radiation, become heated by work performed on it by viscous stresses, undergo compression in pressure gradients and change its potential energy due to displacement in the gravitational field, etc. Neglecting the contributions from radiation and external forces, an energy balance (i.e., the First Law of Thermodynamics) leads to the *energy (transport) equation* [5.21, 22]

$$\frac{\partial}{\partial t}\,\rho(u + 1/2\ v^2) = -\nabla\cdot\rho\underline{v}(u + 1/2\ v^2) + \nabla\cdot(k\nabla T) - p\nabla\cdot\underline{v} + \nabla\cdot[\underline{\underline{\tau}}\cdot\underline{v}]$$

$$\rho\,\frac{D}{Dt}\,(u + 1/2\ v^2) = \mathrm{div}(k\ \mathrm{grad}\ T) - p\ \mathrm{div}\ \underline{v} - \mathrm{div}[\underline{\underline{\tau}}\cdot\underline{v}] \qquad (5.147)$$

for the energy interchanges in a moving fluid as seen by a stationary observer and an observer moving with the fluid, respectively. Eq.(5.147) is physically identical with (5.3). The left side represents the rate of change in the sum of internal and kinetic energy of a stationary control volume and a moving unit mass, respectively. On the right side, in the heat conduction term, Fourier's law $\underline{q} = -k\nabla T$ is inserted where k is the thermal conductivity. The last two terms correspond to the $\nabla\cdot(\underline{\underline{\Pi}}\cdot\underline{v})$ term in (5.3), separating the contributions from compression work and frictional heating, with $\underline{\underline{\tau}}$ the stress tensor.

For a general solution with respect to the temperature field (5.147) must be supplemented with a thermal equation of state, e.g., $p/\rho = RT/M$; a caloric equation of state, $u = u(T,\rho)$; empirical relations for $\rho(T)$, $k(T)$; and it must be simultaneously solved with the Navier-Stokes and continuity equations, unless a constant property fluid is assumed.

Fortunately in most practical cases it is possible to considerably simplify the general energy equation. For instance, for "incompressible" fluids with div $\underline{v} = 0$, (5.74), obviously the compression-work term vanishes. However, having in mind the application of (5.147) to buoyancy-driven convective phenomena, where $\nabla\rho \neq 0$ (in connection with gravity) supplies the motional drive, it is comforting to know that ρ = const. is only a sufficient but not a necessary condition for the neglection of that term. In fact, as a general similarity consideration shows [5.21], *both* terms that account for the conversion of mechanical into thermal energy are relatively insignificant if

1) the free-stream velocity of a given fluid is much smaller than its velocity of sound, and
2) pressure transients, $\partial p/\partial t$ can be neglected.

Hence the following simplified steady-state form is by no means restricted to liquids, where $\rho \simeq$ const. is usually well met. It holds also for gaseous flows under most practical crystal growth conditions. It is useful to know that, for instance, in stagnation flow, even with free-stream velocities as high as 50 m/s typical gases heat up only a few °C.

Rewriting (5.147) under the above assumptions, i.e., neglecting the velocity-dependent last two terms and the kinetic energy term, for a constant property fluid one obtains the *simplified energy equation*

$$\rho C_p\left[\frac{\partial T}{\partial t} + (\underline{v}\cdot\mathrm{grad})T\right] = k\nabla^2 T \qquad (5.148)$$

with C_p[cal g^{-1} deg^{-1}] the specific heat at constant pressure or, collecting the property coefficients

$$\frac{\partial T}{\partial t} + (\underline{v}\cdot\mathrm{grad})T = \kappa\nabla^2 T \qquad (5.149)$$

with $k = \kappa/\rho C_p$[cm^2 s^{-1}] the *thermal diffusivity*. Of course, for multicomponent fluids, C_p and k must be appropriately averaged. For the special case that $\underline{v} \to 0$ one obtains the *heat conduction equation*

$$\frac{\partial T}{\partial t} = \kappa\nabla^2 T \quad . \qquad (5.150)$$

For similarity considerations it is important to know which dimensionless group governs the solutions of the energy equation. Hence, as exercised earlier on several occasions we refer lengths and velocities to L and V [see, for instance, (5.100)] and introduce a dimensionless ratio for the density

$\bar{\rho} = \rho/\rho_\infty$ by reference to a bulk fluid density ρ_∞. Thus, one obtains the *dimensionless energy equation*

$$\bar{\rho}[\frac{\partial T}{\partial t} + (\underline{W}\ \mathrm{grad})T] = \frac{k}{\rho_\infty C_p VL} \nabla^2 T \quad . \tag{5.151}$$

The dimensionless group which characterizes the relative weight of conductive and convective heat flux

$$N_{Pe}^T = \frac{\rho_\infty C_p VL}{k} = \frac{VL}{\kappa} \tag{5.152}$$

is known as the *thermal Peclet number*. In analogy to the solutal N_{Pe}, (5.130), which according to (5.132) can be written as product of N_{Re} and N_{Sc}, one can express N_{Pe}^T as

$$N_{Pe}^T = \frac{VL}{\nu} \cdot \frac{\nu}{\kappa} = N_{Re} \cdot N_{Pr} \tag{5.153}$$

with the *Prandtl number* N_{Pr}.

The remaining development with respect to heat transfer at fluid-solid interfaces, the formulation of a thermal boundary layer equation and the characterization of a thermal boundary layer width δ_T are strictly analogous to the corresponding mass transfer considerations in Sect. 5.4. Therefore, we confine their presentation to Table 5.15. Correlations of distinct boundary layer behaviors to groups of fluids will be made in Sect. 5.5.4 after a survey of typical values for the governing fluid properties.

For comparisons of interfacial heat transfer data and correlations with mass transfer it is useful to introduce the thermal analogue to the Sherwood number (5.141), the dimensionless temperature gradient

$$N_{Nu} \equiv -L \left[\frac{\partial}{\partial y} \left(\frac{T - T_\infty}{T_0 - T_\infty} \right) \right]_{y=0} = f\left(N_{Re}, \frac{v_0}{V}, N_{Pr} \right) \tag{5.154}$$

known as the *Nusselt number*. L is again a characteristic length; T_0 and T_∞ are the interfacial and bulk fluid temperature, respectively; v_0 is the mass average velocity normal to the interface (zero for zero mass flow through the phase boundary) and V is a characteristic free-stream velocity outside the momentum boundary layer. The actual temperature gradient at y=0 normal to the

Table 5.15. Analogy between momentum, mass, and non-radiative heat transfer

	Momentum Transfer	Mass Transfer	Heat Transfer
Steady-state transport equation for constant property fluid	$(\underline{v}\ \mathrm{grad})\underline{v} = -\nabla p + \eta\nabla^2\underline{v} + \underline{f}$	$(\underline{v}\ \mathrm{grad})W_A = \mathcal{D}_{AB}\nabla^2 W_A$	$(\underline{v}\ \mathrm{grad})T = \frac{k}{\rho C_p}\nabla^2 T$
Restrictions	$\rho = \mathrm{const} \rightarrow \mathrm{div}\ \underline{v} = 0$	$\rho = \mathrm{const}$	$\lvert\underline{v}\rvert \ll v_{sound}$
Dimensionless transport equation	$(\underline{W}\ \mathrm{grad})\underline{W} = -\frac{p_0}{\rho V^2}\nabla p + \frac{1}{N_{Re}}\nabla^2\underline{W}$	$(\underline{W}\ \mathrm{grad})W_A = \frac{1}{N_{Pe}}\nabla^2 W_A$	$\bar{\rho}(\underline{W}\ \mathrm{grad})\Theta = \frac{1}{N_{Pe}^T}\nabla^2\Theta$
Dimensionless groups	$N_{Re} = \frac{VL}{\nu}\ , \quad \nu = \frac{\eta}{\rho}$	$N_{Pe} = \frac{VL}{\mathcal{D}_{AB}}$	$N_{Pe}^T = \frac{VL}{\kappa}\ , \quad \kappa = \frac{k}{\rho C_p}$
		$= \frac{\nu}{\mathcal{D}_{AB}} \cdot \frac{VL}{\nu}$	$= \frac{\nu}{\kappa} \cdot \frac{VL}{\nu}$
		$= N_{Sc} \cdot N_{Re}$	$= N_{Pr} \cdot N_{Re}$
Boundary layer widths	$\delta_0 \propto (\nu)^{1/2}$	$\delta = C\left(\frac{N_{Re}}{N_{Pe}}\right)^n \delta_0$	$\delta_T = C\left(\frac{N_{Re}}{N_{Pe}^T}\right)^n \delta_0$
		$\delta \simeq N_{Sc}^{-1/3}\delta_0$	$\delta_T \simeq N_{Pr}^{-1/3}\delta_0$
		$\delta \simeq \left(\frac{N_{Pr}}{N_{Sc}}\right)^{1/3}\delta_T =$	$\left(\frac{\mathcal{D}_{AB}}{\kappa}\right)^{1/3}\delta_T$
Boundary layer equations	$u\frac{\partial u}{\partial x} + v\frac{\partial u}{\partial y} = \frac{1}{\rho}\frac{dp}{dx} + \nu\frac{\partial^2 u}{\partial y^2}$	$u\frac{\partial W_A}{\partial x} + v\frac{\partial W_A}{\partial y} = \mathcal{D}_{AB}\frac{\partial^2 W_A}{\partial y^2}$	$u\frac{\partial T}{\partial x} + v\frac{\partial T}{\partial y} = \kappa\frac{\partial^2 T}{\partial y^2}$

interface is then

$$\left(\frac{\partial T}{\partial y} \right)_{y=0} = \frac{T_\infty - T_0}{L} N_{Nu} \quad . \tag{5.155}$$

Now, interfacial heat transfer at zero or very low mass transfer across the solid-fluid boundary is exclusively or dominantly conductive (diffusive). Here, in contrast to mass transport, we don't have to be concerned about a diffusion-induced advective term. Consequently, with $v_0 = 0$ the temperature profile in a (stagnant) fluid adjacent to the interface is linear, $T = T_\infty + (T_0 - T_\infty)(y/L)$, which, upon insertion into (5.154), leads to a Nusselt number equal to unity. Hence, in general, $N_{Nu} > 1$ indicates in addition to conduction a convective heat flux contribution. In other words, for zero mass transfer across an interface N_{Nu} can be viewed as the ratio of total to conductive interfacial heat flux.

Employing (5.155) the interfacial heat flux is

$$q_0 = -k\left(\frac{\partial T}{\partial y} \right)_{y=0} = -k \frac{T_\infty - T_0}{L} N_{Nu} \quad . \tag{5.156}$$

Introducing a *heat transfer coefficient* h by

$$q_0 = -h(T_\infty - T_0) \tag{5.157}$$

we obtain for this case without interfacial mass flux the relation

$$N_{Nu} = \frac{hL}{k} \tag{5.158}$$

which is often given as the "definition" of the Nusselt number.

5.5.2 Coupled interfacial heat and mass transfer

Simultaneous interfacial mass transfer introduces an advective term into (5.156), as we have seen, e.g., in (5.149). As depicted in Fig.5.26a, for a rotating disk and a fluid with $N_{Pr} = N_{Sc} = 0.7$, the resulting deformation of the steady-state temperature profile depends strongly on the direction of the interfacial mass flux. The action of "blowing" (crystal dissolution) is to fill the space immediately adjacent to the disk with fluid having nearly

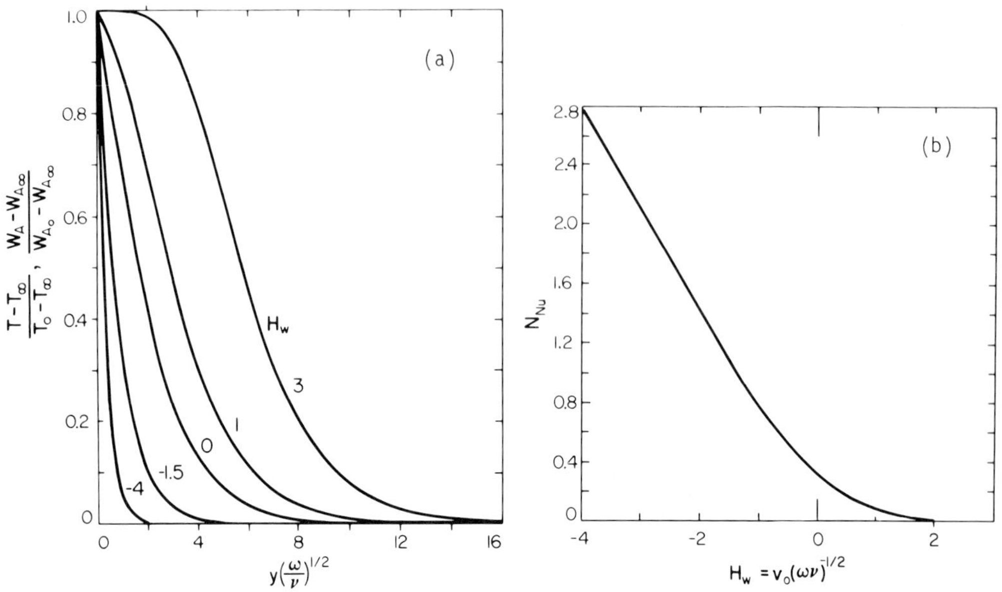

Fig.5.26. Heat and mass transfer about a rotating disk for $N_{Pr} = N_{Sc} = 0.7$. (a) Dimensionless temperature and weight fraction profiles for various dimensionless interfacial mass average velocities: $H_W = v_0(\omega\nu)^{-1/2}$. (b) Nusselt number, (5.154), with $L = (\nu/\omega)^{1/2}$, versus H_W. From [5.44] by permission of ASME

the same temperature as that of the disk and hence to reduce the interfacial temperature gradient. "Suction" (crystal growth), on the other hand, brings ambient fluid closer to the interface thereby steepening the gradient.

Let us emphasize that the change in N_{Nu} with interfacial mass flux (Fig. 5.26b) characterizes the change in the *conductive* interfacial heat flux (Problem 5.13). In order to obtain the *total* interfacial heat flux, the convective flux $C_pT_0v_0$ must be added to q_0(Nu) of (5.156), where v_0 is again the mass average velocity normal to the interface. Of course, if one uses (5.158) rather than (5.154) as the definition of N_{Nu}, the Nusselt number obtained (e.g., from a measured h) does not correspond to the actual interfacial temperature gradient. The definition of N_{Nu} by (5.158) must be seen as a remnant of the stagnant-film model in which the total heat flux is envisioned to be purely conductive, driven by a fictitiously increased or reduced temperature gradient for suction or blowing, respectively.

The magnitude of the convective flux contribution can be expected to scale with the specific heat content of the substance transported. This is confirmed

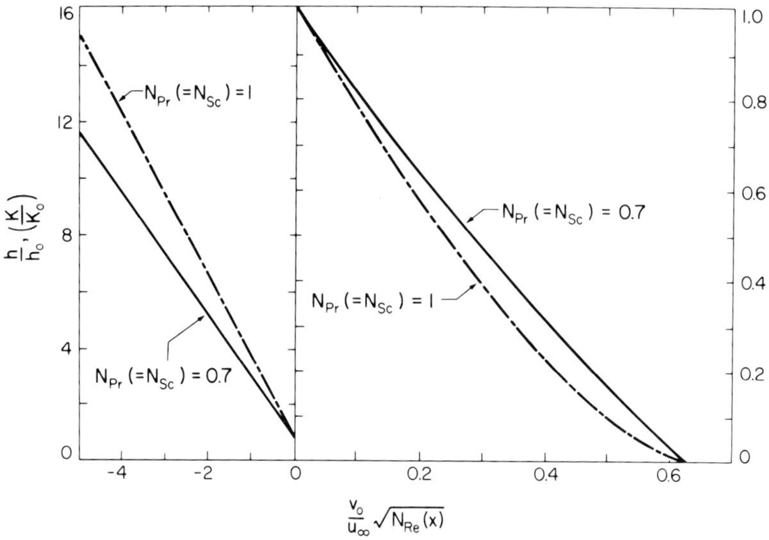

Fig.5.27. Heat and mass transfer parameters for laminar flow over a flat plate. The subscript "0" indicates transfer coefficients with vanishing net interfacial mass flux. N_{Pr} belongs to h and N_{Sc} to K. u_∞ is the free-stream velocity (parallel to plate) and $N_{Re}(x)$ the Reynolds number based upon the distance from the leading edge, x. Adapted from [5.61]

by Fig.5.27 where one sees that the ratio of the heat transfer coefficients with and without interfacial mass flux, h/h_0, increases with N_{Pr}.

We have seen that, in principle, crystal growth and melting increase and decrease, respectively the interfacial temperature gradient that exists under no-growth conditions. However, as Problem 5.12 shows, this effect is rather small even at growth rates of several cm/hour. Furthermore, in reality this effect will be buried under the deformation of the temperature profile from the latent heat delivered or consumed at the interface, a parameter which we have ignored at this point. The above material is presented here mostly to supply a sound basis for deductions of interfacial mass fluxes from solutions to truly analogous heat transfer configurations.

5.5.3 Correlations of solutions to mass and heat transfer problems

As the reader will have noticed, there is a full correspondence between the simplified energy transport equation (5.149) and the mass transport equation *for systems with constant* ρ, (5.120). Direct transformation is achieved by exchanging T and κ with W_A and D_{AB}, respectively. This analogy between convective-diffusive mass and heat transfer is, of course, also observed in situations with vanishing bulk flow, as is reflected in the mathematical equivalence of Fick's second law (5.128) and the heat conduction equation (5.150). This mathematical correspondence has led to the custom of deducing solutions to mass transfer problems from solutions to "analogous" heat transfer configurations found in the extensive heat transfer literature (e.g., [5.47-50]).

This can be a very efficient approach. Yet in drawing such analogies one must not lose sight of the fundamental difference between diffusive heat and mass transfer. As heat diffuses, neglecting self diffusion, the molecules remain "around" their time-averaged coordinates. Hence the mass-connected reference frame, with respect to which the conductive heat flux is expressed, remains stationary, unless, of course, a convective motion is imposed by external means. As mass diffuses, however, under most circumstances the (same) mass-centered reference frame with respect to which the diffusive mass flux is expressed is moving itself merely due to the diffusive mass flux. This diffusion-induced advective flux is discussed at length in Sects. 5.2.2 and 5.4.

Consequently, a correlation between a merely conductive heat transfer system and a geometrically similar mass transport configuration will only be reasonably accurate if the mass transfer rates are so low that the advective reference frame velocity is insignificant. With higher concentrations and/or transfer rates for the species that, for instance, cross an interface, predictions of the mass flux from heat-transfer solutions for $v_0 = 0$ become increasingly erroneous [see also (6.30)]. Full correspondence can only be expected if N_{Pe} and N_{Pe}^T (i.e., N_{Sc} and N_{Pr} if N_{Re} is the same) *and* V_0 have the same value. This is illustrated in Fig.5.27 for the heat and mass fluxes onto a flat plate from a (mainly) parallel flowing fluid. Besides the geometric and dynamic similarity (contained in the factor $\sqrt{N_{Re}(x)}/u_\infty$ that governs the boundary layer widths) also v_0 must be similar, i.e., the product $(v_0/u_\infty)\sqrt{N_{Re}(x)}$ must be the same, to obtain the same heat and mass transport coefficients upon exchange of N_{Sc} with N_{Pr}.

It must also be realized that (5.120), strictly speaking, applies only to flows in which the mass density does not change along a streamline. Thus rigorous analogies between mass and heat transfer along the traditional line sketched above are restricted to cases where the fluid components have comparable mass density, such as mixtures of gases of similar molecular weight. For the description of mass transport in systems with uniform total molar concentration (e.g., gas mixtures in general) one may be tempted to exploit the mathematical similarity between (5.124) and (5.120). Note, however, that $\underline{v}^*$ is used as bulk flow velocity in (5.124) in contrast to the $\underline{v}$ in (5.120), and these average velocities are only equal if all components possess the same molecular weihgt, i.e., (with C = const.) if ρ = const.

5.5.4 Thermal diffusivities and boundary layers

Thermal diffusivities can be calculated from thermal conductivity, specific heat and density data according to $\kappa = k/C_p\rho$. Theoretical concepts, experimental methods and data for thermal conductivities of gases and liquids have been reviewed by TOULOUKIAN et al. [5.51], BRETSZNAJDER [5.15], TSEDERBERG [5.52] and REID and SHERWOOD [5.27]. Estimates of k in gases at high temperatures were tabulated by SVEHLA [5.26].

For gases the kinetic theory shows that at moderate pressures the thermal conductivity k is related to the viscosity by [5.15]

$$k = KC_v\eta \tag{5.159}$$

where K, Maxwell's modulus, is almost constant over a fairly wide range of temperatures and pressures, but it depends on the number of atoms in the gas molecule. For monoatomic gases K is found to be around 2.5. For polyatomic gases K is given by

$$K = \frac{9(n + 2) - 5n}{4n} \quad , \tag{5.160}$$

where n is the number of degrees of freedom of the considered gas molecules. When strongly polar molecules are considered, (5.159) and (5.160) may lead to errors of up to 15%.

The thermal conductivities of liquids are much greater than those of gases. Analogous to the temperature dependence of viscosities, the thermal conductivity in gases and liquids increase and decrease, respectively, with temperature. As pointed out recently [5.42] most data for thermal conductivities of liquids lie within the range $(0.5 \pm 0.2)k^S$ and $(0.6 \pm 0.16)k^S$ for covalent and metallic liquids, respectively, with k^S being the thermal conductivity of the cor-

Table 5.16. Thermal diffusivity data and Prandtl numbers for crystal growth fluids

Fluid	T_m[°C]	T_κ[°C]	$\kappa[\frac{cm^2}{s}]$	$N_{Pr} = \frac{\nu}{\kappa}$	References for κ"
H_2	-259	700	10.8	0.68	
		1000	16.9	0.68	
		1300	24.2	0.69	[5.51, 53] (calcul. from η and C_p)
O_2	-218	700	1.5	0.73	
		1000	2.36	0.76	
Air		50	0.26	0.71	[5.21]
H_2O	0	0	$\approx 10^{-3}$	≈15	[5.54]
Hg	- 38.4	20	5×10^{-2}	2×10^{-2}	[5.21]
Ag	960	1000	7×10^{-2}	5×10^{-2}	
Zn	420	425	0.2	2.2×10^{-2}	
Ga	30	31	0.15	2.2×10^{-2}	
Ge	937	940	4×10^{-2}	3.2×10^{-2}	
Si	1410	1420	5×10^{-2}	5.4×10^{-2}	
NaCl	801	801	1.3×10^{-2}	0.56	[5.54]
$MgAl_2O_4$	2135	2150	$\approx 5 \times 10^{-3}$	≈8	
UO_2	2500	2600	$\approx 4 \times 10^{-3}$	≈12	
Al_2O_3	2045	2050	$\approx 10^{-2}$	≈40	
Te	452	452	7×10^{-3}	7	
Se	220	220	10^{-2}	63	

responding solid near the melting point. Specific heat data are compiled in TOULOUKIAN et al. [5.53]. BRICE [5.42] has put together a particularly comprehensive and informative section on the estimation of specific heats and heats of transitions.

Thermal diffusivities can also be directly determined, based upon the heat conduction equation, (5.150) in transient determinations of properties the temperature dependence of which is clearly understood. TOULOUKIAN et al. [5.54] have compiled directly determined thermal diffusivities and reviewed the experimental methods.

Table 5.16 lists values for κ of various fluids frequently used in crystal growth and in basic mass and heat transfer studies. The data for the liquids were chosen for temperatures close to T_m. Employing the corresponding kinematic viscosity data in Table 5.12 we have also calculated the Prandtl numbers for the various fluids. One finds that $N_{Pr} = 0.7$ is rather representative for gases over a wide range of temperatures. Molten metals and elemental semiconductors, due to their high thermal conductivities lie typically between $N_{Pr} = (1 \text{ to } 5)10^{-2}$. In contrast, oxide melts, due to their high viscosities, have $N_{Pr} = 5$ to 50. From Table 5.15 one sees that, in strict analogy to $\delta \approx N_{Sc}^{-1/3}\delta_0$, the *thermal* and momentum *boundary layer* widths are related through

$$\delta_T \approx N_{Pr}^{-1/3}\delta_0 = \left(\frac{\kappa}{\nu}\right)^{1/3}\delta_0 \quad . \tag{5.161}$$

Consequently, with respect to characteristically different thermal boundary layer extent one cannot simply distinguish between liquids and vapors as is possible concerning mass transfer boundary layers. As schematically indicated in Fig.5.28, in liquid metals the thermal boundary usually extends three times further into the fluid than the velocity transition region. However, in molten oxides and aqueous solutions δ_0 is somewhat wider than δ_T. In gases thermal and momentum boundary layers are of comparable width.

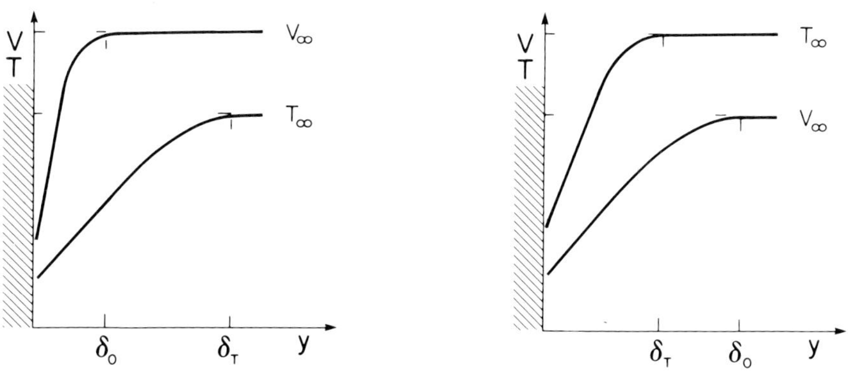

(a) $N_{Pr} < 1$ (liquid metals, semiconductors)

(b) $N_{Pr} \gtrsim 1$ (aqueous solutions, molten oxides)

Fig.5.28. Comparison between velocity and temperature distribution at solid-fluid interfaces for different fluid groups

At last we are in a position to relate the widths of concentration and temperature boundary layers for a given situation in terms of a characteristic dimensionless group. Such a relation will prove particularly useful in considerations of morphological stability [1.1]. It also allows for estimates of the concentration gradient at an interface from the often much easier to measure temperature profile. From a comparison of the equations $\delta = \delta(\delta_0)$ and $\delta_T = \delta_T(\delta_0)$ in Table 5.15 one obtains

$$\delta_T \simeq \left(\frac{N_{Sc}}{N_{Pr}}\right)^{1/3} \delta = \left(\frac{\kappa}{D_{AB}}\right)^{1/3} \delta \quad , \qquad (5.162)$$

where the ratio N_{Sc}/N_{Pr} is sometimes referred to as *Lewis number* N_{Le}. Here we can consider gases and liquids again as dinstinctly differently behaving groups of fluids. We see that for gases, where representative values for κ and D_{AB} are both around 0.7 cm^2/s, $\delta_T \simeq \delta$. This is in excellent accord with the prediction of kinetic gas theory which yields the same value for the self-diffusion coefficient and the thermal diffusivity of an ideal gas. The thermal boundary layer in liquids, however, will always be wider than the concentration boundary layer. In molten oxides their ratio is smaller than in liquid metals where, due to the high k, δ_T may extend ten times as far into the fluid as δ. This general distinction between heat transfer in gases and liquids is intuitively obvious. In liquids, heat can be transferred by collision without large displacement of the molecules, whereas in gases (neglecting radiation) a molecule has to travel through the free mean path distance before it can pass on thermal energy.

5.6 Radiative Heat Transfer [5.55-59]

Radiative heat transfer differs in two important aspects from conduction and convection of heat. Firstly, no medium is required for the transport. Secondly, the rate of heat dissipation by radiation varies approximately as the fourth power of the absolute temperature as compared to the first power dependence approximated by the other two modes. For these reasons radiation becomes the dominant heat transfer mode at high temperatures and/or in the absence of an atmosphere.

In general, the (optical) geometries employed in materials preparation are rather complex. In addition, the thermal radiative properties of materials, in particular at the high temperatures and under the (impure) surface conditions encountered in real laboratory situations, are often poorly characterized. As a consequence of these typically difficult to quantify conditions radiation is widely neglected in crystal growth discussions. But even (by necessity) semi-quantitative considerations of radiative transfer can prove very valuable in the crystal growth practice as we shall experience in later sections. Hence, we will review here the fundamental laws and definitions associated with radiative energy exchange between surfaces and radiation transfer through participating, i.e., (weakly) absorbing media.

All matter is continually emitting *thermal radiant energy* as a result of the thermal motion of its particles (electrons, ions, atoms and molecules). Consequently all matter is also continuously being bombarded by radiant energy from its surroundings. Hence, already for mere geometrical reasons the analysis of radiative transfer is considerably more intricate than, e.g., that of conduction. The propagation of radiation at any point can typically not be represented by a single vector as in the case of heat flow by conduction. To specify the radiation, for instance, incident at a given point, it is necessary to know the radiation from all directions. This leads to rather complex formulations of the radiation exchange relations in general. Hence, first we will introduce the basic parameters with a minimum of geometrical considerations and be concerned with geometrical refinements in later parts of Sect. 5.6.1. Actual "view factors" are discussed in Sect. 5.6.2 together with general aspects of shielding and temperature measurements in radiant heat transfer. In Sect. 5.6.3 a radiation term is added to the energy transport equation (5.148).

5.6.1 Fundamental concepts

Radiant energy can be alternatively envisioned as being transported by either electromagnetic waves or by photons. This dualism is necessary since neither model completely describes the nature of all radiation phenomena. The radiation properties of certain classes of surfaces and materials may be predicted based merely upon electromagnetic theory. The domain of quantum mechanics lies more with the description of spectral properties of matter.

Whether considered as electromagnetic wave or photon flow, radiant energy travels within homogeneous media in straight lines. The frequency ν is constant regardless of the medium through which the radiation is propagated. The speed of propagation c and the wavelength λ, related to c via

$$c = \lambda \nu \ , \qquad (5.163)$$

change with the *index of refraction n* of the medium according to

$$c = c_0/n \quad \text{and} \quad \lambda = \lambda_0/n \qquad (5.164)$$

where the speed of light in cacuum, $c_0 = 2.9979 \times 10^{10}$[cm/s] is a fundamental constant of nature and λ_0 is the wavelength in vacuum, which according to (5.163) is $\lambda_0 = c_0/\nu$. Note that n is, in general, a function of frequency (see Fig.5.29) and temperature.

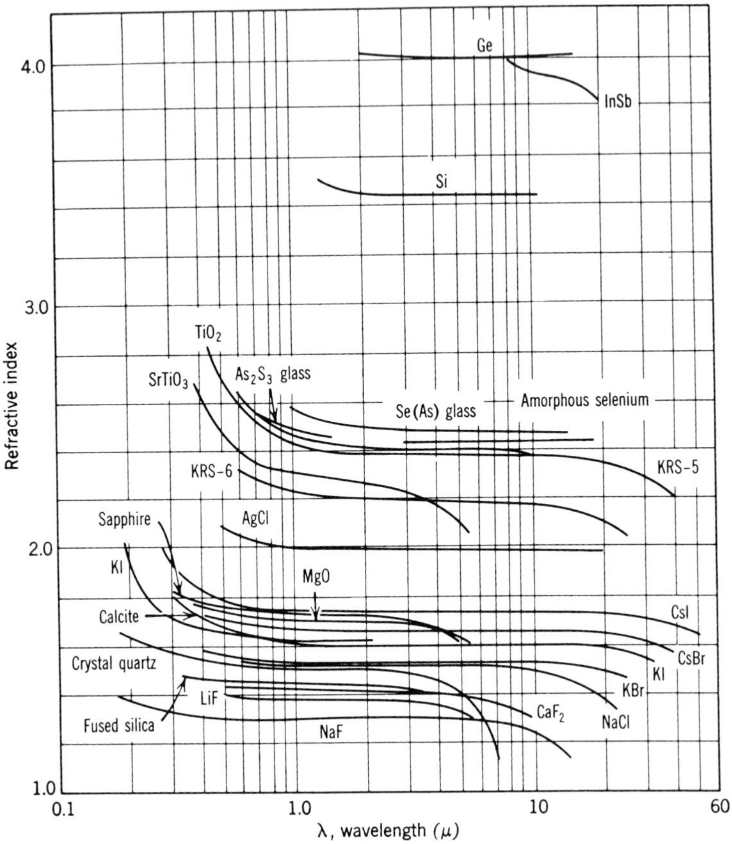

Fig.5.29. Refractive indices of various transparent materials as function of wavelength. From [5.70] by permission of McGraw-Hill

The wavelength range of technologically utilized thermal radiation lies approximately between 0.3 μm and 50 μm (where 1 μm = 10^{-4}cm) or in other words, it ranges from the "near ultraviolet" to the "far infrared".

On propagation through the interface between two media of different n, a beam of radiant energy is refracted (deviated) toward the normal to the interface in the medium of higher index of refraction, as indicated in Fig. 5.30. The angles β_1 and β_2 of the incident and refracted beam, respectively

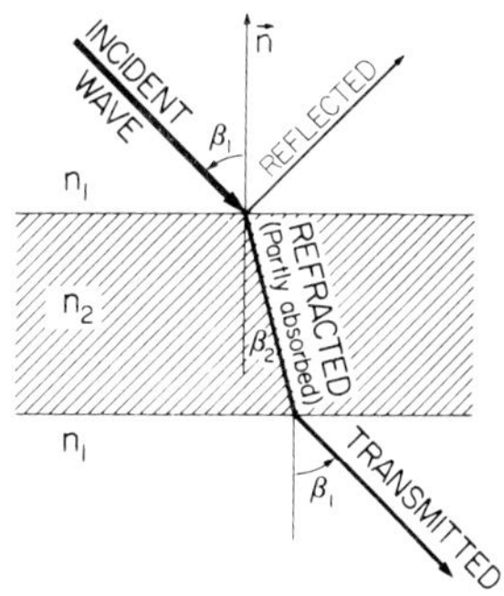

Fig.5.30. Definition sketch for optical parameters

with the normal obey the relation

$$\frac{\sin\beta_2}{\sin\beta_1} = \frac{n_1}{n_2} \tag{5.165}$$

known as *Snell's law*. (See Problem 5.17).

Radiant energy incident on an optically flat and homogeneous medium, as schematically shown in Fig.5.30, is partly *reflected*, partly *absorbed* and the rest is *transmitted*. In reality, with media containing inhomogeneities and microscopic foreign particles, *scattering* (changes in the direction and energy of photons due to collisions) can also play an important (although often difficult to quantify) role in radiation energy transfer. In particular in crystal growth, contributions to "reflection" and "absorption" due to scattering on rough and/or contaminated surfaces or occlusions and grain-boundaries in bulk materials must often be considered beyond the optical behavior that would result from the idealized (and tabulated) parameters for homogeneous materials and optically flat surfaces.

Denoting the thermal radiant energy by Q, one defines the *radiant* "flux" ϕ as

$$\phi = \frac{dQ}{dt} \quad . \tag{5.166}$$

Commonly used units for this time rate of flow of radiant energy are [erg/s] and [Watt]. Note that ϕ, in contrast to the transport fluxes used earlier, is *not* expressed in reference to a unit area.

Neglecting scattering at this point one can draw an energy balance and write the incident flux as the sum of the reflected, absorbed and transmitted fluxes,

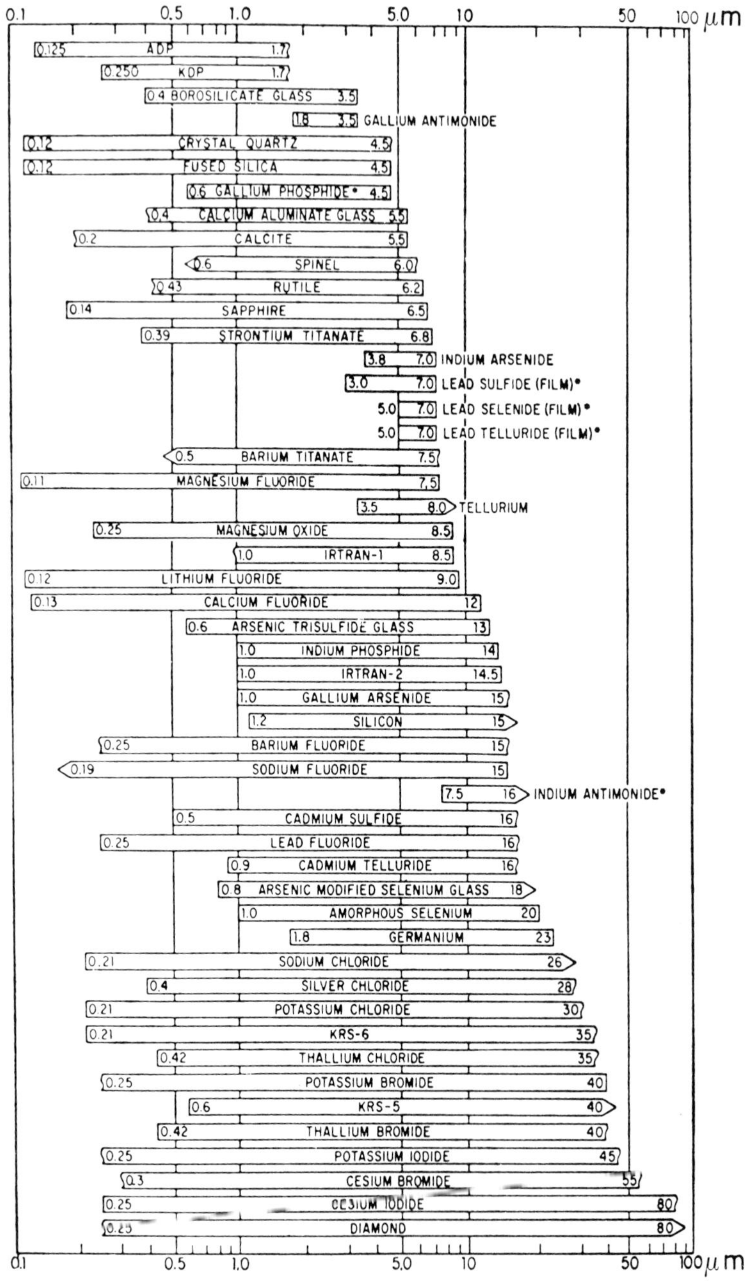

Fig.5.30a. Transmission regions. The limiting wavelengths for both long and short cutoffs have been chosen as the wavelengths at which a sample 2 mm thick has a transmissivity of 0.1. Materials marked with an asteriks have a maximum τ < 0.1. From [5.70] by permission of McGraw-Hill

$$\phi_i = \phi_r + \phi_a + \phi_t \quad . \tag{5.167}$$

Defining the *reflectivity* ρ as the ratio of reflected to incident flux ϕ_r/ϕ_i, the *absorptivity* α as ϕ_a/ϕ_i and the *transmittivity* τ as ϕ_t/ϕ_i, (5.167) can be rewritten as

$$\rho + \alpha + \tau = 1 \quad . \tag{5.168}$$

The transmission ranges of various crystalline materials and glasses are given in Fig.5.30a. For *opaque* materials $\tau = 0$ and thus $\rho + \alpha = 1$, i.e., of the three radiation properties only is independent. Radiation incident on the the three radiation properties only one is independent. Radiation incident on the surface of an opaque material never penetrates into greater depths. Hence, for opaque materials, the absorption, reflection and emission (see below) or radiation are surface phenomena and, as such, depend sensitively on the chemical and structural surface conditions. Systems in which, in addition to $\tau = 0$, $\rho = 0$ and hence $\alpha = 1$, are called *black bodies*.

It should be noted that the radiative heat transfer literature contains a great variety of inconsistent systems of definitions, notations and units used. A number of suggestions have been made to standardize the nomenclature of radiation [5.55]. Encouraged by the National Bureau of Standards, some authors use the ending "*ivity*" (i.e., reflectivity, absorptivity, etc.) to characterize the radiation properties of ideal surfaces (i.e., those optically smooth and perfectly clean). The ending "*ance*" (e.g., reflectance) is then reserved for real surfaces, i.e., for measured properties where there is a need to specify surface conditions. We will not use this (experimentally rather difficult to achieve) distinction. We will follow the custom of using the "*-ivity*" ending for intensive properties (such as electrical resistivity and thermal conductivity) and the "-ance" ending for extensive properties as in resistance and conductance.

Before we continue discussing relationships between optical quantities, we need to consider the emissive properties of a system and also concern ourselves more specifically with spectral parameters. Using the Second Law of thermodynamics and quantum-statistical considerations PLANCK [5.60] has shown that there is a maximum amount of radiant energy that can be emitted by a finite system at a given temperature and wavelength. It turns out that only a black body (which was defined above by the condition $\alpha = 1$, i.e., as absorbing radiation quantitatively) emits such maximum energy throughout the whole spectral range. All real surfaces emit less radiation than a black-body radiator at the same temperature and, at the same time, reflect part of the energy incident upon them. Although highly idealized, the black body concept is very useful as a "reference state" in radiation transport discussions. Experimentally one can approximate black body behavior very well with a very small hole in an otherwise enclosed isothermal cavity *of any material*.

The energy emitted into the whole hemispherical space by a black body (surface) per unit of time and per unit area in a frequency range $d\nu$ is $e_{\nu b}d\nu$ with

$$e_{\nu b}(T) = \frac{2\pi h\nu^3 n^2}{c_0^2[\exp(h\nu/kT) - 1]} , \tag{5.169}$$

known as *Planck's equation*. Values for Boltzmann's constant k and Planck's constant h are given in Table 5.17. It is apparent from (5.169) that the radiation flux emitted by a black body at temperature T into a medium having a real refractive index n is n^2 times the flux emitted at the same temperature into vacuum, where by definition n = 1, or into gases where $n \approx 1$.

Table 5.17. Black-body radiation constants

Boltzmann's constant: $k = 1.380 \times 10^{-16}$ erg K^{-1} = 1.380×10^{-23} Watt s K^{-1}
Planck's constant: $h = 6.625 \times 10^{-27}$ erg s = 6.625×10^{-34} Watt s^2
C_1 in (5.171): $C_1 = 3.741 \times 10^{-5}$ erg cm^2 s^{-1} = 3.741×10^{-16} Watt m^2
C_2 in (5.171): C_2 = 1.439 cmK = 1.439×10^{-2} mK
Stefan-Boltzmann constant: $\sigma = 5.669 \times 10^{-5}$ erg s^{-1} $cm^{-2}K^{-4}$
= 5.669×10^{-8} Watt m^2K^{-4}

If n is constant in the frequency range considered, it is often useful to recast (5.169) in terms of wavelength. Employing the relations $e_{\lambda b}d\lambda = -e_{b\nu}d\nu$, (5.163) and (5.164) one obtains (Problem 5.14) for the spectral distribution of the hemispherical black-body emission

$$e_{\lambda b}(T) = \frac{2\pi hc_0^2}{n^2\lambda^5[\exp(hc_0/n\lambda kT) - 1]} \tag{5.170}$$

or, collecting the constant terms

$$e_{\lambda b}(T) = \frac{C_1}{n^2\lambda^5[\exp(C_2/n\lambda T) - 1]} \tag{5.171}$$

with C_1 and C_2 as given in Table 5.17.

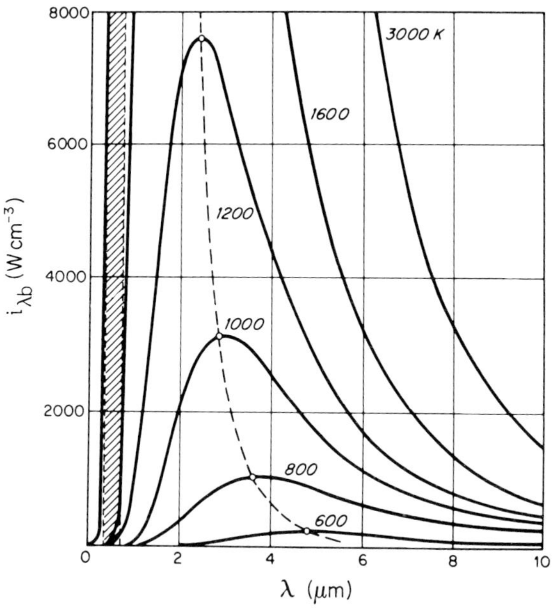

Fig.5.31. Spectral intensity distribution ("monochromatic intensity") of the radiation emitted by a black body into vacuum (n = 1) for a number of temperatures. From [5.61] by permission of McGraw-Hill

Figure 5.31 represents curves for the spectral distribution of black-body radiation as a function of wavelength for a number of temperatures. The dashed area indicates the visible part of the spectrum. One recognizes that for temperatures with which one is generally concerned in materials preparation most of the emitted radiation lies in the infrared, i.e., at wavelengths larger than the visible ones. The human eye, however, is so sensitive that we can readily see a surface at 500°C (i.e., the eye responds to the minute energy flux from the short-wavelength tail of the distribution) as dull red in an otherwise dark environment. Even at 5600 K where the black-body wavelength distribution approximates the spectral distribution coming from the sun, only about one-third of the total energy radiated is in the visible range. For numerical evaluations of Fig.5.31 note that the intensity (defined below) rather than the hemispherical spectral energy employed in (5.169-171) is plotted versus λ.

Rewriting and differentiating (5.170) one can show (Problem 5.15) that the maximum of the emitted spectral power $e_{\lambda b,max}(T)$ lies at a wavelength λ_{max} which is related to T by *Wien's displacement law*

$$(n\lambda T)_{max} = 2898[\mu m\ K] \quad . \tag{5.172}$$

This shift of the peak emissive power with T is also indicated by the dashed line in Fig.5.31.

The total hemispherical emissive power of a black body, $e_b(T)$, i.e., the energy emitted per unit time and area into an adjacent medium over all frequencies (or wavelengths), can be obtained from (5.169) by the integration

$$e_b(T) = \int_0^\infty e_{\nu b} d\nu = n^2 \sigma T^4 \quad , \tag{5.173}$$

known as the *Stefan-Boltzmann equation* with σ the Stefan-Boltzmann constant, the value of which is given in Table 5.17. Note that "total" is used to characterize the summation over all ν's or λ's whereas hemispherical denotes the summation over all directions into the half-space. It is this T^4-dependence of the emitted energy that makes radiation the dominant heat transfer mode at elevated temperatures. Eq.(5.173) has considerable practical consequences which are discussed in various later sections.

The spectral and total emissive power of a real (non-black) surface is advantageously expressed in terms of the black-body emissive power. One defines the total and spectral *emissivity* of a real surface as

$$\varepsilon(T) = e(T)/e_b(T) \quad \text{and} \quad \varepsilon_\lambda(T) = e_\lambda(T)/e_{\lambda b}(T) \quad , \tag{5.174}$$

respectively, where e and e_λ are the total and spectral emissive power per unit area of real surface taken at the same temperature (and, for e_λ, at the same wavelength interval) as that of the black body used for reference. Note that only in special cases (e.g., with some electrical insulators) does $\varepsilon_\lambda = \varepsilon$ over a significant spectral range. Such special systems with wavelength-independent emissivity are termed *grey surfaces* or *bodies*. Obviously, by definition, $\varepsilon = \varepsilon_\lambda = 1$ for a black body.

An important relation between ε (or ε_λ) and the absorptivity α (or α_λ) is given by Kirchhoff's law which in its most general form is derived, e.g., in [5.60]. In essence *Kirchhoff's law* states that for a system in thermodynamic equilibrium

$$\alpha = \varepsilon \quad \text{and} \quad \alpha_\lambda = \varepsilon_\lambda \tag{5.175}$$

for each participating surface. Hence, in general, an absorbing surface (body) emits more energy than a transmitting surface at the same temperature. With $\varepsilon_b = 1$, (5.175) yields $\alpha_b = 1$, correctly characterizing the black body as a perfect absorber.

In most texts (5.175) is derived for thermodynamic equilibrium in the most restrictive sense, i.e., for systems uniform in all state variables and hence no net radiative transport. There is, however, ample experimental evidence that Kirchhoff's law holds well also in situations with net transfer as long as the gradients are not so steep that the *local equilibrium* concept is jeopardized. This, as indicated in Sect. 2.10, implies that the population of energy states participating in the absorption and emission processes must still be well approximated by their (thermal) equilibrium distribution. In certain fluorescence and in laser phenomena (overpopulation of a level with respect to its regular thermal population) the local thermodynamic equilibrium assumption and hence (5.175) will not hold. Other restrictions on Kirchhoff's law may arise from the polarized nature and the directionality of radiation. For a thorough discussion of these see [Ref. 5.55, Chap. 3]. For our semi-quantitative discussions, however, we can (fortunately) ignore these refinements and assume the general validity of (5.175).

The data presented in Table 5.18 and Fig.5.32 reveal some interesting trends for the total emissivity of groups of materials and its dependence on surface conditions. Clean polished metal surfaces possess in general low emissivities. Their ε's increase with temperature; note that for the cases in Table 5.18 in which a temperature range is shown, the corresponding range in ε is given. Roughness and contamination increase the emissivity. Heavy oxide layers may result in emissivity values as high as those of surfaces of electrical bulk insulators.

In compilations of this sort, the directional characteristics of the emissivitiy (not to be confused with the strong directionality of the emitted flux discussed below) is often accounted for. However, the total hemispherical emissivity of smooth metals is at most 30% higher than their total emissivity nor-

Table 5.18. Total emissivity data for various metals and insulators. Compiled from [5.53, 54, 57]

Surface	Temperature [°C]	ε
Aluminum, polished	200 - 600	0.04 - 0.06
rough	20	0.07
heavily oxidized	100 - 550	0.20 - 0.33
Chrominum, polished	20 - 550	0.08 - 0.27
Copper, highly polished	20	0.02
black oxidized	20	0.76
Inconel, stably oxidized	250 - 900	0.55 - 0.78
Platinum, polished	500 - 1500	0.1 - 0.22
Silver, polished or deposited	20 - 550	0.01 - 0.03
oxidized	20 - 550	0.02 - 0.04
Asbestos	20	0.95
Alumina (Al_2O_3), sintered	200 - 1900	0.75 - 0.35
sapphire	20 - 200	0.5 - 0.55
Carbon, rough plate	800 - 1800	0.8
lamp black	20	0.95
Porcelain, glazed	20	0.93
Quartz, glass (3 mm)	20	0.75
Silicon carbide	200 - 1300	0.85 - 0.83
Water (0.1 mm or more)	20	0.96

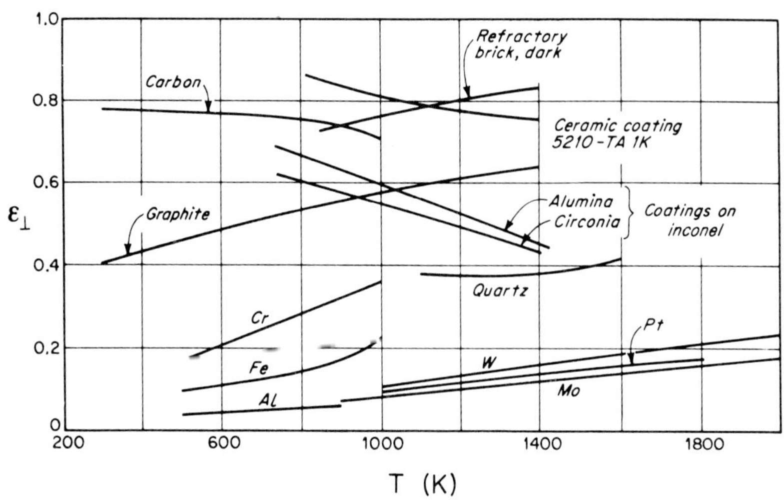

Fig.5.32. Total emissivity of various materials as a function of temperature. From [5.61] by permission of McGraw-Hill

mal to the surface; for insulators this anisotropy is typically even less pronounced [Ref. 5.61, p. 605]. Thus, it is felt here, that the uncertainties in tabulated data from the rather subjectively characterized surface conditions ("highly polished, polished", etc.) make such distinctions irrelevant for most practical purposes. Note the considerable scatter in published data by comparing some of the values in Fig.5.32 with those in Table 5.18!

Total absorptivity and reflectivity values for opaque systems ($\tau = 0$) can, in principle, be deduced from tabulated emissivity data through (5.175) and (5.168) *provided that* the grey-body condition applies and that the flux densities are not extremely high where absorptivity and reflectivity become dependent on the amount of incident radiant energy. With systems with strong absorption bands rather than continuous spectral absorption (such as gases and certain "transparent" solids; see below) additional sources of information must be found.

For many purposes it is advantageous to know the spectral details of the emissivity of a material (and the related absorptivity and reflectivity) rather than only its total emissivity. For some groups of materials the qualitative variations of ε_λ with the wavelength is well established. For metals the "monochromatic" (i.e., spectral) emissivity decreases with increasing wavelength. (This is one of the reasons why the total emissivity of metallic surfaces increases with increasing temperature where greater portions of the emitted radiation stems from shorter wavelength ranges.) For insulators, on the other hand, there is a general tendency for ε_λ to increase with wavelength. Owing to the presence of local (with respect to λ) emission bands in the infrared, however, the variation of $\varepsilon_\lambda(\lambda)$ is quite irregular. These trends are well illustrated in Figs.5.33a and b.

Up to this point we have discussed radiation properties in terms of radiant energy emitted into or absorbed from the entire hemispherical space above a surface. For applications, however, one is typically more concerned with the properties that characterize radiation travelling in specific angular directions with respect to surfaces. Such properties are designated as *directional*, or angular, and distinguished by a prime from the corresponding hemispherical property. Directional transport is then expressed by referring to a small solid angle $d\omega$ (see Fig.5.34) around the direction of propagation (β, θ) where β is the angle of inclination with respect to the surface-normal $\underline{n}$ and θ is the azimuthal angle with $\theta = 0$ chosen arbitrarily. [Remember that the solid angle subtended by a sphere is 4π steradian (ster) and that $d\omega$ can be regarded as

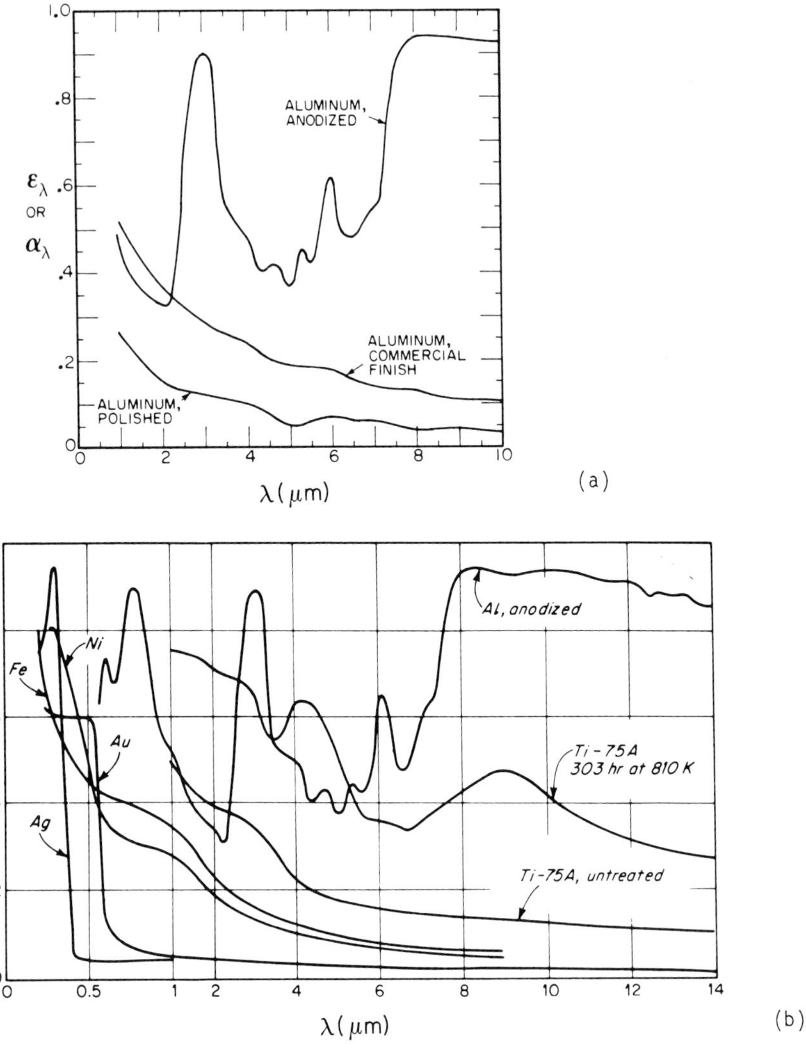

Fig.5.33. Spectral emissivity versus wavelength for various materials; (a) from [5.57] and (b) from [5.61] by permission of Brooks/Cole and McGraw-Hill, Publishing Companies, respectively

the dimensionless surface element dA_s/r^2 where dA_s is the area of intersection between the cone ("pencil") originating at the center of the sphere, and the sphere itself).] Total or spectral *intensity*, respectively, are then defined as the total or spectral energy passing per time through $d\omega$ per *projected* surface area $dA_p = dA \cos\beta$ normal to the direction (β, θ), i.e.,

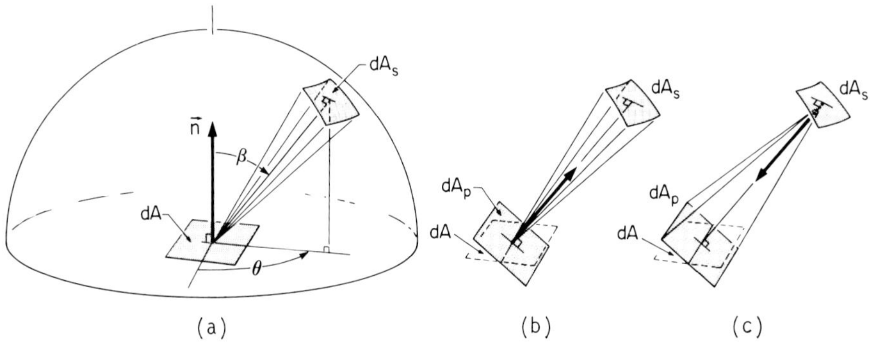

Fig.5.34. Defining sketch for radiative intensity. (a) Surface element dA within hemisphere of unit radius. Energy transfer (b) from dA_p to dA_s in emission and reflection, (c) from dA_s to dA_p for incident radiation

$$i'(\beta,\theta) = \frac{d\phi(\theta)}{dA\ \cos\beta\ d\omega} = \frac{e'(\beta,\theta)}{\cos\beta} \quad \text{and} \quad i'_\lambda(\beta,\theta,\lambda) = \frac{d\phi_\lambda(\theta)}{dA\ \cos\beta\ d\omega} = \frac{e'_\lambda(\beta,\theta)}{\cos\beta} \ , \tag{5.176}$$

respectively. Equation (5.176) can be employed to characterize emitted, absorbed and reflected intensity depending on the origin and direction (inward or outward) of the flux considered; see Fig.5.34. (See Problem 5.16.)

The radiation field within a black body is completely isotropic. Hence the intensity of radiation from a black body, as defined on the basis of projected area, is independent of the direction of emission. Thus $i'_{\lambda b}(\beta,\theta,\lambda) = i'_{\lambda b}(\lambda)$, is a function of λ only. For applications one is generally more interested in the energy emitted in a given direction [i.e., into the solid angle $d\omega$ centered around the direction (β,θ)] per unit of actual (unprojected) surface area, e' or e'_λ. Inserting the direction-independent i'_b and $i'_{\lambda b}(\lambda)$ into (5.176) we obtain *Lambert's cosine law*

$$e'_b(\beta) = i'_b \cos\beta \quad \text{and} \quad e'_{\lambda b}(\lambda,\beta) = i'_{\lambda b}(\lambda) \cos\beta \ , \tag{5.177}$$

where the e'_b and $e'_{\lambda b}$ are called the black-body *directional total* and *directional spectral emissive power*, respectively. Hence the directional power emitted by a black body (surface) has a maximum normal to the surface ($\beta=0$) and is zero parallel to the surface as indicated in the "circular characteristic" of Fig.5.35 (see Problem 5.18). Surfaces with a ($\cos\beta$)-dependence of the directional emissive power as in (5.177) are called *diffusive surfaces*.

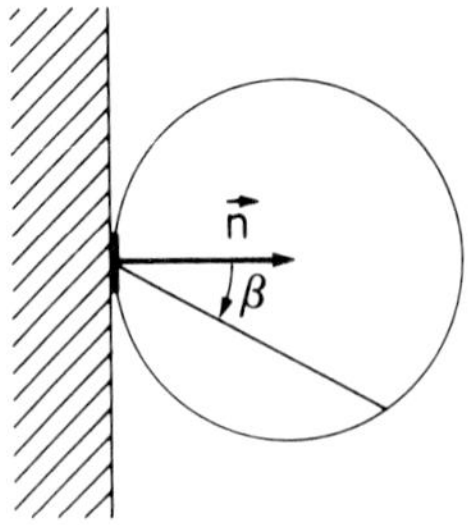

Fig.5.35. Angular dependence of the directional emissive power of a black-body surface element (Lambert's cosine law) schematic

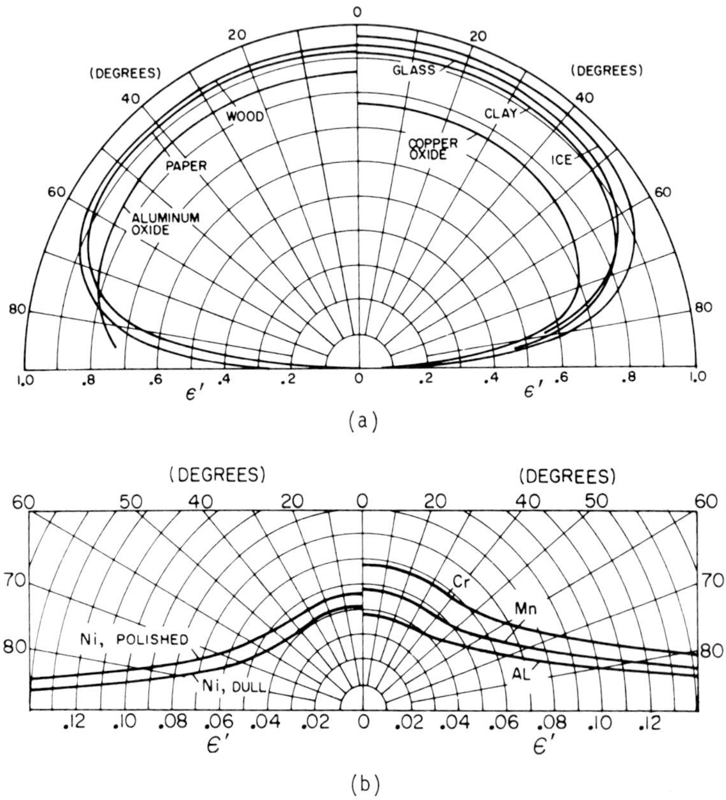

Fig.5.36. Angular distribution of the total directional emissivity for various (a) insulators, and (b) metals. From [5.57] by permission of Brooks/Cole Publishing Company

Non-black surfaces can in general only approximate (5.177), as shown in Fig. 5.36, and often possess some dependence on θ.

The isotropy of the black-body emission intensity results in very simple relations between the directional emissive powers and their hemispherical counterparts. From $e = \int_{\beta=0}^{2\pi} \int_{\theta=0}^{\pi/2} e'(\beta,\theta) \sin\beta \, d\beta d\theta$, [5.55], with (5.176) and $i' = i'_b = \text{const.}$, one obtains readily

$$e_b = \pi i'_b = \pi e'_b(\beta=0) \quad \text{and} \quad e_{\lambda b} = \pi i'_{\lambda b} = \pi e'_{\lambda b}(\beta=0) \quad . \tag{5.178}$$

For non-black surfaces, since $i' = i'(\beta,\theta)$, this integration cannot be carried out as readily. Equation (5.178) however, even if often only approximated, can be very helpful for estimates.

In analogy to the definition of the hemispherical emissivity in (5.174) it is useful to relate the directional emissive power of a non-black surface to the maximum possible directional emissive power, i.e., to that of the black body in the same direction by

$$\varepsilon'(\beta,\theta,T) = \frac{e'(\beta,\theta,T)}{e'_b(\beta,T)} \quad \text{and} \quad \varepsilon'_\lambda(\lambda,\beta,\theta,T) = \frac{e'(\lambda,\beta,\theta,T)}{e'_b(\lambda,\beta,T)} \quad . \tag{5.179}$$

Figure 5.36 displays the angular distribution of the total directional emissivity ε' for various dielectrics and metals. Note that for an ideally diffusive surface that follows Lambert's cosine law, ε' would be unity from $-90° < \beta < 90°$, i.e., the ε' values would lie all on the outermost circle — which must not be confused with the "circular characteristics" for e' depicted in Fig.5.35. One sees that for non-metals ε' remains essentially uniform, at the typically large values, for inclination angles β between 0° and 50-60°; then it drops off quite rapidly to zero. Metals, however, behave quite differently with $\varepsilon'(\beta > 40°)$ increasing sharply. This, at a first glance, seems to be at variance with our earlier statement that the hemispherical emissivity of metals deviates only slightly from the emissivity normal to the surface. One must consider, however, that at large angles where ε' increases sharply, a relatively small portion of the hemispherical power is emitted, as indicated in Fig.5.35 for a black surface.

Before we leave the discussion of optical surface properties and proceed to radiative transfer through "optically participating" media (i.e., media with absorption of radiation in and emission from volume elements) let us pause and summarize some of the above material by applying it to a practical problem. Consider a furnace used in materials preparation. Various ceramic and metal parts may constitute the inner working space ("cavity"). This space may, furthermore, contain some powders and "transparent" crystal pieces. Let us also assume that we have succeeded, via an appropriate heating and insulation geometry to heat all inner surfaces and parts in the cavity to very much the same temperature. Since we have chosen, say 1000° C, where there is already considerable emission in the part of the visible spectrum, we may be tempted to assume that we can conveniently observe the process that we want to monitor. However, looking through a small hole into the furnace cavity we cannot discern "a thing" — inspite of the large differences in the emissivities of the materials used! With our naive expectation we have ignored (5.167), (5.175), i.e., Kirchhoff's law, and (5.176). In an isothermal cavity, no *net* radiative transfer occurs and hence all (projected) surface elements "radiate" with the same intensity. Surface areas with large absorptivity *and hence* large emissivity emit much radiation and reflect little of the radiation received from other areas. For areas with little absorptivity the opposite is true. They emit relatively little, yet they reflect more of the radiation impinging from other areas. (See also Problem 5.16.)

In the definition of the intensity i' above, specific reference was made to radiation *leaving* a surface. In discussions of radiative transfer *through* absorbing, emitting (and possibly scattering) media, one will be concerned with the *net* rate at which radiative energy is locally transferred within the medium. It is useful to distinguish, in this instance, the intensity from i' of (5.176), by using the symbol I, (see, e.g., [5.57]). The intensity I is then defined as the local net flux of radiant energy per unit area normal to the "pencil" of rays (see Fig.5.34) per unit solid angle.

Consider a monochromatic beam of radiation of initial intensity $I_{\lambda 0}$. As the beam traverses a medium of thickness s it will be attenuated as the result of local absorption and scattering within the medium. The change in intensity has been found to be proportional to the local intensity I_λ, and hence the decrease is given by

$$dI_\lambda = -K_\lambda I_\lambda ds \tag{5.180}$$

where the proportionality "constant" K_λ is called the *extinction coefficient**, which, in general is a function of temperature, pressure and composition of the material. The extinction coefficient, corresponding to the two mechanisms of attentuation, is the sum of the *absorption coefficient* a_λ and the *scattering coefficient* σ_λ,

$$K_\lambda = a_\lambda + \sigma_\lambda \quad , \tag{5.181}$$

---

* Note that $k_\lambda = (\lambda a_\lambda)/4\pi$ is also sometimes called extinction coefficient.

with all having units of reciprocal length. Therefore, they are often called linear coefficients, in contrast to the mass coefficients, e.g., $a_{\lambda,m} = a_\lambda/\rho$ (where ρ is the local mass density of the absorbing and scattering species) preferred by some workers. Integration of (5.180) over the path of length s, assuming uniformity of the medium, gives the relation for the transmitted intensity at s

$$I_\lambda = I_{\lambda 0}\exp(-K_\lambda s) \tag{5.182}$$

which is known as *Bouguer's* or *Lambert's law*. Equation (5.182) has also been referred to as Beer's law. More properly, however, Beer's law should be understood as the special form of (5.182) in which $K_\lambda = a_\lambda$, i.e., when the attenuation is only due to absorption. From the photon point of view the quantity $F(s) = \exp(-K_\lambda s)$ in (5.182) represents the fraction of photons whose free path exceeds a given length s. In other words, F(s) is the fraction of photons that have neither been scattered nor absorbed in transversing s. (This, of course, assumes that the contribution of "forward scattered" photons to I_λ is negligible.) Hence, in analogy to the molecular free path one calls very descriptively the reciprocal of the extinction coefficient the *mean free path for photons* of wavelength λ or the *radiation mean penetration distance*

$$L_m = \frac{1}{K_\lambda} \quad \text{with} \quad L_m = \frac{1}{a_\lambda} \quad , \tag{5.183}$$

for non-scattering media.

One of the most important dimensionless parameters associated with "radiation-participating" media is the *optical thickness* of a system defined as

$$\tau_\lambda = K_\lambda L = \ln(I_{\lambda 0}/I_\lambda) \tag{5.184}$$

where the I_λ in the very right part of (5.184), obtained from (5.182), is measured after the radiation has traversed the characteristic distance L of the system. (Note that the *optical density* of a system, a term commonly used by spectroscopists, is defined, very similarly to (5.184), as $OD = \log(I_{\lambda 0}/I_\lambda)$ and hence $OD \approx \tau_\lambda/2.3$.

Now, with (5.183) we can relate the optical thickness with the radiation mean free path and obtain

$$\tau_\lambda = \frac{L}{L_m} \quad , \tag{5.185}$$

which is very useful in classifying basically different radiation transfer regimes. In systems with $\tau_\lambda >> 1$, referred to as *optically thick*, every element of the medium interacts radiatively only with its neighbors. Hence, as in the case of continuum molecular conduction the radiation transfer within the medium becomes a diffusion process. One can show ([5.57], etc.) that in this limit the momochromatic radiation flux is given by

$$q_{r\lambda} = - \frac{4}{3a_\lambda} \, \mathrm{grad}(e_{b\lambda}) \tag{5.186}$$

which is known as the *Rosseland diffusion equation*. In the case of *optically thin* systems, i.e., for $\tau_\lambda << 1$, practically all photons emitted by a given element within the medium will travel directly to the bounding surfaces. In the limit that $\tau_\lambda \to 0$ we have a "non-participating" medium (at the given wavelength; it could strongly participate at other wavelengths!) which has its parallel in free molecular flow where molecules only interact with the bounding surface of a conduit. The most frequently encountered case, of course, involves intermediate values of τ_λ where radiation exchange occurs between all (volume) elements of a medium. Consequently, the local radiation flux can only be obtained from an integro-differential equation which we will not discuss here.

Expressions for the monochromatic absorption coefficient a_λ can, in principle, be derived from quantum mechanical calculations. However, with the exception of monatomic and diatomic gases, the complexity of the calculations has made this approach not very attractive. Hence, most information consists of experimental data or combinations of experiment and analysis. In general, monatomic and symmetric diatomic molecules, since they have no electric dipole moment, cannot significantly absorb [or emit, see (5.175)] by means of vibration-rotation bands. Such gases are transparent to thermal radiation at moderate temperatures. At high temperatures, however, even monatomic and (originally) symmetric diatomic gases can significantly participate in radiation exchange mostly due to electronic transitions or, as in the case of air, dissociation of the symmetrical O_2 and N_2 molecules with partical recombination to form the unsymmetrical NO molecules. Since the energy levels associated with the rotation and vibration modes in gases are well separated, the absorption spectra consist typically of discrete bands (Fig.5.37) in contrast to the fairly smooth radiative property variation with wavelength encountered in condensed phases (see, e.g., Fig.5.33). We see that for thick gas layers α_λ goes

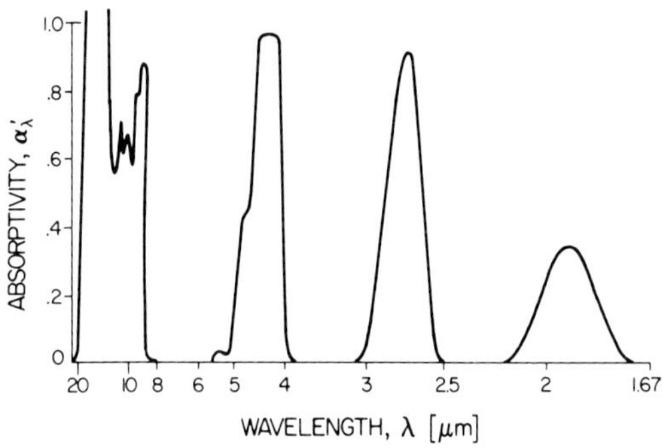

Fig.5.37. Low-resolution absorption spectrum for CO_2 gas at 830K, 10 atm, and a path length through gas of 38.8 cm. Adapted from [5.55]

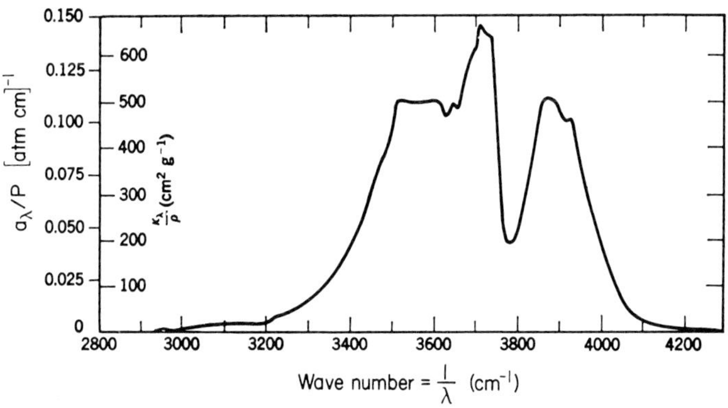

Fig.5.38. Spectral absorption coefficient for water vapor at 1000K in the 2.7 μm region. Adapted from [5.56]

to unity in the region of the main bands (fundamental modes). Hence, providing Kirchhoff's law applies, the gas will also emit like a black body in the spectral region of these bands.

In gases, provided that the temperature is too low to cause significant dissociation, the absorption coefficient a_λ is proportional to the mass density and hence to the pressure P at a given temperature. Thus for comparisons it is convenient to plot the (density-independent) ratio a_λ/P rather than a_λ for individual pressures. An example is given in Fig.5.38.

Now, let us tie the emission and absorption parameters of a system to its reflection properties. We have seen in (5.168) that for an opaque material $\rho + \alpha = 1$. Hence, if the hemispherical absorptivity α is known, the *hemispherical reflectivity* ρ (hemispherical incidence and hemispherical collection) can be regarded as known. To the extent that Kirchoff's law holds for a specific case, ρ may then also be deduced from

$$\rho = 1 - \varepsilon \quad . \tag{5.187}$$

For instance we can predict from Fig.5.33 that gold will reflect infrared radiation extremely well while its reflectivity throughout most of the visible part of the spectrum is rather poor. This, indeed, has been made use of very elegantly in various industrial high-temperature viewing port configurations and in crystal growth for the design of "transparent" high-temperature furnaces [5.65]. Typically in these applications glass is coated with a gold film that is thin enough to give some transparency in the visible, yet resulting in almost quantitative reflection of heat.

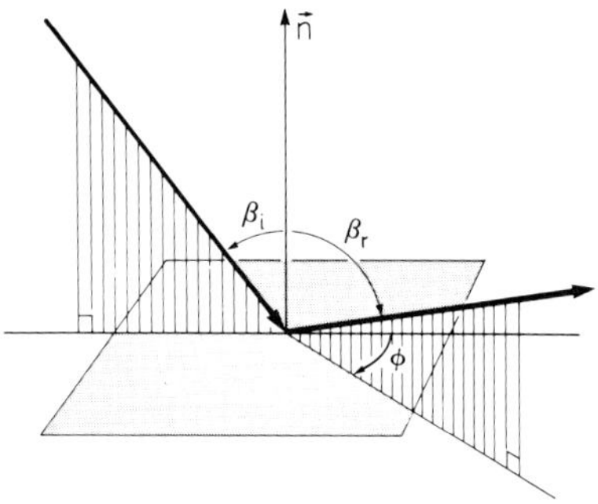

Fig.5.39. Definition sketch pertinent to directional reflection

Actually, the directionality aspects connected with reflective radiant transfer are much more intricate than (5.187) seems to suggest. Recall that the directional emissivity and directional absorptivity pertain, respectively, to emitted and to incident beams of radiation. The reflectivity, however, involves both incoming and outgoing radiation, both of which may have a directional characteristic, as schematically indicated in Fig.5.39. One can define three distinct directional reflectivities: 1) the angular-hemispherical reflectivity (incident beam with specific β_i, reflected radiation collected in entire half space); 2) the hemispherical-angular reflectivity (indicent radiation from entire hemisphere, collection in a specific direction β_r); and 3) the biangular reflectivity with incident radiation and collection of radiation in specific directions, β_i and β_r respectively. In our semi-quantitative discussions we will mostly be concerned with the biangular reflectivity, which in many practical cases has a strong *specular* component, i.e., most of the reflected energy is collected around $\beta_r = \beta_i$ and $\phi = 0$. The directionality of the reflected energy is a sensitive function of the morphology/roughness of the reflecting surface. Perfectly *diffuse reflection,* i.e., the intensity of the reflected radiation is uniform in all angular directions, irrespective of the angle of incidence, will only occur when the wavelength is smaller than the characteristic length of the surface roughness. This and the strong specular effects for smooth (relative to λ) surfaces is nicely illustrated in Fig.5.40. The ordinate of each graph is the biangular reflectance $\rho ba(\beta_i, \beta_r, \phi)$ normalized by the corresponding biangular reflectance in specular-ray direction $\rho_{ba}(\beta_i, \beta_r = \beta_i, 0)$. (Note that for simplicity's sake we are not carrying the subscript λ notation.) Alternately, the ordinate also represents the reflected intensity in the direction (β_r, ϕ) relative to the reflected intensity at the specular angle. The abscissa is the angle β_r. Points to the left of $\beta_r = 0$ are for $\phi = 180°$ (see Fig.5.39), and to the right data are given for $\phi = 0$, 45 and 90°. The open and full circles correspond to $\beta_i = 10$ and 45°, respectively. In such a representation a diffusive surface would lead to a horizontal line at an ordinate value of unity.

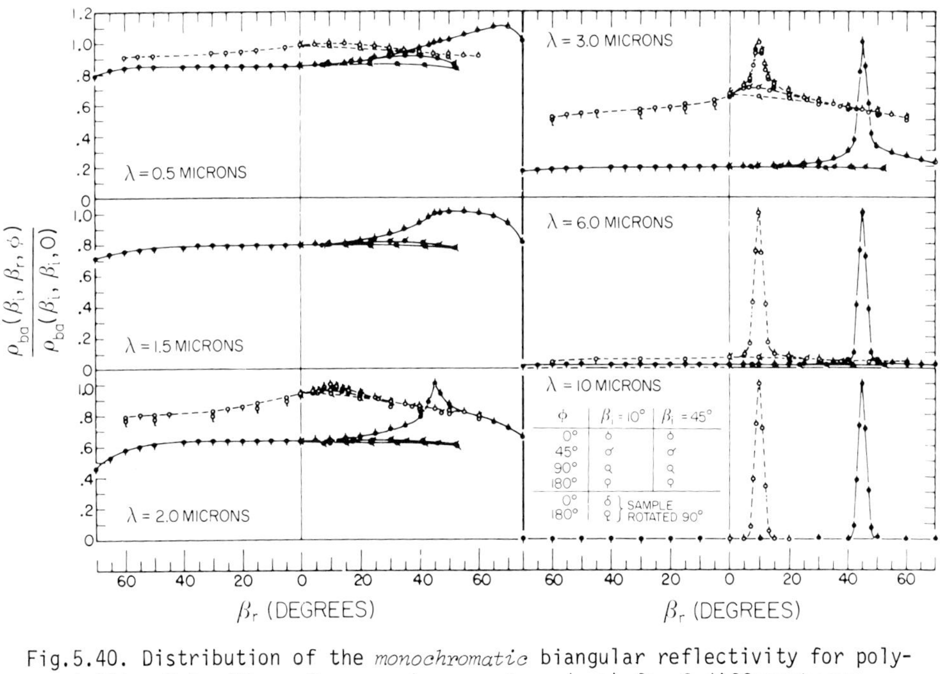

Fig.5.40. Distribution of the *monochromatic* biangular reflectivity for polycrystalline MgO with surface roughness = 1 μm (rms) for 6 different wavelengths. See text. Adapted from [5.57]

Specular reflection, on the other hand, would be represented by a sharp spike at $\beta_r = \beta_i$ with a zero ordinate at all other β_r.

Similar although considerably less detailed results illustrating the dominance of specular reflection are given for various other materials in Fig. 5.41.

The morphology of a surface can also have a substantial effect on the reflectivity, as shown in Fig.5.42 for crystalline lead sulfide coating of same mass per unit surface area but different "structure".

For optically smooth and chemically clean surfaces the monochromatic reflectivity in the specular direction can be predicted from electromagnetic theory. The general relations, accounting for polarization, directionality and phase angle between incident and reflected waves are very complex. We will only point out trends and restrict ourselves to unpolarized reflection. Thermal radiation is in general only weakly polarized. Fig.5.43 illustrates the dependence of the specular reflectivity on the optical parameters of the reflecting (interface) material and $\beta_i = (\beta_r)$. We see that ρ'_λ, for a given refractive index n_2 of the reflecting medium increases with the extinction coefficient K_2. The reader is urged to carefully discuss this seemingly contradictory fact ("increasing reflection with increasing absorption") in Problem 5.19. For all curves of different K_2 there is a broad range in β for

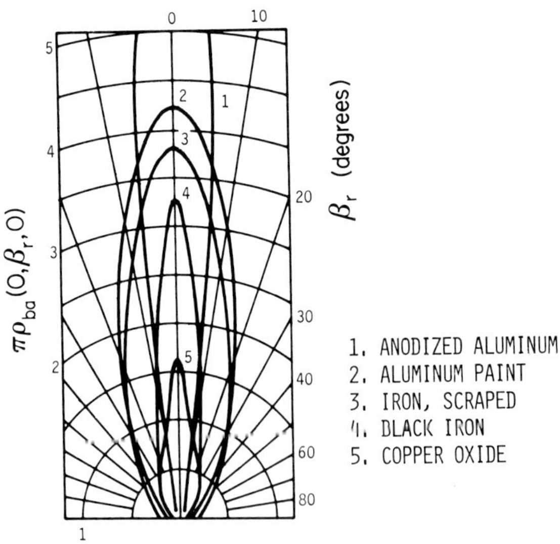

Fig.5.41. Distribution of the *total* biangular reflectivity of various materials in the plane of incidence for normal incidence. From [5.57] by permission of Brooks/Cole Publishing Company

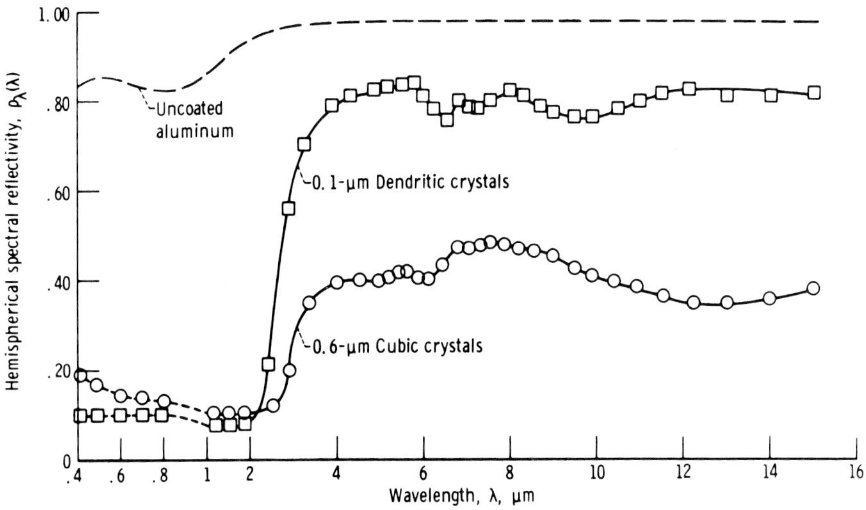

Fig.5.42. Hemispherical spectral reflectivity for normal incidence on lead sulfide coated aluminum. Coating per unit area for both sets of data 0.68 mg/cm^2. From [5.55] by permission of McGraw-Hill

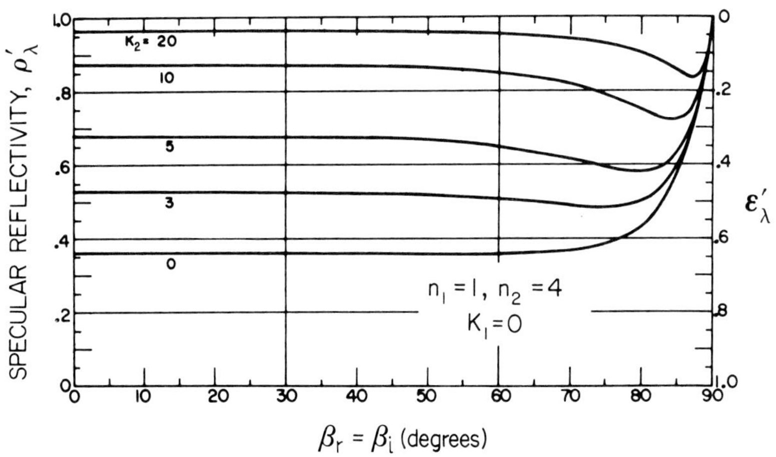

Fig.5.43. Predictions of the spectral specular reflectivity and the spectral emissivity from electromagnetic theory. After [5.57]

which the reflectivity is essentially independent of angle. Furthermore, as β approaches 90° *(grazing angle)*, the specular reflectivity approaches unity regardless of the value of K.

For normal incidence the electromagnetic discussion yields for the unpolarized specular reflectivity

$$\rho'_{\lambda n} = \frac{(n_2 - n_1)^2 + (K_2 - K_1)^2}{(n_2 + n_1)^2 + (K_2 + K_1)^2} , \tag{5.188}$$

which for the special case of incidence through a non-refracting and non-absorbing medium ($n_1 = 1$, $K_1 = 0$, as in Fig.5.43) reduces to

$$\rho'_{\lambda n} = \frac{(n_2 - 1)^2 + K_2^2}{(n_2 + 1)^2 + K_2^2} . \tag{5.189}$$

Further reduction occurs for transparent media where with $K_2 \to 0$

$$\rho'_{\lambda n} = \left(\frac{n_2 - n_1}{n_2 + n_1}\right)^2 , \tag{5.190}$$

known as the (simplified) *Fresnel relation*. Evaluating (5.190) for radiation transfer through a fused silica wall ($n_1 = 1$, $K_1 = 0$, $n_2 = 1.5$, $K_2 \gtrsim 0$), for instance, we see that on each surface (two per wall; see Fig.5.30) about 4% of the incident energy (between $0.2\ \mu m < \lambda < 3\ \mu m$) is reflected even under ideal circumstances. (Note that quartz glass begins to absorb considerably at wavelengths above 2.5-3μm, depending on its purity.)

At this point we have defined all essential parameters needed in discussions of radiative heat transfer in crystal growth. For applications to actual systems we need to consider specific, finite geometries (Sect. 5.6.2) and allow for combinations of radiant energy transport with convective-conductive heat transfer (Sect. 5.6.3).

5.6.2 Applications of the radiative transfer concepts

For a grey surface of area A and emissivity ε at temperature T we know, by combining (5.172) and (5.174), that the rate with which thermal energy is emitted into the hemispherical space is

$$\phi = \sigma\varepsilon A T^4 \quad . \tag{5.191}$$

In order to calculate actual heat fluxes between finite surfaces, however, we must take the specific geometry into account. The essential parameter is obviously the fraction of radiant energy leaving one surface that reaches another surface in a given situation. This fraction F is termed *geometric configuration factor* or view factor, also angle factor. When the configuration factor between each pair of all surfaces (of known ε's and T's) of an enclosure are known one can, in principle, develop a set of equations that governs the radiation exchange within the enclosure. This, in general, is a formidable mathematical task. Hence, in the following we will restrict ourselves to a few simple geometries which suffice to illustrate the essential trends. References for configuration factors of various complex geometries encountered in materials preparation are then given in the text.

The basic configuration for the derivation of geometric view factors is the radiative interchange between two differential area elements of arbitrary orientation, Fig.5.44. The total energy per unit time leaving dA_1 and incident upon dA_2 [see also Fig.5.34 and (5.176)] is

$$d^2\phi'_{d1\to d2} = i'_{1b} dA_1 \cos\beta_1 \, d\omega_1 \tag{5.192}$$

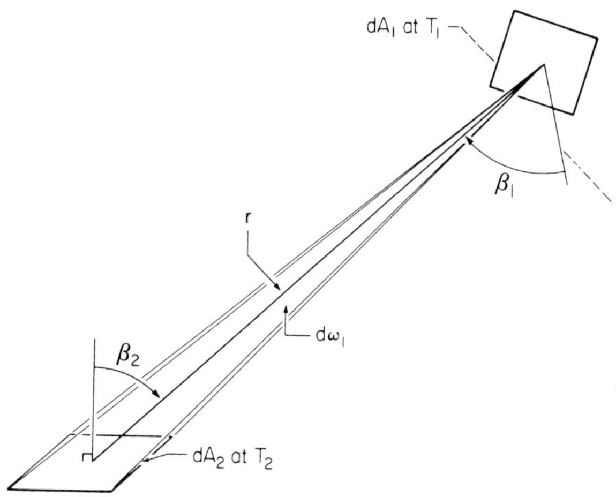

Fig.5.44. Radiation exchange between differential area elements

where we have assumed that both elements are black and hence i' does not depend on direction (diffusive emission). With $d\omega_1 = dA_2 \cos\beta_2/r^2$, (5.192) becomes

$$d^2\phi'_{d1\to d2} = \frac{i'_{1b} dA_1 \cos\beta_1 \, dA_2 \cos\beta_2}{r^2} \quad . \tag{5.193}$$

A symmetrical relation is obtained for $d^2\phi'_{d2\to d1}$. (See also Problem 5.20.) From (5.178) and (5.173) $i'_b = e_b/\pi = \sigma T^4/\pi$. Inserting this into (5.193) and dividing by the total energy $e_b = \sigma T^4$ leaving dA_1 one arrives at the differential geometric configuration factor

$$dF_{d1\to d2} = \frac{\cos\beta_1 \cos\beta_2}{\pi r^2} \, dA_2 \quad , \tag{5.194}$$

which depends only on the size of dA_2 and its orientation with respect to dA_1. Configuration factors between differential and finite areas and between finite and finite areas can then be obtained by integration of (5.194) [5.55]. (Problem 5.21.) Table 5.19 summarizes the energy exchange-equations and the

Table 5.19. Summary of radiative energy-exchange relations, configuration factor definitions and reciprocity relations. After [5.55]

Geometry	Energy Exchange	Configuration factor	Reciprocity
Elemental area to elemental area	$d^2\phi'_{d1\rightleftarrows d2} = \sigma(T_1^4 - T_2^4) dA_1 dF_{d1\to d2}$	$dF_{d1\to d2} = \frac{\cos\beta_1 \cos\beta_2}{\pi r^2} dA_2$	$dA_1 dF_{d1\to d2} = dA_2 dF_{d2\to d1}$
Elemental area to finite area	$d\phi_{d1\rightleftarrows 2} = \sigma(T_1^4 - T_2^4) dA_1 F_{d1\to 2}$	$F_{d1\to 2} = \int_{A_2} \frac{\cos\beta_1 \cos\beta_2}{\pi r^2} dA_2$	$dA_1 F_{d1\to 2} = A_2 dF_{2\to d1}$
Finite area to finite area	$\phi_{1\rightleftarrows 2} = \sigma(T_1^4 - T_2^4) A_1 F_{1\to 2}$	$F_{1\to 2} = \frac{1}{A_1} \int_{A_1} \int_{A_2} \frac{\cos\beta_1 \cos\beta_2}{\pi r^2} dA_2 dA_1$	$A_1 F_{1\to 2} = A_2 F_{2\to 1}$

integral definitions of the configuration factors. Note in particular the reciprocity relations in the last column, which allow for convenient deductions of supplementary configuration factors and cross-checking of results.

An extensive compilation of references for solved configuration factors is given in [Ref. 5.55, App. B]. Table 5.20 lists a few selected results that will be useful in this and other sections.

Let us apply the above to a configuration often found, at least in approximated form, in materials preparation. Assume that a body of surface area A_1 at temperature T_1 and with $\varepsilon_1 = \varepsilon(T_1)$ is completely surrounded by another body of area A_2 and emissivity ε_2 at T_2. Let the geometry be such that the inner body cannot "see" any part of itself so that all radiation emitted by it is either absorbed or reflected by the outer body, i.e., $F_{1\to 2} = 1$. Also let the inner body intercept a fraction A_1/A_2 of the radiation emitted by the outer body, i.e., $F_{2\to 1} = A_1/A_2$ as, for instance, for the two long coaxial cylinders in Table 5.20.

The energy emitted per unit time by the outer body is from (5.191) $\sigma\varepsilon_2 A_2 T_2^4$. Of that energy the inner body absorbs a fraction $(A_1/A_2)\varepsilon_1$ and reflects a fraction $(1 - \varepsilon_1 A_1/A_2)$ to the outer body which intercepts all of it and reflects a fraction $(1 - \varepsilon_2)$ of which a further fraction $(A_1/A_2)\varepsilon_1$ is absorbed by the inner body and so on. Summing over all the terms of this series (Problem 5.22), one obtains for the total radiation received per time by the inner body

$$\phi_{2\to 1} = \frac{\sigma A_1 T_2^4}{1/\varepsilon_1 + A_1/A_2(1/\varepsilon_2 - 1)} \quad . \tag{5.195}$$

The corresponding expression for $\phi_{1\to 2}$ differs only in the T_1^4 term that replaces T_2^4 in (5.195). Consequently, the net rate of heat transfer between the two bodies is

$$\phi_{1\rightleftarrows 2} = \frac{\sigma A_1 (T_1^4 - T_2^4)}{1/\varepsilon_1 + A_1/A_2(1/\varepsilon_2 - 1)} \quad . \tag{5.196}$$

For $A_2 \gg A_1$ (e.g., a small sample or crucible in a big furnace) the denominator in (5.196) reduces to $1/\varepsilon_1$, that is to say that the heat flux becomes

Table 5.20. Selected diffusive configuration factors. From [5.55] by permission of McGraw-Hill

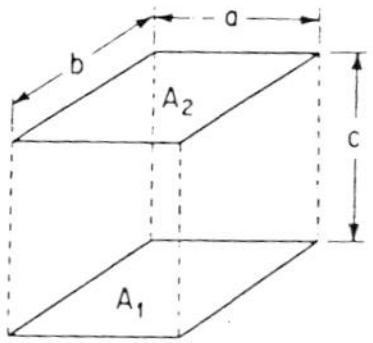

Identical, parallel, directly opposed rectangles.

$$X = \frac{a}{c} \qquad Y = \frac{b}{c}$$

$$F_{1-2} = \frac{2}{\pi XY}\left\{\ln\left[\frac{(1+X^2)(1+Y^2)}{1+X^2+Y^2}\right]^{1/2} + X\sqrt{1+Y^2}\tan^{-1}\frac{X}{\sqrt{1+Y^2}} + Y\sqrt{1+X^2}\tan^{-1}\frac{Y}{\sqrt{1+X^2}} - X\tan^{-1}X - Y\tan^{-1}Y\right\}$$

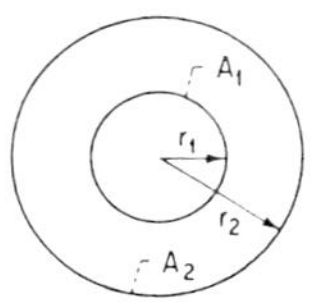

Concentric cylinders of infinite length.

$$F_{1-2} = 1$$

$$F_{2-1} = \frac{r_1}{r_2}$$

$$F_{2-2} = 1 - \frac{r_1}{r_2}$$

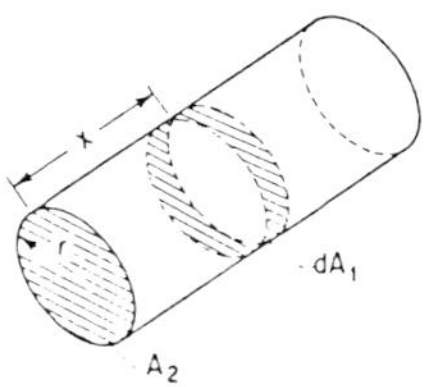

Ring element dA_1 on interior of right circular cylinder to circular disk A_2 at end of cylinder.

$$X = \frac{x}{2r}$$

$$F_{d1-2} = \frac{X^2 + 1/2}{\sqrt{X^2 + 1}} - X$$

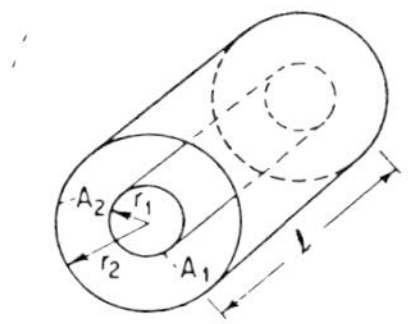

Two concentric cylinders of same finite length.

$$R = \frac{r_2}{r_1} \qquad L = \frac{l}{r_1}$$

$$A = L^2 + R^2 - 1$$

$$B = L^2 - R^2 + 1$$

$$F_{2-1} = \frac{1}{R} - \frac{1}{\pi R}\left\{\cos^{-1}\left(\frac{B}{A}\right) - \frac{1}{2L}\left[\sqrt{(A+2)^2-(2R)^2}\cos^{-1}\left(\frac{B}{RA}\right) + B\sin^{-1}\left(\frac{1}{R}\right) - \frac{\pi A}{2}\right]\right\}$$

$$F_{2-2} = 1 - \frac{1}{R} + \frac{2}{\pi R}\tan^{-1}\left(\frac{2\sqrt{R^2-1}}{L}\right) - \frac{L}{2\pi R}\left\{\frac{\sqrt{4R^2+L^2}}{L}\sin^{-1}\left[\frac{4(R^2-1)+(L^2/R^2)(R^2-2)}{L^2+4(R^2-1)}\right] - \sin^{-1}\left(\frac{R^2-2}{R^2}\right) + \frac{\pi}{2}\left(\frac{\sqrt{4R^2+L^2}}{L} - 1\right)\right\}$$

where for any argument ξ: $-\frac{\pi}{2} \leqslant \sin^{-1}\xi \leqslant \frac{\pi}{2}$

$$0 \leqslant \cos^{-1}\xi \leqslant \pi$$

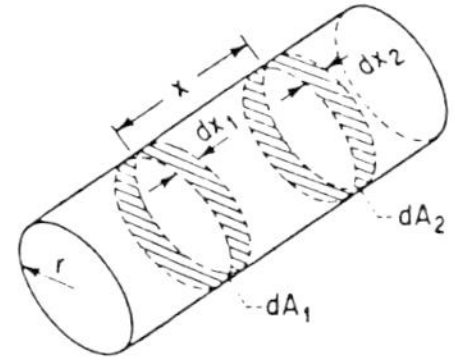

Two ring elements on the interior of a right circular cylinder.

$$X = \frac{x}{2r}$$

$$dF_{d1-d2} = \left[1 - \frac{2X^3+3X}{2(X^2+1)^{1/2}}\right]dX_2$$

independent of ε_2. In the other extreme in which $A_2 \to A_1$ (or $F_{2\to1} \to 1$) (e.g., the heating source fits closely the shape of a sample) ε_1 and ε_2 have the same weight in the denominator of (5.196). Hence, given a choice one will employ a material with high ε_2 for good radiation coupling. This case with $A_2 \to A_1$ with only a small distance between the surfaces can also be viewed as transfer between two "infinite planes".

Now, if we consider the inner body as the heat source, then (5.196) represents the heat loss to the environment. Since most refractory materials have comparably high ε's (metals under oxidizing conditions!) little can be done to reduce these losses through the choice of materials. A more efficient approach utilizes "floating" radiation shields between the two bounding surfaces of fixed temperature. Assume n shields have been inserted between T_1 and T_{n+2} (the former T_2). In steady state the total heat flux will be constant through all shields. Summing over all equations of the form of (5.196) that hold between pairs of neighboring surfaces (inner surface and first shield, first shield and second shield, etc.) the rate of heat transfer becomes

$$\phi_{1\rightleftarrows n+2} = \frac{\sigma(T_1^4 - T_{n+2}^4)}{\sum\limits_{i=1}^{i=n+1} \frac{1}{A_i}\left[\frac{1}{\varepsilon_i} + \frac{A_i}{A_{i+1}}\left(\frac{1}{\varepsilon_{i+1}} - 1\right)\right]} \quad , \tag{5.197}$$

which for a set of "infinite parallel planes" (i.e., spacing of the shields and surfaces small as compared to their diameter, and length large as compared to diameter so that end-losses can be ignored) with equal emissivity reduces to

$$\phi_{1\rightleftarrows n+2} = \sigma A(T_1^4 - T_{n+2}^4)\left(\frac{1}{n+1}\right)\left(\frac{\varepsilon}{2-\varepsilon}\right) \quad . \tag{5.198}$$

Comparing (5.198) and (5.196) it is seen that the introduction of n floating shields reduces the radiation losses from the inner body by a factor (n+1). This can be further improved upon by choosing low-ε material for the shields, i.e., $\varepsilon_{shield} < \varepsilon_1$ and ε_{n+2}. An elegant heat shielding design consisting of a close-spaced multiturn spiral has been described in [5.118].

To this point we have only considered diffusively emitting and consequently diffusively reflecting surfaces. Yet most real surfaces have strong specular reflection components, as shown in the foregoing section. Specular reflection complicates heat transfer discussions considerably. For the most simple case, the two parallel infinite plances, of course, nothing changes as compared to diffusive transfer. All emitted and reflected radiation leaving surface 1 will directly reach surface 2 and vice versa, as indicated in Fig.5.45a. For specular exchange between two concentric cylinders, for instance, this is not the case. Any radiation leaving A_1 after impingement and partial absorption on A_2 will always be reflected onto A_1 and can never be reflected from an element on A_2 onto another element on A_2 (see Fig.5.45b, path a_1 and a_2). However, the radiation emitted from A_2 can travel along either of two types of paths b or c (with the latter never reaching A_1) as shown in Fig.5.45b. Hence the fraction $F_{2\to 2}$ will not contribute to the interchange. Only the fraction $F_{2\to 1}$ will be reflected back and forth between A_1 and A_2. Consequently $\phi_{2\to 1} = A_2\varepsilon_2 F_{2\to 1}T_2^4$, with $F_{2\to 1} = A_1/A_2$, becomes $\phi_{2\to 1} = A_1\varepsilon_2\sigma T_2^4$ and thus depends only on A_1 and not on A_2. Hence, specular exchange between concentric cylinders corresponds to the "infinite" parallel plate problem with *both* surfaces taking on the area of the *inner* body.

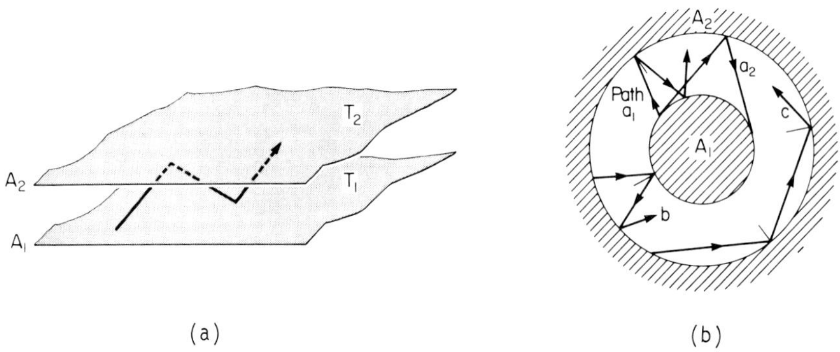

Fig.5.45. Radiation exchange between specular surfaces of simple geometry; (a) Infinite parallel planes, (b) paths in gap between concentric cylinders or spheres

Radiative exchange must also be carefully considered in certain types of temperature measurement. Assume that a temperature sensor is positioned in a transparent medium, such as a gas, of temperature T_g in an enclosure, the walls of which have a temperature T_w. Due to radiative interchange with the wall the temperature sensor will acquire a different temperature T_s. The radiation error $T_s - T_g$ can be estimated from a simple heat balance [5.61], describing the heat transferred per unit surface area and time from the gas to the sensor by convection and from the sensor to the walls by radiation, in the form

$$h_c(T_g - T_s) = \varepsilon_s \sigma (T_s^4 - T_w^4) \quad , \qquad (5.199)$$

where h_c is a convective heat transfer coefficient and ε_s the emissivity of the temperature sensor surface. For the right-hand term we have taken advantage of the fact that the sensor surface is typically much smaller than the area of the surrounding walls [see (5.196)]. When the difference between T_w and T_s is small compared to either temperature one can linearize the right-hand term and write $\varepsilon_s \sigma (T_s^4 - T_w^4) \approx 4\varepsilon_s \sigma T_w^3 (T_s - T_w) = h_r(T_s - T_w)$, thus introducing a radiation heat transfer coefficient h_r. The relative radiation error E is then

$$E = \frac{T_g - T_s}{T_g - T_w} = \frac{1}{1 + h_c/h_r} \quad . \qquad (5.200)$$

As has been pointed out before [5.61], even under conditions that are typical around room temperature E can become as big as 20%. At elevated temperatures E can become still more pronounced. In order to reduce h_c/h_r and, hence, the radiation error, one will

(a) employ a sensor surface with high reflectivity (low emissivity) in the IR, and

(b) increase the convective-conductive heat transfer by decreasing the diameter of the sensor (very thin wire thermocouple or thermistors).

In opaque media, on the other hand, such problems do not arise, since radiative coupling between a sensor and the surrounding opaque medium will only aid in acquiring the correct T_s.

In pyrometric temperature measurements significant errors can arise when some of the above considerations are ignored. The two major problems one faces here are:

(a) Pyrometers are calibrated for an emissivity of the surface equal to unity. Hence, rather than taking readings from the surface of a material with $\varepsilon < 1$, one should use a (cylindrical) hole in the material. Such a hole is a reasonably good black body if the pyrometer can be focused several diameters deep into the hole.

(b) Reflections of much hotter objects on the surface of the sample or transmission through transparent samples (melts!). These problems are difficult to deal with and solutions (thermocouples vs. pyrometer, shielding, etc.) must be "tailored" to the individual case.

For an extensive discussion of heat transfer in semi-transparent solids, specifically including some crystal growth aspects, the reader is referred to a recent review by VISKANTA and ANDERSON [5.66].

5.6.3 Combined conductive-convective-radiative heat transfer

The presence of a thermal radiation field within a medium alters the energy transport (energy conservation) equation (5.3, 147), in principle, as the result of three effects. 1) In the presence of temperature gradients a net radiation energy flux contributes to the heat transfer. 2) Each element of the medium contains additional radiant energy. 3) The radiation pressure augments the conventional fluid dynamic pressure. Unless extremely high temperatures prevail, the last two effects are generally insignificant. Hence, for our purposes we will only include the radiative heat flux $\underline{q}_R$ in the general heat flux term $\underline{q}$ in (5.3). Under the provisions for the validity of (5.148) the simplified energy equation becomes then

$$\rho C_p\left[\frac{\partial T}{\partial t} + (\underline{v}\cdot \mathrm{grad})T\right] = k\nabla^2 T - \mathrm{div}(\underline{q}_R) \quad . \tag{5.201}$$

Before (5.201) can be applied to any problem involving a radiation participating medium, a relation for $\underline{q}_R$ as function of the temperature must be obtained. For instance, for optically thick media [see (5.185)] where only neighboring elements interact radiation-wise, one can employ the Rosseland diffusion equation (5.186). For intermediate optical thickness, where radiation exchange takes place between all elements of a system, an integral expression is obtained for the radiation flux vector. Consequently, the energy equation (5.201) becomes an integro-differential equation.

5.7 Stefan Problems

In most of the foregoing discussions it is assumed that

(a) the shape (typically chosen planar);

(b) the temperature and/or concentration (typically the equilibrium values); and

(c) the growth rate or velocity (typically chosen constant)

of a propagating phase boundary are known – without concern for the transport conditions necessary in the adjacent phases to realize a specific choice of (a-c). In nature and in the laboratory the situation is in general quite the opposite. One will be able to specify the temperatures and concentrations in the phases at some distance from the interface (outer boundary conditions) together with the transport parameters (diffusivities, conductivities, etc.) in the phases. In addition, if the latent heat and the segregation coefficients are known, one can set up a heat and mass balance, respectively, for the interface (inner boundary condition). The location and velocity of the boundary, however, will be a function of all these conditions, and of the interfacial kinetics parameters, which we ignore at this point by assuming equilibrium throughout. For instance, during the pulling of a crystal from a congruent melt (see Fig.6.41 and [1.2]) under fixed seed and melt temperature, assuming dominant cooling through the seed, the growth rate will decrease with increasing crystal length. Also, under otherwise identical conditions a material with lower heat of solidification will grow faster since it takes higher growth rates to liberate heat at the interface with the same rate.

Mathematically, in order to obtain a relation for the changing location of an interface, one must solve the time-dependent heat and/or mass diffusion equation subject to the existence of a *free (inner) boundary*. The balance for the diffusant (heat, mass) across this boundary adds to the complexity of the mathematics. The first solution to such a free boundary problem was given by Lamé and Clapeyron in 1831. These authors calculated the thickness of the solid crust generated by the cooling of a liquid filling the half-space $x > 0$ under the influence of a constant temperature at the plane $x = 0$. They found that the thickness of the solid is proportional to the square root of the time but did not determine the proportionality constant [5.67]. A rigorous solution to this and similar solutions was then put forward by Stefan in 1889. Hence, the general class of transport problems involving the continuity (diffusion) equation and energy equation with a moving boundary or free boundary are called *Stefan problems*. In the Russian literature these problems are sometimes referred to with the rather lengthy name of Lamé-Clapeyron-Stefan problems.

Stefan problems are among the more difficult boundary value problems in mathematical physics. We will use solutions to Stefan problems in the Sects. on non-steady segregation (6.5.1) and on dynamic morphological stability [1.1], as well as in discussions of various transport phenomena in specific crystal growth arrangements [1.2]. Detailed derivations of solutions to Stefan problems, however, go beyond the scope of this text.

The extensive literature on phase change problems with a moving boundary has recently been surveyed in a comprehensive monograph by RUBINSTEIN [5.67] and in reviews by BANKOFF [5.68] and MUEHLBAUER and SUNDERLAND [5.69]. Most recent developments in the treatment of moving boundary problems have been reported in [5.150]. PARKER [Ref. 5.1, Chap.3] has emphasized the crystal growth aspects of Stefan problems, including interfacial kinetics and surface energy considerations.

5.8 Free Convection

Convective flow reduces the diffusional barrier to interfacial heat and mass transport. Convection is therefore beneficial to crystal growth provided that the remaining diffusional barrier is *uniform* across an interface and is *not temporally fluctuating* due to instabilities (non-steadiness) in the convective flow. This is discussed in Section 6.5.1 on non-steady segregation. The convective phenomena encountered in crystal growth, due to often very extreme boundary conditions and complex composition of the fluids, are among the most complicated ones observed. (As a crystal grower one has full sympathy for the difficulties faced by meteorologists in making reliable weather forecasts!) Consequently, our understanding of convective phenomena, in particular in crystal growth from low Prandtl number fluids (metals and some semiconductors; see Table 5.16), is rather limited at this point. Fortunately, considerable efforts are being made currently towards a better understanding of the fluid dynamics of crystal growth processes. In fact, a good part of the recent development concerning non-steady convection is due to crystal growers.

As pointed out in the introduction to this chapter, convective flow may be forced by (dominating) contact forces. Or flow may be driven by body forces.

Then convection is termed free or "natural." Forced convection is treated in Sect. 5.3. In this following section various aspects of *free convection from (heated) planes* are reviewed. Trends that arise from additional (lateral) confinement and less idealized boundary conditions are also indicated here. Free convection in fully confined geometries relevant to crystal growth, however, will be discussed in [1.2] following the detailed presentations of the respective crystal growth techniques.

Large parts of this section have greatly benefitted from authoritative reviews by VELARDE [5.71] and CARRUTHERS [5.81, 82]. The reader is also refered to a book on convective stability by GERSHUNI and ZHUKOVITSKII [5.151] and a review by NORMAND and coworkers [5.152].

5.8.1 Introductory Remarks

There is a multitude of driving mechanisms for free convection. There is also a large number of schemes for the categorization of these mechanisms. Some schemes that were very useful at some time have become more confusing than helpful due to the rapid development in this area, yet they are still in use. For instance, it is still customary to talk about "Bênard convection" if, as we will see, actually "Rayleigh convection" in a "Bênard geometry" is meant. Lacking a commonly agreed upon terminology we will simply identify the various types of convection according to their underlying mechanisms. Yet, in order to facilitate "cross-reference" and to do justice to the pioneering achievements of the founding fathers, we will also use parenthetically the "conventional" names.

Free convective flow may be driven by *buoyancy* and/or *surface tension* (obviously this jeopardizes our above distinction between free and forced convection according to the predominance of body and "surface", respectively). Various mechanisms found in crystal growth, which result in *potentially* convectively unstable situations, are listed in Table 5.21. Gravitationally unstable situations (dense "on top" of less dense) in which convection tries to overturn a fluid to lower the free energy of a system can arise from temperature and/or concentration gradients. Surface-tension-driven convection may also originate in temperature and concentration gradients. Note the emphasis on "potential" above. This is to indicate that it may take a system-dependent critical value of the destabilizing gradient to actually obtain convective flow. Or, a density gradient that is stable in one configuration may be un-

Table 5.21. Free (natural) convection mechanisms

<u>Buoyancy-driven</u>
thermal expansion
thermal diffusion
temperature dependence of solubility
temperature dependence of chemical equilibrium
externally imposed, intrinsically unstable solute gradient
externally imposed, intrinsically stable solute gradient plus heat flow
electro-diffusion
<u>Surface-tension-driven</u>
temperature dependence of surface tension
composition dependence of surface tension

stable in another one. Hence, whether these mechanisms or *combinations thereof* actually lead to convective motion must be decided from case to case by a detailed stability analysis. Note also that the term "convective instability" is used in the literature for any transition between different convective states. The transition from the quiescent state to the lowest motional (steady or time-dependent) state is only one transition out of a whole spectrum of possible "instabilities".

Convection due to thermal expansion was qualitatively correctly explained by Lomonosov in 1753 (for this and other historical references see [5.71]). The first convection studies in the laboratory were performed by BÉNARD in 1900 [5.72]. Very shallow layers (h < 1mm) of molten spermaceti wax, heated from below through a metal plate and with the upper surface freely exposed to air, displayed a convection pattern of dominantly hexagonal cells. Fig. 5.46 shows two of Bénard's original photographs together with a particularly regular pattern obtained recently with silicone oil [5.73]. RAYLEIGH, in a fundamental theoretical paper in 1916 [5.74] interpreted Bénard's observation as instability due to buoyancy resulting from expansion of the heated fluid. Since then, it has been recognized for a long time that Bénard's studies were mostly concerned with surface-tension-driven convection rather than the thermal expansion-buoyancy mechanism. BLOCK in a still widely unknown paper published in 1956 [5.75], showed that Bénard cells can be obtained in layers as shallow as 50 μm and even with heating from above, i.e.,

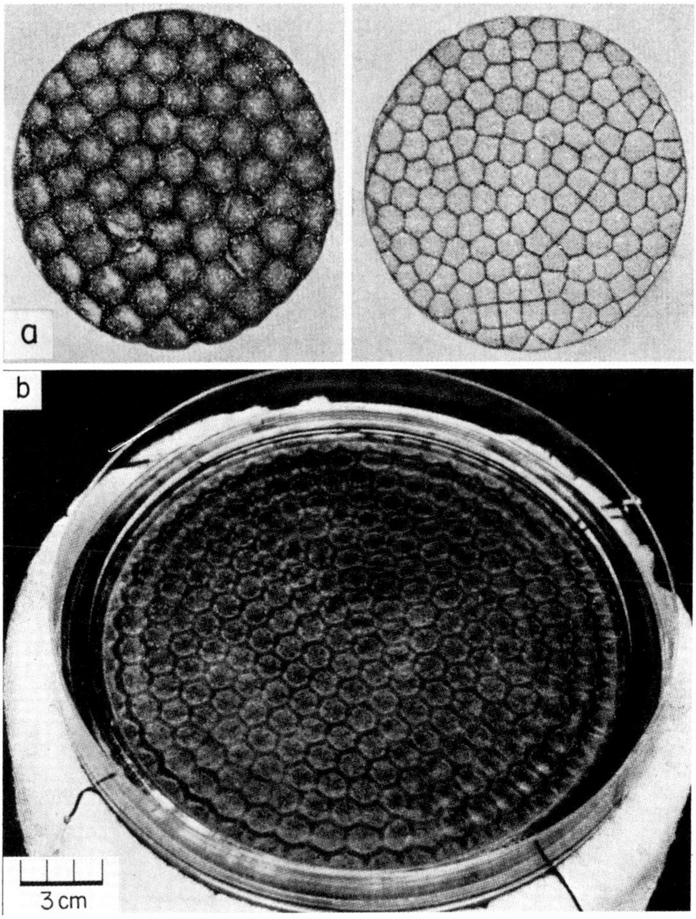

(a) (b)

Fig.5.46. Hexagonal convection pattern (Bénard cells) in liquid-air interfaces. (a) Molten spermaceti wax [5.72] and (b) silicone oil [5.73]

with a "stabilizing" temperature gradient. Independently, PEARSON (1958), stimulated by observations made on drying paint films, put forward the first stability analysis based upon the temperature dependence of surface tension [5.76]. He was also the first one to show *quantitatively* that Bénard's experiments were dominated by surface tension rather than buoyancy. Unfortunately the convection terminology has not followed suit yet. "Bénard-Rayleigh convection" is still a commonly used term for the description of instabilities in fluids that are heated from below. We will rather refer to the underlying mechanisms to minimize confusion. The short term *expansive convection* ("Rayleigh convection" if the temperature gradient is parallel to the gravitational acceleration) will be used in this text for buoyancy-driven flow due to thermal expansion. "Thermal convection", often employed for this mechanism, is probably the least specific term possible, considering all the other buoyancy-driven mechanisms that also require a temperature gradient. For clear distinction and brevity, surface-tension-driven flow (Bênard convection, Marangoni convection) will be referred to as *σ-convection*. Following common custom *Bênard problem* or *geometry* will be used for a bottom-heated fluid configuration of *small aspect ratio* a; where a is defined as

$$a = \frac{\text{height of fluid}}{\text{lateral extent}} \quad . \tag{5.202}$$

Of course, for some fluid geometries (5.202) does not supply an unambiguous definition and must be supplemented correspondingly.

A considerable fraction of the voluminous work on convective instabilities has been concerned with establishing the proper *planform* (flow pattern) for the Bênard problem. In the beginning, polygonal cellular flow was considered the only true eigenform for flow caused by bottom heating. This was so strongly believed that any manifestations of a periodicity in flow patterns was referred to as "cellular" — although it could have been interpreted as a parallel roll pattern as well (see, e.g., [5.96]). More careful analysis of recent experiments (see, e.g., [5.73, 77]) performed on fluid layers between horizontal plates to exclude surface tension effects, revealed almost exclusively rolls (Fig.5.47). Polygonal cells were only found for a fluid with extremely temperature-dependent viscosity [5.78], or when the mean temperature was changed during the experiment [5.79]. Note that the latter condition may readily occur in crystal growth. In rectangular cylindrically bound layers one has found straight two-dimensional rolls with axis parallel to the shorter wall as well as square "cells" as superposition of two rolls with perpendicular axis. Circularly bound layers showed concentric ring rolls. These findings have lead to the widespread opinion "that for standard fluids, hexagons (or in general polygonal-like structures) appear when surface tension is involved in the problem, whereas rolls and only rolls are the structure in Rayleigh convection" [5.71]. Yet the controversy about the appropriate planforms for expansive and σ-convection cannot be considered as resolved. Most recent convection studies performed under micro-gravity conditions on board Apollo XVII have unambiguously established that cellular motion, setting in at a critical temperature gradient, develops also in the absence of gravity [5.80]. However, practically simultaneously with the polygonal cells in the center of the fluid, concentric side rolls developed near the bounding walls. This tends to indicate that rolls are dominantly side-wall effects

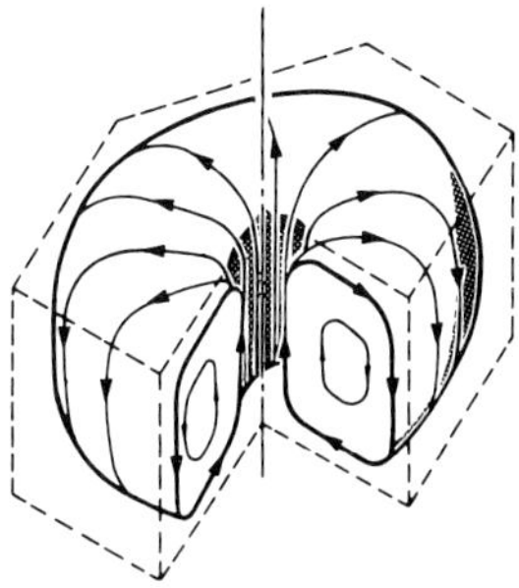

(a)

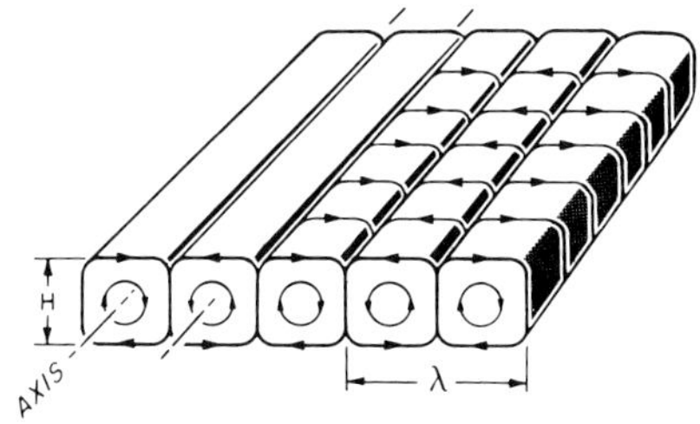

(b)

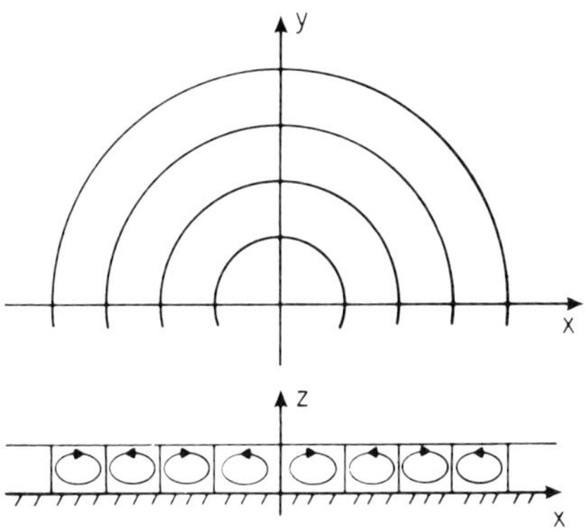

(c)

Fig.5.47. Convective flow in Bénard geometry. (a) Single Bénard cell, (b) Straight two-dimensional rolls; from [5.81]. (c) Concentric ring rolls, from [5.77] by permission of Wiley and Sons, and Marcel Dekker, respectively

and not particularly characteristic of the driving mechanism. For a detailed review of the theoretical and experimental work on this question see [5.77].

It is interesting to note a distinctly different cellular flow behavior between gases and liquids. In liquids, as indicated in Fig.5.47a, the flow is ascending in the center and descending at the outer periphery of an individual (hexagonal) cell. In gases, however, the flow direction is opposite. This has been explained in terms of the temperature dependence of the viscosity of the fluid (for references see [5.77]). The flow in the cell center follows the direction of increasing kinematic viscosity (see also Sect. 5.3.3). This model was very elegantly confirmed with studies on liquid sulfur where $d\nu/dT$ is first negative, as with most other liquids, but between 153-180°C $d\nu/dT$ becomes sharply positive. Correspondingly, a reversal in the cellular flow direction was observed in that range [5.83].

5.8.2 Buoyancy-driven convection

This section is concerned with theoretical and experimental investigations of buoyancy-driven convection from heated horizontal and vertical planes. As a basis for the later discussion of monocomponent expansive, solutal and double-diffusive convection we will first outline the fundamentals of linear (small perturbation) stability theory, which will also be helpful for the treatment of dynamic morphological stability theory [1.1].

Linear stability analysis [5.84, 88]

Convective stability theory is concerned with the response against disturbances of arbitrary size of a fluid with a potentially unstable density and/or surface-tension distribution. Linear stability theory restricts itself to infinitesimal perturbations. The basic objective is to determine the conditions under which the system takes on another state different from the previous one. *Given the appropriate perturbation*, when will a transition from, say, the quiescent state to a state of steady convection occur?

The flow velocities under conditions not too far above those required for the onset of convection are in general small. Therefore density changes that may result from inertial forces are negligible. Furthermore, it is assumed that the spatial variations of the fluid properties (from dependence on temperature, concentration, etc.) are small as compared to their average value. These assumptions allow us to base the description of motion on the simplified Navier-Stokes equation (5.75) which, considering gravity as the only driving force for flow, takes on the form

$$\rho \frac{D\underline{v}}{Dt} = -\nabla p + \eta\nabla^2\underline{v} + \rho\underline{g} \quad , \tag{5.203}$$

where $\underline{g}$ in the body force term is the vector of the gravitational acceleration (0, 0, -g). Let us now consider the case that a potentially unstable density gradient is established in a uniform *vertical* temperature gradient resulting from conduction between a lower horizontal plane which is heated and an upper horizontal plane which is cooled. We don't have to specify the lateral extent of the planes at this point. Let T_0 and ρ_0, respectively, indicate the temperature and density of the fluid at the point chosen as origin, for instance, at the lower plate. Let the subscript u denote *local* quantities in the *unperturbed* (here quiescent) state. The superscript will be used for the change in these quantities due to the disturbance. Hence in the *perturbed* state the real local quantities are

$$T = T_u + T', \quad \rho = \rho_u + \rho', \quad p = p_u + p' \quad . \tag{5.204}$$

Note that even in an incompressible fluid, gravity (i.e., the weight of higher fluid layers) induces a pressure gradient. For *subcritical* conditions, that is in the quiescent fluid, $T' = 0$, $\rho' = 0$ and $p' = 0$. The density distribution is then

$$\rho = \rho_u = \rho_0[1 - \beta(T_u - T_0)] \quad , \tag{5.205}$$

with the thermal expansion coefficient, $\beta = V^{-1}(\partial V/\partial T)_p$. Also, the linear temperature profile assumed above can be written as

$$T = T_u = T_0 + \nabla Tz \quad , \tag{5.206}$$

where ∇T is the unperturbed temperature gradient. Combining (5.205) and (5.206) one gets

$$\rho_u = \rho_0(1 - \beta\nabla Tz) \quad . \tag{5.207}$$

Hence for the unperturbed, motionless state the Navier-Stokes equation (5.203) reduces to

$$0 = -\nabla p_u + \rho_u \underline{g} \quad . \tag{5.208}$$

Thus, a *necessary* criterion for static equilibrium in fluids is that the hydrostatic pressure and the mass density are constant in horizontal planes.

For the perturbed state (5.203) takes on the form

$$(\rho_u + \rho')\frac{D\underline{v}}{Dt} = -\nabla(p_u + p') + (\rho_u + \rho')\underline{g} + \mu\nabla^2\underline{v} \quad . \tag{5.209}$$

The first two terms on the right-hand side can be regrouped to $(-\nabla p_u + \rho_u\underline{g}) - \nabla p' + \rho'\underline{g}$, where the term in parentheses is zero due to (5.208). Thus (5.209) can be reduced to

$$(\rho_u + \rho')\frac{D\underline{v}}{Dt} = -\nabla p' + \rho'\underline{g} + \mu\nabla^2\underline{v} \quad . \tag{5.210}$$

After dividing both sides by ρ_0, the left-hand side of (5.210), using (5.207) becomes

$$\frac{\rho_u + \rho'}{\rho_0}\frac{D\underline{v}}{Dt} = \left(1 - \beta\nabla Tz + \frac{\rho'}{\rho_0}\right)\frac{D\underline{v}}{Dt} \quad . \tag{5.211}$$

In order to compare the relative weight of the terms in (5.211) for usual experimental conditions we must express ρ' in terms of T'. Employing (5.204) and (5.207) one obtains (the derivation is left to the reader as Problem 5.23)

$$\rho' = \rho_0 \beta T' \quad . \tag{5.212}$$

Realizing that typical values for β are a few 10^{-3} for gases and a few 10^{-4} for liquids, one sees that $\rho'/\rho_0 << 1$ for all practical conditions. The second term in (5.211) can also be neglected as compared to unity as long as the temperature gradient does not take on extreme values. This assumption was also made above in form of negligible property variations ($\Delta\mu$, $\Delta\beta$, $\Delta\kappa$) throughout the fluid. Hence one may disregard in this problem the change in density due to temperature "except in so far as it modifies the operation of gravity", as RAYLEIGH [5.74] has phrased it. This approach is commonly attributed to BOUSSINESQ [5.85]. However, MIHALJAN [5.86], in a careful analysis of the "Boussinesq approximation", has pointed out that Oberbeck [5.87] has made use of it in meteorological studies already in 1879. In fact, Oberbeck's derivation in terms of power series appears more profound than the list of assertions with which Boussinesq justified the approximation.

Now, with (5.212) the *Oberbeck-Boussinesq approximation* of the Navier-Stokes equation, i.e., the equation of motion for the perturbed state, can be written as

$$\frac{D\underline{v}}{Dt} = - \frac{\nabla p'}{\rho_0} - \beta T' \underline{g} + \nu \nabla^2 \underline{v} \quad . \tag{5.213}$$

For the slow flow velocities that prevail not too far from critical conditions, one may further simplify (5.213) by neglecting the second-order terms of the velocities. The resulting linearized Boussinesq equation reads then

$$\frac{\partial \underline{v}}{\partial t} = - \frac{\nabla p'}{\rho_0} - \beta T' \underline{g} + \nu \nabla^2 \underline{v} \quad . \tag{5.214}$$

Of course, for the transition from (5.213) to (5.214) one must, strictly speaking, assume that not only the velocity but also the velocity *gradients* in the fluid are small; note the $\partial u/\partial x$... terms in Dv/Dt, (5.76). Hence the linearized equation holds well for fluids with $N_{Pr} > 1$ only, as we will discuss below.

For a derivation of the Boussinesq approximation for a compressible fluid (ideal gas) where the temperature and density in (5.213) must be replaced by "adiabatically corrected" values, see [5.95].

Analogous to the above derivation of the equation of motion for the (lowest) convection state one can obtain the corresponding linearized heat transfer equation (see, e.g., [5.88]) as

$$-\underline{v}\nabla T = \left(\frac{\partial}{\partial T} - \kappa\nabla^2\right) T' \quad . \tag{5.215}$$

Combining (5.214) and (5.215) and the continuity equation (5.74) so as to eliminate all other variables, one obtains for the vertical velocity component of the perturbation

$$\left[\left(\frac{\partial}{\partial t} - \nu\nabla^2\right)\left(\frac{\partial}{\partial t} - \kappa\nabla^2\right) + \beta\nabla Tg\nabla_2^2\right]w = 0 \tag{5.216}$$

with the two-dimensional Laplace operator $\nabla_2^2 = (\partial^2/\partial x^2) + (\partial^2/\partial y^2)$. Alternatively, with somewhat different combinations of the linear equations one can eliminate w and obtain a like equation in T'.

In order to assess the stability of a system one must investigate its reaction to all possible disturbances. To make this task mathematically manageable one assumes that an arbitrary infinitesimal perturbation can be decomposed into a basic set of components (modes), for instance into (spatial) Fourier components. Then instability to one single mode out of the whole set implies instability. Note, however, that this supplies a sufficient but not a necessary criterion. Deviations from the simplifying assumptions of a Boussinesq fluid may result in instabilities under conditions that are stable from the point of view of linear theory. As we will see below, for instance, such *subcritical* instabilities can originate in the temperature dependence of fluid properties.

For the problem of infinite lateral extent of the fluid (Bénard geometry) one may build up the velocity perturbation from discrete Fourier components in the form

$$w(x,y,z;t) = \sum_k A_k(z,t)\exp(i\underline{k}\,\underline{r}) \tag{5.217}$$

with $\underline{r} = (x,y)$ and the wave vector $\underline{k} = (k_x, k_y)$ and solve (5.216) for given boundary conditions with the product trial function

$$w_k = A_k(z)F_k(x,y)\exp(\sigma_k t) \tag{5.218}$$

where k stands for the particular Fourier mode used for the stability test. The time factor $\exp(\sigma_k t)$ decides the stability question. For $\sigma_k < 0$ the considered perturbation with wave vector k will be damped out, and the state of the fluid is stable against perturbations of that spatial periodicity. At $\sigma_k = 0$, that is, at marginal stability, the component and hence a perturbation containing this component will just be retained. However, if $\sigma_k > 0$, the perturbation increases exponentially with time and the unperturbed fluid state is unstable. Under certain conditions, but not for those of the ideal Bénard configuration [5.88], σ_k can become complex and oscillatory fluid motion will set in. This is sometimes referred to as "overstability". What EDDINGTON [5.89] has coined overstability is best explained in his own words: "In the usual kinds of instability a slight displacement provokes restoring forces tending away from equilibrium; in overstability it provokes restoring forces so strong as to over-shoot the *corresponding* position on the other side of equilibrium." Thus "overstability" exists only if, in addition to being complex, σ_k has a positive real part Re and oscillatory motion of increasing amplitude develops. There are, however, various situations that lead to oscillatory convective flow with decreasing [damped, $\mathrm{Re}\{\sigma_k\} < 1$] or perfectly constant amplitude. Hence, "overstability", as introduced into hydrodynamic stability discussions in [5.84] is only a special case and should not be used for oscillatory modes in general.

The amplitude factor $A_k(z)$ in the trial function (5.218) is arbitrary, i.e., in contrast to σ, independent of the boundary conditions. Consequently, neither the amplitude nor the direction (upward or downward) of the vertical velocity component can be obtained from linear theory. A determination of the actual planform is also beyond the reach of the linear approximation. Any solution with the characteristic periodicity (wave vector) of the critical state (rolls, hexagons, rectangles,...) are equally allowed. The planform as well as the vertical velocity component can, in principle, be decided upon with nonlinear, finite amplitude perturbation theory [5.90-92].

In this context, it is also important to note that the threshold condition obtained from linear theory for the Bénard problem of constant property fluids characterizes indeed the lowest *necessary* density gradient for instability, even for finite amplitude perturbations [5.93].

Expansive convection

Flows near critical condition:
Although the results for convective instabilities between *infinite horizontal boundaries* (Bénard problem) have little *direct* bearing on crystal growth, they well illustrate the fundamental trends. Figure 5.48 represents the result of

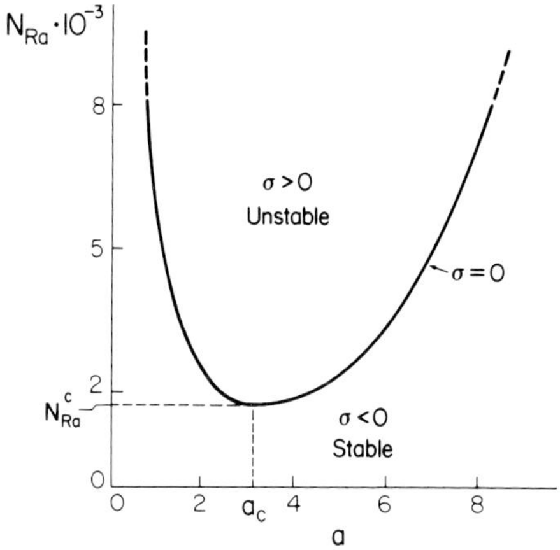

Fig.5.48. Marginal stability curve (σ = 0) for expansive convection in Bénard geometry for rigid-rigid boundaries. N_{Ra} versus normalized wave number a. After [5.77]

linear stability theory for a configuration with rigid boundaries of much higher thermal conductivity than the fluid. One sees that instability (onset of steady convection) is predicted when the governing dimensionless group

$$N_{Ra} = \frac{\beta g \Delta T d^3}{\nu\kappa} \quad , \tag{5.219}$$

the *Rayleigh number*, exceeds a critical value N_{Ra}^c, around 1700 for rigid-rigid conducting boundaries. In (5.219), $\beta = (1/V_0)\partial V/\partial T$ is the thermal expansion coefficient and ΔT the temperature difference between the horizontal surfaces that are separated by d. The Rayleigh number can be understood as a dimensionless temperature gradient. The ratio of N_{Ra} and N_{Pr} defines the *Grashof number*, i.e.,

$$N_{Ra} = N_{Pr}\cdot N_{Gr} = \frac{\nu}{\kappa}\cdot\frac{\beta g \Delta T d^3}{\nu^2} \tag{5.220}$$

Table 5.22. Critical Rayleigh numbers for the onset of steady convection in infinite horizontal fluid layer heated from below. After [5.97]

Bounding Surfaces	Both conducting	Upper insulating Lower conducting	Both insulating
Both free	658	-	120
Upper free	1101	669	320
Both rigid	1708	1296	720

which in turn is the dimensionless product of the ratios of buoyant to viscous forces and inertial to viscous forces, respectively;

$$N_{Gr} = \frac{F_b F_{in}}{(F_{vis})^2} \quad . \tag{5.221}$$

The abscissa in Fig.5.48 represents the wave number $a = 2\pi/\lambda$, where λ is the wavelength of the perturbation measured in units of the total fluid height d.

Table 5.22 summarizes the results of linear stability theory for N_{Ra}^{c} for various combinations of the thermal and mechanical properties of the boundaries. The stabilizing effect of rigid (no slip) as well as thermally conducting surfaces is clearly reflected. "Conducting" is used for the situation in which the thermal conductivity of the fluid, k_f, is much smaller than that of the bounding material, k_b, so that the temperature perturbations vanish on the boundaries. The dependence of N_{Ra}^{c} on the ratio k_f/k_b is depicted in Fig.5.49 [5.94].

Evaluating (5.219) for specific systems (see Problem 5.24) one notes the following trends:

(a) in liquids much larger temperature gradients are needed than in gases to sustain convection;

(b) gases, in contrast to liquids, are convectively considerably more stable at elevated temperatures than at room temperature.

Due to the mathematical complexity of such overall stability considerations there is a strong desire for heuristic arguments. Hence, qualitative discussions of convective behavior are often based on local criteria. In interpretations of the expansive Bénard problem, for instance, one can encounter the following arguments: In order to trigger *sustained* convection,

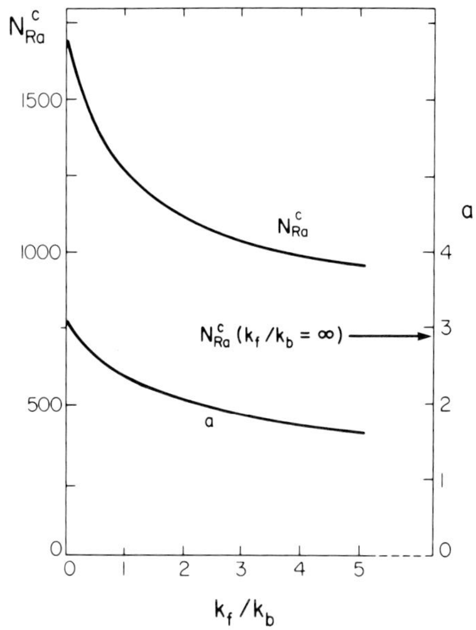

Fig.5.49. Expansive convection in Bénard geometry. Dependence of N^c_{Ra} and the corresponding critical wave number a on the ratio k_f/k_b. Rigid-rigid boundaries. From [5.94] by permission of the Royal Society, London

a perturbation must displace a buoyant fluid element, say upwards, in a manner that the element *remains* buoyant. Hence, the fluid element must be displaced rapidly enough so that it remains at a higher temperature than the new environment for a significant amount of time. Viscosity impedes such a displacement. High thermal conductivity of the fluid favors rapid thermal adjustment. Consequently, the viscosity and the thermal conductivity are stabilizing factors [see (5.219)]. From these arguments one can even envision why instability sets in for a specific size (lateral periodicity) of a perturbation (Fig.5.48). Very small displaced fluid elements ($a \gg 0$), due to their large surface to volume ratio, lose heat more readily than larger elements. Very large elements, on the other hand, act self-stabilizing by significantly changing the temperature of the environment.

We see that such heuristic arguments have some value in pointing out trends. However, threshold phenomena cannot be explained based on such local descriptions. A local consideration is rarely representative for the system*.

---

* G.H. Westphal has helped the author considerably in clarifying this point.

Referring to the Bénard problem, the above simplifying approach ignores that there are extended areas with horizontal and hence practically isothermal flow, besides the up or down displacements considered. There are even elements which remain stationary during steady convection. *All* these elements, due to their *relative* motion, will participate in the viscous dissipation of energy. Yet, only part of them will, via their motion (and energy transport), *tend* to lower the energy of the system. Thus an instability criterion must be derived from a *global* consideration of the fluid, i.e., a consideration of heat and mass transfer in a volume that is representative for the overall problem. For the Bénard configuration, for instance, that representative fluid volume would contain *at least* one (hexagonal) cell or roll plus the corresponding upper and lower boundary areas. For less symmetric configurations one may have to include the whole system.

Based on rigorous global considerations CHANDRASEKAR has derived a very revealing interpretation of the threshold in the expansive Bénard problem: "Instability occurs at the minimum temperature gradient at which a balance can be readily maintained between the kinetic energy dissipated by viscosity and the internal energy released by the buoyancy force" [Ref. 5.84, p. 34].

At the threshold, N_{Ra}^{c}, as briefly mentioned before, linear theory [5.88] as well as finite amplitude (nonlinear) analysis [5.93] predict the onset of a steady non-oscillatory convection mode. Overstability *at the onset* cannot occur in the buoyancy-driven monocomponent Bénard problem. Observations to the contrary can be traced back to the presence of horizontal ∇T-components. Such a situation of σ_k in (5.218) being real and equal zero for the marginal state is customarily referred to as "validity of the principle of the exchange of stabilities (PES)" *for that case*. It is frequently assumed that the PES is generally valid; this assumption is wrong. Below we will encounter various more complex convection configurations for which the marginal state is oscillatory.

Most theoretical predictions concerning expanisve convection in the Bénard geometry are well borne out by experiment. One must keep in mind, however, that in calculations for the Bénard geometry one assumes an infinite lateral extent of the fluid. In practice, this condition can only be approximated. One can keep the aspect ratio very low, yet the diameter of containers used in convection studies seldom exceeds 20 cm. Large diameters impose problems with respect to uniform heating of the bounding plates. Consequently, the effect of lateral walls, though often ignored, is present in any experiment.

(a) (b)

(c) (d)

(e) (f)

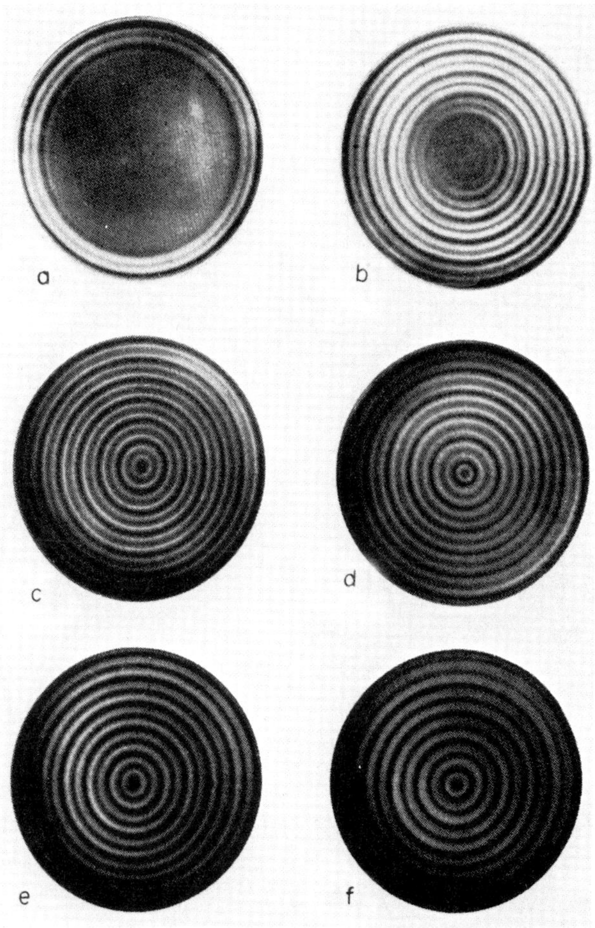

Fig.5.50. Development of subcritical rolls and increase of wavelengths with N_{Ra}. For details see text. From [5.98] by permission of Pergamon Press

It has been shown [5.98] that the presence of lateral walls and of (associated) small horizontal temperature gradients may cause *subcritical rolls*. The first roll originated near the circular cylindrical wall already at about 0.6 N_{Ra}^c (see the discussion of [5.98] in [5.71]). The second and third roll (Fig.5.50a) developed at 0.75 and 0.85 N_{Ra}^c, respectively. Further increase in ΔT causes more rolls to form until at N_{Ra}^c the entire cross section is filled with concentric rolls (Fig.5.50b). At *supercritical* Rayleigh numbers (i.e., $N_{Ra} > N_{Ra}^c$) the innermost ring began to shrink and disappeared at sufficiently high ΔT. With further increase in N_{Ra} consecutive disappearance of rings (Fig.5.50d-f) was followed by irregularities in the flow

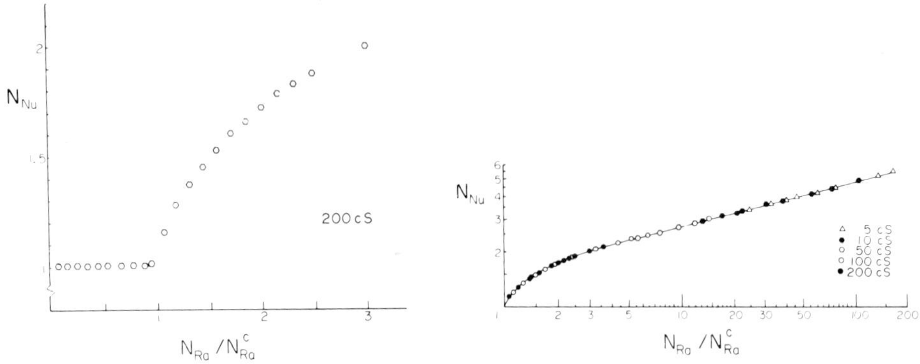

Fig.5.51. Heat transfer curves for various silicone oils. N_{Nu} vs. N_{Ra}/N_{Ra}^{c}. Bénard geometry, height-diameter ratio =4/100. From [5.98] by permission of Pergamon Press

and finally by time-dependent flow. It is interesting to note that in the subcritical range, inspite of the presence of several rolls, the heat flux across the layer remained practically unchanged as depicted in Fig.5.51. Only at and above N_{Ra}^{c} a measurable increase in heat transfer was observed!

The various critical N_{Ra}'s and wavelengths of the flow pattern predicted by linear theory for the onset of instability have been experimentally confirmed with high accuracy for a large variety of fluids with relatively temperature-independent properties. (Silicone oils resemble particularly well such "Boussinesq fluids".) For fluids with significant temperature dependence of the relevant properties (mostly k and η), the Boussinesq equation cannot yield accurate results. Indeed for such "Non-Boussinesq" fluids nonlinear analysis shows that "subcritical" instabilities (with respect to N_{Ra}^{c} from linear theory) may arise in the presence of *finite amplitude* disturbances even though infinitesimal disturbances decay. Also, the analysis shows that the pattern of the steady subcritical motion consist of regular hexagonal cells. (See [Ref. 5.90, Sect. 7]. This combination of fluids with strongly temperature-dependent properties (melts near freezing point!) and finite amplitude perturbations (from mechanical vibrations or non-ideal temperature control) is likely to occur in crystal growth.

A few more words on the flow pattern since it is of particular importance for interfacial mass transfer. For instance, the effective segregation (Sect.

6.1) depends strongly on the flow velocity distribution adjacent to an interface. For $N_{Ra} > N_{Ra}^c$, linear theory allows equally for all wavelengths that are within the marginal stability curve (Fig.5.48). In practice, however, the convection rolls observed in shallow fluid layers (i.e., $a \lesssim 1$) always possess widths which are comparable to the layer height. Hence, the number of rolls is a function of the aspect ratio. During a crystal growth process the aspect ratio and thus the number of rolls may change, as, for instance, during the pulling of a crystal from its melt. Thus, there may be discrete jumps in the convection pattern that "flushes" an interface as a crystal grows and the remaining melt layer becomes shallower. In addition most experiments indicate that the number of rolls is a (weak) linear function of N_{Ra} up to about $N_{Ra} = 10\, N_{Ra}^c$ [5.77]. The number of rolls decreases with N_{Ra}. This is theoretically not understood yet. In low Prandtl number fluids this increase in wavelength is more pronounced than in fluids with $N_{Pr} \gtrsim 1$. Also of practical importance is the fact that the wavelength depends on the physical character of the bounding horizontal planes. As predicted theoretically (Fig.5.49) and experimentally confirmed the wavelength decreases with increasing thermal conductivity of a boundary. Also λ is larger at free boundaries as compared to rigid ones. The influence of surface roughness on convective stability is taken up below.

Radiative heat transfer has tacitly been ignored in the stability considerations to this point. Yet radiation transfer may be significant in high-temperature crystal growth from (semi-) transparent fluids (melts, fluxes). Radiative exchange modifies the linear temperature profile obtained from conduction only. The temperature of the fluid layers adjacent to a boundary will be closer to the temperature of the boundary than without radiation. This reduces the effective layer height d in (5.219) and hence should act stabilizing. A recent theoretical treatment of the expansive Bénard problem including radiative heat transfer confirms this expected increase in N_{Ra}^c [5.99]. Well-defined experiments are still lacking in this area.

The above derivation of the stability criterion was based upon the simplified energy equation (5.148), i.e., viscous conversion of motional into thermal energy was not taken into account in the energy balance. In most practical crystal growth situations this is well justified. However, as a most recent treatment of the expansive Bénard problem including viscous dissipation shows, under certain conditions major errors must be expected [5.100]. The key parameter is the non-dimensional dissipation number $D_i = \beta gd/C_p$, with C_p the specific heat at constant pressure and all other parameters as in (5.219). As soon as the D_i of a specific system (not restricted to Bénard configuration) is not small as compared to one the viscous dissipation acts stabilizing. That is to say that for $D_i \geqslant 0.1$ the convective stability will be significantly increased or, in the case of flow, the convection velocity will be reduced. For typical parameter values the group $\beta g/C_p$ has values of $10^{-4} - 10^{-7}$ cm and hence viscous dissipation will only play a role in systems with large length scale. However, at very low temperatures where β/C_p can become very big, or at very high body forces, even typical container dimensions suffice to obtain $D_i \approx 1$. In crystal growth these extreme conditions are, for instance, found in the solidification of rare gas crystals [1.2] or in crystallization in a centrifuge [5.20].

N^c_{Ra} is independent of N_{Pr}. However, once free convective flow has set in there are very important differences in the flow behavior between fluids of small and large Prandtl numbers. For forced flow near solid fluid interfaces we have seen in Fig.5.28 that for $N_{Pr} \gtrsim 1$ the momentum boundary layer is wider than or comparable to the thermal boundary layer. Consequently, as convection sets in, the initial linear temperature profile is more readily deformed by the flow than in fluids with $N_{Pr} \ll 1$. This trend is schematically indicated in Fig.5.52 for the temperature distribution in the vertical center plane of an individual convection roll, for instance. (Asymmetries that will arise from the preferential direction given by $\underline{g}$ are ignored.) Consequently, the temperature gradient in the "core" of the roll is sizeably reduced. That, in turn, causes the convective streamlines to be more concentrated adjacent to the ("convection-sustaining") horizontal planes. The flow becomes increasingly confined to the boundary regions. For fluids with $N_{Pr} < 1$ (Figs. 5.28a and 52b) quite different behavior can be expected. Since convection contributes to a lesser extent to heat transfer in these flows the linear temperature profile is less deformed upon onset of convection. Consequently, more of the original gradient can *remain* active over a greater volume of the fluid for driving the flow. Even with further increase in N_{Ra} (or ΔT for that matter) the flow maintains more of a circulatory bulk character.

The flow velocities can also be expected to be quite different for these two groups of fluids. We know that convection sets in at the same N_{Ra}, independent of N_{Pr}. This, however, according to (5.220) and (5.221) implies that for low-N_{Pr} fluids the Grashof number and hence the flow velocity (in

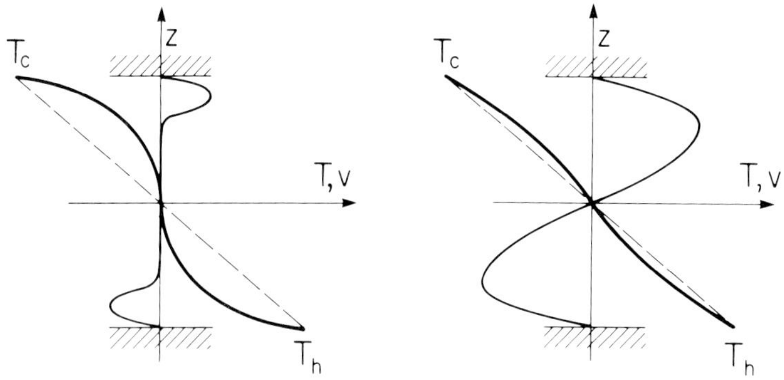

(a) (b)

Fig.5.52. Schematic temperature and velocity profiles (heavy and light curves, respectively) in the vertical center plane of a convection roll between horizontal planes. (a) $N_{Pr} > 1$, (b) $N_{Pr} < 1$. Assymmetry from gravity ignored. Dashed lines represent temperature profile from conduction only

the inertial force term) must be much higher than for high-N_{Pr} fluids where the same N_{Ra} leads to a smaller N_{Gr}. (Triumph of dimensionless analysis!) These qualitative deductions are well confirmed by theoretical and experimental results, some of which are presented in later sections. Up to two orders of magnitude higher flow velocities can be expected in, say, liquid metals as compared to molten oxides for the same N_{Ra}. The high free-flow velocities encountered in low-N_{Pr} fluids, as indicated without justification in connection with (5.214), invalidate the linearized Boussinesq equation in these systems as soon as convection sets in. On the other hand, for $N_{Pr} > 1$ linear theory yields often reasonable results over a range of several N_{Ra}^{c} beyond the marginal condition.

The difference in flow patterns and velocities of the two groups of fluids has various practical implications. For instance, the resulting mixing behavior in melts can be expected to be quite different. Also, the surface roughness of a container will have a quite different influence on the convective stability of the fluid. Surface roughness can be important for high-N_{Pr} fluids where the flow is concentrated in a "boundary layer". For low-N_{Pr} fluids, however, where the flow is more distributed throughout the bulk, surface roughness will be of little influence.

Convection on vertical boundaries in contrast to the Bénard geometry possesses no threshold. Any finite horizontal temperature gradient leads to buoyancy-driven flow unless there are other, stabilizing forces present. This can readily be seen from the *necessary* hydrostatic equilibrium condition (5.208) which, dropping the subscript u and realizing that ρ may be a function of position is

$$\nabla p = \rho(\underline{r})\underline{g} \quad . \tag{5.222}$$

Following [5.101], we take the curl of (5.222) and obtain (Problem 5.25)

$$\nabla\rho \times \underline{g} = 0 \quad . \tag{5.223}$$

With $\rho = \rho(T)$, (5.223) becomes

$$\frac{d\rho}{dT} \nabla T \times \underline{g} = 0 \quad , \tag{5.224}$$

which is only fulfilled if the temperature gradient is parallel (or antiparallel) to the gravity vector. In the presence of *any* horizontal ∇T component, (5.224) is non-zero, and the quiescent state is not a solution of the Navier-Stokes equation (5.203) for that system; that is to say, flow will set in. The vertical layers of different density can simply slide or fold over one another. No threshold has to be overcome since no horizontal layers have to intermix, as in the Bênard geometry, to move towards a state of lower energy.

For a treatment of the free convective flow at a vertical plane the N_{Pr} of the fluid plays again a decisive role. If $N_{Pr} \gtrsim 1$, as discussed in connection with horizontal boundaries, the flow will be confined to a region close to the vertical plate, and solutions to the boundary-layer-form of the governing equations will yield realistic results. For $N_{Pr} \ll 1$, however, where the flow extends far into the bulk of the fluid, the higher-order terms in the vertical (z) direction in the equation of motion cannot be neglected anymore. There are numerous solutions and experiments in the literature of the high-N_{Pr} case of free flow due to a vertical plate (either heated or cooled) in an infinite fluid since the first solution was presented by Pohlhausen in 1930. For references see, e.g., [5.21, 61, 92]. More recently OSTRACH (e.g., [5.102]) has derived solutions for the whole range of N_{Pr}. Figure 5.53 presents the dimensionless temperature and velocity profiles he obtained. The N_{Ra} used in these similarity plots is based upon the distance z from the

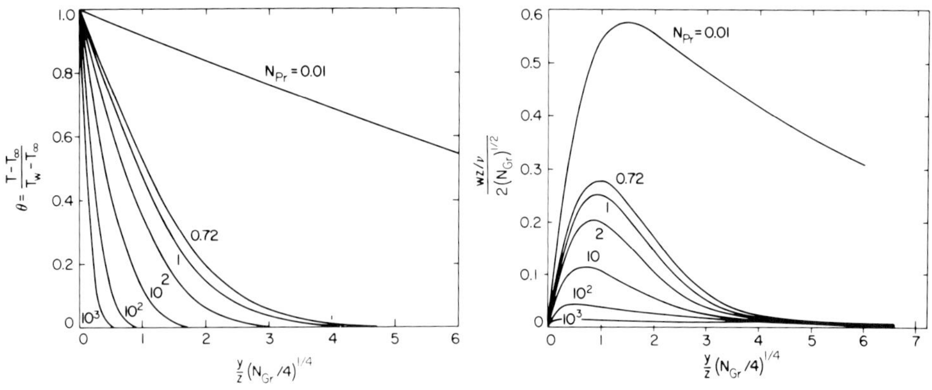

(a) (b)

Fig.5.53. Dimensionless temperature profile (a), and velocity profile (b) for free convection on a vertical plate. After [5.101]

leading edge; y is the horizontal distance from the plate; T_W and T_∞ are the temperature of the plate and the quiescent bulk fluid, respectively. The previously expected dependence of the flow character on N_{Pr} is well confirmed in these plots. Also, note that the velocity and thermal boundary layer thickness grow with $z^{1/4}$ as opposed to the $x^{1/2}$ dependence found for forced flow parallel to a plane (5.95).

The non-radiative heat flux between the isothermal vertical plane and the fluid can be computed from the dimensionless interfacial temperature gradients in Fig.5.53a. For the Nusselt number, (5.154), one obtains [5.92]

$$N_{Nu} = -F(N_{Pr}) \cdot (N_{Gr})^{1/4} \quad , \qquad (5.225)$$

where $F(N_{Pr})$ is given in Fig.5.54.

Radiative exchange has been included in the vertical-plane convection problem in recent work [5.103]. With increasing optical thickness of the fluid, (5.184), the (additional) radiative heat flux to the fluid adjacent to the plane, causes an increase in the flow velocity as well as in the velocity boundary layer thickness beyond the corresponding non-radiative values. Results for a system with $N_{Pr} = 0.7$, a conduction-to-radiation parameter $N = k\beta/4\sigma T_W^3 = 1$ [k:thermal conductivity; β:extinction coefficient K

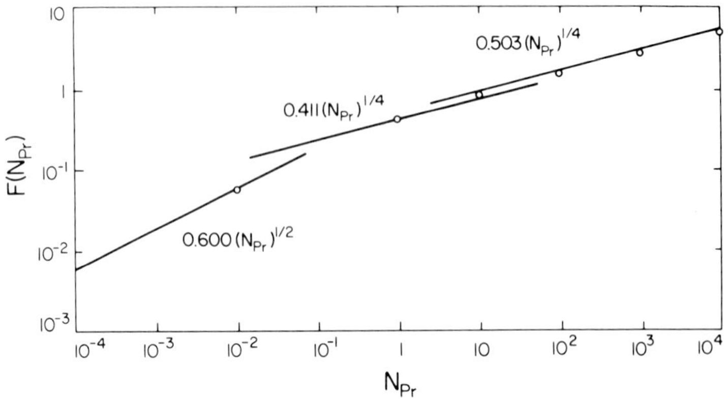

Fig.5.54. Heat transfer characteristics [$F(N_{Pr})$ in (5.225)] for free convection from an isothermal vertical plane. After [5.92]

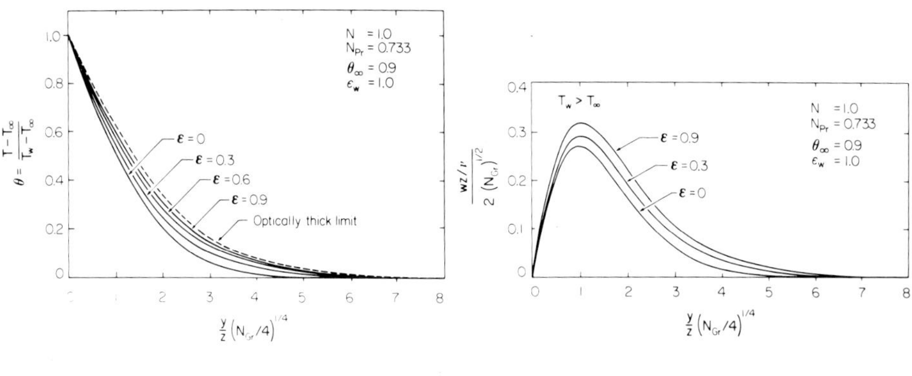

(a) (b)

Fig.5.55. Effect of radiative heat exchange on convective flow on a vertical, isothermal plane. For parameters see text. (a) dimensionless temperature profiles; (b) dimensionless velocity profiles. From [5.103] by permission of Pergamon Press

in (5.180)], $T_\infty = 0.9\ T_w$, and a wall emmissivity of 1, are given in Figs. 5.55a and b for various values of the Bouguer number $\xi \equiv \beta z(N_{Gr}/4)^{-1/4}$.

Supercritical flows:

Upon continued increase of N_{Ra} beyond the marginal state (which may be at $N_{Ra}^c = 0$ when a horizontal ∇T is involved), flow patterns of higher and higher complexity appear in a fluid, assuming that the appropriate wavelength perturbation is present. The transitions from certain planforms or modes to the next higher modes occur again at discrete values of N_{Ra}. Hence, the necessity arises for additional indexing of the critical Rayleigh numbers: N_{Ra}^{c1} for the marginal state, N_{Ra}^{c2}, etc. (There is, however, no commonly agreed upon notation yet.) The specific values of these higher critical Rayleigh numbers depend on the specific boundary conditions (shape and physical properties of boundaries) *and*, in contrast to N_{Ra}^{c1}, also on N_{Pr}. This latter dependence follows from the distinctly different flow behavior of low- and high-N_{Pr} fluids discussed above. Characteristic differences in convective stability can also be expected for a given fluid depending on whether the driving ∇T is strictly parallel to $\underline{g}$ or whether it possesses a horizontal component. The shear flows and rolls induced by horizontal temperature gradients are inherently gravitationally stable ("light on top of heavy") as opposed to the rolls obtained in vertical destabilizing temperature gradients. In the crystal growth practice one encounters typically a combination of horizontal and vertical heat fluxes. The resulting complexity of the convection phenomena, though intellectually fascinating, has been impeding crystal growth developments considerably.

In the following we will discuss higher convective modes with particular concern for oscillatory instabilities. First, strictly vertical upward heat flow between parallel planes (Bênard configuration) is considered. Then we discuss combinations with horizontal heat flows that result from the tilting of the parallel planes. Finally two-dimensional heat flow is added to the parallel-plate problem in the horizontal and vertical configuration. Trends that arise from additional lateral confinement are indicated throughout.

In Fig.5.56 KRISHNAMURTI [5.104] has compiled the results of her extensive studies of the expansive Bênard problem (circles) and data of other workers (other symbols). One sees that only N_{Ra}^{c1} (transition from quiescent fluid to steady rolls, curve I) is independent of N_{Pr}. All other regime boundaries show a strong dependence on the fluid properties for $N_{Pr} \lesssim 50$. At curve I the previously discussed steady straight rolls develop. In laterally confined fluid layers (with $a < 1$) the rolls prefer to meet most of the boundaries at right angles. In circularly bound layers this leads to bent steady roles [5.104]. In rectangular enclosures the rolls are typically parallel to the short side of the "cell". In high Prandtl number fluids this two-dimensional flow becomes unstable with respect to a steady three-dimensional flow when

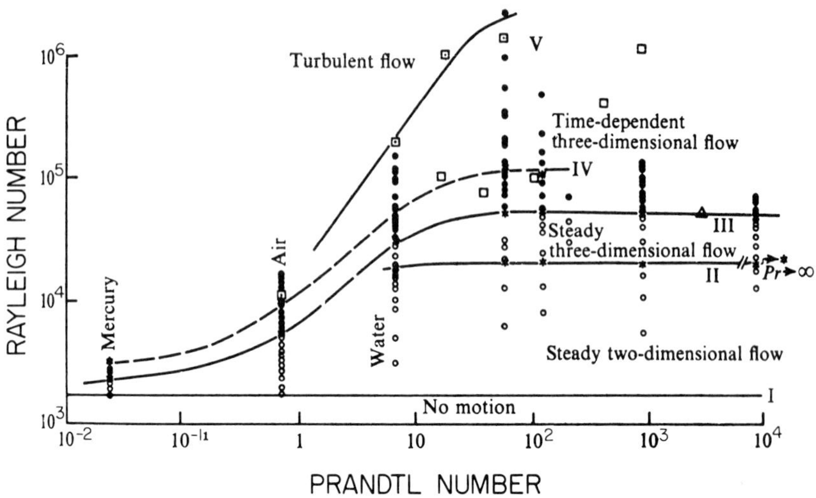

Fig.5.56. Regime diagram for expansive convection between horizontal infinite planes. From [5.104] by permission of Cambridge University Press

the conditions for curve II are reached. Two types of time-independent three-dimensional flows have been observed in the Bénard geometry: zigzag instabilities and cross-roll instabilities. The zigzag instability corresponds to a bending of the original rolls (Fig.5.57a). The cross-roll instability rolls emerge at right angles to the original rolls (Fig.5.57b). Upon further increase of N_{Ra} a steady regular bimodal flow evolves from the cross-roll instability. A particularly well-developed example for this regular cell pattern is given in Fig.5.58. Cross-roll cells show a considerable hysteresis [5.104, 105].

Theoretical information about the stability ranges of these three-dimensional flows can only be obtained when the *advection* of buoyancy and momentum *by the flow generated by buoyancy* is taken into account, which, by definition requires the nonlinear transport equations. The theoretical models, mostly developed by BUSSE and coworkers (see [5.106]) agree rather well with the observed data. Fig.5.59 depicts the stability regions calculated for N_{Pr} = 7. The lowest curve corresponds to the marginal stability curve in Fig. 5.48. For wave number-N_{Ra} combinations below this curve, the fluid is quiescent. Inside the "pear-shaped" region two-dimensional flow (i.e., parallel rolls) is stable. Outside of this region, above R_c (our N_{Ra}^{cl}), of course, in-

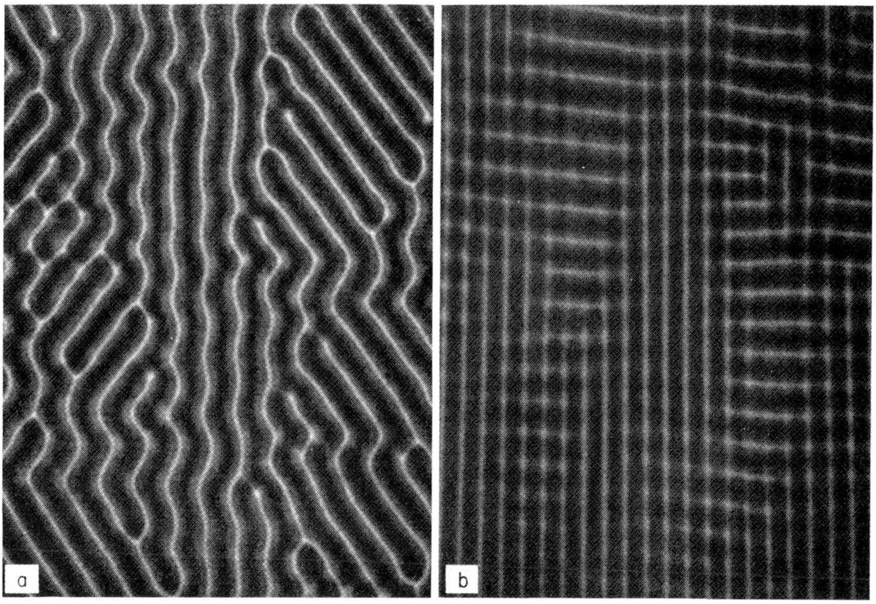

(a) (b)

Fig.5.57. Steady three-dimensional flows observed in Bênard geometry at moderate N_{Ra} of high-N_{Pr} fluids. (a) Zigzag instability; (b) cross roll instability. From [5.105] by permission of Cambridge University Press

stabilities in the rolls develop. At low N_{Ra} zigzags become stable if the wave number a of the original rolls is smaller than the critical value of about 3.2; cross-rolls will prevail for larger a's. Above $N_{Ra} = 2.3 \times 10^4$ rolls of all wave numbers are unstable and bimodal convection will dominate the flow.

Figure 5.56 shows that at low N_{Pr} three-dimensional steady instabilities lose importance. This is also well understood and reflects stabilizing influence of the momentum advection terms, which are inverse proportinal to N_{Pr}, and of the viscous dissipation terms.

Above curve III (Fig.5.56) the flow is (periodic) time dependent. Considerable efforts have been made recently to understand the origin of the oscillatory heat and mass transfer observed in this region. At this point it appears that $N_{Pr} \approx 5$ forms the borderline between two different mechanisms. In high-N_{Pr} fluids it is a gravitational boundary layer instability that leads to oscillations. Buoyant fluid accumulates at the heating bound-

Fig.5.58. Bimodal convection. $N_{Ra} = 2.5 \times 10^4$, $N_{Pr} = 450$. Photographed through top glass plate. From [5.105] by permission of Cambridge University Press

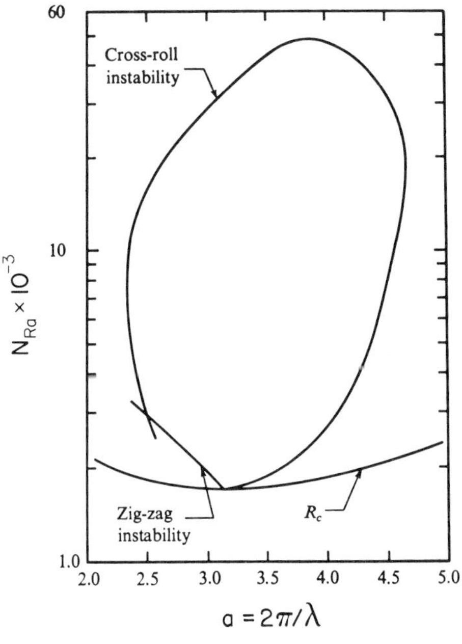

Fig.5.59. The stability of two-dimensional convection in a water layer ($N_{Pr} = 7$). Rigid boundaries, infinite thermal conductivity. From [4.106] by permission of Cambridge University Press

ary (the same argument applies, of course, to cooling upper boundaries) and grows until a local critical N_{Ra} is exceeded. Then "thermals" (Fig.5.60)

Fig.5.60. Development of thermals near a heated horizontal boundary. From [5.108] by permission of Cambridge University Press

break away and rise. A theoretical model for this process by HOWARD [5.107] explains most of the observations at least semi-quantitatively. Strong support comes from photographic studies of the development of thermals [5.108]. Unequivocal evidence, however, that the temperature oscillations observed in layers of high viscosity silicone oil originate in the periodic sweeping-by of thermals (there called "plumes") is only very recent [5.104]. The theoretical treatment of thermals is deferred to the section on melt dynamics [1.2]. The approximate time scale of oscillations due to thermals can be deduced from heuristic arguments: Most of the time during the formation of a thermal, heat transfer near the boundary is dominated by conduction. The period after a breakaway until a new thermal begins to form is observed to be short as compared to the "growth time". Hence the periodicity will be essentially governed by the *thermal diffusion time* $\tau_\kappa = d^2/\kappa$ [for the mass diffusion analogon see (6.55)] where d is comparable to the thermal boundary layer widths. For typical conditions we see that the period of these oscillations will be of the order of minutes.

For low-N_{Pr} fluids one would expect for reasons discussed previously that oscillatory instabilities result from a "bulk-mechanism" rather than the boundary layer instability observed for fluids with high Prandtl numbers. Indeed nonlinear analysis [5.109] shows that with decreasing N_{Pr}, the terms in the equation of motion which are responsible for the generation of vertical vorticity become significant at $N_{Pr} \approx 5$. The resulting instability to the original steady rolls resembles a wave propagating along a rope. This "snaking" motion is schematically depicted in Fig.5.61. "The roll pattern is shifted perpendicular to its axis periodically along the axis and in time" [5.109].

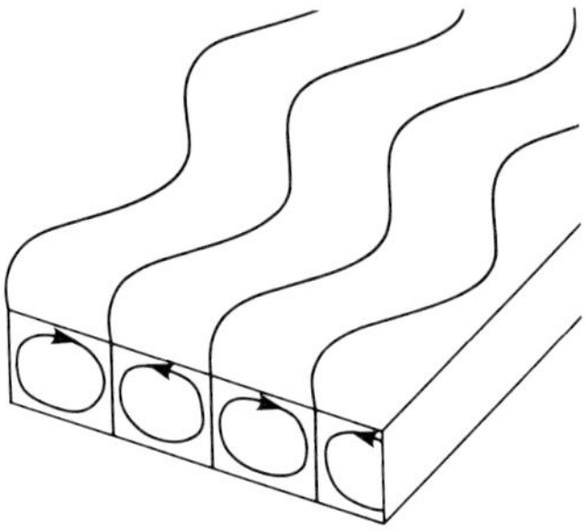

Fig.5.61. Oscillating convection rolls (schematic). The bending of the rolls propagates along the roll axis in time. From [5.106] by permission of Cambridge University Press

The period of the oscillations due to the development of vertical vorticity for a roll spacing of $\sqrt{2}$ times the layer height has been given (for *free* boundaries) to be [5.109]

$$\tau = 0.65(h^2/\kappa)[(N_{Ra} - N_{Ra}^{cl})/N_{Ra}^{cl}]^{-1/2} \quad . \tag{5.226}$$

Thus τ is of the order of the *viscous diffusion time* $\tau_\nu = d^2/\nu$ [for the mass diffusion analogue see (6.55)]. For $\nu = 3 \times 10^{-3}$cm/s (typical of liquid metals) and d = 1cm one obtains $\tau_\nu \approx 5$ min. Hence, both mechanisms result in comparable oscillation frequencies. A distinguishing feature, however, is the proportionality between the square of the frequency and N_{Ra}. This functional dependence has meanwhile been confirmed so well, even for completely confined fluids, that it can be used, in turn, for an identification of the vertical vorticity mechanism. This is convenient since low-N_{Pr} fluids are in general not transparent and hence don't allow for visual observation of the flow pattern.

It is particularly noteworthy that for fluids with very low N_{Pr} oscillations set in at Rayleigh numbers which are only about 10% in excess of N_{Ra}^{cl}.

This, at some points, had lead to the belief that convection in liquid metals is always unsteady. Unfortunately, very little is known about convection in low-N_{Pr} fluids at high N_{Ra}'s. Qualitatively though it is probably safe to assume that thermal boundary layer instabilities will probably not occur in these systems even at very high N_{Ra}'s. As discussed in context with Fig.5.52, the flow is well distributed throughout the cross section of a roll or cell and, thus, should continuously dissipate the potential build up of buoyant energy caused by adverse vertical temperature gradients.

At this point it is also important to realize that the consequences of a transition from one instability (convection mode) to another for the interfacial transport in crystal growth are twofold:

1) As the flow pattern changes, the supply pattern for dopants *and* thermal energy changes, which leads to a change in the interfacial segregation (Sects. 6.3 and 6.5) and a change in the heat flux parallel to the interface.
2) At a transition point the slope of the heat flux versus N_{Ra} changes typically rather abruptly. This is depicted for a fluid with N_{Pr} = 100, for instance, in Fig.5.62 for all transitions given in Fig.5.56. Hence, as

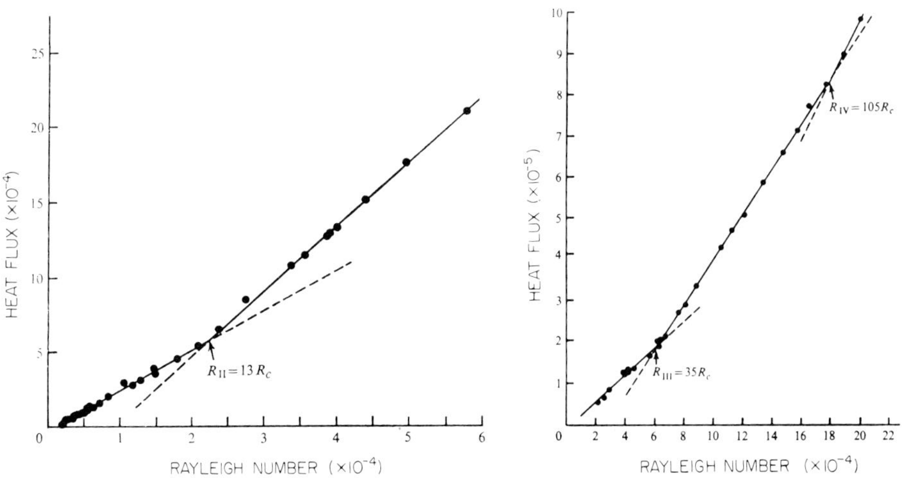

(a) (b)

Fig.5.62. Dimensionless heat flux (i.e., $N_{Nu} \times N_{Ra}$) versus N_{Ra} showing the various transitions to higher convective instabilities. $N_{Pr} = 10^2$. (a) from [5.110] and (b) from [5.111] by permission of Cambridge University Press

the nutrient for a growing crystal undergoes such a mode change, the heat flux across the interface and hence the growth rate may change significantly. This, again, may change the segregation behavior.

When *lateral boundaries* are added to the Bénard problem, the resulting additional viscous dissipation on these walls acts stabilizing. The critical Rayleigh numbers increase with the aspect ratio [height to width, (5.202)] of the confining geometry. If, however, a horizontal temperature gradient is introduced with these lateral bounds (which in practice is difficult to avoid) the overall effect can be destabilizing as discussed below.

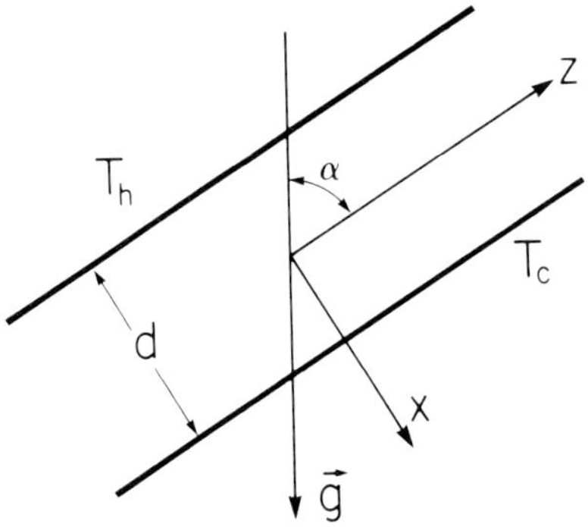

Fig.5.63. Defining sketch for convection between parallel infinite planes tilted out of the vertical position

Having treated the Bénard geometry and the vertical plane, we shall now proceed to the convective phenomena that arise between parallel planes that are *tilted* at various orientations with respect to $\underline{g}$. The geometrical parameters are characterized in Fig.5.63. For a tilt angle of $\alpha = -90°$ the system represents the expansive Bénard problem. Convective flow instabilities are of thermal-gravitational origin. For $\alpha = 0°$, the basic shear flow pattern will be symmetric about the (imaginary) center plane with upward flow on the heated plate and downward flow on the cooled plate. Instabilities in this orientation will be of hydrodynamic nature, i.e., the result from the instability of the two opposing flows. For angles between these extreme positions one will expect a complicated interplay between the two mechanisms in which the Prandtl number (i.e., the boundary layer behavior) of the fluid plays a decisive role. Numerical solutions to this complex problem were recently obtained by Russian workers [5.112, 113] for the range $5 \geq N_{Pr} \geq 0.2$

and were extended to considerably larger and smaller Prandtl numbers by [5.114]. The limited experimental data available at this point confirm the theoretical analyses very well. The results of these studies can be summarized as follows [5.114]:

1) $10^3 \geq N_{Pr} \geq 12.7$: In the vertical orientation ($\alpha = 0$) at a critical N_{Gr} instabilities to the basic shear flow set in as transverse oscillatory rolls (see, e.g., Fig.5.61, axes in y-direction). The specific value of N_{Gr}^c decreases strongly with increasing N_{Pr}. As α is changed to a certain small negative value transition to steady longitudinal rolls (axes in direction of basic flow, i.e., z-direction) occurs. As depicted in Fig.5.64, this angle of transition α_t changes from -1° up to a maximum of -2° as N_{Pr} varies from 12.7 to 27. For $N_{Pr} > 27$, α_t diminishes again. Inclination acts destabilizing, i.e., instabilities set in at decreasing values of N_{Gr} as α increases.

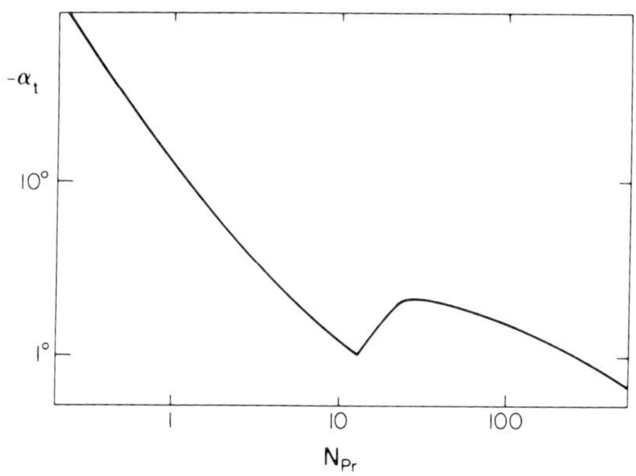

Fig.5.64. Angle of transition from transverse to longitudinal convective instabilities between inclined infinite plates. After [5.114]

2) $0.24 \leq N_{Pr} \leq 12.7$: Results calculated for $N_{Pr} = 0.2$, 1 and 5 are shown in Fig.5.65. The values obtaind for $\alpha = -90°$ must, of course, agree with the results of the Rayleigh-Bénard stability analysis. For a comparison one must note that the N_{Gr} for vertical and inclined slot problems are generally expressed in terms of the half-width d/2, and half the temperature difference between the plates, $\Delta T/2$ (see Fig.5.63). Hence, N_{Gr} is multiplied by the factor 1/16 in Fig.5.65. As mentioned previously, "transverse" and "longitu-

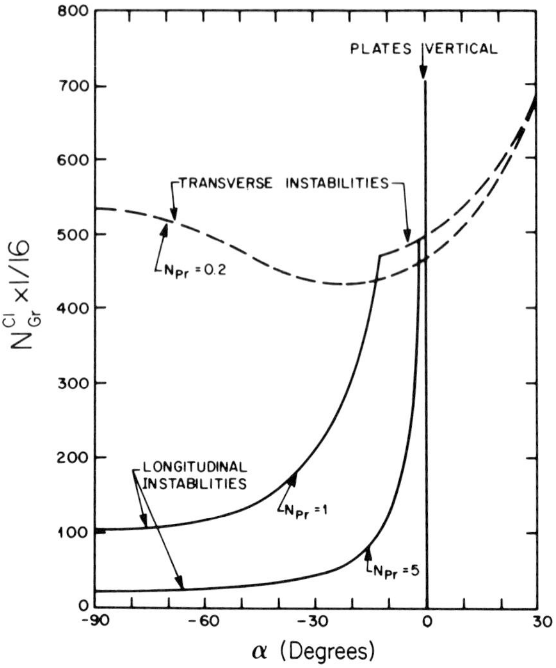

Fig.5.65. Variation of N_{Gr}^{cl} for instabilities to the basic shear flow between differentially heated inclined parallel planes for various N_{Pr}. $\alpha = -90°$: Bénard problem, $\alpha = 0°$: infinite vertical slot; [5.112]. From [5.81] by permission of Marcel Dekker

dinal" rolls are degenerate for perfectly horizontal planes with a strictly vertical ∇T. The slightest tilting, however, will lift this degeneracy. In high-N_{Pr} fluids the rolls orient themselves parallel to the incipient basic shear flow. For $N_{Pr} \approx 0.2$, on the other hand, we see that the rolls will be oriented at right angles to the basic flow. Overall the *basic flow acts stabilizing*. The more the plates are tilted out of the horizontal the larger is the N_{Ra}^{cl} (or N_{Gr}^{cl}) needed to start longitudinal rolls in fluids with $N_{Pr} \lesssim 1$. Quantitatively,

$$N_{Ra}^{cl} = \frac{1708}{\sin\alpha} \tag{5.227}$$

for rigid conducting boundaries, for instance. Eq.(5.227) simply reflects the reduced effective gravitational acceleration normal to the planes upon tilting.

With decreasing negative α a transition from the steady longitudinal instabilities to *steady* transverse instabilities (horizontal rolls) occurs at a discrete α_t, as opposed to the behavior of fluids with $N_{Pr} > 12.7$ where the transverse mode is time dependent. The value of α_t depends again on N_{Pr}. Starting at $-1°$ for $N_{Pr} = 12.7$, α_t increases to about $-10°$ for $N_{Pr} = 1$. For $N_{Pr} < 1$, $\alpha_t = -90°$ is rapidly approached (Fig.5.64).

Note also the regime for $0° < \alpha < 30°$ in which transverse rolls are established as instabilities to the basic flow by "heating from the top".

3) $N_{Pr} < 0.24$: In this range, which is representative of liquid metals, only horizontal (i.e., transverse) rolls can occur as instabilities to the basic flow.

As last step in this section let us consider the effects of *non-uniform heating* of the parallel plates, i.e., let us add a horizontal ∇T to the Bénard problem and a vertical ∇T to the differentially heated vertical slot geometry.

The Bénard problem with imposed small horizontal temperature gradient ∇T_h has been treated for $N_{Pr} \gtrsim 1$ and conducting free boundaries and rigid boundaries by [5.115] and [5.116], respectively. Similar to the results for the tilted Bénard problem above, it is found that the critical Rayleigh number for instabilities to the basic flow is increased with increasing ∇T_h. Steady instabilities are preferred in the form of longitudinal rolls and transverse rolls depending on whether N_{Pr} is larger or less than 5.1 and 1.7 for free and rigid boundaries, respectively. For low-N_{Pr} fluids, a most recent analysis [5.116] shows that with increasing ∇T_h the primary flow soon becomes stable against steady longitudinal rolls but becomes increasingly unstable to oscillatory transverse perturbations. Theoretical results for $N_{Pr} = 0.05$ are depicted in Fig.5.66.

It becomes apparent from this plot that horizontal temperature gradients as small as *a few percent* of the destabilizing vertical gradient suffice to develop oscillatory convection. Hence in the laboratory practice it is very difficult to avoid oscillations in low-N_{Pr} fluids upon "heating from the bottom".

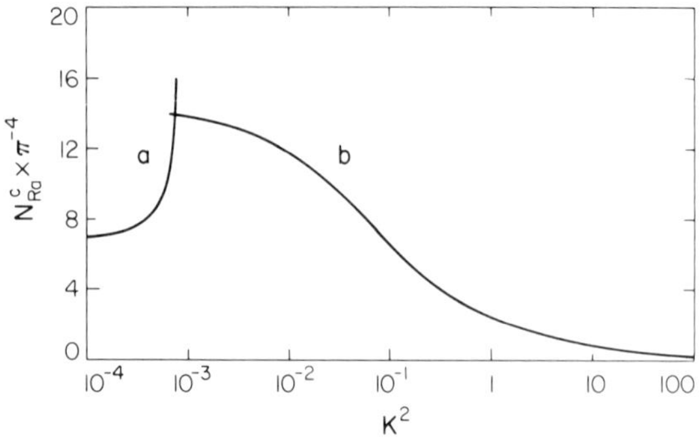

Fig.5.66. Critical Rayleigh number for non-uniformly heated horizontal planes as function of the ratio of the applied horizontal and vertical temperature gradients ($K^2 \approx \nabla T_g/4\nabla T_v$). $N_{Pr} = 0.05$. (a) steady convection, (b) oscillatory convection. After [5.116] by permission of the American Institute of Physics

Convection resulting from two-dimensional heat flow in a vertical slot can be expected to be particularly complex since the horizontal temperature gradient generates flows which are colder at the bottom and hotter at the top, while the vertical temperature gradient (from bottom heating) supports opposite behavior. This situation has been analyzed for a wide range in $\nabla T_h/\nabla T_v$. Theoretical results for $N_{Pr} = 1$ are given in the regime diagram depicted in Fig.5.67 in terms of the horizontal Grashof number $N_{Gr} = \beta g d^4 \nabla T_h/\nu^2$ and the vertical Rayleigh number $N_{Ra} = \beta g d^4 \nabla T_v/\nu\kappa$, (5.219) and (5.220), where d is the distance between the semi-infinite parallel plates. The case of $N_{Ra} = 0$ corresponds to the vertical slot case ($\alpha = 0$) in Fig.5.65. Hydrodynamic instability between the two vertical counter flows sets in at $N_{Gr}^{c1} \times 1/16 \approx 500$ (note the poor agreement between the two presentations). As bottom heating sets in (i.e., $N_{Ra} > 0$) the velocity of the basic shear flow will increase. This reduces the hydrodynamic stability and the steady horizontal (i.e., transverse) rolls will be sustained at lower and lower horizontal N_{Gr}^{c1} as the imposed vertical temperature gradient is increased (curve 1). Eventually, for $N_{Pr} = 1$ at $N_{Ra}^{c2} = 1550$, the upward flow becomes equally possible along either the hotter or colder plate, and the velocity of the basic flow becomes so high that any arbitrarily small N_{Gr} results in instability. Passage through this critical point is associated with an inversion and decrease of the shear flow velocity. Hence the hydrodynamic stability increases again, and larger N_{Ra}'s are required to sustain instabilities (curve 2). In the absence of a ∇T_h, buoyancy-driven convection sets in at $N_{Ra} \approx 2100$ (for $N_{Pr} = 1$). A ∇T_h alters that critical N_{Ra} until at a combination of $N_{Ra}^{c2} = 2880$ and $N_{Gr}^{c1} = 640$ the transverse steady rolls become unstable with respect to oscillatory modes.

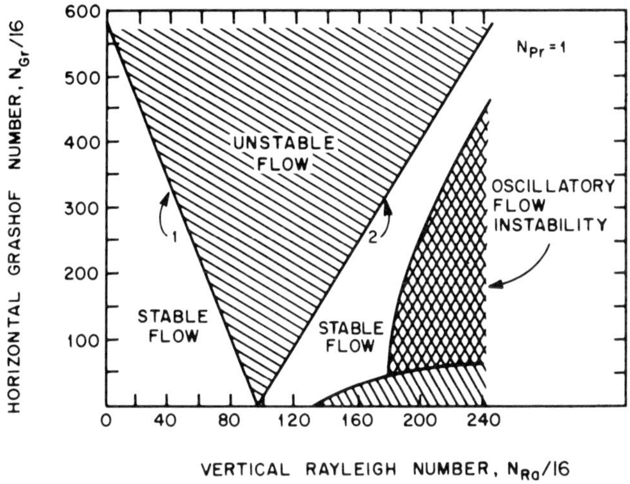

Fig.5.67. Stability diagram for convective flow between parallel vertical plates of semi-infinite extent with combined heating from below and from the side [5.117]. From [5.81] by permission of Marcel Dekker

Solutal or chemical convection

Density gradients in monocomponent ("pure") fluids can only be due to thermal expansion. In multicomponent fluids, however, non-uniform density can result from various other mechanisms. Most simply a density gradient may be sustained by external means (e.g., solute addition) or by an interfacial segregation process. Examples are concentration layers that may arise upon the confluence of salty water and fresh water, or the solute gradients that arise adjacent to growing crystals.

For the analysis of *solutal* convection that may be driven by these solutal gradients one may take advantage of the analogy between heat transfer and mass transfer *at low concentrations* where the advective velocity normal to the interface can be neglected. Hence the above thermal (expansive) convective stability analysis (based upon simultaneous solutions to the simplified momentum and heat transfer equations) is completely analogous to a solutal convective stability analysis in which the momentum and mass transfer equations (for "uniform" density systems) are solved simultaneously. In both situations density changes are only considered in the buoyancy term. The former problem becomes the latter when the Prandtl number is replaced by the *Schmidt number*, the thermal expansion coefficient $\beta = (1/V_0)\partial V/\partial T$ by the *solutal density coefficient* $\gamma = (1/\rho_0)\partial\rho/\partial C$ and the temperature gradient by the concentration gradient ∇C. Thus solutal convective instabilities should occur when the *solutal Rayleigh number*

$$N_{Ra,c} = \frac{g\gamma\nabla C d^4}{\nu D_{AB}} \tag{5.228}$$

reaches a critical value of the analogous expansive convective problem, e.g., 1707 for the Bênard geometry with rigid boundaries that supply or extract the destabilizing solute species readily (the analogue to "conducting"). The validity of (5.228) for the onset of solutal convection has been demonstrated experimentally. In one set of experiments layers of polyethylene glycol (PEG) solutions were destabilized by water that was given off by an osmotic membrane used as lower boundary [5.119]. Other workers studied convection resulting from diffusion of SO_2 into water layers through the top surface [5.120]. Most interestingly, convection was also observed in the experiments on PEG when the wet membrane was placed on top [5.119]. Convection was then connected to the evaporation process through the membrane. When the evaporation was prevented, no convection was observed. This suggests that the destabilizing density gradient was established by the evaporative cooling at the upper boundary.

Values of the solutal density coefficient γ in liquids are typically between 0.1-1 [atomic fraction]$^{-1}$ as opposed to the thermal expansion coefficients which range from 10^{-4} - 10^{-3}[deg]$^{-1}$. In addition D_{AB}'s in liquids are in general considerably smaller than κ's. Hence solutal convection can be sustained by *extremely low* concentration gradients. In fact, as Problem 5.26

demonstrates, in aqueous solutions the critical ∇C is of the order of 10^{-6} ppm cm^{-1} and in liquid metals 10^{-5} ppm cm^{-1} [5.121]. Hence in liquids the small concentration gradients that are unavoidable in practical crystal growth can readily drive non-steady convection, or effectively counteract expansive convection (see next section) depending on the orientation of ∇C. In gases, where the kinematic viscosities *and* diffusion coefficients are on the average 3 orders of magnitude higher than in liquids (Table 5.12 and Sect. 5.3.3) the required concentration gradients are significant. However, they can exist at growing interfaces or as a result of a temperature-dependent chemical equilibrium in the presence of a temperature gradient. This latter mechanism, which plays an important role in chemical vapor transport and deposition processes [1.2], is referred to as *chemical convection*. Since the simultaneous transport of heat and mass is involved in this situation we defer its discussion to the following section on double-diffusive convective phenomena.

Non-steady solutal convection in the Bénard geometry has not yet been investigated in great detail. (Results for laterally confined geometries will be discussed in the solution-growth Chapter [1.2]. However, in analogy to the previous discussion of the expansive convective phenomena one can expect the following: Depending on the Schmidt number of the fluid, instabilities to steady rolls will either occur as boundary layer instabilities ($N_{Sc} > 1$) or in form of the development of vertical vorticity ($N_{Sc} < 1$). Hence the solutal analogue to the thermally induced "meandering" (Fig.5.61) observed in liquid metals ($N_{Pr} \ll 1$) will probably occur in gases only. (See also Fig.5.20.) In liquids, however, oscillatory solutal convection will likely result from the periodic release of blobs of buoyant liquid from the concentration boundary layer, in analogy to the "thermals" depicted in Fig. 5.60. The presence of such *"solutals"* was deduced indirectly from mass transfer results obtained in the previously mentioned SO_2-water experiments [5.120]. The frequency of oscillations from solutals will be governed by the concentration diffusion time $\tau_c \approx \delta'^2/D_{AB}$ (see Eq.6.55) where δ' is comparable to the concentration boundary layer width. Thus for typical crystal growth conditions the period will be of the order of a few minutes, as has been observed, e.g. in high-temperature solution growth [5.122]. A more detailed discussion of non-steady solutal convection is deferred to the mass transfer sections of melt and solution growth systems [1.2].

In the absence of thermal gradients and of intrinsic concentration gradients, convectively destabilizing density gradients can arise if a body force other than gravity acts selectively on the fluid components. For instance, very recent work demonstrates that in electrolyte solutions convection sets in when a critical electrical potential gradient is exceeded [5.123]. Similar behavior can be expected in electrophoretic systems, for magnetic fluids in magnetic fields, or in centrifugal fields.

In principle, one can employ all these "non-thermal" mechanisms for the suppression of expansive convection. However, as the following discussion will show, such double-diffusive situations may readily result in "subcritical" (with respect to expansive convection) oscillatory motion.

Double-diffusive convection [5.124-126]

In many crystal growth systems two or more components of often quite different diffusivities are transported simultaneously. The term component is used here in the widest sense and may refer to solutes as well as heat. Such a double-diffusive transport can lead to a whole range of new convective phenomena which have attracted considerable interest only recently. Probably the most striking feature is that instabilities can develop even when the overall density decreases upward, i.e., in a *statically* stable configuration. This leads to various convective curiosities in nature, such as "salt fountains" and "salt fingers". For a qualitative introduction to these surprising phenomena and some simple, very instructive experiments see [5.153, 154].

Linear stability criteria for double-diffusive convection have been developed directly from the linear stability calculations for (simple) expansive convection discussed above. Summaries of these efforts have been given in [5.127, 128]. With the introduction of two "concentration" differences ΔT and ΔS (which need not be restricted to temperature and solute concentrations) between the (horizontal) boundaries and two different molecular diffusivities κ_T and κ_S, four non-dimensional parameters are required to specify the system. For instance in a *thermosolutal* or thermohaline system one uses the thermal N_{Ra}, (5.219), the solutal $N_{Ra,c}$, (5.228), the Prandtl number $N_{Pr} = \nu/\kappa_T$ and the ratio of the diffusivities, i.e., the Lewis number

$N_{Le} = \kappa_T/D_{AB}$ [see also (5.162)]. For given N_{Pr} and N_{Le} the results of the stability analysis may be presented as contours in a Rayleigh number (N_{Ra}, $N_{Ra,c}$) plane, each point on this plane representing a system with given temperature and salinity stratification. Fig.5.68 depicts semiquantitatively the various convection regimes obtained theoretically for systems with $N_{Pr} = 10$ and $N_{Le} = 100$, resembling the transfer of salt and heat in

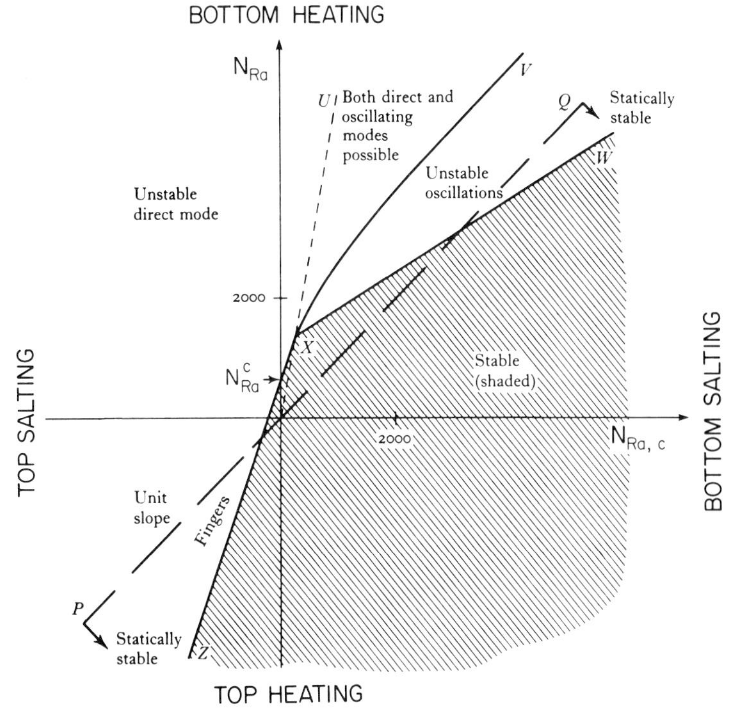

Fig.5.68. Diagram of the various thermosolutal convection regimes in the N_{Ra}, $N_{Ra,c}$ plane. Semi-quantitative presentation for $N_{Pr} = 10$ and $N_{Le} = 100$. After [5.126] and [5.125]

water. The sign convention is such that positive N_{Ra} correspond to a destabilizing ∇T (heated from bottom) and positive $N_{Ra,c}$ to stabilizing ∇C (salted from bottom). The line $N_{Ra} = N_{Ra,c}$ represents an equilibrium state of constant density. Hence if $N_{Ra} \leq N_{Ra,c}$ the fluid is *statically* stable. In the quadrant $N_{Ra} < 0$ and $N_{Ra,c} > 0$ (top heating, bottom salting) perturbations cannot grow. When $N_{Ra} > 0$ and $N_{Ra,c} < 0$ (bottom heating, top salting) all points in this quadrant above the line XZ are unstable (direct mode region). For

$N_{Ra,c} = 0$ this regime boundary must go through N_{Ra}^{cl} of the purely thermal problem. The equation for XZ is

$$N_{Ra} = N_{Ra,c} \cdot N_{Le} + N_{Ra}^{cl} \quad . \tag{5.229}$$

This allows one to reduce the condition for thermosolutal marginal stability to that of the monocomponent case by expressing an effective Rayleigh number

$$N_{Ra,eff} = N_{Ra} - N_{Ra,c} \cdot N_{Le} \tag{5.230}$$

with the critical effective Rayleigh number being numerically equivalent to the N_{Ra}^{cl} of the corresponding Rayleigh-Bénard system (see Table 5.22). Also the critical wave number and the flow pattern induced by a combination of thermal and solutal gradients in this regime are the same as those in the mere thermal situation. From (5.230) we see again that concentration gradients are much more effective toward destabilization than temperature gradients. (Actually the slope of XZ has been strongly reduced in Fig.5.68 for a better distinction from the N_{Ra}-axis.) Very small concentration gradients suffice to either stabilize or destabilize a fluid with respect to N_{Ra}^{cl}, depending on the relative orientation of ∇T and ∇C.

There are two regions in the Rayleigh number plane (Fig.5.68) in which a thermosolutal system behaves fundamentally different from a purely thermally stratified system: the "overstable" region and the "salt finger" region. Overstable, i.e., oscillatory conditions can arise when the temperature gradient is destabilizing and the solute gradient is stabilizing (bottom heating and salting, so called diffusive regime) — even when the overall density gradient is stable (below $N_{Ra} = N_{Ra,c}$ at large $N_{Ra,c}$)! Suppose that a fluid element is displaced upward in this situation. Since the parcel cools off much more rapidly than it loses its salinity ($\approx$ 100 times), the remaining buoyancy forces will drive it back toward the initial position. However the lag in temperature between the parcel and its surrounding will cause the parcel to return to its initial position faster than it leaves. It will overshoot its equilibrium position, buoyancy will drive it upward again, etc. Thus the energy to sustain motion comes dominantly from the component with the *higher diffusivity*, here the heat. Note that in this situation the *onset* of convection is oscillatory. Also, note again that the presentation in Fig. 5.68 is intentionally distorted and the "triple point" X is actually very close to the vertical axis. Hence very small ("stabilizing") impurity concentrations in liquids can cause oscillatory motion way below the oscillatory instability conditions for the pure fluid.

In the "finger" region (top heating and salting, below $N_{Ra} = N_{Ra,c}$) a basically different mechanism prevails and only steady modes are found. As an element is displaced upward (or downward), due to the slow salinity gain (or loss), it remains buoyant inspite of rapid adjustment in temperature. Thus the motion draws its energy from the potential energy in the component of *lower diffusivity*. At the onset the convective motion in this instability region is very similar to the familiar "direct mode". The widths of the rolls or cells is comparable to the layer height. But at larger N_{Ra} thin columns ("finger") form which evidently can release the potential energy of the salt field more efficiently.

In the above heuristic discussion we have tacitly assumed that higher solute concentrations lead to an increase in the fluid density ($\gamma > 0$) and higher temperatures to a decrease. Yet we know that this is not necessarily the case. Thermal expansion anomalies (e.g., water) as well as systems with $\gamma < 0$ are known. Hence in combination with the two possible orientations for each ∇T and ∇C one can devise *twelve* different situations for thermosolutal convective instabilities [5.129], an exercise in combinatoric left to the reader.

The stability calculations have recently been extended into the nonlinear regime by various workers. For references see [Ref. 5.126, pp. 259] and [5.130]. Considerable insight has been obtained into supercritical motions for various boundary conditions in the Bénard geometry and for the inclined slot [5.131, 132]. The material is too extensive to be reviewed here. It should be pointed out however that at higher N_{Ra} the development of a horizontal layer structure (each layer well mixed, with sharp boundaries to neighboring layers) is expected in the bottom heated and salted configuration. Such layer formation is particularly favored in the presence of additional horizontal temperature gradients.

Most of the theoretical predictions for the onset of instability, the global features of the flow patterns and the oscillation periods have been reasonable well confirmed experimentally. However the complex details seen experimentally in the convection states can only be explained qualitatively. One must note that all theoretical models, at this point, are based upon the assumption that the undisturbed temperature *and* concentration gradients are constant throughout

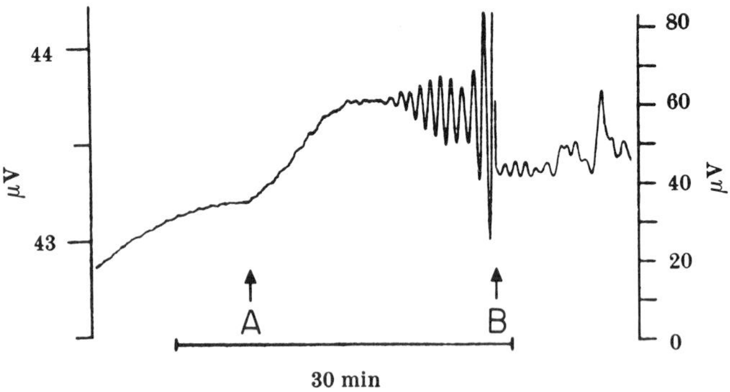

Fig.5.69. Section of differential thermocouple output showing the oscillatory breakdown of stably stratified layer of sugar solution heated from below. A: change of heating rate. B: refer to right-hand scale after B. From [5.133] by permission of Macmillan Journals

the fluid. The latter is difficult to realize experimentally.because of the impermeability for solutes of most materials of which boundaries might be constructed. Also, as discussed earlier, even if the boundaries are perfectly permeable for the solute (as in crystal growth) linear gradients can only be approximated if the solute concentration and its (steady-state) flux are very low (Sect. 5.2.2). Hence the current state of theoretical description of double-diffusive convection, though very valuable for the interpretation of trends and possible phenomena, cannot be expected to yield more than semi-quantitative results for the typical crystal growth situation.

Figure 5.69 depicts a particularly well-developed example for the onset of an overstable convective mode in a thermosolutal system [5.133]. The temperature was obtained from the differential output of two thermocouples placed 0.5 and 1.0 cm above the heated bottom plate. Note the reduction of the recorder sensitivity by a factor of 50 at B.

To this point we have specified the concentration and temperature gradients independently. In other words, boundaries with specific constant temperature *and* concentration values were assumed. With respect to the concentration boundary values, however, practical systems are typically more complex. In crystal growth one may (at best) be able to specify a solute concentration at a growing interface. But the concentration values at the container walls of a closed system will have to be considered as the *result* of the "internal" chemical potential distribution of the solute. In the presence of a temperature gradient the solute distribution may thus be governed by *thermal diffusion* (Sect. 5.2.4), for instance.

A convective stability analysis including thermal diffusion is very similar to the discussion of thermosolutal convection. The momentum (Navier-Stokes) and energy transport equation in the Boussinesq simplification are solved together with the continuity (convective-diffusion) equation which now contains an additional term for the separating effect of thermal diffusion, see e.g., (5.49) and (5.57). The extensive work in this area, almost exlusively performed during the last few years, cannot be done justice in a short section within this text. Hence we will restrict ourselves to a qualitative discussion, with emphasis on the trends that are important for crystal growth. For a more quantitative study of this fascinating area the reader is referred to [5.121, 134-136].

The effect of thermal diffusion on convective stability depends on the sign of the "Soret coefficient" and the solutal expansion coefficient γ. (The reader should be aware that the nomenclature in the thermal diffusion literature is far from uniform; see also Sect. 5.2.4 and, for a comparison of the various coefficients used in recent work, [Ref. 5.137, pp. 348-9]. Considering the Bénard problem (with bottom heating) a positive Soret coefficient S_T implies that the (initially homogeneous) fluid becomes stratified such that the (heavier) solute is enrichted in colder layers. Now it depends on the sign of γ whether this represents a stabilizing or destabilizing situation. For the more common case that increased solute concentration leads to higher mass density, thermal diffusion acts destabilizing in analogy to the "top salting" situation of the mere thermosolutal problem. Also, note that in real experiments for a given system the N_{Ra} necessary to achieve instability [i.e., a certain $N_{Ra,eff}$ in (5.230)] will depend on how rapidly the convective state is approached! *Thermal diffusion is very slow*. In particular, in liquids heating rates orders of magnitudes below those employed in ordinary convection studies must be used if steady-state behavior in thermal diffusion systems is to be obtained. If heated too rapidly as compared to the thermal-diffusion relaxation time (hours!), the fluid behaves essentially as though it were pure. Recent numerical experiments have also demonstrated the possibility of *transient* oscillations [5.136].

Negative Soret coefficients (the heavier solute tends towards hot) act convectively stabilizing. For large S_T this can be rather pronounced. For instance in aqueous 1N-LiI solutions N_{Ra}^{cl} as large as 196 times the value of 1708 were observed [5.138]. In analogy to the mere thermosolutal behavior this "bottom heated and salted" situation results in *oscillatory modes* at marginal conditions. Reflecting again the low thermal diffusion coefficients, the periods of these oscillations are of the order of minutes (gases, ionic solutions) but are expected to be as long as several hours in certain liquids [5.121]. Even more complex situations can arise when the sign of S_T is temperature dependent as in aqueous NaCl solutions, for instance, where with 0.5N-NaCl below 12°C the salt reverses its flow direction from hot→cold to cold→hot [5.137].

Most of the limited experimental data on thermal-diffusion-modified convective instabilities available at present were obtained under (idealized) conditions that are not representative of crystal growth situations. *Much* more work is needed in this area before its significance for crystal growth can be fully appraised. One must keep in mind, however, that thermal diffusion *can* readily lead to density gradients that are considerably larger than those typically obtained through thermal expansion.

Chemical reactions and their temperature dependence offer another possibility for establishing solutal gradients in a (crystal growth) fluid that may influence the convective behavior. Extensive theoretical work of the effect on convective stability of temperature and concentration gradients created by surface catalyzed reactions has been put forth recently; for references see [5.139]. These calculations show that *chemical convective instabilities* can occur also in heating from above, in analogy to the thermosolutal problem.

In this context it should be mentioned that interfacial heat and mass transfer can be oscillatory for mere kinetic reasons without requiring buoyancy forces, as has been demonstrated experimentally and theoretically for the hot wall pyrolysis of gaseous reactants [5.140, 141]. Oscillatory chemical reactions in general have been reviewed in [5.142].

The addition of a concentration gradient to the previously discussed expansive convection problem on a *vertical wall* complicates matters considerably. Thermosolutal convection on vertical walls, which is important in various crystal growth situations, has been reviewed in [Ref. 5.92, pp. 304]. The relative thickness of the velocity and temperature boundary layer is not a simple function of N_{Pr} and N_{Sc} anymore. The combined buoyancy effect of the heat and solute fields alters the velocity distribution of Fig.5.53 significantly. For the case that the vertical plane is a heat source and solute sink even negative values of velocity are predicted in the *outer* boundary region for smaller values of N_{Sc} (high diffusivities). This results from the incoming mass flux first experiencing the negative buoyancy effect of the wider concentration boundary layer. For this it is assumed that the fluid density increases with solute concentration so that the fluid is heavier in the outer boundary region where buoyancy from thermal expansion is insignificant; see also Fig.5.20, 28 and (5.162).

5.8.3 Surface-Tension-Driven Convection

Surface-tension-driven convection (σ-convection) has become of particular interest in connection with materials preparation efforts in low-gravity environments such as the Apollo missions, and the just beginning Skylab program. This is not to say that σ-convection is not important in earthbound processes. In fact, as already indicated in Sect. 5.8.1, it has been recognized at latest since PEARSON's analysis [5.76] that most of Bénards classical convection experiments on bottom-heated, shallow horizontal layers [5.72] were governed by surface tension rather than by buoyancy.

The molecular origin and physical content of the concept "surface tension" will be discussed in detail in the chapter on nucleation in [1.1]. Here we will view surface tension simply as the force that acts on the interface *between two fluid phases* tending to minimize the interfacial area. Quantitatively, it is the force that appears to act tangentially on an interface across a line of unit length in the interface. The surface tension of liquids decreases with temperature "since as the temperature rises the difference between a liquid and its vapor phase decreases" [5.143]. Note, however, that in general the surface tension changes by no more than 5% in a 100°C temperature interval. Surface tension depends also on composition and electric field. Hence gradients of any of these *along* the interface cause surface tension gradients. Those, in turn, act like shear stresses applied "by the interface" on the adjoining bulk fluids and can induce flows and/or affect existing one. From this it is obvious that horizontal temperature gradients (i.e., ∇T *parallel* to the free surface) can sustain flow even in the absence of gravity. Note that this *thermocapillary flow* has the same direction as the basic buoyancy driven shear flow. However, in the Bénard-geometry, where ∇T is *normal* to the upper free surface all points on the *unperturbed* surface possess the same σ! How can motion set in there? A realistic analysis of this case [5.144] shows that the deformation of the interface (by perturbations) plays an essential role in obtaining a tangential $\nabla\sigma$. The resulting (local) flow, in turn, sets up a horizontal temperature gradient which *tends* to sustain the flow even after the initiating perturbation has ceased. In the presence of gravity, however, the process, in close analogy to the purely buoyancy-driven instability in the Bénard geometry will only be self-sustaining if the sum of the rates at which potential energy is liberated by capillarity and buoyancy exceeds the rate at which kinetic energy is dissipated by viscosity. Thus convection driven by combined buoyancy and capillary forces from a vertical ∇T will only set in after a certain threshold is overcome. The value of this barrier must be obtained from stability analysis. Qualitatively though, we can expect from our heuristic discussion that this threshold will decrease

(a) as gravity is reduced for a given geometry, and

(b) as the layer height is reduced under given gravity conditions.

Extensions of Pearson's stability analysis support these expectations. For the Bénard geometry instability from combined buoyancy and capillarity is predicted [5.144] when a critical value of the *Marangoni number*

$$N_{Ma} = -\frac{\nabla T h^2}{\eta\kappa}\frac{\partial\sigma}{\partial T} \tag{5.231}$$

is exceeded, where ∇T and h are the vertical temperature gradient and layer height, respectively. The critical N_{Ma} on earth and for an insulated free surface is $N_{Ma}^c = 80$. In the absence of gravity [5.145] there is no critical N_{Ma} for the onset of steady flow and the critical wavelength (cell diameter) of a *Marangoni flow* pattern is very similar to that of buoyancy-driven flow. This leads to such a strong coupling of the two mechanisms under combined conditions that one can write in good approximation [5.144]

$$N_{Ra}^{(1)}/N_{Ra}^{c1} + N_{Ma}^{(1)}/N_{Ma}^{c} = 1 \tag{5.232}$$

where $N_{Ra}^{(1)}$ and $N_{Ma}^{(1)}$ are the values at marginal stability for the two agencies that reinforce one another. The actual functional relationship (very close to a straight line) is depicted in Fig.5.70. As also shown in this figure, a liquid layer that is stable to buoyant forces can still be destabilized by surface-tension gradients.

In comparing (5.231) and (5.219) we note for a given ∇T that the Rayleigh number scales as the fourth power of the fluid layer height whereas the Maran-

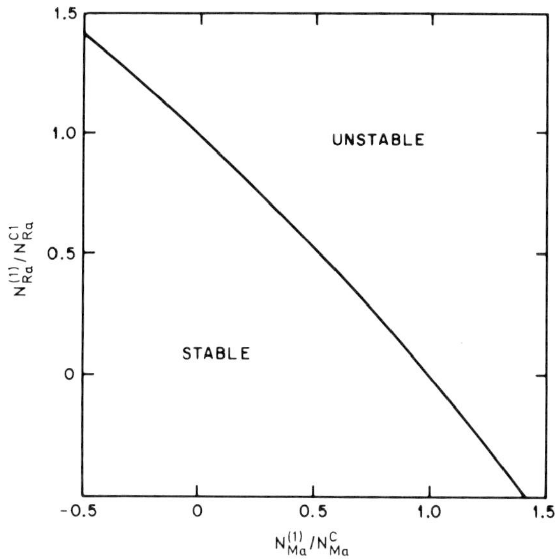

Fig.5.70. Influence of surface-tension gradients on the critical Rayleigh number for steady flow for heating a liquid from below. After [5.143]. From [5.81] by permission of Marcel Dekker

goni number is only proportional to the square of h. On earth, therefore, σ-convection will dominate gravity-driven convection only if the characteristic length is "small". As estimated in Problem 5.27, this "crossover length" in liquids of interest in crystal growth may still be of the order of a centimeter!

Thermocapillary convection, i.e., from a horizontal ∇T, in an (earth-bound) horizontal fluid layer has also been discussed [5.147]. For other references see [5.146, 148].

Traditionally the relative strength of buoyancy and surface tension forces is deduced from the *Bond number*

$$N_{B_0} = \rho g h^2/\sigma \quad . \tag{5.233}$$

However, as has been pointed out most recently in a critical review of the literature by OSTRACH [5.148], the limits of the applicability of the Bond number are uncertain and a more general criterion is required. He derived, based on a physically more relevant velocity normalization, the following dimensionless groups (modified Bond numbers) as criteria for the relative weight of buoyancy to σ-convection terms

$$\frac{N_{Gr}A}{N_\sigma} = \frac{\rho\beta g L^2}{(\partial\sigma/\partial T)} \equiv \overline{N}_{B_0} \quad (\text{for } N_{Pr} < 1) \tag{5.234}$$

and

$$\frac{\overline{N}_{B_0}}{(N_\sigma A^2)^{1/3}} \quad (\text{for } N_{Pr} > 1) \tag{5.235}$$

where A = h/L is the aspect ratio of the confined, laterally heated layer (height h, width L) used for the derivation and

$$N_\sigma = \frac{\Delta Th}{\eta\nu}\frac{\partial\sigma}{\partial T} \quad , \tag{5.236}$$

i.e., $N_\sigma = N_{Ma}/N_{Pr}$ and hence (5.233) is the ratio N_{Ra}/N_{Ma}. Equation (5.234), which applies to "viscous flow" situations ($N_{Pr} < 1$, Fig.5.52b) had been used before by [5.147] in reciprocal form and with a different length scale. Eq.(5.235) applies to boundary layer flow ($N_{Pr} > 1$, Fig.5.52a), see Problem 5.27.

One sees from (5.233)-(5.235) that under reduced gravity, in a spacecraft at, say 10^{-4}g, surface tension is *likely* to dominate buoyancy. Experiments on board Apollo 14 [5.146] showed indeed a significant reduction of the critical Marangoni number, less though than theoretically predicted. On the other hand, even if gravity is reduced by several orders of magnitude, buoyancy *can* still be predominant. But even the rapid near-turbulent convection found on earth in a typical high-temperature crystal growth environment is likely to be replaced by a slow (steady) flow.

One of the major difficulties in performing defined surface tension measurements and thermocapillary or Marangoni flow experiments results from the strong dependence of σ on "surfactants" (surface-active impurities). An example is given in Fig.5.71 where we note that the surface tension of liquid iron is reduced by 25% by impurity concentrations as low as 200 ppm!

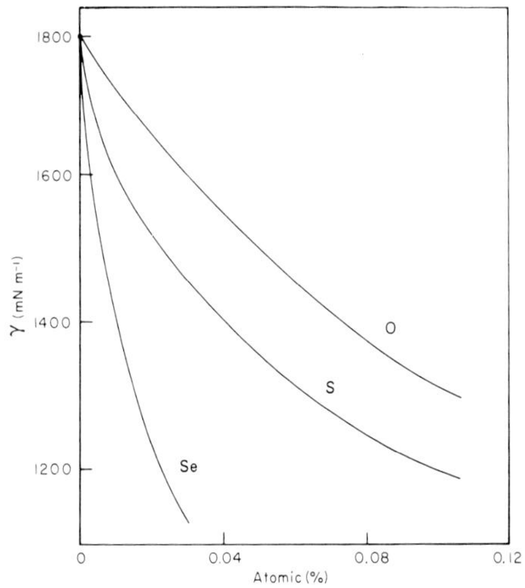

Fig.5.71. Dependence of the surface tension of liquid iron at 1500°C on the content of O, S and Se. From [5.149] by permission of Academic Press

Problems

5.1 Prove (5.15) for a binary system.

Hints: $\rho = \rho_A + \rho_B$, $C = C_A + C_B$, $C_A\bar{V}_A + C_B\bar{V}_B = 1$.

5.2 In a reputable textbook on the mathematics of diffusion one finds the following statement: "...in the early measurements it is usually assumed that the total volume of the system remains constant as diffusion proceeds *and so* the mutual diffusion coefficient D^V is measured". Discuss why this statement is (at best) misleading. Rephrase the statement so it becomes generally valid.

5.3 Derive an expression for $N_A(\nabla X_A)$ for diffusion of A through stagnant B analogous to (5.42).

Answer: $N_A = -\dfrac{C}{1 - X_A}\mathcal{D}_{AB}\nabla X_A$.

5.4 Derive analytical expressions for the steady-state concentration distributions for diffusion of a component A through a stagnant component B (see Fig.5.2). Assume constant $\bar{V}_A$, $\bar{V}_B$, and $\mathcal{D}_{AB}$.

(a) Given are: C_{Ao} and C_{Al}.

What is $C_A(x)$ and $C_B(x)$ in terms of C_{Ao}, C_{Al}, $\mathcal{D}_{AB}$, and $\bar{V}_A$ and $\bar{V}_B$, respectively?

(b) What is the total molar flux $N_A(C_{Ao}, C_{Al}, \mathcal{D}_{AB}, \overline{V}_A$ and $l)$?

(c) What is $C_A(x) = f(N_A, C_{Ao}, \mathcal{D}_{AB}, \overline{V}_A)$?

Answers: a) for A: $$\frac{1 - C_A(x)\overline{V}_A}{1 - C_{Ao}\overline{V}_A} = \left(\frac{1 - C_{Al}\overline{V}_A}{1 - C_{Ao}\overline{V}_A}\right)^{x/l}$$

b) $$N_A = -\frac{\mathcal{D}_{AB}}{\overline{V}_A l}\ln\left(\frac{1 - C_{Al}\overline{V}_A}{1 - C_{Ao}\overline{V}_A}\right)$$

c) $$C_A(x) = \frac{1}{\overline{V}_A}\left[1 - (1 - C_{Ao}\overline{V}_A)\exp\left(\frac{N_A\overline{V}_A}{\mathcal{D}_{AB}}x\right)\right] .$$

5.5 STEFAN [5.11] stated in his original paper in 1882 on the diffusive evaporation of water through a stagnant column of air that "... the volume of vapor passing through a unit area per unit time is

$$V = -\frac{D}{P-p}\frac{dp}{dx} \; \ldots"$$

where P and p are the total pressure and partial pressure of water, respectively. Show that this statement is equivalent to (5.42).

5.6 Derive the Navier-Stokes equation for the x-component from (5.75). Hints: Carry out the detailed steps of differentiation, collect terms and note that $\nabla\cdot\underline{v} = 0$ for an "incompressible" fluid.

5.7 Hydrodynamic boundary layer width in crystal growth from the melt:

(a) Estimate δ_0 for a non-rotated KCl crystal which is pulled from its melt, based upon the stagnation in plane flow model. Assume a typical convective flow velocity of 5 cm/sec normal to the interface and a density of the molten salt of about 2.1 g/cm^3.

(b) How fast would one have to rotate a crystal in order to obtain the same δ_0 in an oxide melt with typically $\rho = 0.5$ poise and $\rho = 5$ g/cm^3?

5.8 A sphere of radius R_1 is suspended in a one-dimensional flow of a fluid, the kinematic viscosity, ν, and flow velocity V_1 of which are given. In order to exert the same acceleration on a sphere of R_2 in a fluid of ν_2, what must V_2 be?

4.9 Derive (5.126) in analogy to the derivation of (5.120).

5.10 Derive (5.125) and (5.127). State the assumptions made and point out the practical cases for which we can expect reasonable accuracy from these equations.

5.11 What minimum stirring velocity is typically required in an aqueous solution in order to significantly enhance the mass flux in flow direction above the mere diffusive range? Justify the various assumptions made.

5.12 If it were possible to keep T_0 constant, independent of growth rate, how fast would one have to grow an NaCl crystal under rotation from its melt (with fixed T_∞) in order to increase the interfacial temperature gradient by 30% over the no-growth value. Use Fig.5.26 and data given in Tables 5.12 and 5.16. Discuss the practicality of the nessesary assumptions and the relevance of the result obtained.

5.13 Derive Fig.5.26b from Fig.5.26a.

5.14 Derive (5.170) from (5.169). For hints see p. 375.

5.15 Derive Wien's displacement law (5.172) by differentiation of Planck's spectral distribution. Hint: Multiply (5.171) with $(n^3T^5)^{-1}$.

5.16 As radiation passes through a participating medium, it is attenuated due to absorption processes and, at the same time, augmented by emission from the volume elements that the radiation passes. Naive question: Since for most systems according to Kirchhoff's law the emissivity is equal to the absorptivity – why, then, is there any net attenuation?

5.17 Calculate the critical angle of incidence for total internal reflection at a boundary between fused silica and air for radiation with λ = 500 nm.

5.18 Show that $e_b'(\beta)$ has indeed the circular characteristic plotted in Fig. 5.35.

5.19 Discuss the (seemingly contradictory) fact that the reflectivity of materials increases with increasing absorption coefficients.

5.20 As shown in the figure below, a black square with side 0.2 cm is at temperature 1000 K and is near a tube 0.2 cm in diameter. The opening of the tube acts as a black surface, and the tube is at 500 K. What is the net radiation exchange along the connecting path S between the square and the tube opening? [5.55].

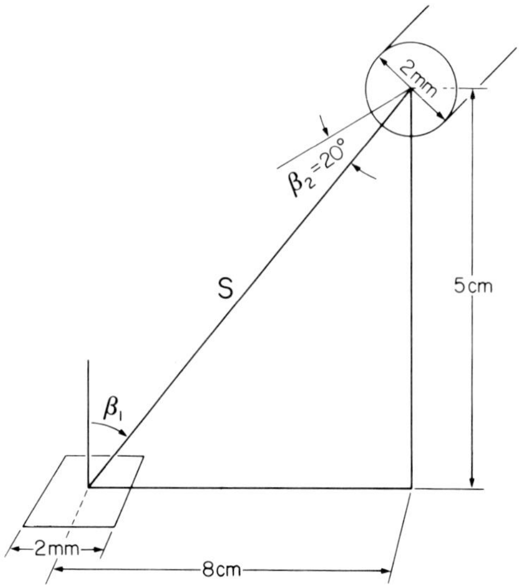

5.21 Derive the diffusive configuration factors and energy exchange relations between a differential element and a finite area, and between a finite area and another finite area, as given in Table 5.19.

5.22 Derive (5.195) by summing over the geometric progression formed by the multiple reflection and absorption terms.

5.23 Derive (5.212).

5.24 Calculate the critical temperature gradient for the onset of expansive convection in the Bênard geometry for

(a) hydrogen and argon at atmospheric pressure for RT and 1000 K (show that $\beta = 1/T$ for gases), and

(b) silicon and Al_2O_3 at their melting point.

Use material parameters from Tables 5.12 and 5.16. Discuss the validity of the Boussinesq-fluid approximation for these systems.

5.25 Verify (5.223).

5.26 Calculate the critical concentration gradient for solutal convection in an aqueous NaCl-solution. Deduce the solutal expansion coefficient, for instance, from the tables on concentration properties of aqueous solutions in the CRC-Handbook. D_{AB} and ν can be found in Tables 5.7 and 5.10.

5.27 Estimate from (5.233) and (5.234) the layer height for which surface tension and buoyancy forces become comparable for

(a) molten silicon ($\partial\sigma/\partial T \approx -0.06$ dynes $cm^{-1}{}^{\circ}C^{-1}$),

(b) molten iron ($\partial\sigma/\partial T \approx -0.4$ dynes $cm^{-1}{}^{\circ}C^{-1}$).

(Data from [5.81]). Justify all other assumptions.

6. Segregation

The discussion of phase transitions in Chapters 3 and 4 has shown that only monocomponent and congruently melting or vaporizing multicomponent systems, respectively, remain unchanged in composition during solidification or crystallization. In all other systems the composition of the solid differs from that of the coexisting fluid (liquid or vapor) even in equilibrium. This phenomenon is called *segregation* or redistribution. In this chapter we will be mainly concerned with "impurities", i.e., concentrations of solutes of, say, a few percent and below. We will, however, emphasize the consequences of adding large concentrations of components (solvents, inert gases) for the formal description of segregation dynamics.

Segregation is important for crystal grwoth for a variety of reasons. Firstly, one is frequently faced with the necessity of purifying starting materials which are inadequate for crystal growth processes. Here the question is how to apply segregation most effectively. Secondly, one may want to predetermine the composition of a nutrient (doped or non-stoichiometric melt, etc.) in order to obtain a specific composition of the growing crystal. Thirdly, one may want to add a major component to a nutrient in order to obtain more advantageous experimental conditions, such as lower temperatures, than those feasible with "pure" starting materials. Then again the question arises how to minimize incorporation of such additives (solvents, transport gases, etc.) in the growing crystal. In any case, whether the problem is purification, doping or engineering of growth parameters, the central question is to what extent the crystal composition reflects that of the nutrient from which it grows.

In this chapter after introducing some new parameters called the segregation coefficients, we will discuss segregation from the thermodynamic point of view of equilibrium between homogeneous bulk phases. Next, the bulk equilibrium conditions will be relaxed step by step and supplemented by dynamic models for realistic macroscopically steady mass transfer conditions. This results in practical concepts for purification, for the achievement of uniform composition and yields insight into microsegregation phenomena that result from non-steady conditions. Finally we will discuss phenomena that occur when even solid and liquid directly at the interface are not in equilibrium.

Our discussions will dominantly be concerned with solid-liquid phenomena since precious little information is available on segregation in crystal growth from vapors. In the presentation of dynamic models, however, we will stress the conceptual differences between segregation from liquids and vapors that originate in distinclty different material properties and consequently different mass transfer conditions.

6.1 Segregation Coefficients

In practice, crystallization is rarely conducted so slowly that a system can readjust its composition homogeneously throughout the *bulk* of its phases according to the phase diagram. At practical crystallization rates equilibrium may still exist between fluid and solid directly at the interface. However, the limited mass transfer rates will typically cause concentration gradients in the bulk phases, as indicated in Fig.6.1.

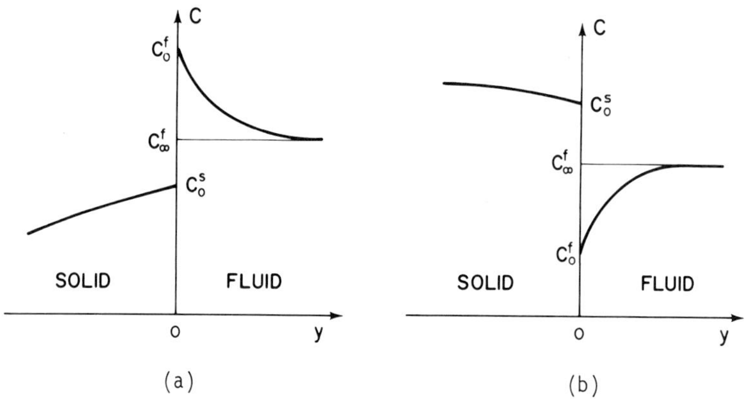

Fig.6.1. Typical dynamic concentration profiles near solid-fluid interfaces at finite growth rates. The considered component is (a) partially rejected (b) preferentially incorporated into the solid

In order to describe segregation quantitatively one defines segregation coefficients. Three different definitions are advantageously used in particular situations. If, for instance, there is no clear evidence that local equilibrium exists between solid and fluid at the interface, one will cautiously base a discussion of the interfacial transfer of a component on the *interfacial segregation coefficient* k_0, which is defined as

$$k_0 = \frac{W_0^s}{W_0^f} = \frac{X_0^s}{X_0^f} = \frac{C_0^s}{C_0^f}\frac{\rho^f(0)}{\rho^s(0)} \qquad (6.1)$$

where the concentration of the considered component in the solid and fluid phase (superscripts s and f, respectively) are taken directly at the interface (subscript zero) as indicated in Fig.6.1. It is useful to memorize that for components which are partly rejected by the progressing interface and hence become accumulated in the residual fluid, $k_0 < 1$ (Fig.6.1a). On the other hand, $k_0 > 1$ corresponds to preferential take-up of the considered component by the solid and depletion in the fluid as depicted in Fig.6.1b.

For a description of the efficiency of, say, a purification process k_0 is not too informative. Whenever segregation between bulk phases is of interest one will refer to the composition of the bulk fluid rather than to the interfacial C_0^f and use the *effective segregation coefficient* k that is defined as

$$k = \frac{C_0^s}{C_\infty^f}\frac{\rho^f(\infty)}{\rho^s(0)} \qquad (6.2)$$

where the subscript ∞ designates "far away from the interface". Note that for $k_0 < 1$ as well as for $k_0 > 1$ the effective segregation coefficient approaches unity upon increase of the growth rate of the solid, V^s and hence, macroscopic segregation ceases eventually at high enough V^s. (See Problem 6.1.)

Most segregation work has been concerned with impurity-in-melt systems where $\rho^f/\rho^s \approx 1$. Hence it has become customary to ignore the density ratio in (6.1) and (6.2). For simplicity we will follow this custom through parts of the text. In discussions of segregation in solution and vapor growth, however, one must resort to the original definitions in terms of weight fraction ratios and (equivalent) mole fraction ratios.

If solid and fluid are in mass transfer equilibrium at the interface, then k_0 becomes equal to the *equilibrium segregation coefficient* k* which depends only on the thermodynamic properties of the system rather than the mass transfer kinetics at the interface. This case of

$$k_0 = k^* \tag{6.3}$$

is frequently approached in practice. For solute concentrations above 1% or so, equilibrium segregation coefficients can easily be deduced from phase diagrams. Figure 6.2a shows that k^*'s for components which lower the solidification temperature of a system are smaller than unity. In systems with upward sloping coexistence lines, as in Fig.6.1b, $k^* > 1$.

If the experimental conditions are such that the concentration gradient in the liquid becomes negligible ("ideal mixing case") and

$$k \approx k_0 \quad , \tag{6.4}$$

maximum efficiency is obtained in a segregation process.

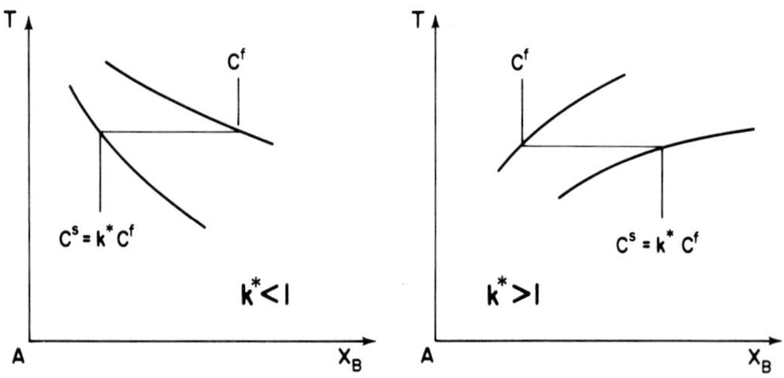

(a) (b)

Fig.6.2. Portions of phase diagrams in which the solidification temperature is (a) lowered, (b) raised by the solute, and corresponding segregation coefficients k_0

As will be discussed below, equilibrium segregation coefficients depend in general on the temperature, the pressure, and on the concentration of the component of interest, say B, as well as on the presence of other components. Therefore, it is not meaningful to call $k^*(T, P, X_B, X_i)$ a segregation "constant" as is customary. At low concentrations, segregation coefficients are often found to be relatively concentration independent. That "constant" range, however, depends on the individual system. For example, in germanium k_0-values for most impurities are found to be concentration independent in a range as wide as $10^{-2} > X_B > 10^{-10}$. However, there are counter examples, $k_{Ca/NaCl}$ is 0.3 at 10^{-2} mole fractions in the melt and becomes 1.0 at 5×10^{-5}, i.e., no segregation occurs at this level. Hence, extrapolation of k*-values from higher concentrations (phase diagrams!) to impurity levels below 10^{-2} can be very misleading.

6.2 Equilibrium Segregation

6.2.1 Thermodynamics of segregation

The equilibrium segregation coefficient k* can be predicted theoretically if one knows certain experimentally accessible parameters such as the heat and entropy of mixing in the liquid and solid solution. Even if some of these parameters are not known and need to be estimated based on idealized models, the thermodynamic approach can yield valuable information. Let us first consider segregation upon solidification from a binary melt and then expand the concept to crystallization from solutions.

Segregation from melts. Neglecting possible impurity-impurity interaction, host and each impurity independently are considered as binary systems. Host and impurity terms will be characterized by subscripts 1 and 2, respectively. Subscripts l and s will be used for the liquid and solid phases, respectively, in contrast to the convention in previous sections, where phases were denoted with superscripts.

As expressed in (2.89), in equilibrium the chemical potential of each of the components is equal in both phases, hence,

$$\mu_{2l} = \mu_{2s} \quad \text{and} \quad \mu_{1l} = \mu_{1s} \quad . \tag{6.5}$$

For the chemical potential of a component in a real solution one can write, as discussed in Chap. 2,

$$\mu_i = \mu_i^* + RT \ln(\gamma_i X_i) \quad , \tag{2.113}$$

where μ_i^* refers to a standard state that is chosen such that μ_i^* depends only on intensive state variables (pressure, temperature, ...), but not on the composition. The activity coefficient γ_i characterizes the deviation from ideality (by its deviation from unity), and X_i is the relative concentration in mole fractions.

In a theory for segregation coefficients k^* in solidification from melts one then obtains from (6.5) and (2.113)

$$k^* = \frac{X_{2s}}{X_{2l}} = \frac{\gamma_{2l}}{\gamma_{2s}} \exp\left(\frac{\mu^*_{2l} - \mu^*_{2s}}{RT}\right) = \frac{\gamma_{2l}}{\gamma_{2s}} \exp\left(\frac{\Delta\mu^*_2}{RT}\right) \quad . \tag{6.6}$$

The chemical potentials μ^*_{2l} and μ^*_{2s}, as well as $\Delta\mu^*_2$, the change in chemical potential upon "melting" of pure component 2, are taken at the equilibrium temperature T, i.e., with dilute solutions approximately at the melting point T_1 of the host.

For an evaluation of (6.6) from experimentally accessible parameters, two idealizations are commonly made:

(a) the heat capacities of solid and liquid component 2 are assumed to be temperature independent between T and T_2, the melting point of pure 2. This allows for a convenient calculation of $\Delta\mu^*_2$ by setting the enthalpy change $\Delta h_2(T)$ of pure 2 equal to the heat of fusion at T_2, in the form

$$\begin{aligned} \Delta\mu^*_2 &= \Delta h_2(T) - T\Delta s_2(T) \\ &= \Delta h_2(T_2) - T\,\frac{\Delta h_2(T_2)}{T_2} \quad , \end{aligned} \tag{6.7}$$

where $\Delta S_2(T_2) = \Delta h_2(T_2)/T_2$ is the entropy change of 2 upon melting at T_2. The error resulting from this approximation is for most systems smaller than that from uncertainties in the other thermochemical parameters in (6.7).

(b) Solid and liquid solutions are assumed to behave regularly, i.e., only a configurational entropy change occurs upon dissolution whereas the vibrational entropy remains unchanged. This allows for computation of the activity coefficients from the respective heats of dissolution in the following way. Rewriting (2.113) shown above one gets

$$\gamma_2 = \frac{1}{X_2} \exp\left(\frac{\mu_2 - \mu^*_2}{RT}\right) = \frac{1}{X_2} \exp\left(\frac{\Delta\mu^d_2}{RT}\right) \quad . \tag{6.8}$$

Note that (6.8) stands for two equations: with indices s for the solid solution and with l for the liquid solution. $\Delta\mu^d_2$, the change of the chemical potential of 2 upon dissolution, can be expressed as

$$\mu_2 - \mu_2^\star = \Delta\mu_2^d = \Delta h_2^d - T\Delta s_2^d \tag{6.9}$$

where Δh_2^d and Δs_2^d are the heat (enthalpy) and entropy of dissolution, respectively. With the regular solution assumption [see (3.35)], (6.9) becomes

$$\Delta\mu_2^d = \Delta h_2^d + RT\ \ln(X_2) \quad , \tag{6.10}$$

and (6.8) is consequently

$$\gamma_2 = \exp\left(\frac{\Delta h_2^d}{RT}\right) \quad . \tag{6.11}$$

Inserting (6.7) and (6.11), one each for solid and liquid solutions, into (6.6) one obtains

$$\ln(k^*) = \frac{1}{R}\left[\frac{\Delta h_{2l}^d}{T} - \frac{\Delta h_{2s}^d}{T} + \frac{\Delta h_2(T_2)}{T} - \frac{\Delta h_2(T_2)}{T_2}\right] \quad . \tag{6.12}$$

The first two terms in this relation contain the heat of dissolution or the energy required to transfer pure molten and solid component 2 into the host lattice and host melt, respectively. Depending on whether $\Delta h_{2l}^d \lesseqgtr \Delta h_{2s}^d$, the segregation coefficient decreases or increases. In other words, the impurity will accumulate in the phase in which it causes less energy increase. The last two terms contain the heat of fusion of the "pure impurity". Depending on whether the melting point of 2, $T_2 \lesseqgtr T$, the segregation coefficient is shifted to lower or higher values, as we have anticipated already from mere graphical arguments in Fig.6.2. Note that all terms contain T or T_2 in the denominator. From this one can expect that, for instance, crystallization from solution around room temperature leads, in general, to stronger segregation than melt crystallization. Certainly the heat of dissolution term, particularly Δh_{2l}^d, will be different from the melt case, but the 1/T dependence forms in most systems an overriding effect, as we will verify later.

This model and equations similar to (6.12) were devised by [6.1] and in a more general form by [6.2]. In these and other papers [6.3, 4] the theory has been applied to segregation from covalent semiconductor melts. The heats of dissolution were calculated based on simple models, and it was shown that the vibrational change of entropy term for most impurities in Ge and Si is

Table 6.1. Values of segregation coefficients k* and k_0 in germanium and silicon [6.3]. From [6.5] by permission of North Holland Publ. Co.

Solute	Germanium		Silicon	
	Calculated	Experimental	Calculated	Experimental
Sn	2×10^{-2}	2×10^{-2}	8×10^{-3}	1.6×10^{-2}
Pb	1×10^{-4}	1.7×10^{-4}	–	–
As	1×10^{-1}	2×10^{-2}	1×10^{-3}	3×10^{-1}
Sb	2×10^{-2}	3×10^{-3}	3×10^{-3}	2.3×10^{-3}
Bi	4×10^{-5}	4.5×10^{-5}	–	–
Al	1×10^{-1}	7.3×10^{-2}	4×10^{-2}	2×10^{-3}
Ga	2×10^{-2}	8.7×10^{-2}	1×10^{-2}	8×10^{-3}
In	8×10^{-4}	1×10^{-3}	3×10^{-4}	4×10^{-4}
Tl	1×10^{-5}	4×10^{-5}	–	–

small as compared to the overriding Δh^d_{2l} term; hence, the regular solution approach is well justified [6.3]. Table 6.1 shows the good agreement of the calculated values with experimental data.

In ionic systems, the calculation of heats of dissolution is considerably more complex than in covalent lattices. Satisfactory results have yet been obtained only for monovalent impurities. Problems encountered with multivalent impurities in polar lattices will be mentioned later. Only recently has thermodynamic theory been applied to alkali halide melt segregation [6.6, 7]. The activity coefficients γ_{2s} were calculated, assuming again regular solutions, from Δh^d_{2s} data computed for monovalent alkali and halogen impurities [6.8]. Very little, however, is known about the impurity host interaction in melts, i.e., the actual γ_{2l}-values. Some workers [6.6] drew qualitative conclusions by comparison of experimental and theoretical results for k*. Others [6.7], calculating activity coefficients based on a simple melt structure model [6.9], obtained quite different results. Thus, the uncertainty in calculated segregation coefficients for monovalent impurities in solidifcation from alkali halide melts is typically ± 0.1.

Segregation from solutions. For crystallization from (aqueous) solutions the theoretical treatment is somewhat more complex than for solidification from melts. The energetic conditions are similar to those in the solidification case. Thus, the chemical potential of the impurity in the solid in

(2.113) above can be written

$$\mu_{2s} = \mu_{2s}^{*} + RT \ln(\gamma_{2s}X_{2s}) \quad , \tag{6.13}$$

where μ_{2s}^{*} is again the chemical potential of the pure solid component 2, although with a different value due to the lower equilibrium temperature. If the solvent is quantitatively rejected at the interface (anhydrous crystallization), γ_{2s} reflects only host-impurity interaction. For the host in the solid, with $X_{1s} \approx 1$, one can assume ideal behavior and, therefore,

$$\mu_{1s} = \mu_{1s}^{0} \quad , \tag{6.14}$$

The main difference between segregation from melts and solutions occurs in the liquid phase. For the impurity

$$\mu_{2l} = \mu_{2l}^{*} + RT \ln(\gamma'_{2l}X_{2l}) \quad , \tag{6.15}$$

where γ'_{2l} now reflects impurity-host interaction and impurity-solvent interaction and, hence, is denoted by a prime. Here the choice of the standard state deserves special attention. Since the mole fraction of the impurity can often not be varied up to unity without phase change (solubility limit), reference to "pure" component 2 becomes rather meaningless. Thus, one refers here more advantageously to a fictitious state in which the impurity is infinitely dilute in the solvent. This is legitimate since a component i of a real solution is found to approach quasi-ideal behavior both as $X_i \to 1$ (Raoult's law) and as $X_i \to 0$ (Henry's law), as discussed in Sect. 2.11. For $X_i \to 0$, of course, this does not imply that interactions can be neglected. It means, however, that for an appropriately chosen μ_{2l}^{*} in (6.15), γ'_{2l} can be set equal to unity in the very dilute solution range.

The same consideration applies to the dissolved host material in the liquid phase. Therefore,

$$\mu_{1l} = \mu_{1l}^{*} + RT \ln(\gamma'_{1l}X_{1l}) \quad , \tag{6.16}$$

where γ'_{1l} accounts for host-solvent *and* host-impurity interaction, the latter of which can be neglected for the host for all practical purposes as $X_{2l} \to 0$. From the equilibrium condition, combining (6.13) through (6.16) one obtains for the segregation coefficient

$$k^* = \frac{X_{2s}/X_{1s}}{X_{2l}/X_{1l}} \times \frac{X_{2s}X_{1l}}{X_{2l}} = \frac{\gamma'_{2l}}{\gamma_{2s}\gamma'_{1l}} \exp\left[- \frac{(\mu^*_{2s} - \mu^*_{2l}) - (\mu^*_{1s} - \mu^*_{1l})}{RT} \right] . \tag{6.17}$$

The parameters μ^*_{2l} and μ^*_{1l} are not directly accessible experimentally. One can, however, express them in terms of the chemical potential of the pure components in equilibrium with their saturated solutions. Since

$$\mu^*_{is} = \mu^*_{il} + RT \ln(\gamma^{sat}_{il} X^{sat}_{il}) ,$$

(6.17) can be rewritten, with $\gamma'_{1l} = \gamma^{sat}_{1l}$, as

$$k^* = \frac{X_{2s}X_{1l}}{X_{2l}} = \frac{\gamma'_{2l}}{\gamma_{2s}\gamma_{2l}{}^{sat}} = \frac{X_{1l}{}^{sat}}{X_{2l}{}^{sat}} \tag{6.18}$$

Resolving (6.13) with respect to γ_{2s} and substitution into (6.18) yields then

$$k^* = \frac{\gamma'_{2l}}{\gamma^{sat}_{2l}} \frac{X^{sat}_{1l}}{X^{sat}_{2l}} X_{2s} \exp\left(- \frac{\Delta\mu^d_{2s}}{RT} \right) . \tag{6.19}$$

The change in chemical potential $\Delta\mu^d_{2s}$ involved in the dissolution of pure component 2 in the host solid can be expressed again as in (6.9) in terms of the heat of dissolution Δh^d_{2s} and the change in entropy Δs_{2s} as $\Delta\mu^d_{2s} = \Delta h^d_{2s} - T\Delta s_{2s}$. Separating the entropy change into its configurational part $\Delta s^{mix}_{2s} = -R \ln(X_{2s})$ and the vibrational part Δs^{vibr}_{2s} one thus obtains from (6.19)

$$k^* = \frac{\gamma'_{2l}}{\gamma^{sat}_{2l}} \frac{X^{sat}_{1l}}{X^{sat}_{2l}} \exp\left(- \frac{\Delta h^d_{2s} + T\Delta s^{vibr}_{2s}}{RT} \right) . \tag{6.20}$$

Similar analyses were put forward by [6.10, 11]. The above derivation [6.12] has the advantage that the relation for k* contains only parameters that are experimentally and/or theoretically accessible.

One sees from (6.20) that the segregation coefficient for crystallization from solution is the product of:

(a) the "solution interaction ratio" of the activity coefficient for the impurity at the trace level and at saturation concentration,

(b) the solubility ratio of host and "pure" impurity, and

(c) the exponential term, the numerator of which represents the "excess chemical potential" (deviation from ideality) for dissolution of the impurity in the solid host [not to be confused with Δg^{xs} in (3.33)].

Little is known concerning activity coefficients of solutes in the presence of high concentrations of other solutes, i.e., for impurities in saturated solutions. For dilute ternary systems (low concentrations of 2 *and* 1 in water) one would estimate γ'_{2l} from the Debye-Hückel theory which is based upon the electrostatic interaction between individual (isolated) ions of completly dissociated solutes. However, the present case, where the activity coefficient for the component 2 will be modified by 1-2 (solute-solute) interactions, is beyond the scope of this classical model. Hence, we are left with scarce experimental information that only allows following up trends for γ'_{2l}. However, as the examples for activity coefficients of ionic compounds in Fig.6.3 suggest, the term (a) will typically be rather close to unity.

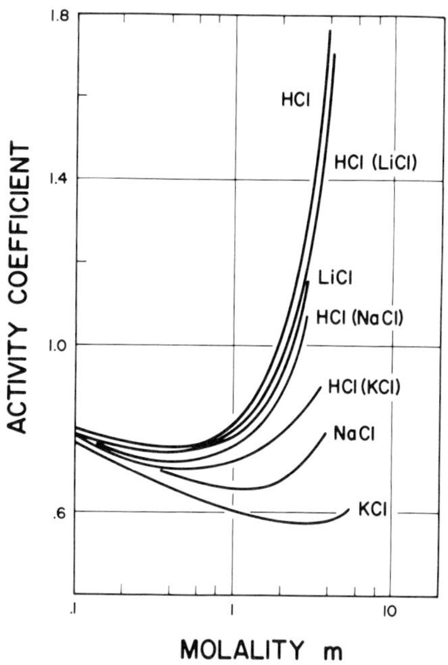

Fig.6.3. Activity coefficients of some ionic compounds in aqueous solutions. For instance: HCl(LiCl) is γ of HCl at concentration = 0.01M in a solution of water and LiCl of concentration m. From [6.12] by permission of American Institute of Physics

The term (b), for which data are easily accessible, can deviate considerably from unity when host and impurity differ sizeably in their solubility in the solvent. The most significant term, however, is (c). Crystallization temperatures from solutions are, in general, considerably lower than solidification temperatures from melts. Hence, neglecting for the moment the vibrational entropy term, with Δh_{2s}^{d} being about the same as in the melt case, one expects from the exponential part a much smaller k* value for segregation from solution.

Another reason for increased segregation in the solution case can be solute-solvent interaction. Such *solvation* effects occur particularly in polar solvents. In water, for instance, solute ions associate in a bond-like manner with "water molecules" (i.e., hydronium ions H_3^+O) and it is these hydrated species, rather than the "nude" ions, that enter the interfacial segregation process. Solvation, due to screening of the ions, leads typically to a reduction of the *effective concentration* of the solute. This results in smaller γ_{2l}'-values in (6.20) and hence in lower k* values.

The above thermodynamical analysis (6.20) has recently been applied to segregation in alkali halide crystallization from aqueous solutions [6.12]. Calculations of the vibrational entropy change based upon the Debye model of solids, employing reststrahlen frequency data, showed that Δs_{2s}^{vibr} cannot be neglected in alkali halides. The resulting calculated k* values are in good agreement with data obtained in carefully performed experiments. Deviations from other data can be traced to non-equilibrium conditions during crystallization; see Sect. 6.5.2.

6.2.2 Experimental observations

Compilations of effective segregation coefficients for various semiconductor and other inorganic systems were given by SHOEMAKER and SMITH [6.42], KRÖGER [6.13] and BRICE [6.14], for alkali halides by ROSENBERGER [6.15]. An excellent review on trace element segregation and application to geological systems by McINTIRE [6.16] should also be mentioned here.

The thermodynamically predicted stronger segregation upon crystallization at lower temperatures [see (6.12) and (6.20)] is in general experimentally verified. This is, for instance, illustrated in Table 6.2 in a comparison of k_0-values for various impurities in alkali halides observed in melt solidification and crystallization from aqueous solutions, respectively.

Figure 6.4 reflects the commonly observed concentration independence of k_0 for impurities that are in the same valence state as the ionic host lattice. Furthermore, it shows a decrease in segregation with increasing

Table 6.2. Selected experimental segregation coefficients for crystallization from melts and aqueous solutions. Concentrations between 10^{-2} and 10^{-3} mole fractions of impurity in nutrient. From [6.12] by permission of the American Institute of Physics

System	Melt	Solution
NaCl:Li	0.2	<.005
NaCl:Ca	0.3	.2
KCl:Na	0.3	.001
KCl:Ba	0.05	.00005
KCl:Rb	0.6	.05
NaCl:I	0.06	<<.0004
KI:Br	0.8	.4

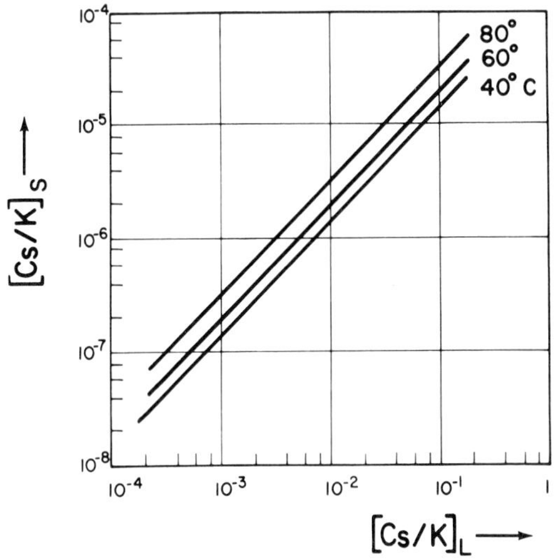

Fig.6.4. Distribution of Cs between aqueous KCl solutions and KCl single crystals. After [6.17]

T as expected from (6.20). Hence the choice of the crystallization temperature can be significant for the effectiveness of purification even in solution growth at rather low temperatures. We will, however, encounter various arguments in connection with dissolvation and interfacial mobility of adatoms that advise against too low temperatures [1.1, 2]. Thus, in practice one can rarely take full advantage of the temperature dependence of k*.

A number of mechanisms have been proposed to account for the concentration dependence of segregation coefficients. Reviews can be found in [6.18] and [6.19]. We will discuss here only a few examples.

In ionic systems one observes frequently a strong concentration dependence in the segregation of solutes that differ in their valence state from the solvent. For instance, divalent ions in alkali halides exhibit progressively weaker segregation with decreasing concentration in the melt. Table 6.3 shows this for the system Ca:NaCl. One sees that segregation ceases completely at 5×10^{-5} mole fractions. This tendency is also observed in crystallization from solution, but probably due to the hydration of the divalent ions less pronounced. What is the mechanism behind this effect?

Table 6.3. Concentration dependence of the melt segregation coefficient for Ca:NaCl. Data from [6.15]

Concentration in melt, X_2	k_0
10^{-2}	0.3
7×10^{-3}	0.4
10^{-3}	0.6
5×10^{-4}	0.8
5×10^{-5}	1.0

When multivalent ions, for instance, M^{++} (M standing for metal) are incorporated into a lattice consisting of monovalent ions, cation vacancies are generally formed for charge neutrality reasons. If one neglects complex formation for the M^{++} in the ionic melt, this reaction can be schematically expressed in the form $M_l^{++} \rightleftarrows M_s^{++} + \boxed{+}$ with an equilibrium quotient [see (2.102)]

$$k_1 = \frac{[M_s^{++}][\boxed{+}]}{[M_l^{++}]} \tag{6.21}$$

where the square brackets indicate concentrations.

The lattice, however, contains also thermally created Schottky defects, the concentration of which is a function of T only (at p = const.). This vacancy formation process AH $\rightleftarrows \boxed{+} + \boxed{-}$ can be characterized by another equilibrium constant

$$k_2 = [\boxed{+}] \cdot [\boxed{-}] \quad . \tag{6.22}$$

The concentration of occupied lattice sites, [AH], which is very large as compared to the vacancy concentration, is considered to be constant and is contained in the constant k_2. Overall charge neutrality requires that

$$[M_s^{++}] + [\boxed{-}] = [\boxed{+}] \quad . \tag{6.23}$$

Since these two mechanisms of vacancy creation are "concentration-coupled", the segregation coefficient,

$$k^* = \frac{[M_s^{++}]}{[M_l^{++}]} = \frac{k_1}{[\boxed{+}]} \quad , \tag{6.24}$$

becomes a function of the vacancy concentration in the crystal. Based upon (6.21) through (6.23), k* can be expressed in terms of k_1, k_2 and the impurity concentration in the melt (Problem 6.2) as

$$k^* = \frac{k_1}{(k_1[M_l^{++}] + k_2)^{1/2}} \quad . \tag{6.25}$$

Since, in general, $k_2 < k_1$, the segregation coefficient increases with decreasing impurity concentration (i.e., with decreasing weight of the $k_1[M_l^{++}]$ term). At high impurity concentrations (6.25) approaches the form

$$k^* = \left(\frac{k_1}{[M_l^{++}]}\right)^{1/2} \quad , \tag{6.26}$$

and hence k* becomes independent of the Schottky defect concentration. In other words, when with decreasing $[M_s^{++}]$ the impurity-induced vacancy concentration becomes comparable to the Schottky defect concentration, the incorporation of divalent impurities become energetically more favorable than at higher concentrations. This mass action law type of analysis of the concentration dependence of segregation was introduced by WAGNER [6.20]. To fit typical experimental data, k_2 has to be chosen such that $[\boxed{+}]$ in the pure material at the melting point of AH is about 10^{-4}. This value agrees very well with the results one obtains for the Schottky vacancy concentration from ionic conductivity measurements.

As this discussion indicates, one finds in general that data for segregation of impurities of different valence states than the host, obtained at concentrations above 100 ppm, do not necessarily reflect k_0 at low concentrations.

Impurity-solvent interaction and complex formation can be another reason for a sometimes rather complicated concentration dependence of the segregation coefficient. For instance, lead when present in low concentrations in the $KCl:H_2O$ system has a $k_0 > 1$. (Lead is also very likely involved in the nucleation mechanism of KCl crystallization from aqueous solution.) With increasing lead concentration, k_0 decreases. Unity is reached at about 10^{-3} mole fractions, and $k_0 = 0.1$ at 4×10^{-3}. However, on further increase of the lead concentration in the solution, k_0 reaches suddenly a value of several hundred. This final increase can be explained by coprecipitation of the complex salt $KPbCl_3$. (For references and more details see [6.15]).

Segregation coefficients in a certain impurity-host system can also depend on the presence of other impurities (solute-solute interaction). For instance the boron segregation on solidification of doped germanium varies considerably with the oxygen concentration of the melt [6.21]. There is some evidence that boron-oxygen complexes form in the melt leading to a reduction of $k_0 = 7$ for the low-oxygen case, to $k_0 = 1$ at higher concentrations.

Another example for this group is presented in Fig.6.5a [6.114]. The iron concentration in hydrothermally grown zincite (ZnO) was found to depend strongly on the concentration of KOH in the nutrient. Also, with decreasing temperature the segregation is more pronounced in this system, as we have exptected, in general, for mere thermodynamical reasons.

A pronounced case of segregation changes due to impurity interaction in the solid is illustrated in Fig.6.5b. It was found [6.22], that k_0 for Nd^{3+} in $CaWO_4$ is a function of the concentration of alkali metal impurities. M^+ ions act as charge compensation for the Nd^{3+} in the solid. Hence, in the presence of M^+ fewer cation vacancies are formed upon incorporation of Nd^{3+} ions, making their dissolution in the solid energetically more favorable – as is reflected in an increase of k_0. The concentration dependence $k_0(M^+)$ has been explained in a mass action law analysis similar to the one presented above.

Although the interaction mechanism is particularly obvious in ionic solids, k-changes due to solute-solute interaction are not restricted to ionic materials. For a discussion and references on interaction effects in covalent systems see, for instance [6.14, 18].

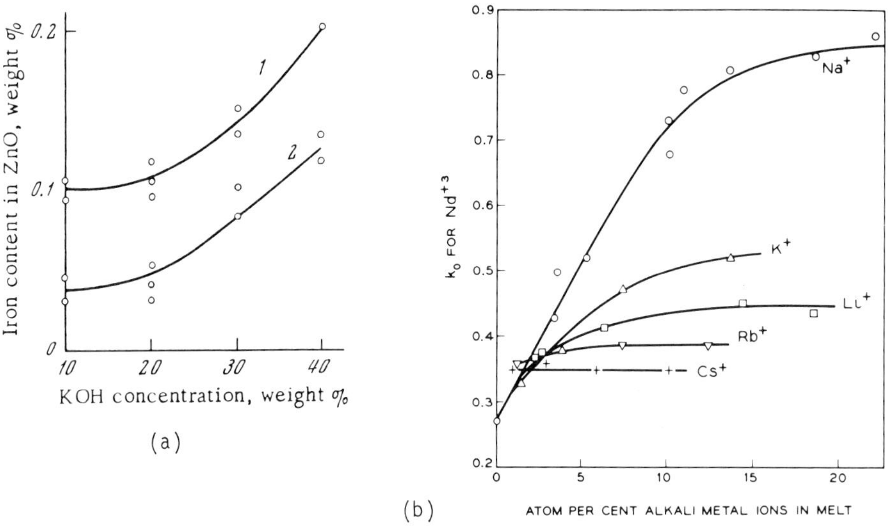

Fig.6.5a. Iron concentration in ZnO crystals as a function of the KOH concentration of the hydrothermal solution. 1) 450°C, 2) 350°C. From [5.104] by permission of The American Institute of Physics.
Fig.6.5b. Experimental segregation coefficients for Nd^{3+} in $CaWO_4$ as a function of concentration of compensating alkali metal ions. Nd^{3+} concentration 1-2 atomic percent in melt. After [6.22] and from [6.23] by permission of Prentice Hall

There have been various efforts to correlate segregation coefficients to the difference in size (misfit) between host and substitutional solute atoms or ions. Such an approach can be very useful if restricted to "similarity considerations", that is, to extrapolations of k-values to electronically very similar solutes, for which the difference in contributions to the terms in (6.12) and (6.20) consists in essence only in a change in the strain energy contribution to Δh^d_{2s}. An ideal group for such similarity considerations are, of course, the rare earths due to their identical outer electron configuration. It has been demonstrated that rare earth segregation data for calcium molybdate can be very satisfactorily correlated to the misfit of the individual ions with respect to Ca^{2+} [6.24]. Assuming that the only change in free energy of the system consists in a change in the (misfit) strain energy $\Delta G_{\Delta r}$ in the solid solution and that none of this contribution is released by dislocation formation, one can expect that

$$k_{\Delta r} = k_0 \exp(-\Delta G_{\Delta r}/RT) \quad , \tag{6.27}$$

where k_0 is the segregation coefficient of a similar solute with radius r_J equal to the host cation radius r_A. Based upon an isotropic elastic continuums model and neglecting the, in this case, small entropy changes, one obtains the proportionality [6.24]

$$\Delta G_{\Delta r} \propto [\frac{1}{2} r_A(r_J - r_A)^2 + \frac{1}{3} (r_J - r_A)^3] \quad . \tag{6.28}$$

Figure 6.6 illustrates the good fit obtained from (6.27) and (6.28) with the experimental data.

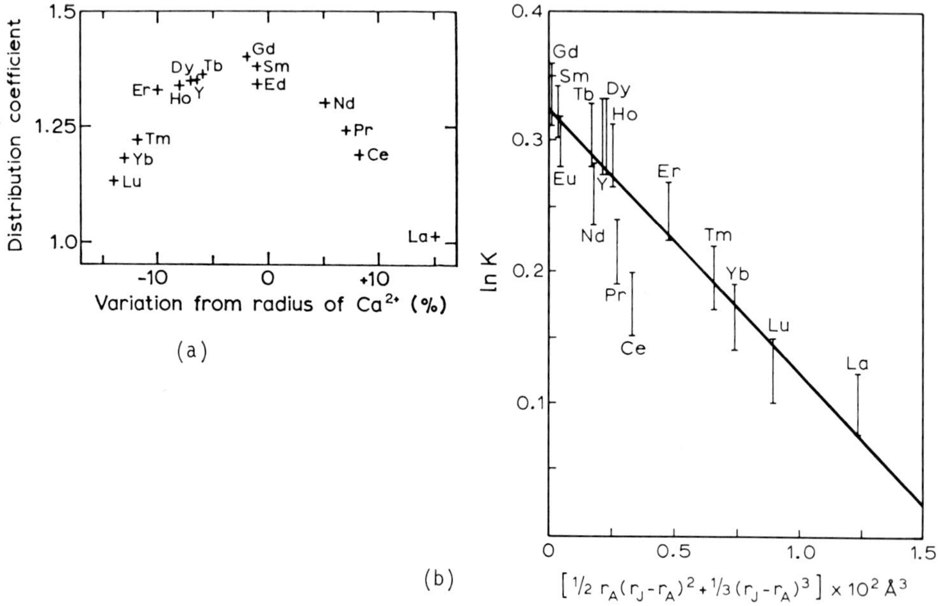

Fig.6.6. Segregation coefficients of rare earths in calcium molybdate. (a) k as a function of misfit. From [6.14]. (b) ln(k) as a function of the excess strain terms (see text). From [6.24]. By permission of North Holland Publ. Co.

Figure 6.7 illustrates a similar exponential behavior of a segregation coefficient on the ionic radius of a lattice component. Plotted are, in contrast to Fig.6.6, the distribution coefficient *ratios*

$$\frac{k_C}{k_R} = \frac{\dfrac{C/(C + R) \text{ in the crystal}}{C/(C + R) \text{ in the flux}}}{\dfrac{R/(C + R) \text{ in the crystal}}{R/(C + R) \text{ in the flux}}} \tag{6.29}$$

where C and R stand for the mole fraction of the considered component and the rare earth, respectively. The abscissa shows the R-ion radii as directly determined for the respective systems. In this particular form (6.29) represents a measure for the probability that C (Y in the garnet case, Pb in the R-ortho ferrite system) substitutionally accopies a R-site. Evaluation of the upper plot shows that the propensity of Y to occupy a larger R-cation site increases approximately with the one-hundredths power of the R-ion radius!

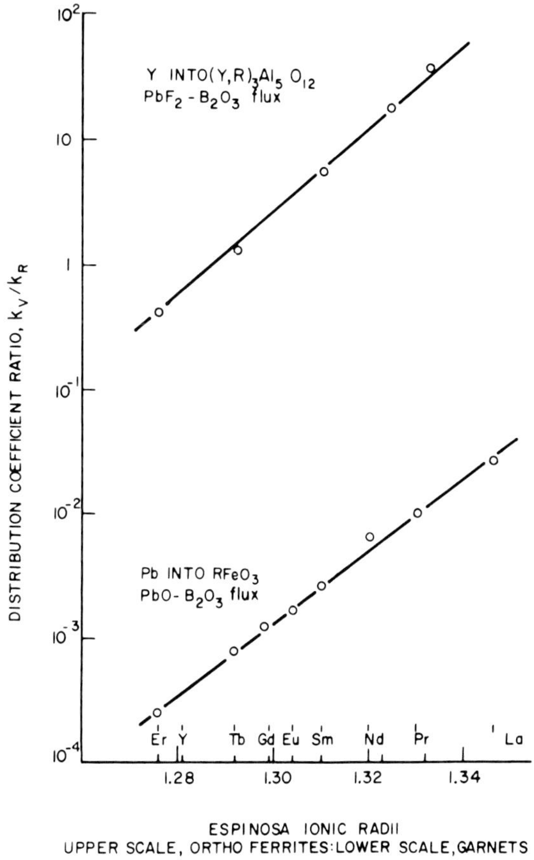

Fig.6.7. Distribution coefficient ratio, k_V/k_R versus rare earth ion radius. $k_V = k_Y$ for yttrium entering the garnet (upper curve); $k_V = k_{Pb}$ for lead entering the rare earth ortho ferrite (lower curve) in flux growth. From [6.55] by permission of Pergamon Press

6.3 Steady-State Segregation

In this section the assumption of bulk equilibrium between the phases will be given up. We will deal with the dynamic nature of segregation and describe the concentration distribution in the fluid adjacent to an interface. Thereby we will employ the mass transfer models and boundary layer approximations treated in Chap. 5. We will assume that diffusive fluxes in solids can be neglected as compared to crystallization rates encountered in practice. In other words, we consider solute concentrations that form in the solid at the interface as "frozen in" during consecutive growth of the solid. However, since diffusivities in liquids and gases are, for most systems, several orders of magnitude higher than in solids, (Sect. 5.2.3) diffusive-convective redistribution in the nutrient, will be taken into account. In addition, we will assume that the governing segregation coefficient, k_0 or k^*, is independent of the prevailing dynamic parameters of the segregation process.

For simplicity we will restrict ourselves to a one-dimensional description, normal to the interface. Remembering, for instance, the uniform flow adjacent to a rotating disk (Fig.5.10) we note that for certain configurations a one-dimensional description can be fully adequate. For other configurations where the interfacial velocity field depends strongly on the specific location on a fluid-solid boundary (e.g., flow parallel to a plate, Fig.5.11) we will simply consider the results obtained from the one-dimensional treatment as a local statement with respect to the locally existing parameters.

The component distribution in the (interfacial) fluid is governed by the prevailing diffusive and convective conditions. Thus simultaneous solutions to the mass conservation (continuity, convective-diffusive) equation (5.1), (5.120) etc. and the momentum conservation (Navier-Stokes, or boundary layer) equation (5.2), (5.77) or (5.90) must be sought. For the calculation of concentration *gradients* immediately at the interface and their relations to growth rates [see (5.42, 112)] we were able to take advantage of the fact that fluid motion from free or forced convection ceases on the solid-fluid interface. Only the diffusion-induced growth flux penetrates the interface which we consider as stationary. For the current problem, going away from the interface we lose this advantage. In the fluid, in order to calculate the concentration *distribution* in addition to the growth rate we must know the actual fluid velocity (in the lab frame) normal to the interface, $v_n(y)$, originating in free or forced convection. Again viewed in the stationary interface frame, the mass average velocity to be used for solving the continuity equation must be taken as the sum of the interfacial flow velocity

(5.114) and $v_n(y)$

$$v = v(0) + v_n(y) = \frac{\rho^s}{\rho^f(0)} V^s + v_n(y) \quad . \tag{6.30}$$

This forms, in general, a formidable mathematical task. Hence, the desire for simplification is great. Simplifying models – and their limitation – are described below. Let us first consider the conditions that lead to steady state adjacent to an advancing (growing) and segregating interface.

Suppose a planar solid-fluid interface begins to move with a growth rate V^s into the fluid. Some components of the fluid (impure melt solution or gaseous mixture) are either partially rejected by or preferentially build into the growing solid. It is assumed that the interfacial growth kinetics proceeds more rapidly than the transport kinetics in the fluid – which is not necessarily the same as assuming interfacial equilibrium. Consequently, individual constitutents of the fluid will either accumulate ($k_0 < 1$) or be depleted ($k_0 > 1$) in front of the propagating interface. Accumulation or depletion proceeds with time as depicted in Fig.6.8 in a reference frame that

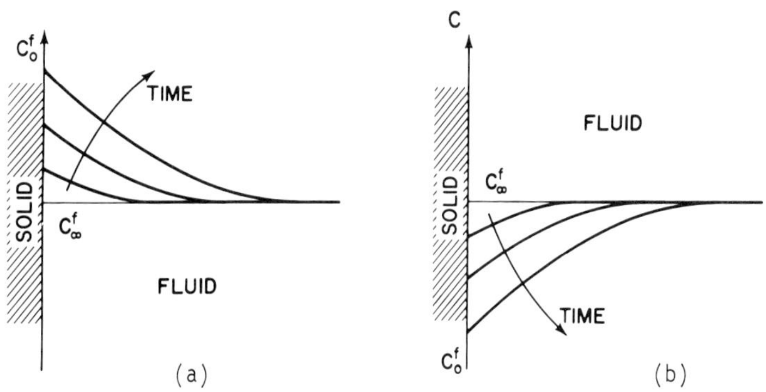

Fig.6.8. Transient concentration profiles in interfacial fluid for (a) $k_0 < 1$ and (b) $k_0 > 1$

moves with V^s. Note that in multicomponent fluids different components may be accumulated and depleted simultaneously. The concentration gradients at the interface will become continuously steeper. Hence the diffusive transport rates, either into the bulk fluid (for $k_0 < 1$) or towards the interface (for $k_0 > 1$), increase with time. Eventually, a quasi-steady state is reached, e.g., for $k_0 < 1$, when the interfacial rejection rate is balanced by the transport rate away from the interface. Note that a *true steady state*, with C_0^f = const. can only occur when either

(a) the fluid volume is infinite, or

(b) for finite fluid volume, when the concentration of the impurity (relative to the host) C_0^s that is accepted into the solid, becomes equal to the impurity concentration (relative to the host)) C_∞^f in the fluid and hence no effective segregation takes place, i.e., $k = 1$, or

(c) for a finite fluid volume, when the component flux into the crystal is balanced by a flux into the fluid of the same components from a source — a scheme encountered in certain vapor and solution growth techniques and in certain melt growth arrangements designed to achieve uniform composition of the solid, which are treated in Sect. 6.4.3.

6.3.1 Segregation in growth from a nutrient of uniform density

There are crystal growth systems in which the nutrient on its way towards the growing interface undergoes, inspite of segregation, only minor total density changes. This will be the case when the nutrient contains a strongly dominating component as, for instance, in growth from an impure, yet essentially congruent melt. There, segregation of the impurities at the interface will change the local fluid density $\rho^f(y)$ only insignificantly over zero-growth conditions. Similar situations arise in growth from liquid or gaseous solutions with very low solubilities of the crystal forming components. At high concentrations, in order to maintain $\rho^f \approx$ uniform, solutes and solvent must have very similar densities.

As indicated before, the situation of constant mass density along streamlines is very advantageous for the theoretical description since the momentum conservation equation (5.1) and the continuity equation (5.2) are thus uncoupled. One can obtain the velocity distribution $\underline{v}$ independent of the actual component distribution. The interfacial concentration distribution is then derived from the continuity equation (for ρ = const.) (5.120) employing the independently derived $\underline{v}$. Hence in steady state and for one dimension we have to solve

$$D_{AB} \frac{d^2 W_B(y)}{dy^2} - v \frac{dW_B(y)}{dy} = 0 \tag{6.31}$$

for the appropriate boundary conditions, where v is given by (6.30).

For a discussion of segregation it is convenient to rewrite (6.31) in terms of molar concentrations [see also the comments following (6.2)]. With $C_B = W_B\rho^f/M_B$ and cancelling out the constant ρ^f/M_B one obtains

$$\frac{d^2C_B(y)}{dy^2} - \frac{v}{\mathcal{D}_{AB}}\frac{dC_B(y)}{dy} = 0 \quad . \tag{6.32}$$

Segregation and conservation of component B leads to the interfacial boundary condition at y=0:

$$C_0^s V^s = C_0^f k_0 V^s = C_0^f v(0) + \mathcal{D}_{AB}\left(\frac{dC^f}{dy}\right)_{y=0}$$

$$= C_0^f \frac{\rho^s}{\rho^f(0)} V^s + \mathcal{D}_{AB}\frac{dC^f}{dy} = C_0^f V_{cf} + \mathcal{D}_{AB}\frac{dC^f}{dy} \tag{6.33}$$

where we have employed (5.114) for v(0) and dropped the subscript B. The V_{cf}, used as abbreviation in (6.33), is called the crystallization flow velocity, following a notation introduced by WILCOX [6.25, 26]. At the outer edge of the concentration boundary layer we assume (by definition) that the concentration takes on the (uniform) bulk value. Hence the second boundary condition is that at $y = \delta$

$$C^f = C_\infty^f \quad . \tag{6.34}$$

Sor far we didn't have to impose any simplifications besides the uniform density assumption. Note, however, that we have still the mass average velocity v in (6.32) which, in general, as expressed in (6.30), is y-dependent. To circumvent this mathematical difficulty $v_n(y)$ in (6.30) is traditionally ignored and only V_{cf} taken into account. This simplification is often justified with the claim that "close to an interface diffusion is the dominant mass transfer mode". There is nothing wrong with this qualitative statement. The question is, however, *how close* to the interface does one have to be to make this a valid claim. Let us analyze this based upon the (local) Peclet number (5.130). Taking the concentration boundary layer width δ as characteristic length and $v = V_{cf} + v_n$ as characteristic velocity, one can discuss the mass transfer situation across the boundary layer in terms of

$$N_{Pe} = \frac{V_{cf}\delta}{\mathcal{D}_{AB}} + \frac{v_n\delta}{\mathcal{D}_{AB}} \quad . \tag{6.35}$$

The first term on the right represents the relative weight of the advective and diffusive fluxes toward the interface, and the second term can be seen as the ratio of the externally caused convective flux to the diffusive flux. Now we see that the key parameter for the decision of whether ignoring v_n is physically realistic is $\mathcal{D}_{AB}/V_{cf} = Y'$ (and not simply the condition that $v_n \ll V_{cf}$). If the experimental situation is such that the so called *characteristic diffusion distance* $Y' < \delta$ and consequently the first right-hand term in (6.35) becomes big, it takes significant values of v_n to influence the transport in the boundary layer, i.e., to contribute significantly to N_{Pe}. In cases where $Y' \gg \delta$ one may still decide, for mathematical convenience, to ignore v_n, or in other words, to replace the real diffusive-convective flux situation within δ by a model in which there is "diffusion-only" within a narrower *diffusion boundary layer* of width δ'. (We recognize here, of course, the stagnant film model, briefly discussed on p. 294.) However, in these cases one should keep the distinction between δ and δ' in mind, i.e., one should not expect the concentration profile obtained under the diffusion-only assumption to closely resemble the actual distribution. This point has caused considerable confusion in the past.

Setting $v = V_{cf}$, the solution to (6.32) for the boundary conditions (6.33) and (6.34) is

$$c^f(y) = c^f_\infty \frac{k_0 \frac{\rho^f}{\rho^s} + \left(1 - k_0 \frac{\rho^f}{\rho^s}\right) \exp\left(-\frac{V_{cf}}{\mathcal{D}_{AB}} y\right)}{k_0 \frac{\rho^f}{\rho^s} + \left(1 - k_0 \frac{\rho^f}{\rho^s}\right) \exp\left(-\frac{V_{cf}}{\mathcal{D}_{AB}} \delta\right)} \tag{6.36}$$

which for $\mathcal{D}_{AB}/V_{cf} \ll \delta$ (diffusion only) reduces to

$$c^f(y) = c^f_\infty \frac{k_0(\rho^f/\rho^s) + [1 - k_0(\rho^f/\rho^s)]\exp[-(V_{cf}/\mathcal{D}_{AB})y]}{k_0(\rho^f/\rho^s)} \quad . \tag{6.37}$$

Recalling that diffusion coefficients in fluids range typically from $D = 10^{-3}$ to 10^{-5} cm^2/s (Sect. 5.2.3) and noting that melt-growth rates may be anywhere between 2×10^{-5} and 2×10^{-3}cm/s, one sees that the characteristic diffusion distance in melts may take on values from 50 to 5×10^{-3}cm. This, as we will see, has to be compared with a range in concentration boundary layer width under melt-growth conditions of $\delta = 0.2$ to 10^{-3}cm. Hence the diffusion-only case of (6.37) may well occur in practice. However, if this condition is not met in practice, then (6.36) with δ' for δ must be taken for the fictitious concentration profile within the diffusion boundary layer.

Now, in order to deduce an expression for the effective segregation coefficient, one sets y = 0 in (6.36) which yields the steady-state concentration at the interface, and, employing (6.2), one obtains

$$k = \frac{C_0^s}{C_\infty^f} = \frac{k_0}{k_0(\rho^f/\rho^s) + [1 - k_0(\rho^f/\rho^s)]\exp[-(V_{cf}/\mathcal{D}_{AB})\delta]} \quad . \tag{6.38}$$

Equation (6.38) was originally derived in 1953 by BURTON, PRIM and SLICHTER (BPS) [6.27] for $\rho^f = \rho^s$ in the form

$$k = \frac{k_0}{k_0 + (1 - k_0)\exp(-\delta V^s/D)} \tag{6.39}$$

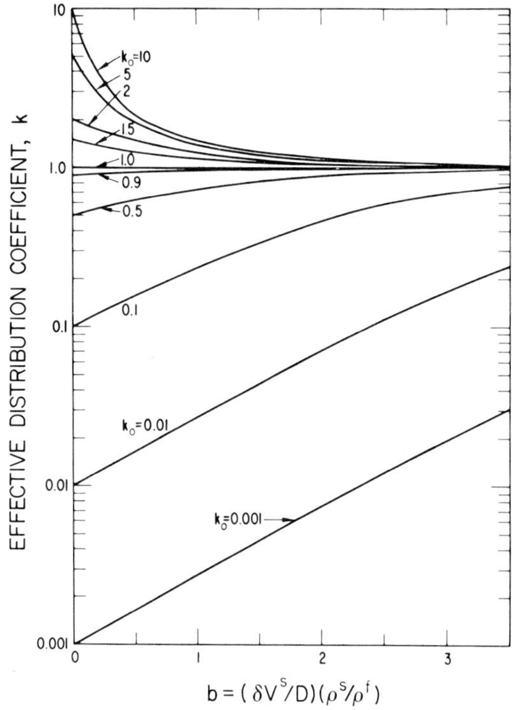

Fig.6.9. Effect of normalized growth rate $\delta V_{cf}/\mathcal{D}_{AB}$ on effective segregation coefficient. From [6.28] by permission of Marcel Dekker

partly based upon the diffusion-only assumption, as we will discuss in Sect. 6.3.3. The exponent terms in (6.38) and (6.39) are often collected in a *dimensionless boundary layer thickness* $b = \delta V_{cf}/D_{AB}$, which is also called *normalized growth rate*.

The functional dependence of k on the normalized growth rate is illustrated in Fig.6.9 for various values of k_0. One sees that for both $k_0 < 1$ and $k_0 > 1$, with increasing growth rate the effective segregation coefficient approaches unity more or less rapidly, depending on δ and D. The central role of δ for the effectiveness of a purification process becomes rather clear here. The smaller δ is, the more efficiently segregation will proceed; hence, the closer k will approach k_0.

6.3.2 Segregation in growth from a nutrient of non-uniform density

There are numerous crystal growth situations in which the nutrient undergoes significant total density changes upon approaching an interface. This can be due to the segregation of components of strongly differing density (molecular weight) present in comparable concentrations. Or non-uniform density may be due to steep interfacial temperature gradients even in situations where the segregating species constitute only a minority. In these cases a rigorous formal description of segregation can only be obtained from simultaneous solutions of the continuity and Navier-Stokes equations — including energy transport when significant temperature gradients are present. This author is not aware of any such general segregation treatments in the literature. The general trends that one will expect for this complex situation are, of course, similar to those contained in the Burton-Prim-Slichter (BPS) description: reduction of the effective segregation with increasing V^s, decreasing D_{AB} and increasing δ. However the quantitative dependences can be expected to be quite different.

6.3.3 Experimental observations

Before we compare the theoretical predictions with experimental results it is useful to summarize the assumptions that are necessary to obtain the steady-state concentration profile in the form of (6.36), i.e., the basis for the BPS relation (6.39):

1) equal and constant cross-sectional area in solid and fluid,
2) planar interface,
3) boundary layer width δ constant across interface,
4) constant crystallization or growth rate V^s,
5) constant diffusion coefficient in the fluid, $\mathcal{D}_{AB}$, independent of concentration, temperature and other impurities,
6) no diffusion in the solid,
7) constant k_0, independent of T and other impurities, and
8) uniform initial concentration $C^f(y)$.

The classical experiments to test the BPS relation (6.39) were performed by BURTON et al. [6.29] with Ge crystals pulled from doped melts. Fig.6.10 represents k-data obtained at varying growth rates at different rotation rates, i.e., at different v_n in (6.30).

One sees that the effective segregation coefficient increases with the growth rate and decreases with the rotation rate, as predicted. It is also noteworthy that for the highest rotation rate, $k \approx k_0$ up to $V^s \approx 4$ cm/hr (≈ 10 µm/s). For a quantitative comparison of the experimental data with the BPS theory let us briefly look at the approach taken in the original paper [6.27]. Initially the authors formulated the segregation problem at a rotating interface as expressed in (6.30) with v taken as the sum of the growth velocity (with ρ^f/ρ^s equal unity) and the normal fluid velocity. For the boundary conditions as in (6.33) and (6.34), and an approximation for v_n given by COCHRAN [6.30] (see also Sect. 5.3.2, Fig.5.10)

$$v_n = 0.5\,\omega^{3/2}\nu^{-1/2}y^2 \qquad \text{for} \qquad y < (\nu/\omega)^{1/2} \tag{6.40}$$

they obtain the solution for the concentration at the interface in the form

$$\frac{C_0^f - C_\infty^f}{C_0^f - C^s} = \int_0^\infty e^{-(X+BX^3)}dX = I \tag{6.41}$$

where $B = 0.17\,\mathcal{D}_{AB}^2\omega^{3/2}/V^{s^3}\nu^{1/2}$ and $X = V^s y/\mathcal{D}_{AB}$. For practical reasons, however, the authors then continue the discussion based on the diffusion boundary layer model, i.e., the assumption that within δ' $v_n = 0$. Beyond δ' the mixing fluid flow is assumed to keep C^f equal C_∞^f. Within this diffusion boundary layer the transport flux is envisioned being merely diffusive-ad-

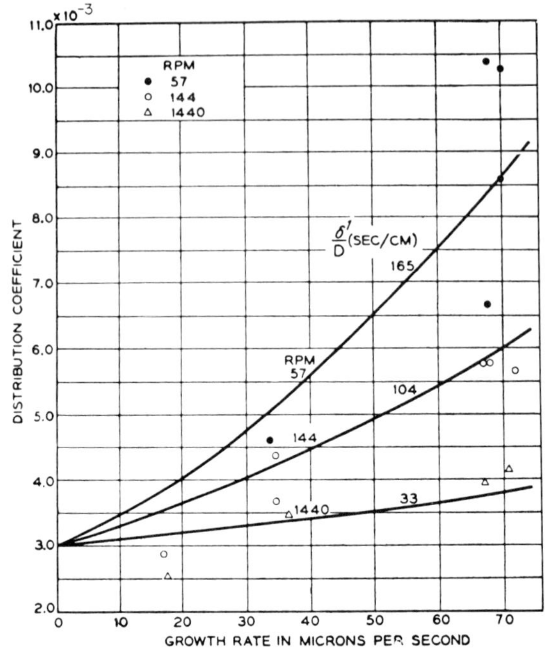

(a)

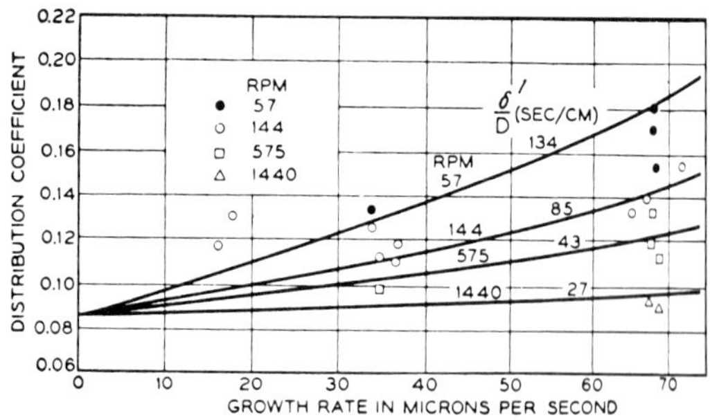

(b)

Fig.6.10. Effective distribution coefficients k for (a) Sb in Ge and (b) Ga in Ge. From [6.29] by permission of the American Institute of Physics

vective. Consequently, the solution for the interfacial concentration C_0^f is obtained from

$$\frac{d^2C(y)}{dy^2} + \frac{V^s}{\mathcal{D}_{AB}}\frac{dC(y)}{dy} = 0 \tag{6.42}$$

and the boundary conditions (6.31) and $C^f(\delta') = C_\infty^f$ in the form

$$\frac{C_0^f - C^s}{C_\infty^f - C^s} = \exp(V^s\delta'/\mathcal{D}_{AB}) \quad , \tag{6.43}$$

which is, of course, identical with (6.36) evaluated at $y = 0$ (Problem 6.3). Equation (6.39) follows then from (6.43) and $k = C^s/C_0^f$, (6.2).

Now, in comparing their experimental data with (6.39) BPS proceeded as follows: After adopting an appropriate value for k_0 (e.g., 0.003 for Sb in Ge, Fig.6.10a) a value for δ'/D_{AB} was chosen by trial and error, i.e., by fitting (6.39) to the data, for one rotation rate. The values for δ'/D_{AB} for the other ω's were then found from this first value by means of the relation

$$\exp(-V^s\delta'/D_{AB}) = 1 - I \quad , \tag{6.44}$$

which follows readily from (6.41) and (6.43). We see that the thus obtained curves fit the experimental data reasonably well.

Summarizing we recognize the BPS theory to be a "hybrid" of the stagnant film model (see p. 294 and Fig.5.24, the δ' is nothing but the d discussed there), and a full-fledged fluid dynamic boundary layer treatment. This is frequently not realized in the literature. As is left to the reader in Problem 6.4, the values for δ' obtained for the experimental conditions used in [6.29] are still about one-half of the true concentration boundary layer widths δ, inspite of $v_n \approx 10V^s$ at $y \approx \delta'$. This illustrates the rapid decrease [quadratic with y, see (6.40)] of v_n and hence of the convective flux upon closer approach to the interface. It also illustrates the usefulness of the diffusion boundary layer model (as long as $v_n(\delta)$ is not $\gg V^s$) which allows for a mathematically drastically simpler treatment of the problem.

More recently, very good agreement with (6.39) was found in a study of the growth rate dependence of k for rhodium in melt-pulled zinc tungstate [6.31]. Fig.6.11 shows the results. The solid line was calculated for $k_0 = 2.38$, $D_{AB} = 1.2 \times 10^{-5}\mathrm{cm}^2\mathrm{s}^{-1}$ and a δ that was derived from an experimentally determined thermal boundary width $\delta_T = 0.16$ mm at 84 rpm crystal rotation rate.

The above discussion may have led to the impression that the BPS description is restricted to segregation at a rotating interface. Yet (6.39) can be applied to any segregation problem, no matter what the specific fluid dynamic configuration (geometry, v_n and V^s) that results in a certain diffusion boundary layer width δ'. No matter how large v_n is as compared to V^s, within some region of width δ' at the interface, diffusion-advection fluxes will always dominate over the convection fluxes. Hence, it has become common practice to use segregation data for estimates of diffusion boundary layer widths;

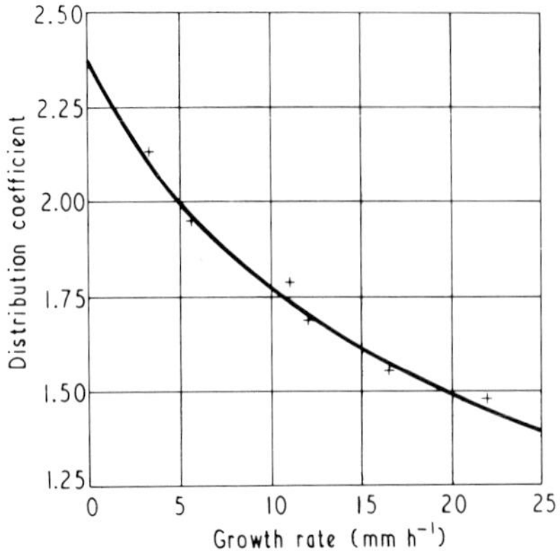

Fig.6.11. Effective segregation coefficient of rhodium in zinc tungstate growth from the melt, as function of growth rate. From [6.31] by permission of the Physical Society, London

Table 6.4. Diffusion boundary layer widths deduced from segregation data observed in melt growth.[6.32]. Data from [6.28]. By permission of McGraw-Hill

Solvent	Solute	δ', cm
	Normal solidification	
Pb	1% Sb	0.05
Mg	0.0016% Fe	0.007
Anthracene	0.1% tetracene	0.2
H_2O	2-10 g/100 ml urea	0.02
H_2O	1-5 g/100 ml urea	0.009
	Czochralski growth	
Ge	$<10^{-3}$% Ta	0.04
Ge	Sb	0.01
Ge	Ga	0.01
Si	Al	0.03
NaCl	0.5 % I	0.0005
	Zone melting	
Pb	10% Sn	0.08
Pb	10% Sn	0.016
Fe	0.2% Ta	<0.015
$GaCl_3$	$1.4 \times 10^{-5} FeCl_3$	0.008
Napthalene	10% benzoic acid	0.07

see Table 6.4..Assuming that V_{cf}, D_{AB}, C_0^s, C_∞^f and k_0 are known, δ' can be calculated from (6.39). We note, however, that the obtained value will only resemble that of the actual concentration boundary layer width if $V^s \approx v_n(\delta)$, i.e., if $D_{AB}/V^s \leqslant \delta$. Other complications resulting from non-steady conditions are discussed on p. 477.

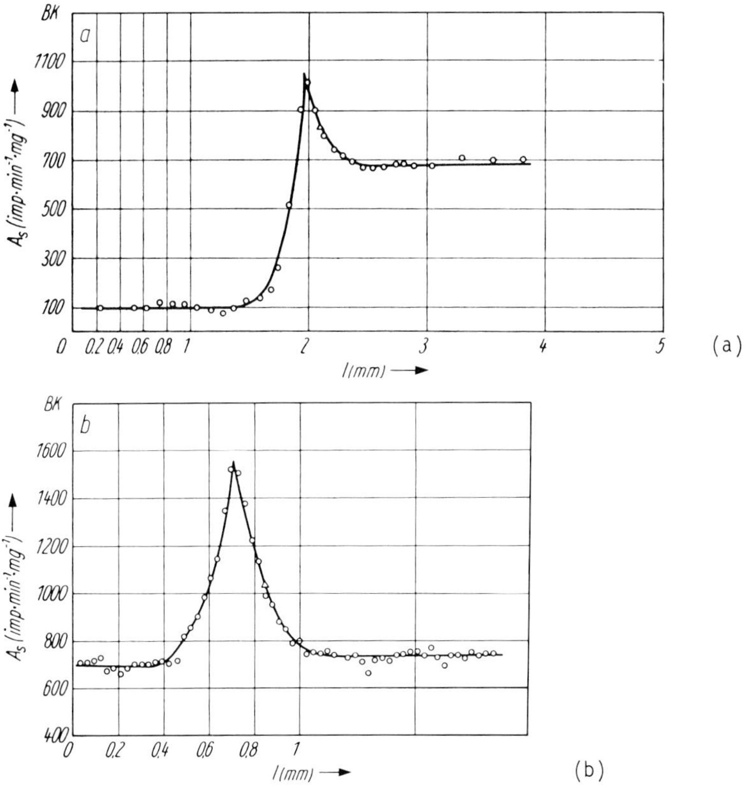

Fig.6.12. Concentration distribution in the region of the interface of 0.05% Zn melt-doped Sn. (a) V^s = 8.8 mm/hr. (b) V^s = 35.4 mm/hr. From [6.36] by permission of Akademie Verlag, Berlin

Experimentally determined interfacial concentration profiles also confirm the BPS description very well [6.36]. Fig.6.12 shows the distribution of Zn in Sn obtained, after quenching of the melt, via a radiotracer analysis of the very finely sliced sample. Fig.6.12a gives the concentration profile about the interface for a growth rate at which k exceeds k_0 = 0.13 only slightly. After a fourfold increase in V^s under otherwise similar conditions, however, one sees that k becomes practically unity (Fig.6.12b). The finite slope of the profile in the "solid region" (to the left of the maximum, C_0^f) resulted from a slight misorientation of the slicing direction with respect to the interfacial plane.

Typically, segregation data are not as well characterized as the aforementioned studies suggest. Hence, only semi-quantitative conclusions can be drawn in most cases that confirm the trends reflected in the BPS relation (6.39). For instance, it has been shown in melt-pulling experiments with alkali halides that at V^S = 0.2 cm/hr the $k_{Sr/KCl}$ is already 10% larger than the extrapolated k_0 [6.33]. On the other hand, other workers found in a similar pulling arrangement that the growth rate had to be increased to 12 cm/hr to obtain the same deviation for $k_{K/NaCl}$ [6.34]. The rotation conditions were not specified in either of the papers. Part of this discrepancy can probably be explained by the lower diffusion coefficient expected for divalent ions in monovalent melts, that is likely due to complex formation. However, for the major part these results illustrate the significance of the individual stirring conditions (free convection, rotation, etc.) in the nutrient and thus the dependence of segregation on the prevailing δ. This is further illustrated by a result obtained also for the K/NaCl system [6.35]. In a Bridgman experiment [1.2], where convection is less significant than in pulling because the cooler crystal is situated below the hotter melt, V^S had to be less than 0.4 cm/hr in order to achieve $k \approx k_0$.

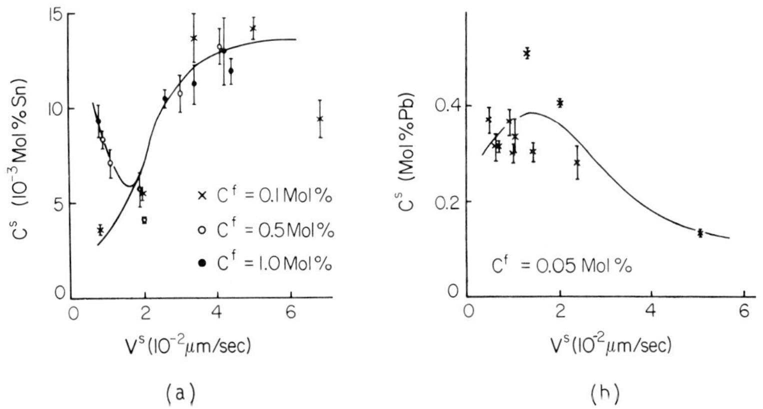

Fig.6.13. Concentration of (a) Sn and (b) Pb in KCl crystals, grown from aqueous solutions, as function of growth rate for various dopant concentrations. From [6.79] by permission of North-Holland Publ. Co.

Only very scarce information is available on the dynamic nature of segregation in growth from solutions. Fig.6.13 shows data obtained from crystallization experiments with lead and tin-doped KCl solutions. The aqueous solutions were unstirred [6.79]. As discussed in Sect. 6.2.2, such complex-forming species display often a rather complicated concentration dependence of k_0. In addition, the above curves show a complicated growth rate dependence of k. There is also strong evidence that part of this complex behavior stems from adsorption (non-equilibrium) phenomena.

With monovalent impurities that, in general, behave in a less complex manner, crystallization rates from aqueous solutions of KCl, KBr and NaCl as low as 15% of the total salt within 3 days were required in order to obtain growth-rate-independent segregation [6.80]. Again, the stirring conditions were unfortunately not specified.

6.4 Macrodistribution

Segregation performed in a controlled manner can advantageously be employed for purification of materials as well as for redistribution of dopants. In the following, concentration profiles in solids will be treated as they result from different geometries used during *fractional solidification* processes. Although the discussion is mostly based on liquid-solid phase transitions, the results apply also to vapor-solid transitions if the appropriate density changes reflected in the crystallization flow velocity V_{cf} are taken into account, as pointed out in Sect. 6.3.1. This section follows quite closely reviews on the subject by PFANN [6.37], and ZIEF and WILCOX [6.38].

6.4.1 Normal freezing

It is assumed that a progressive solidification process of an initially completely molten amount of material can be described in terms of only one coordinate. It is customary to use the solidified fraction g of the material for that purpose, as indicated in Fig.6.14. The advantages of such a dimensionless coordinate are obvious.

Depending on the mixing conditions in the bulk melt (i.e., on k), one will obtain different impurity concentration profiles in the solid. For the "no-mixing, diffusion-only" case we know already that after an initial transient, segregation will cease as $k \to 1$ for any practical growth rate. This is illus-

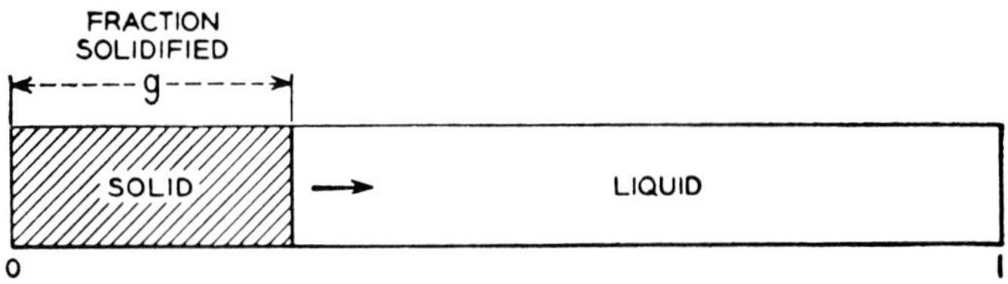

Fig.6.14. Solidification by normal freezing. From [6.37] by permission of John Wiley and Sons

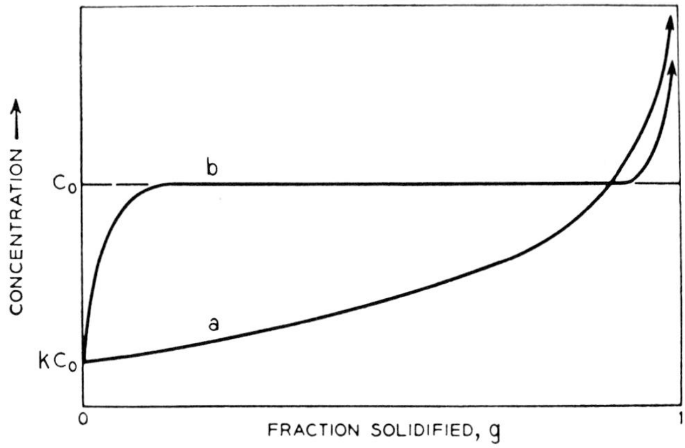

Fig.6.15. Concentration of solute after normal freezing versus fraction solidified for $k < 1$: (a) partial mixing in liquid, (b) transport in liquid by diffusion only. From [6.37] by permission of John Wiley and Sons

trated with curve (b) in Fig.6.15. Mass conservation, of course, requires a final transient at the end where the impurities that were removed from the purer part of the ingot have to be frozen out. Yet this case is rather rare under typical "earthbound" conditions. For the case of partial mixing in the melt, i.e., where diffusion dominates only in a boundary layer at the progressing interface, quite different profiles are obtained as indicated by curve (a) in Fig.6.15. The first layer to solidify from the melt contains a concentration $C_0^S = kC_0$, with C_0 the initial, uniform concentration in the melt and k the effective segregation coefficient. This is the same as for the no-mixing case. But then, with increasing g a much longer zone of lowered impurity content results due to more efficient impurity transport from the segregating interface into the residual melt. The zone of $C^S < C_0$ will be longer the better the macroscopic fluid mixing process is. In other words: The purification efficiency will increase with $k \to k_0$ ("complete mixing"). However, even with complete mixing, C in the residual melt will eventually

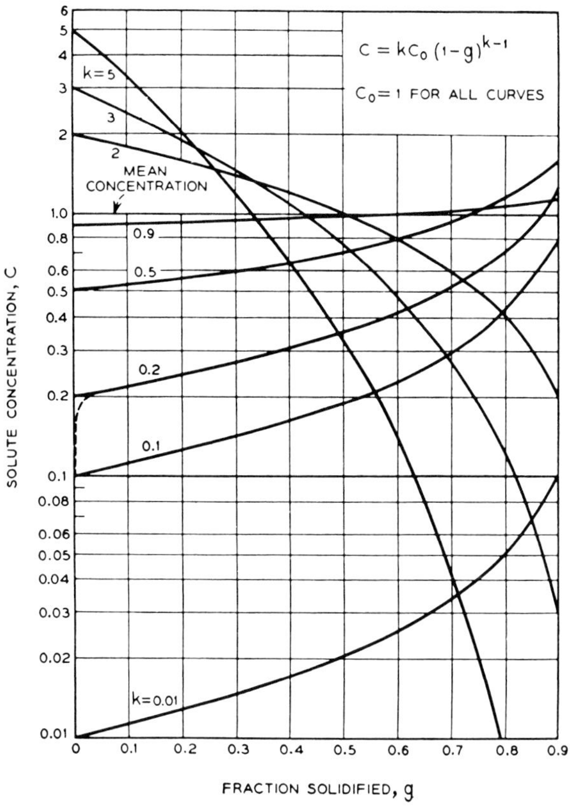

Fig.6.16. Concentration profiles in solid for normal freezing calculated from (6.45) for various effective segregation coefficients k. From [6.37] by permission of John Wiley and Sons

reach C_0/k_0, where effective segregation can no longer occur. Mass conservation again requires a zone with $C^S > C_0$ at the end of the solid ingot.

Assuming that the initial impurity level C_0 is uniform, that k is independent of g and that solid state diffusion can be neglected, and that the considered component is only transported through the melt, one can express the concentration profile in the solid as

$$C = kC_0(1 - g)^{k-1} \quad . \tag{6.45}$$

known as *Pfann relation*. (For derivations see [6.39].) Normal freezing curves for various k values according to (6.45) are plotted in Fig.6.16.

Strictly speaking, the statement that the first layer to solidify from the melt is of composition C_0k is wrong. The effective segregation coefficient applies only after the solid had advanced far enough to accumulate the steady-state amount of impurity in the enriched melt layer. However, as will be discussed in Sect. 6.5.1, compositional transients in melts with any significant convection and hence not too wide diffusion boundary layer width, require typically a few seconds. Hence, the transient region (suggested by the dashed line in Fig.6.16) for solidification in a normal gravity environment is rather narrow.

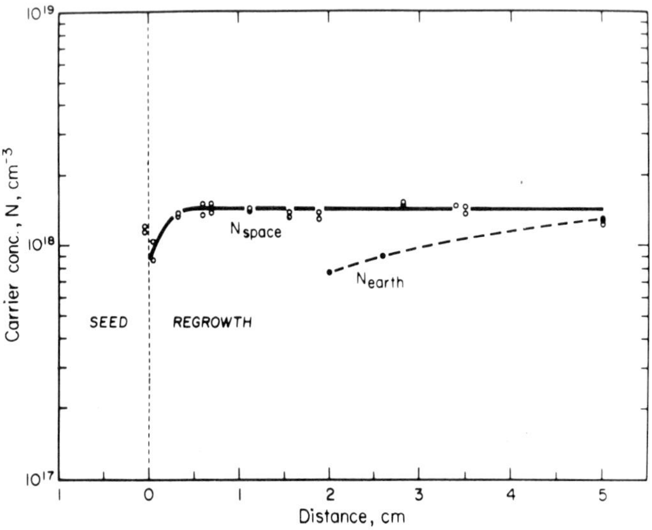

Fig.6.17. Concentration profile obtained from Hall-effect measurements for Te-doped InSb crystal grown under low gravity. Dashed line: Profile obtained under identical conditions on earth. From [6.40] by permission of The Electrochemical Society

In solidification experiments under "zero-gravity", the transient region may take on significant dimensions. There density gradients do not stimulate convective fluid flow and, besides the possibility of surface-tension-driven flow (Sect. 5.8.3), mass transfer is by diffusion only. Figure 6.17 shows a concentration profile obtained in a normal freezing process of Te-doped InSb conducted during the Skylab-III mission [6.40]. As theoretically expected, this experiment beautifully verifies that in a low-g environment ($\approx 10^{-5}$g)

a) the transient region takes on macroscopic dimensions, here about 0.5 cm, and

b) after the transient a true steady state (diffusion-only case, curve b in Fig.6.15) is reached. Due to always present convection, this is practically impossible to realize on earth for growth rates that yield structurally satisfactory single crystals. The microsegregation aspects of this experiment will be taken up in Sect. 6.5.

A normal freezing distribution obtained for partial mixing in the melt is depicted in Fig.6.18. The solid line is calculated from Pfann's relation (6.45).

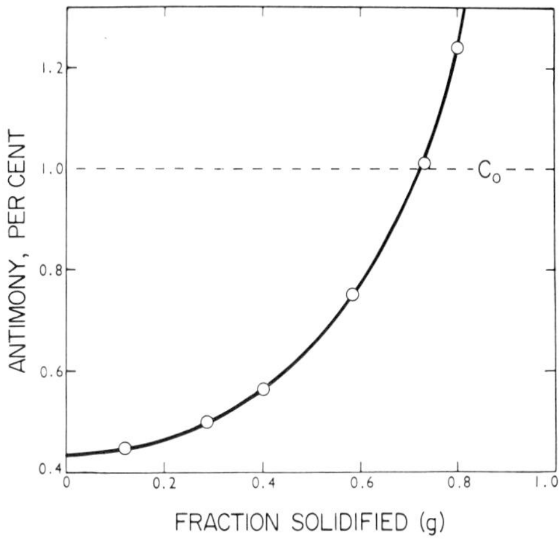

Fig.6.18. Solute distribution in Sb-doped Pb plotted against fraction solidified. From [6.41] by permission of Institute of Metals, London

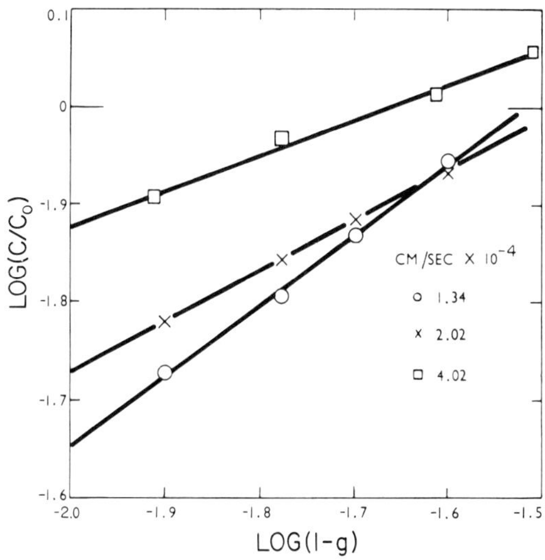

Fig.6.19. Logarithmic plot of solute distribution in Sb-doped lead for various growth rates. From [6.41] by permission of Institute of Metals, London

Normal freezing experiments are frequently employed for determinations of effective segregation coefficients. In plots of $\log(C/C_0)$ versus $\log(1-g)$ one obtains straight lines; see (6.45) and Fig.6.19. Both the slope and the intercept at $g = 0$ yield k.

Normal freezing ("fractional solidification") is a powerful method for material purification if k deviates enough from 1 and if high impurity levels have to be dealt with. In principle, by using smaller and smaller fractions of the initial amount of material (purer ends), each consecutive normal freezing step should result in an even purer substance. In practice, the result of such a chain of processes is limited by the necessity of in-between handling of the material that eventually introduces as many impurities as are removed. Thus, purification limits for multiple normal freezing processes are typically found around solute-to-host concentrations of 10^{-4}.

6.4.2 Zone melting

Zone melting avoids recontamination of the material as it results from in-between handling and use of different containers for each fractional step. It was first employed for crystal growth by ANDRADE and ROSCOE in 1937 [6.54]. However its potential for the ultrapurification of materials was not recognized until W.G. Pfann at the Bell Laboratories applied it in 1952 to germanium. In this technique only part of the material is molten in the form of a zone that is passed through the ingot, as depicted in Fig.6.20. Neglecting again the short transient at the very beginning of the ingot, the solid which first forms from the melt after passage of the molten zone has the composition kC_0. Due to the limited volume of melt into which impurities can be rejected, the condition of "no segregation" ($C^S = C_0$) occurs at a shorter

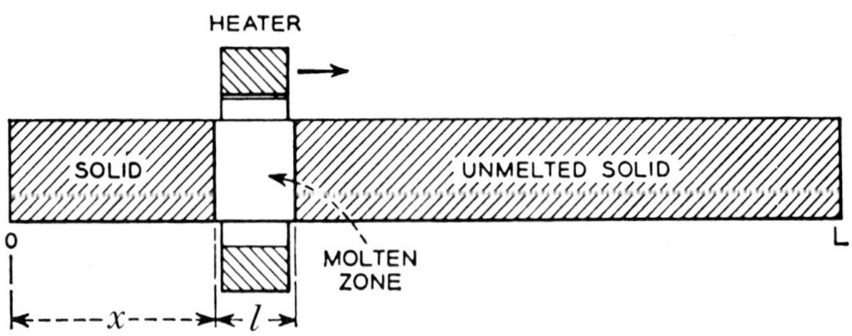

Fig.6.20. Zone melting: molten zone of length l traversing a cylindrical ingot of length L. From [6.37] by permission of John Wiley and Sons

distance from the front end of the ingot than in normal freezing. Figure 6.21 indicates schematically the profile in the solid as it results from different mixing conditions in the liquid zone after one zone pass.

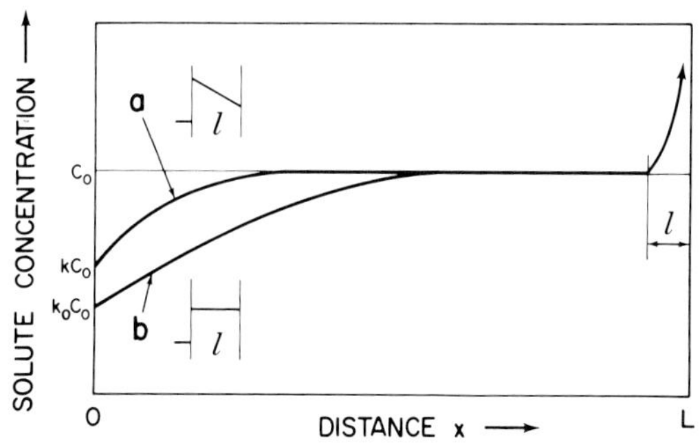

Fig.6.21. Schematic concentration profiles after passage of one molten zone through an ingot of uniform initial C_0. For (a) partial mixing ($k > k_0$) and (b) complete mixing ($k = k_0$), in the melt

Assuming that the zone length l is constant, that material transfer occurs only in the liquid, and that k is independent of ingot coordinate x, the concentration profile after one zone pass can be expressed for the region $0 \leqslant x \leqslant L - l$ in the form

$$C = C_0[1 - (1-k)\exp(-kx/l)] \quad . \tag{6.46}$$

Solidification in the last zone length represents a normal freezing process. Thus, the concentration profile in the region $L - l \leqslant x \leqslant L$ is given by (6.45).

Curves calculated from (6.46) for various k's are plotted in Fig.6.22. One sees that the purification effect from a single zone pass is inferior to that of a normal freezing process as indicated by the dashed line in Fig. 6.22. However, the advantage of zone melting becomes evident when multiple passes are performed. Note that the first solid to freeze from the melt after passage of the second zone contains somewhat more impurities than C_0k^2. As illustrated in Fig.6.23, this comes from the averaging effect remelting has on the profile that resulted from the first zone pass. Correspondingly, the area of increased concentration, $C > C_0$, at the end of the ingot becomes shifted to somewhat lower x-values.

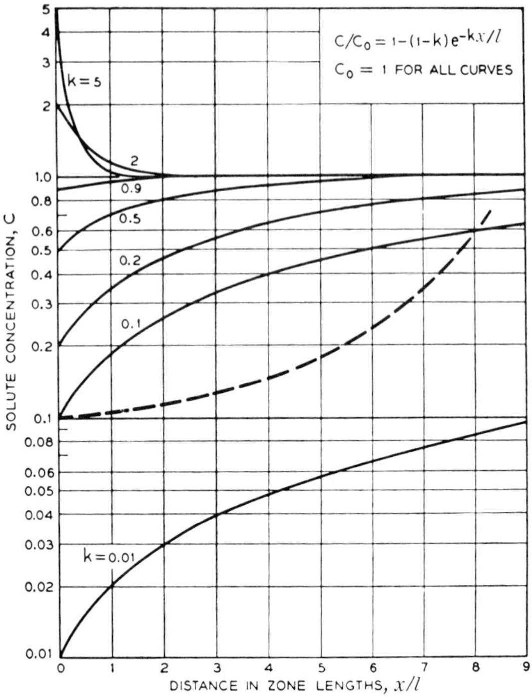

Fig.6.22. Curves for single-pass zone melting showing solute concentration in the solid versus distance in zone lengths from beginning of charge, for various values of the distribution coefficient k. From [6.37] by permission of John Wiley and sons

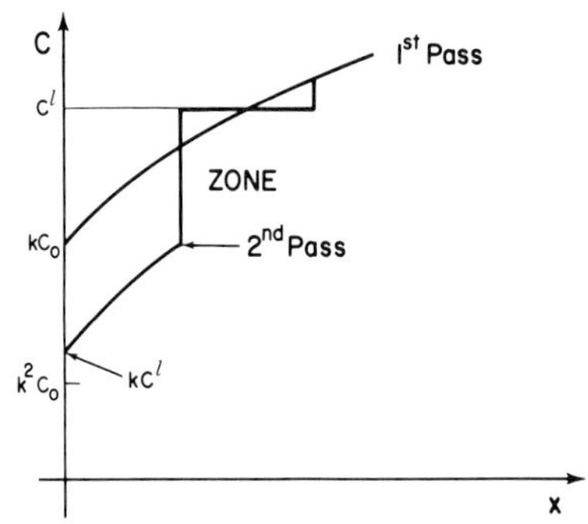

Fig.6.23. Concentration profile before and after pass of second zone at beginning of ingot. C^l: concentration in molten zone

From Fig.6.23 one can deduce that the purification efficiency will increase with decreasing zone length l. The wider the zone the more of the initial concentration gradient is "averaged out" in a consecutive zone and the more the range with $C > C_0$ will be moved towards the beginning of the ingot. Figure 6.25 illustrates this further: compare the profiles after 5 and 10 passes for $L/l = 5$ (dashed) and $L/l = 10$.

Impurity concentration profiles calculated for consecutive zone passes are plotted in Fig.6.24. For the calculations see [6.37].

Comparison of Figs.6.24 and 6.25 demonstrates the importance that the effective segregation coefficient k has for the purification result. In a system with $k = 0.1$, a $C/C_0 = 7 \times 10^{-7}$ is obtained after $n = 8$. Yet only $C/C_0 = 10^{-4}$ can be achieved from $k = 0.2$ after the same number of zone passes.

What is the ultimate distribution that can be reached theoretically as n goes toward infinity? The ultimate profile is reached when the flux of impurities into the zone (from melting the next solid layer) equals the flux of impurities into the freezing solid. Here one expects again that a narrower zone and a smaller value for k allow for a steeper ultimate gradient. Comparison of the ultimate distribution in Figs.6.25 and 26 verifies this clearly.

The ultimate concentration profile for $0 \leqslant x \leqslant (L-l)$ can be approximated by [6.37]

$$C(x) = Ae^{Bx} \tag{6.47}$$

where A and B are constants obtainable from

$$A = \frac{C_0 BL}{e^{BL} - 1} \quad \text{and} \quad k = \frac{Bl}{e^{Bl} - 1} \tag{6.48a and b}$$

For convenience, some calculated B-values are given in Table 6.5. See also Problem 6.5.

The theoretical and technical details of zone refining have been reviewed by PFANN [6.37], SCHILDKNECHT [6.43], ZIEF and WILCOX [6.38] and SHAH [6.44]. Selected technical aspects will be taken up in the chapter on melt growth [1.2]. In the following we will reconsider some of the idealized assumptions that lead to (6.46) and (6.47) and discuss deviations that occur frequently in practice.

a) Solute transport in the solid and vapor

In high quality crystals, the neglect of solid state diffusion is in general a good approximation in comparison to diffusion in fluids (Sect. 5.2.3), although there are some systems known where this condition is not fulfilled [6.28]. However, under typical zone refining conditions (growth rate,

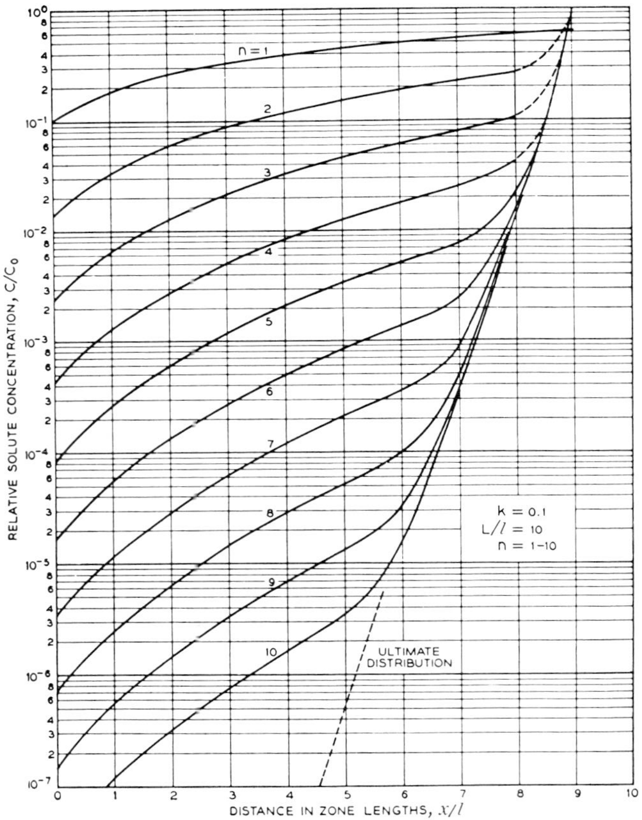

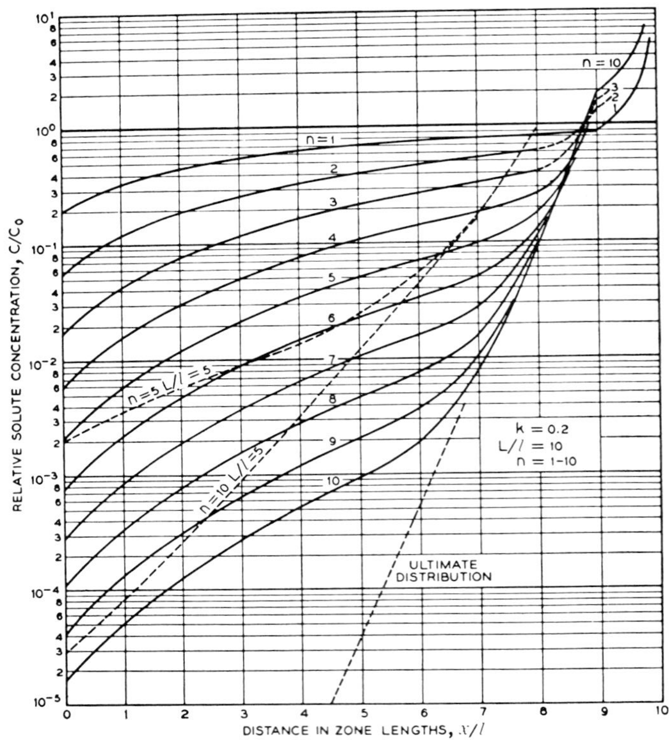

Fig.6.24. Relative impurity concentration C/C_0 (logarithmic scale) versus distance in zone lengths x/l from beginning of ingot for various numbers of passes n and k = 0.1. From [6.37] by permission of John Wiley and Sons

Fig.6.25. Concentration profiles for multiple zone passes and k = 0.2. From [6.37] by permission of John Wiley and Sons

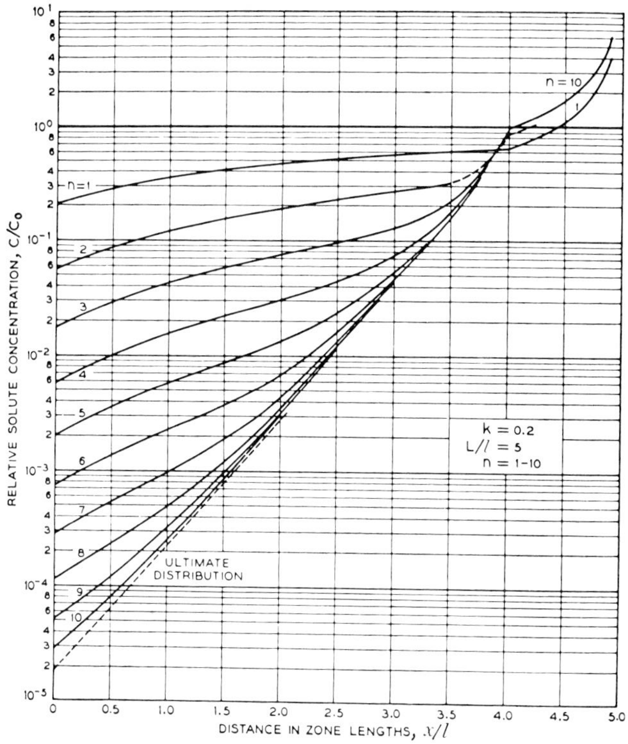

Fig.6.26. Concentration profiles for multiple zone passes, k = 0.2 and $L/l = 5$. From [6.37] by permission of John Wiley and Sons

Table 6.5. The constant B as a function of k calculated from (6.48b) for $l = 1$. From [6.37] by permission of John Wiley and Sons

k	B	k	B	k	B
0.0010	+9.10	0.10	+3.61	0.80	+0.430
0.010	+6.47	0.20	+2.66	1.00	0
0.020	+5.65	0.40	+1.62	1.50	-0.86
0.040	+4.79	0.50	+1.25	2.0	-1.58
0.070	+4.08	0.60	+0.945	3.0	-2.82

T-gradients, etc.) the resulting crystal quality is rather poor, which, in the extreme, may result in macroscopic trapping of solute as we will discuss in the chapter on morphological stability [1.1]. However, even in the absence of such gross effects one obtains usually high densities of defects, such as dislocations, grain boundaries and even macroscopic cracks, which via lowering of the activation energy for diffusion lead to high-diffusivity paths. For instance, for certain aromatic hydrocarbons it has been demonstrated that back diffusion in the solid ingot limits the zone refining result considerably [6.45]. Whereas in these systems in unperturbed lattices D^S ranges from 10^{-9} - 10^{-11} cm^2/s near the melting point, increases by factors of 10^5 - 10^6 have been found for diffusion along dislocations and grain boundaries. This makes the resulting diffusion fluxes comparable to those obtained in the molten state. Fig.6.27 depicts some of the results obtained for anthracene in carbazole [6.45]. Repeated cropping, i.e., removal of the very end of the ingot in order to remove the steepest concentration gradient that contributes most to back diffusion, was found very effective under these circumstances.

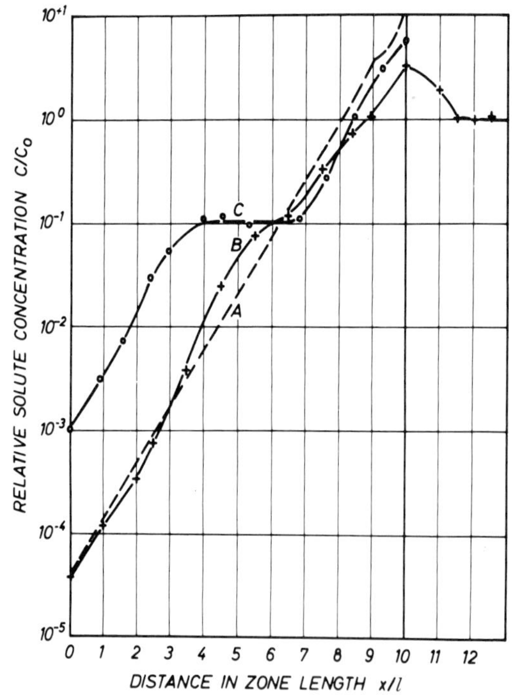

Fig.6.27. Experimental evidence of the influence of backward diffusion on zone refining in carbazole doped with 0.5% anthracene. L/l = 10, k = 0.5, n = 50. Curve A: calculated ultimate distribution. Curve C: standard zoning procedure. Curve B: the length of the solid bar exceeds $10l$. The decrease in the zone length (in A) in the region $L - l < x < L$ was simulated by steady heating-power reduction. Note the concentration increase in the residual, unzoned ingot. From [6.45] by permission of American Institute of Physics

Another possibility for back transport of solute exists when a material has significant vapor pressure, say $p \geqslant 10^{-3}$ torr, at the zone temperature. As schematically indicated in Fig.6.28 convective vapor transport can result in deposition of impurities on the free surface of the ingot on both sides of the zone. The next zone pass will melt these impurities again. Hence, the purification effect at the purer end is reduced. Free solid surfaces exist in crucible-less melting processes (floating zone) as well as in horizontal open boat arrangements. In vertical zone melting in crucibles, where the solid parts act as tightly fitting plugs, vapor transfer is strongly reduced (see Fig.6.30).

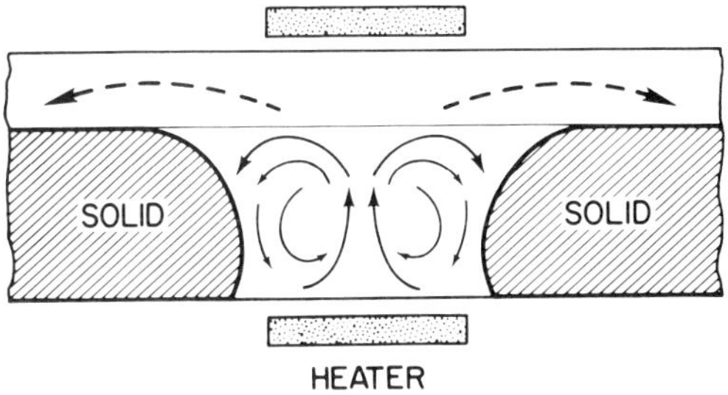

Fig.6.28. Melt convection and vapor transfer from zone to free solid surfaces in horizontal zone melting

Analytical relations for concentration distributions in various normal freezing and zoning modes including redistribution by evaporation have recently been presented in [6.46].

Impurities can, of course, also be transferred from the container material. The limited purity of containers is, as a matter of fact, the limitation to purification in most systems. With melts that have sufficient surface tension, one will therefore try to employ floating zone purification [1.2].

b) Varying zone length

The zone width in a certain heating setup will be such that heat transferred to and from the zone balance. Since these conditions change with the position of an ingot with respect to the zone heater, so will the zone length. This can be particularly pronounced at lower temperatures where heat conduction, as compared to radiation, still plays an essential role. The zone will be considerably wider at both ends, (unless the thermal conductivity of the container is high) since it is only in thermal contact with the solid ingot on one side. In principle, one can compensate for this end effect by programming the heat input. This, however, is a considerable experimental task. Note that zone widening reduces the purification effect.

Reduced segregation at the beginning of an ingot may also have kinetic reasons. In order to obtain maximum purification one will in general melt the whole initial portion of the ingot without having any solid to act as a seed. Depending on the material and crucible surface, considerable supercooling may be required before a new solid nucleates. Hence the initial portion of the contained material may become undercooled and then, after nucleation, freeze much more rapidly then the remainder of the ingot.

c) Dependence of the effective segregation coefficient on zone motion

The necessary conditions for a constant k were already discussed in connection with the experimental observations; Sec. 6.3.3. There are some special aspects to zone melting, however, that make k = const. difficult to achieve in practice. First, the above mentioned zone widening effect at the beginning and end of an ingot causes a change in k, since (assuming a constant heater travel rate) the growth rate will be modified. Furthermore, it is not trivial to design equipment which ensures a linear travel rate that is constant on a microscopic scale. There is always a trend for oscillations between static and dynamic friction. "Jerky" mechanical drives can cause growth rate fluctuations and hence, reduce the purification efficiency.

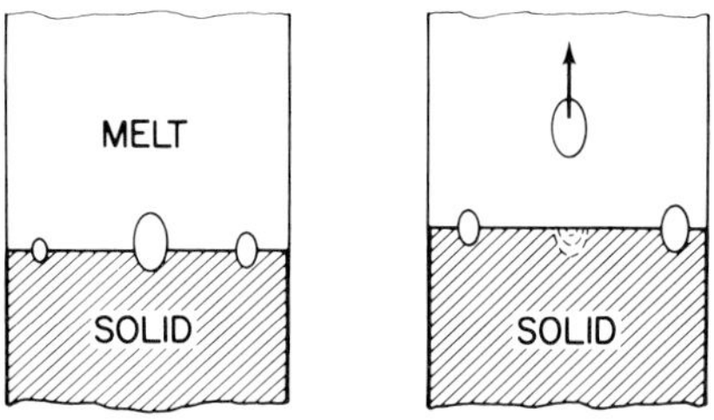

Fig.6.29. Modification of interface shape by gas bubbles and rapid growth after bubble detachment

Growth rate discontinuities can also result from gas bubbles that form at interfaces if the solubility for a gas in the solid is much lower than in the liquid. As schematically indicated in Fig.6.29, bubbles change the interface shape due to modifications of heat and mass transfer. If buoyancy "desorbs" such a bubble from the interface, the deformed part of the interface "catches up" rapidly causing a local reduction of segregation via the growth-rate discontinuity. A practical example of this type is mentioned in connection with Fig.6.31.

Growth rate and hence, segregation fluctuations can also be due to convective instability, i.e., time-dependent convective heat and mass transfer in the fluid, the topic of Sects. 5.8.2 and [1.2]. On the other hand, convection can be used to advantage for reducing the concentration boundary layer width δ and hence, for improvement of macrosegregation. For instance, in a vertical zone melting arrangement, convection at the upper interface in the destabilizing temperature gradient is much stronger than at the lower phase boundary. Most recent investigations [6.48] show that there is even less mixing between the upper and lower halves of the zone than Fig.6.30 suggests. For the consequences of oscillatory microsegregation on the macroscopically averaging k, see Sect. 6.5.

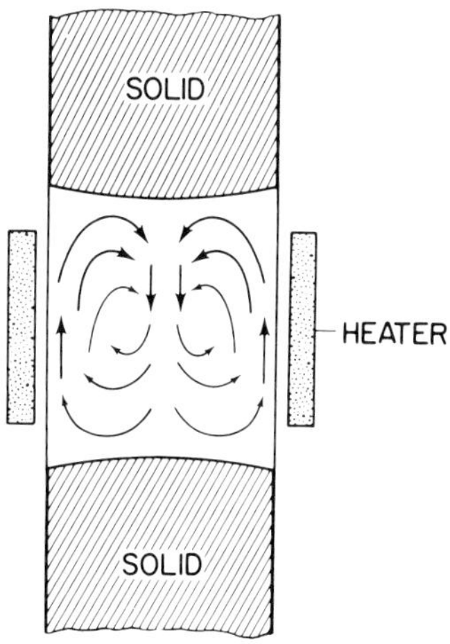

Fig.6.30. Convective mixing in vertical melt zone. After [6.47]

The invention of zone refining was a prerequisite for the development of the early semiconductor systems. The most impressive zone refining results have been obtained with Ge and Si where certain impurity concentrations were reduced to below 10^{-10} mole fractions. Great efforts have also been made with alkali halides. The purest KCl and KBr single crystals to date with 4×10^{-10} and 10^{-9} divalent cations were grown with a zone refining technique [6.49, 50]. Fig.6.31 shows the measured impurity profiles in zone refined KCl ingots, illustrating some of the aforementioned practical limitations.

Limitations of the zone refining process can also result from solubility limits as well as invariant reactions. Consider a simple eutectic system as represented by the phase diagram in Fig.6.32. With an initial charge of com-

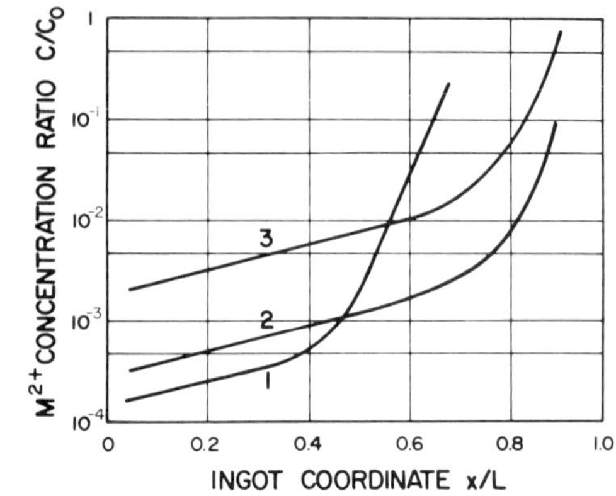

Fig.6.31. Impurity concentration profiles in zone refined KCl ingots. C_0 is the initial impurity level of about 1 ppm. After [6.50]. (1): Polycrystalline into. (L = 28cm, l = 1cm, n = 20). Higher purity in front part due to ultra-pure graphite container, lower purity in rear due to horizontal boat arrangement allowing for vapor transfer. (2): Single crystal. (L = 16.5cm, l = 1.5cm, n = 23). Somewhat lower purity in front from quartz container. Higher efficiency in rear from reduction of vapor transfer in vertical arrangement that is also more advantageous for single crystal growth. (3): Identical to (2) except for chlorine bubble formation at growing interface

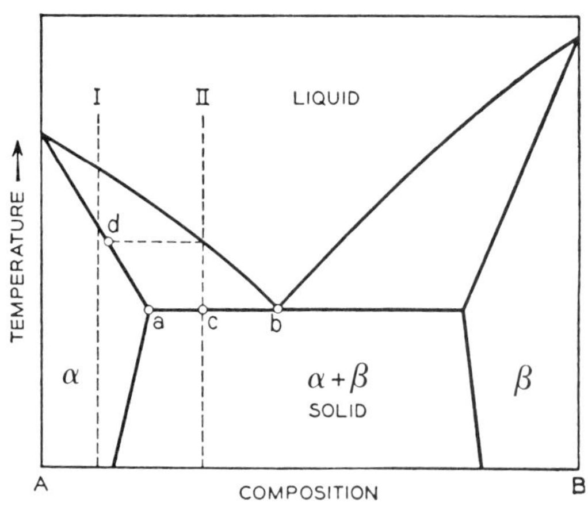

Fig.6.32. Zone refining of eutectic. From [6.37] by permission of John Wiley and Sons

position I, after some zone passes the ingot will eventually consist of phase α ranging in composition from some value close to pure A at the beginning to concentration a at the end. Upon passage of additional zones an increasing amount of β phase will form at the end of the charge. If the initial composition of the charge is at II, zone melting causes a phenomenon that is not found in normal freezing. The beginning of the ingot, after the first zone has passed, has composition d. Then with further travel of the zone, composition a is eventually reached in the solid and b, the eutectic, in the liquid. Here the reaction becomes invariant in a new sense: α of composition a *plus* eutectic ($\alpha + \beta$) is frozen out until the zone reaches the end of the charge where only eutectic will be solidified. This is because the zone is constantly fed with solid of composition c and therefore has to freeze out solid of the same average composition. After repeated passes, more and more of the charge end will consist of mere eutectic while the front will get closer in its composition to pure A. Hence, the proportion of eutectic will be greater than one would get from the lever rule for equilibrium freezing.

In systems with compound formation, the concentration of excess components can be reduced in zone melting processes. As illustrated in Fig.6.33, charges that initially deviate in composition from the congruently melting solid (maximum melting point T_m) will be shifted closer and closer to X_m in the front of an ingot by successive zone passes.

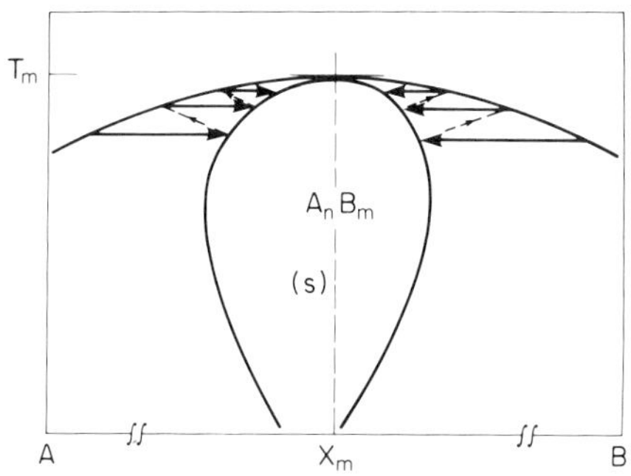

Fig.6.33. Expanded part of binary phase diagram with compound formation. Successive "purification" of off-stoichiometric material by zone melting

Zone refining is by no means restricted to melt-solid segregation. It can in principle be based on any phase equilibrium in which the phases in mass-transfer contact have different solubilities for a considered component. Zone refining has been successfully employed in solid-solid and solid-vapor systems as well. For references see [Ref. 6.37, p. 65]. Zone purification can be aided by an electric field gradient across the zone which causes changes in the solute concentrations at the interface. Or zone purification can be conducted in a mode in which a zone of solvent is moved through an ingot. For references on these two groups see [6.44].

For a survey of zone refining work on inorganic and organic materials performed up to 1966 see ZIEF and SHOEMAKER, and ZIEF in [6.38]. A more recent compilation is found in WILKE [6.51].

6.4.3 Uniform composition

The above discussion has shown that macrosegregation, unless specifically modified, causes a non-uniform solute distribution in fractional solidification processes. Let us now look into the possibilities of modifying fractional solidification to obtain uniform impurity distribution.

First, one can make use of the dynamic nature of segregation. According to the BPS relation (6.39) continuous reduction of an initially high growth rate will result in a more homogeneous plateau-like profile if $k < 1$. For $k > 1$, continuous increase is required. However, this method is rather limited by interface stability requirements [1.1]. Growth rates that are too high lead to unstable growth. Thus, one can only make partial use of the full growth rate dependent range of k.

Another possibility is the creation of a steady-state condition in which material of composition C_0 is solidified from a limited melt volume with $C^f = C_0/k$. Meanwhile material of composition C_0 is fed with the same rate into the melt, as shown in the scheme below:

$$\underset{\text{solid added}}{C_0} \rightarrow \underset{\text{melt}}{C_0/k} \rightarrow \underset{\text{solid withdrawn}}{C_0} \quad . \tag{6.49}$$

We have encountered this condition already in the diffusion-only case in normal freezing (Fig.6.15, curve b) and in zone melting after the initial transient (Fig.6.21). In both cases impurities were accumulated until the concentration in the diffusion boundary layer directly at the interface was C_0/k. From then on effective segregation no longer occured. In this way major portions of a charge could be homogeneously doped. Note, particularly, that the normal freezing process is not well suited for this purpose because it is difficult to avoid convection in order to steadily establish the diffusion-only condition.

A more efficient technique of achieving uniform doping is zone leveling. In this case one creates the steady-state condition right from the beginning by adding an appropriate amount of dopant to the melt as illustrated in Fig.6.34. Except for a final transient, a uniformly doped ingot results from this process after a single zone pass. Note, however, that in reality the conditions are not as simple as suggested in the figure. The depicted case of ideal mixing can only be approximated for very slow zone travel rates, and the concentration profile in the zone will correspond rather to insert (a) in Fig.6.21 (see also Fig.6.40, right side). Therefore, the amount of dopant, initially added to the melt in order to achieve a specific concentration in the solid, will depend on the dynamic condition of the experiment, and a short initial transient can hardly be avoided.

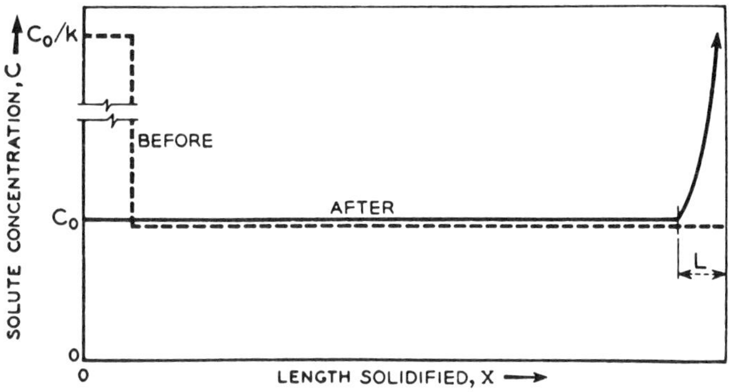

Fig.6.34. Zone leveling: addition of dopant to molten zone at beginning of charge. From [6.37] by permission of John Wiley and Sons

A good approximation of a uniform distribution can also be obtained by repeated back-and-forth passes of a zone through a charge of C_0. thus more or less eliminating the initial and final transients that occur in one-directional zone melting.

Another "straightforward" scheme to fulfill the steady-state condition (6.49) is the so-called push-pull method. In this method a melt is continuously replenished with a melting solid rod while a crystal of the same composition is pulled with the same rate from the melt, as illustrated in Fig.6.35. Rather stringent feed control has to be maintained in this technique.

Figure 6.36 depicts an experimental setup used by Petrow and Semskow for the growth of uniformly doped Ge crystals by the push-pull technique. Still another scheme to obey (6.49) is employed in the two-container method, shown in Fig.6.37. In this method the only critical condition is that the material

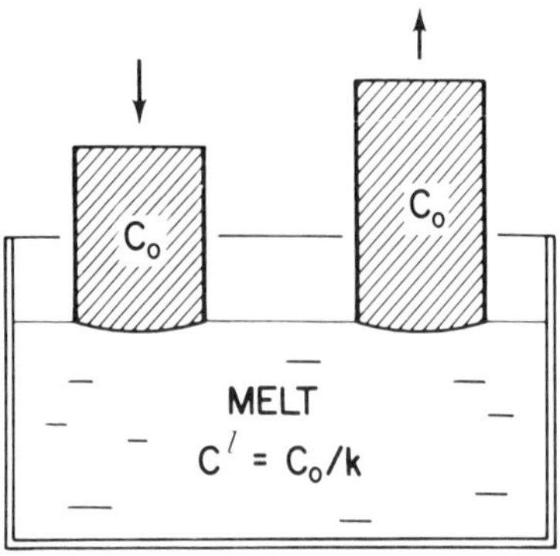

Fig.6.35. Push-pull method for growth of uniformly doped crystals

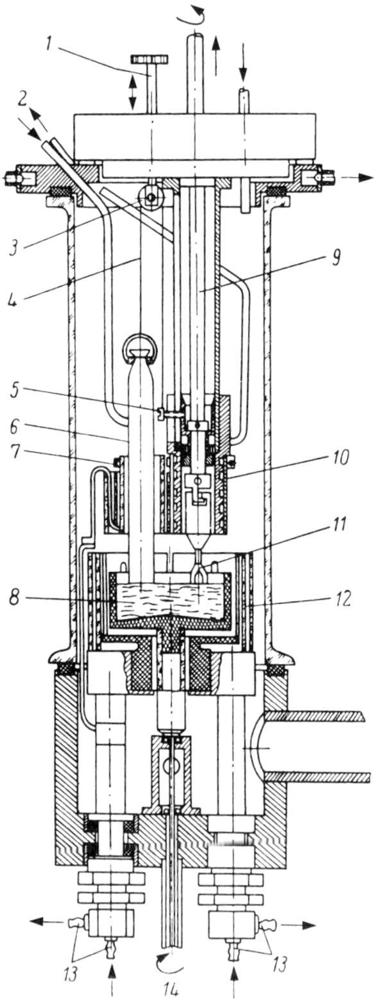

(1) height adjustment for feed rod,
(2) cooling water,
(3) pulley,
(4) feed rod suspension,
(5) quartz vacuum envelope,
(6) melting feed rod,
(7) auxiliary heater,
(8) melt,
(9) pulling rod (note that feed rod is automatically lowered during pulling),
(10) auxiliary crystal cooler,
(11) growing crystal,
(12) graphite heater,
(13) cooled current feedthrough,
(14) crucible rotation

Fig.6.36. Push-pull setup employed for Ge single crystals by Petrow and Semskow. From [6.51] by permission of VEB Deutscher Verlag der Wissenschaften

flow velocity through the interconnecting channel has to be large enough to override the diffusive material exchange between the two containers. In version (a), which requires a rather complex heating setup, the diffusion distance can be kept long; hence, the bore diameter does not impose any technical problem. A drawback of this version is, however, that the immediate melt volume from which the crystal is pulled diminishes with time. The floating crucible technique (b) avoids this problem. The inner volume is automatically kept constant as long as the crucible floats freely. Due to the limited length of the channel, however, rather small bore diameters are required (see Problem 6.6).

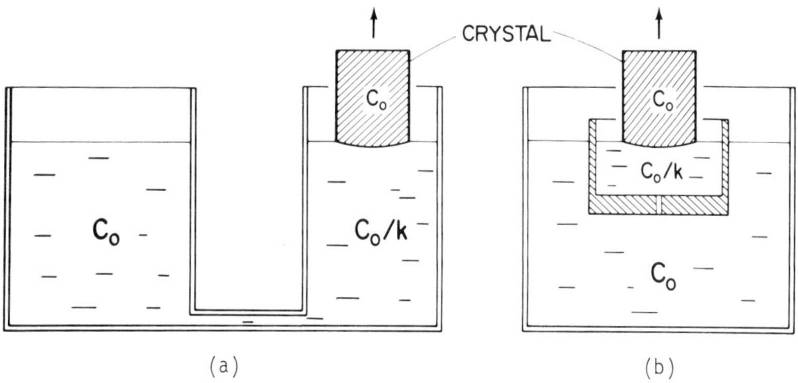

Fig.6.37. Two-container method for achievement of uniform solute concentration in melt-grown crystals (a) principle, (b) floating crucible version

A similar arrangement for the uniform physical vapor growth of incongruently vaporizing materials has been reported recently [6.52] and will be discussed in detail in [1.2].

Figures 6.38 and 39 illustrate schematically some experimental details used in the floating crucible technique.

Fractional solidification in a peritectic system results, in general, in a non-uniform composition. For instance a melt of composition 2 in Fig.6.40 leads initially to solid of compositon a. With progressing solidification, as the melt is enriched with B, the solid composition follows the solvus line a-b. A zone leveling procedure can be used here to obtain solid of uniform composition. Assume that the whole solid charge has a mean composition 4 when the first zone is molten. If melting is performed rapidly, i.e., so that solid state diffusion and, hence, a compositional change according to

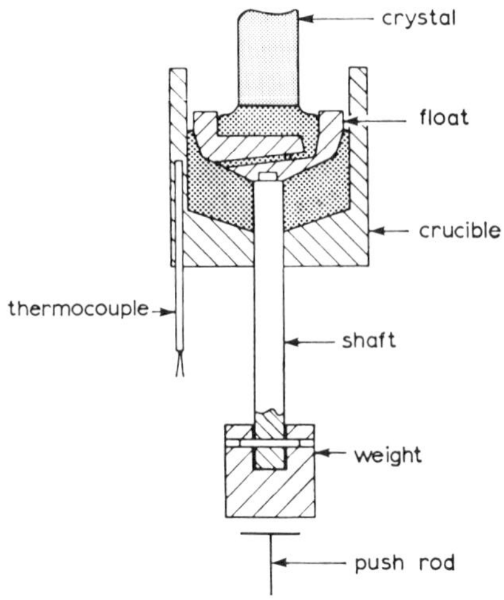

Fig.6.38. A floating crucible system. From [6.14] by permission of North-Holland Publ. Co.

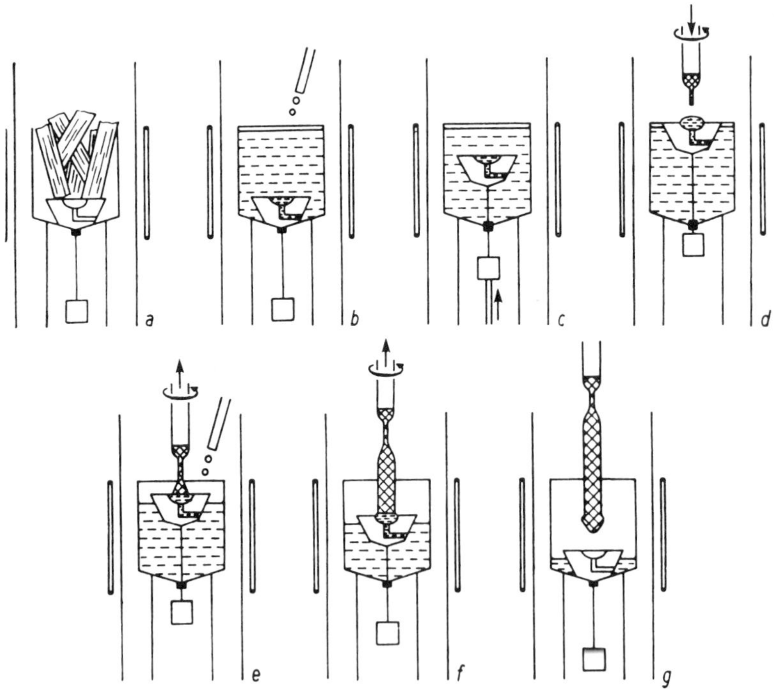

Fig.6.39. Procedure to grow uniformly doped Ge from a floating crucible as used by Goorisen and Karstensen. From [6.51] by permission of VEB Deutscher Verlag der Wissenschaften. (a) Filling with Ge, (b) melt is doped to level C_0, melt flows into capillary when hydrostatic pressure exceeds surface tension, (c) raising of float, (d) seeding of melt in float, (e) doping of float content to C_0/k, (f-g) pulling of crystal

the equilibrium diagram can be neglected, the melt will have the same composition 4. (Note that the melting point of 4 under these dynamic conditions is T_4'.) If one adds B "until the zone composition reaches 3", lowers the temperature close to 3 and starts passing the zone through the charge, then (except for the last zone length) an ingot of uniform composition 4 results. The quotation marks were used since it is not essential that the entire zone be homogenously doped. Composition 3 is only required in the concentration boundary layer at the growing interface. As indicated in the concentration profiles of Fig.6.40, this will require addition of different amounts of B to the zone, depending on growth rate and mixing conditions in the melt, i.e., depending on the effective segregation coefficient k.

For more extensive discussions of theoretical and practical problems in obtaining uniform solute distributions refer to the works of AIRAPETYANTS and SHMELEV [6.53] and PFANN [6.37].

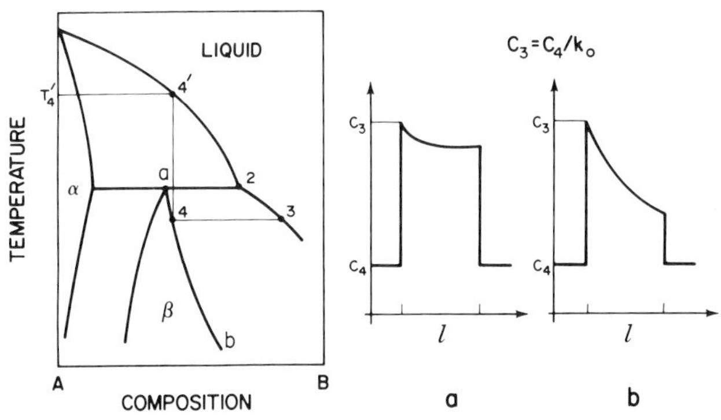

Fig.6.40. Left side: Binary system with peritectic reaction. After [6.37]. Right side: Dynamic concentration profile in molten zone for (a) low and (b) high growth rate — both resulting in the same solid composition

6.5 Microdistribution

In order to obtain a one-dimensional, time-independent description of dynamic segregation in form of the BPS relation (6.38), the above discussion of mass transfer at a solid-fluid interface has been based upon a series of idealized assumptions. In practice, however, these conditions are met only to a limited extent. Some deviations and consequences for macrosegregation have been dealt with in connection with zone melting (Sect. 6.4.2). Here we will go further and treat segregation phenomena that result in a "fine-structure" of the macrodistribution with characteristic distances or spacings ranging from $1\text{-}10^3 \mu m$. Such compositional short range inhomogeneities can seriously limit the usefulness of single crystals for applications. Hence, considerable efforts have been made to understand the origin of these micro-heterogeneities. Only recently insight has grown to the extent where more homogeneous materials can be obtained through control of the pertinent parameters.

Let us divide possible deviations from the idealized assumptions into two groups:

1) Time dependence: *non-steady* conditions for growth rate V^s, solute concentration boundary layer width δ and solute concentration C^f_δ.
2) Kinetics and geometry: *non-equilibrium* in composition between fluid and solid at the interface; non-planar interface; non-uniform solute distribution across the interface.

A division into these two groups is convenient for the discussion's sake, but it is somewhat arbitrary. In dividing them so one has to keep in mind that time-dependent and kinetic effects can be strongly coupled. For instance, a certain non-equilibrium segregation effect may be proportional to the growth rate. Hence, it will be more pronounced in periods of rapid growth and may become insignificant during slow growth parts of the fluctuations. In other words, a non-uniformity in interfacial solute distribution, if caused by convective phenomena, is also likely to be time dependent.

In the following we will discuss non-steady and non-equilibrium phenomena. Geometrical conditions such as actual concentration fields in the fluid and non-planar interfaces as well as severe non-equilibrium phenomena (interfacial breakdown, fluid trapping, etc.) will be taken up in later sections, since they require detailed knowledge of the mass and heat transfer mechanisms that prevail in the specific techniques [1.1, 1.2].

6.5.1 Non-steady segregation

The BPS relation (6.38) applies, strictly speaking, only to steady-state conditions. Yet it allows to *qualitatively* predict changes in the solute concentration C^s_0 in a growing solid whenever the growth rate, boundary layer width, diffusivity and solute concentration in the fluid change. Such changes in C^s_0 will only be detectable if

a) the interface propagates over a measurable distance during a characteristic time of a fluctuation in the segregation conditions,
b) the rate of change, for instance in V^s, exceeds a certain minimum that depends on k* and on the sensitivity of the compositional analytical technique used, and
c) diffusion of the solute in the solid is too slow to equalize growth-induced inhomogeneities during cooling of the crystal to room temperature.

The growth rate V^s is a complicated function of many parameters, as will be discussed in detail in [1.1]. But we know from Sect. 2.10 that interfacial mass transfer is governed by the difference in chemical potential of the phases in contact. Consequently, we can expect from the Gibbs-Duhem equation (2.50) that non-steady segregation will occur when there are changes in any of the extensive state variables in a system. The most effective changes, however, are changes in the temperature since their "scaling factors", the molar entropies, are typically the ones that are the most different in the nutrient and the crystal. However, from (2.50) we see that changes in pressure, concentration, magnetic and electric field may also cause differences in the chemical potentials of components in the (contacting) phases and hence result in growth rate changes and consequently in changes in the effective segregation coefficient. In actuality "things" are even more complex than just indicated. Before, we have (tacitly) looked only at the consequences of $\Delta\mu$-changes of the host material for growth rates and hence effective segretation. Yet upon closer inspection one realizes that when intensive state variables change, so do the chemical potential of all components. Hence, we can even expect the equilibrium segregation coefficients to change. Under most circumstances this will be a second-order effect, mostly buried under the "more responsive" consequences from mere growth rate changes. Yet one must keep this possibility in mind.

The concentration boundary layer width δ is governed by the diffusivity *and* the convective conditions in the fluid. As such it is subject to externally imposed changes (e.g. rotational stirring) and internal changes of the free convective conditions. In addition, the diffusivity may change, for instance, in the presence of an electric field. Yet is should be recalled that changes in δ are only of significance for segregation if the characteristic diffusion distance $Y' = D/V_{cf}$ defined in connection with (6.35) is comparable to or larger than δ.

In reality, many of these phenomena are coupled. For instance, changes in convection can cause the local temperature to vary *and* can modify the boundary layer width, thus being "doubly" efficient with respect to non-steady segregation. Or consider "all" possible consequences of electric fields that are often used in segregation studies. The resulting currents may change the interfacial temperature (Peltier effect, see also p. 469) *and*, if the considered species are ionic, change the effective diffusivity (due to electromigration) *as well as* the interfacial segregation coefficient!

This complex picture becomes even more complicated when evidence is introduced from recent investigations which show that in certain vapor-solid processes the interplay between interfacial mass and heat transfer can cause oscillatory phenomena [6.71, 72]. These findings, in turn, illustrate the arbitrariness of uncoupling non-steady and non-equilibrium phenomena in segregation.

To date the experimental material still outweighs the theoretical understanding of observed non-steady segregation phenomena. Hence, we will first discuss a selection of experimental results before we relate to available models. Thereby, and in particular in the theoretical part, we will follow quite closely an authoritative review by CARRUTHERS and WITT [6.56] on transient segregation effects in Czochralski growth (pulling). Measures to avoid or, at least, reduce, non-steady segregation will be discussed in the specialized chapters on melt, vapor and solution growth [1.2].

Experimental observations

Solute segregation striae and their origin in melt-grown crystals have been investigated for quite some time. Growth rate fluctuations have been linked to two types of temperature fluctuations:

a) temperature oscillations resulting from convective instability in the melt; and

b) temperature modulation from rotation commonly used in pulling from the melt [1.2]. Since it is practically not feasible to obtain an ideally rotational symmetric temperature distribution, rotation of a crystal during growth leads to local periods of slow and rapid V^S in this technique. With pronounced heating assymetry one may even encounter a succession of local growth and back-melting with the periodicity of the imposed rotation, as illustrated in Fig.6.41.

The above separation into (a) and (b) is useful at this point and is favored by the typically quite different periods involved. Yet we will see later that there can be significant coupling between forced convection (here from rotation) and free convection leading to much more complicated flows than the simple pattern in Fig.6.41 suggests.

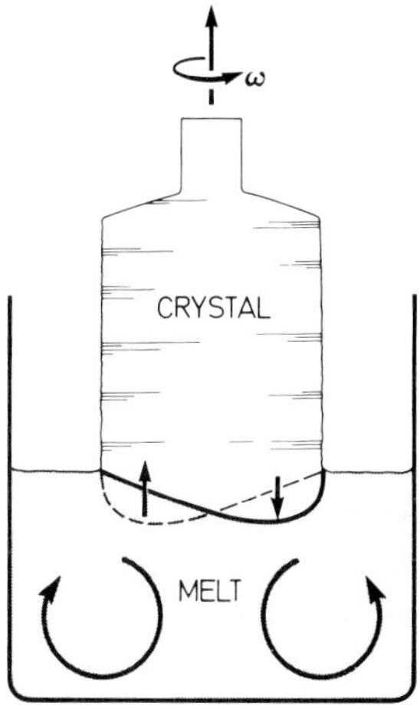

Fig.6.41. Schematic presentation of assymmetric thermal condition in pulling from melt under rotation. Dashed line indicates interface position after rotation by π; zones of potential back-melting and rapid growth are indicated by arrows

Various authors have successfully correlated distances between peaks in the solute distribution to the periodicity of measured temperature fluctions in the melt. For instance, the spacing between rhodium striae in melt-pulled zinc tungstate agreed quantitatively with the product of the (average) growth rate and the 2 min period of intentionally imposed large temperature changes, as shown in Fig.6.42 [6.31]. Superimposed on the imposed T-modulation, rather regular temperature oscillations with an 8 second period and 2°C amplitude were found and tentatively ascribed to convective instability. To some extent these faster fluctuations are also reflected in the rhodium distribution, although not in a simple manner. Microscopic investigation of the solid revealed an even finer periodic striae spacing (not reflected in Fig.6.42) that agreed quantitatively with the product of rotation period and growth rate.

Causality between these mechanism and the micro solute distribution in the growing crystal has been established only very recently. WITT et al. [6.58], after years of development of the required high resolution techniques, were able to demonstrate direct time correspondence between temperature, growth rate and solute fluctuations in crystals grown from the melt. Figure 6.43 demonstrates the correspondence of convective temperature oscillations in the

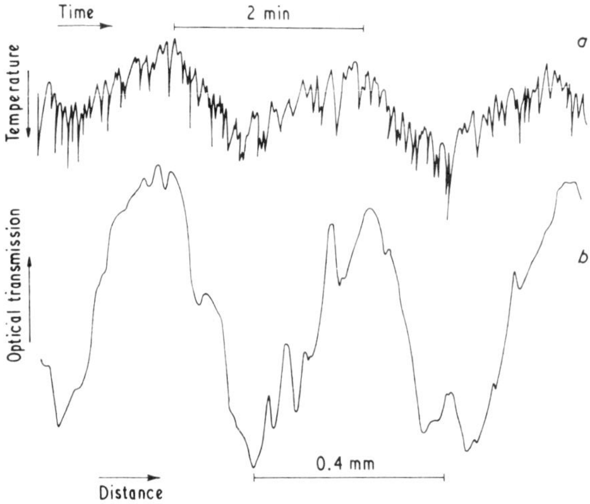

Fig.6.42. (a) Variations of temperature at a fixed point in a zinc tungstate melt vs. time. Externally caused temperature oscillations (2 min period) and superimposed convective fluctuations. (b) Variation of optical transmission (Rh concentration) vs. distance in crystal in growth direction. From [6.31] by permission of The Physical Society, London

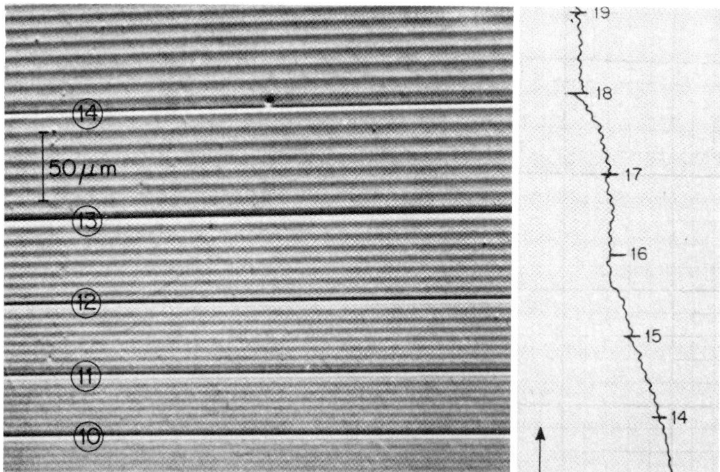

Fig.6.43. (a) Etched section of a Te-doped InSb crystal grown in the presence of (b) thermal oscillations in the melt. Note the constant frequency of the thermal perturbation over the whole region displayed as established by the periodic time markers (21 second intervals). The growth rate decreases from top to bottom. From [6.57] by permission of The Electrochemical Society

melt and solute striation. The interference contrast photomicrograph (Fig. 6.43a), obtained after selective etching, shows a small segment of a Te-doped InSb crystal that had been grown from the melt without rotation. Two features can be distinguished. Firstly, one sees pronounced striations that had been introduced as time markers by Peltier cooling (short period of growth rate increase by current pulse, see the end of this section) at periodic intervals of 21 seconds. Secondly, one recognizes somewhat fainter sub-striae that correspond to dopant heterogeneities corresponding to a period of 2.6 s. Fig. 6.43b represents the melt temperature as recorded from a thermocouple close to the solid-melt interface, also displaying time markers caused by the Peltier current. Direct time correspondence between temperature recording and solute distribution was achieved by identifying the corresponding time markers. One sees that the striations between time markers 14 and 8 correspond to temperature fluctuations which lie below the detection limit of ±0.01° C of the sensing circuit!

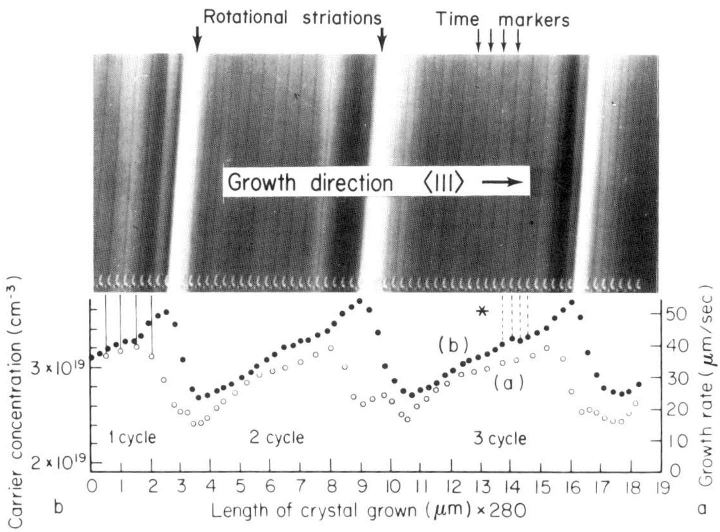

Fig.6.44. Upper part: Differentially etched segment of a Ga-doped Ge crystal pulled at 21.2 μm/s (7.6 cm/hr) with seed rotation (6.16 rpm)
Lower part: Dopant distribution analysis: (a) microscopic growth rate as obtained from periodic time markers (0.5 s intervals); (b) dopant (carrier) concentration distribution as obtained from spreading resistance measurements. From [6.58] by permission of the Electrochemical Society

By increasing the time marker frequency [6.58], it was possible to demonstrate quantitative correspondence between growth rate fluctuations and solute striations. The upper part of Fig.6.44, again an interference contrast photomicrograph, depicts a differentially etched small section of a Ga-doped Ge crystal that had been pulled from the melt. The darker areas correspond to higher Ga concentrations. The following features can be distinguished: 1) rotational striations, 2) growth rate markers introduced by Peltier cooling at time intervals of 0.5 s, 3) impact traces of a spreading resistance probe used in equal spacings of 10 μm in order to determine the Ga concentration. The lower part of Fig.6.44 shows the results of an analysis based upon the time markers and the spreading resistance measurements. One sees that during each rotational cycle the microscopic growth rate varies continuously, thereby changing its value by almost a factor of three. (Microscopic growth rate of up to *thirty times* the macroscopic rates have been observed by the same group in earlier work!) The microscopic dopant concentration fluctuates by about 50%. A detailed analysis shows that during the segments with growth rate acceleration the BPS relation (6.38) holds still surprisingly well; yet it fails during the deceleration periods. Note also the considerable phase shift between maxima of solute distribution and microscopic growth rate, which cannot be accounted for by the BPS steady-state theory. This is not surprising considering that with such drastic growth rate changes, solute steady states in the melt will only be re-established after periods that correspond here to several marker intervals, as will be discussed quantitatively below.

This investigation gives also unambiguous evidence for the correspondence between topographic features deduced from selectively etched surfaces and the solute microdistribution.

The significance of crystal rotation for striae formation becomes particularly obvious in Fig.6.45 which corresponds to a different segment of the same crystal. All data are the same as for the sample in Fig.6.44, except for the slower pulling rate. The left part of this segment grew under seed rotation as specified above and, due to the slower growth rate, even underwent periodic remelting. The right part was pulled without rotation and displays a microdistribution caused by convectively induced growth rate fluctuations only. Note that in the rotated part the phase shift between solute peaks and microscopic growth rate maxima is considerably less pronounced than in Fig.6.44.

Temperature oscillations of rather strict periodicity that are encountered in certain fluid configurations (Sect. 5.8) are illustrated in Fig.6.46, together with the resulting micro-segregation pattern.

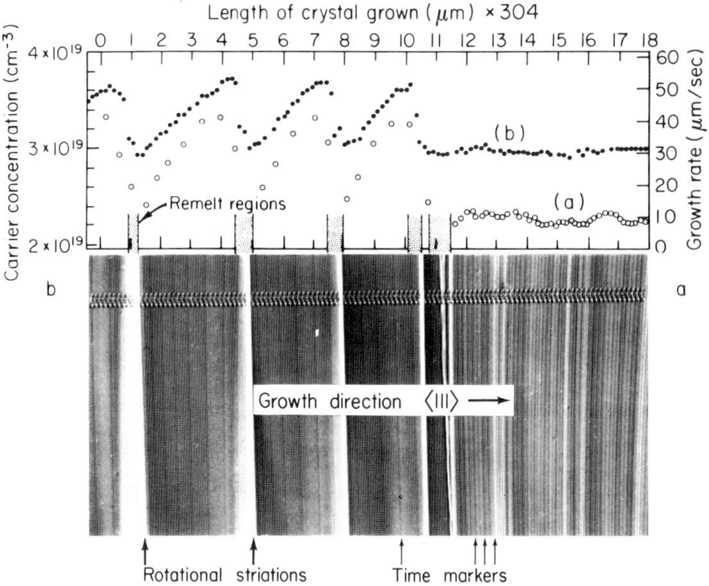

Fig.6.45. Segment of a Ga-doped Ge crystal grown from the melt with (left part) and without (right part) rotation. Data same as for Fig.6.44, except pulling rate reduced to 8.7 μm/s (3.0 cm/hr). From [6.58] by permission of The Electrochemical Society

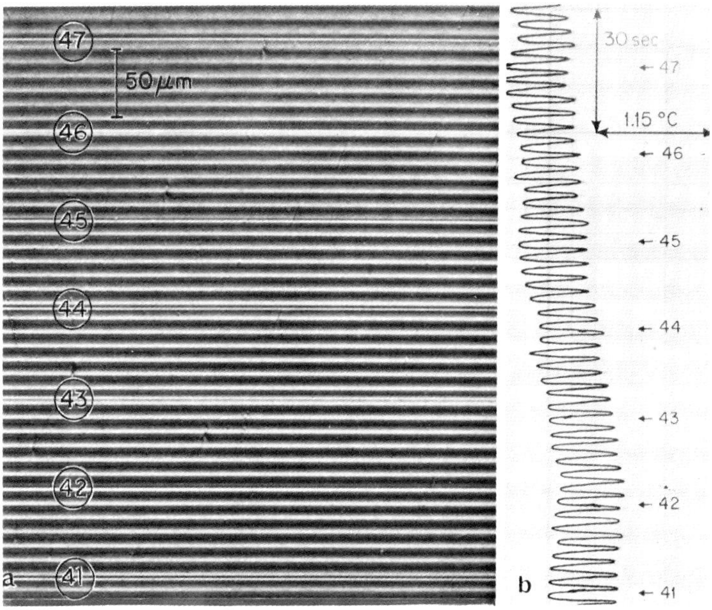

Fig.6.46. (a) Section of Te-doped InSb crystal grown in the presence of thermal oscillations in the melt. Comparison with Fig.6.43 shows that the dopant heterogeneities are here more pronounced. Incipient remelting can be observed. (b) Recording of temperature in melt during growth of the section. From [6.57] by permission of The Electrochemical Society

Most recently, detailed solidification studies were carried out under zero-gravity conditions in order to unambiguously identify the role of buoyancy-driven convection in micro-segregation. Although these studies yielded some unexpected results with respect to surface interactions between melts and their containers, it was clearly established that most of the micro-inhomogeneities found in terrestrial solidification are gravity-induced and that surface-tension-driven convection is considerably less significant under most practical circumstances.

Fig.6.47. Dopant segregation characteristics near the central region of initial regrowth interface of Te-doped InSb crystal grown during Skylab-IV mission. Magnification 580 X. From [6.40] by permission of The Electrochemical Society

Figure 6.47, depicting an etched section of a Te-doped InSb crystal, partly regrown aboard Skylab-IV illustrates these findings [6.40]. The upper part shown was grown on earth by pulling from the melt and shows the familiar rotational and convective striation pattern. Part of the earth-grown crystal was molten in the space experiment and resolidified under zero gravity.

The solidification of Ga-doped Ge was also investigated in parallel experiments on earth and during the Skylab-II mission [6.59]. In Fig.6.48 typical radial resistivity profiles (measured normal to the solidification direction of the normal freezing geometry) for space and terrestrial samples are compared. Neglecting the high rise in resistivity (corresponding to lower Ga concentrations) near the edges of the ingot, one sees that in the absence of gravity the dopant homogeneity is drastically improved.

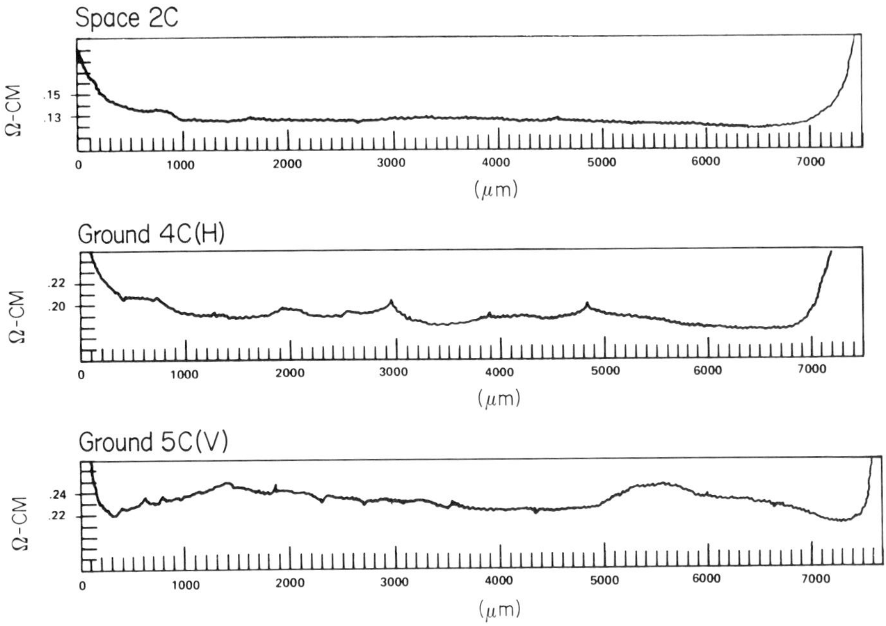

Fig.6.48. Comparison of radial resistivity profiles for Ga-doped Ge measured on sections across the samples for crystals grown on earth in vertically (V) and horizontally (H) oriented crucibles and during Skylab-II mission (top). From [6.59] by permission of North-Holland Publ. Co.

There are, however, remaining compositional fluctuations of ±1.5% about the average concentration. It also appears that the enhanced convective flux expected for the vertical terrestrial solidification indeed introduces a somewhat more curved macro-profile than is observed in the horizontal case. The steep concentration drop near the edges of the samples is not understood at this point. One may be tempted to interpret this observation in terms of surface-tension-driven flow (Marangoni effect, Sect. 5.8.3). Yet both space and earth grown samples show too similar a behavior; at least in the vertical mode where the melt has supposedly no free surface one would expect a different profile. The authors assign this effect qualitatively to the (moderately) curved interface shape that prevailed in the experiments [6.59] or possibly to the resistance measurement process [6.60]. Fig.6.49, representing an enlarged section of the data in Fig.6.48, allows for a better evaluation of the increased homogeneity obtained in the absence of gravity.

It has been found that melt-pulled alkali halides show essentially the same micro-segregation phenomena as the semiconductors discussed above [6.61]. Solute inhomogeneities in KCl were detected through a small addition of

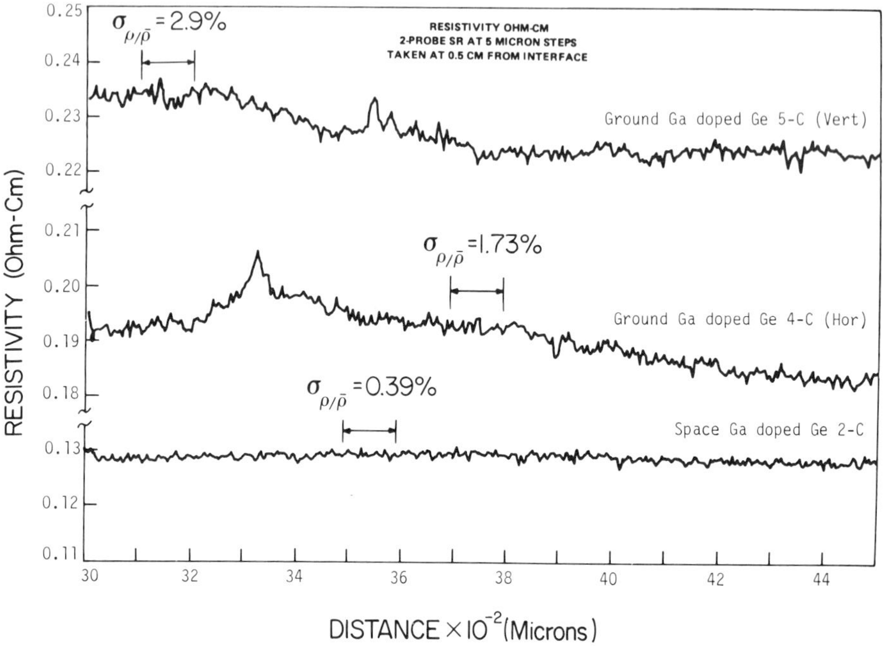

Fig.6.49. Enlarged part of the resistivity variations depicted in Fig.6.48. From [6.59] by permission of North-Holland Publ. Co.

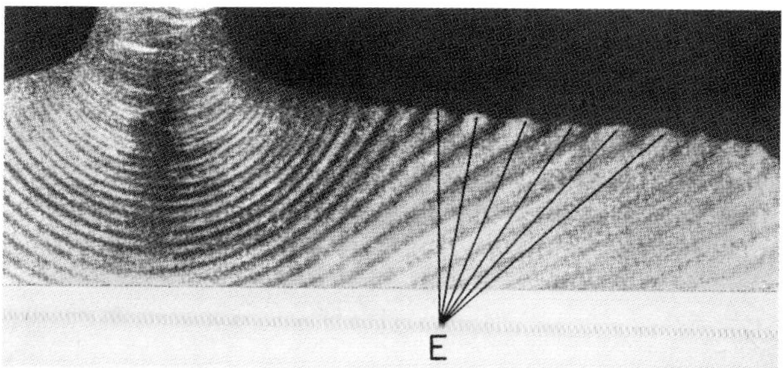

Fig.6.50. Radiograph of a longitudinal cleavage plane of the upper part of a melt pulled KCl crystal doped with Ca-45. Crystal diameter about 3 cm, rotation rate 1 rpm, growth rate increased at half the final diameter. E, enrichment zones. From [6.61] by permission Akademie-Verlag Berlin

β-emitting Ca-45 to the $CaCl_2$ used as dopant in the melt. Fig.6.50 represents an autoradiograph of a cleavage plane through the rotation axis of the crystal. Calcium enriched areas lead to light regions in the radiograph. The calcium concentration was found to fluctuate by a factor of 2 between maxima and minima of the rotational striations. One-to-one relation between the number of rotations and the number of turns in the "segregation sprial" was established. For comparison an analog experiment was performed with Ca-doped NaCl where the segregation coefficient at the chosen concentration is unity. As expected, uniform distribution was obtained throughout the crystal.

These experiments [6.61] revealed also micro-heterogeneties that could be related to convective temperature oscillations in the melt, as illustrated in Fig.6.51.

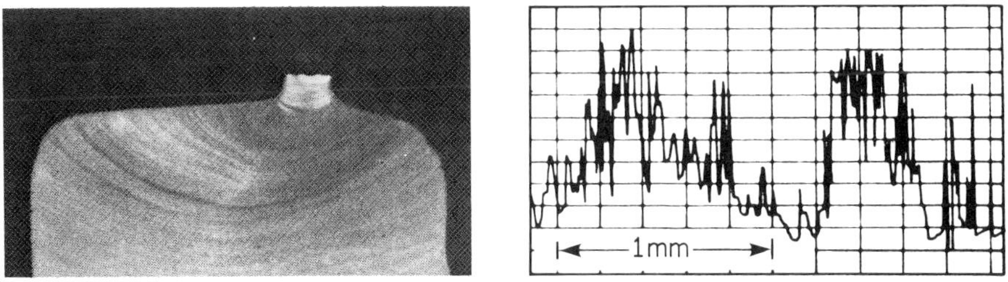

(a) (b)

Fig.6.51. (a) Radiograph of Ca-45 doped KCl crystal that had been pulled from the melt without rotation, natural size. (b) Liminescence intensity of Ag doped KCl crystal under electron irradiation, monitored in growth direction. The 1 mm periodicity corresponds to the rotation rate. Typical periodicity in the "fine structure": 0.05 mm. From [6.61] by permission of Akademie-Verlag Berlin

In addition to rotaticnal modulations of the growth rate in non-uniform temperature distributions, rotation can cause heterogeneities in growing solids for mere fluid dynamic reasons. In [1.2], we will see that rotation creates very complex convection pattenrs [6.88]. For instance, cells can develop that mix very little with the rest of the melt. The hydrodynamic conditions that stabilize these "solute pockets" are rather fragile. Perturbations in the form of rotation rate changes, mechanical vibrations or from free convection may lead to unsteady solute release from ($k > 1$) or take-up into ($k < 1$) these cells and modulate the solute concentration in the solid.

As a well characterized example of roation-induced striations in oxide growth one should mention results obtained on the double niobate

$Ba_xSr_{1-x}Nb_2O_6$ [6.62]. Fig. 6.52 depicts striations resulting from non-steady segregation of, likely K and Ca impurities and their suppression upon stopping of the rotation during pulling.

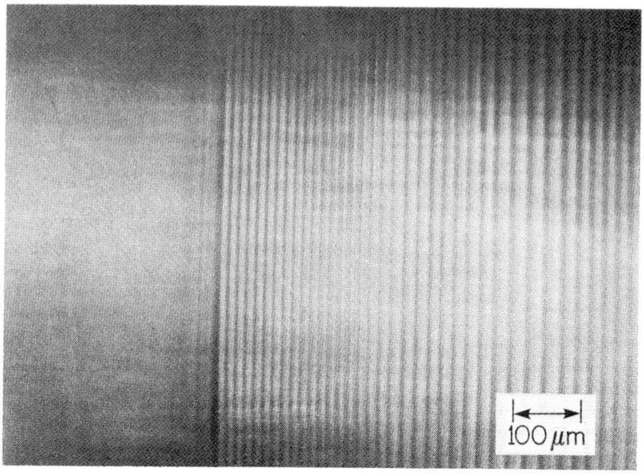

Fig.6.52. Rotation striations in barium-strontium niobate and their suppression upon stopping of the rotation. Microphotograph between crossed polarizers. From [6.62] by permission of North-Holland Publ. Co.

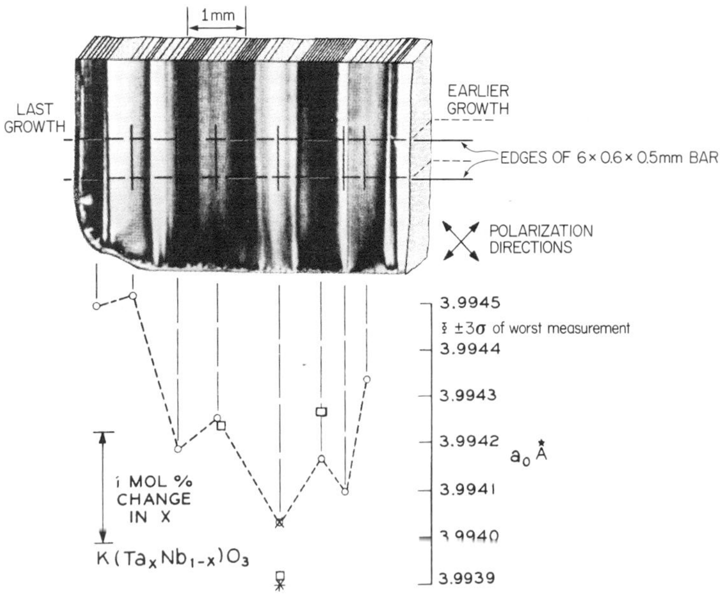

Fig.6.53. Compositional striations in KTN. Upper part: Photograph of a crystal between crossed polarizers. Below: Lattice parameter shown in vertical register with the photo. From [6.63] by permission of Pergamon Press

Micro-segregation phenomena, of course, are not restricted to impure melt solidification. In principle, one will expect such phenomena in any situation where the solid has a different composition than its nutrient. Although the findings are not as numerous and well characterized as in growth from melts, in the following we will give some examples for growth induced micro-heterogeneities for crystallization from solutions (incongruent melts) and vapors. Because of the low growth rates typically encountered in multicomponent processes (Sect. 4.2), convective short period fluctuations are expected to lead to inhomogeneity spacings that are either not resolvable and/or not of practical concern. Hence, observed striations in these systems will likely be due to low frequency convective modes and externally caused long period changes such as power line fluctuations, temperature control variations and room temperature fluctuations.

Figure 6.53 shows one of the few available examples for striations in crystals grown from non-stoichiometric melts. KTN ($K(Ta_xNb_{1-x})O_3$), used as an optical modulator and beam deflector, is typically grown as $K(Ta_{0.63}Nb_{0.37})O_3$ from molten $KNbO_3$ - $KTaO_3$ solutions of the "corresponding" liquidus composition. Yet, in practice the composition change between solidus and liquidus is governed by the dynamics of the mass transfer in the interfacial liquid solution, just as in solute segregation in melts. Any variation in growth rate and/or diffusion boundary layer width will lead to compositional fluctuations about the desired solid composition ratio. Hence, physical properties, such as lattice parameters, refractive indices, etc. that depend upon that ratio will be non-uniform as well. The average increase in the lattice parameter a_0 with progressive growth depicted in Fig.6.53, can be understood from the phase diagram of the $KNbO_3$-$KTaO_3$ system. As the residual solution and hence the growing crystal is progressively enriched in Nb, the lattice parameter increases.

Examples for striations in flux grown crystals are shown in Figs.6.54 and 6.55. In both cases the origin was traced to periodic temperature fluctuations from inadequate furnace control during programmed cooling of the flux.

Impurity striations are also found in hydrothermally grown quartz. Strong banding is probably caused by temperature irregularities from powerline fluctuations, whereas finer banding can be associated with convective fluctuations [Ref. 6.23, p. 288]. A particularly "impressively" striated SiO_2 sample is shown in [6.66].

In crystal growth from low-temperature solutions one can, in principle, also expect non-steady convection-induced segregation phenomena. However, to this author's knowledge, no systematic investigations have been made yet. Solution growth is in general conducted rather isothermally. Particularly in aqueous solution growth, typical temperature gradients are too low to drive

Fig.6.54. Striations in In_2O_3 crystals grown from $PbO-B_2O_3$ flux by temperature lowering. From [6.64] by permission of American Ceramic Society

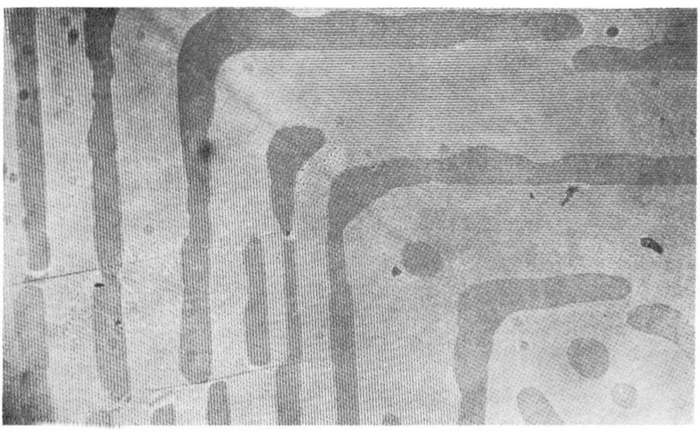

Fig.6.55. Banding in $DyFeO_3$ (Dysprosium iron oxide) grown from $PbO-B_2O_3-PbF$ flux. The striation-free region (lower right corner) was grown by flux evaporation, the striated part by cooling. Note the alignment of the magnetic domains parallel to the bands. Magnification 83 X. From [6.65] by permission of Pergamon Press

non-steady convection. However, as discussed in Sect. 5.8.2, concentration gradients which exist in solutions for various other "non-thermal" reasons are amply sufficient to drive time-dependent convective flow. Very little, though, is yet known about frequencies of such *solutal* convective instabilities. In order to cause detectable heterogeneities in connection with the typical slow solution growth rates, fluctuations in the fluid would have to have rather low frequencies.

Compositional striations have also been observed in crystal growth from the vapor phase and reviewed in [6.67, 68]. One can correlate characteristic striae spacings with typical periodicities of oscillatory convective instabilities recently found in vapor crystal growth systems [1.2]. However, direct causality between these compositional heterogeneities and convective temperature oscillations remains to be demonstrated.

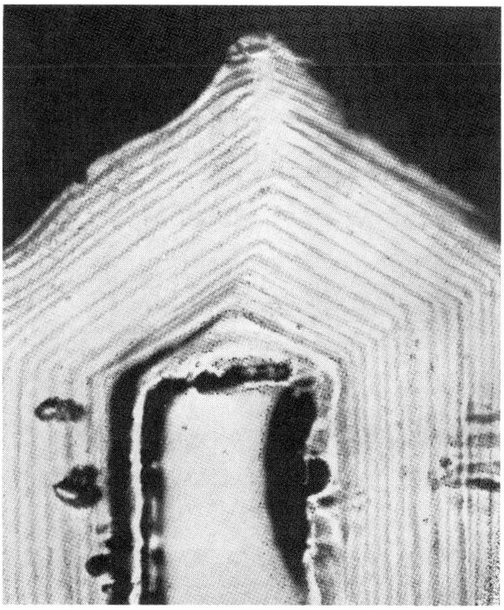

Fig.6.56. Vapor transported Ge wafer with growth striae in {111} planes. ~250 X. From [6.69] by permission of IBM Corporation

Figure 6.56 depicts one of the few available examples for striae in vapor grown crystals that are likely due to a periodic segregation mechanism. Germanium crystals, grown by chemical vapor transport [1.2] with iodine, revealed under the microscope the shown striae with a typical spacing of a few micrometers [6.69]. No efforts were made to determine the origin or chemical nature of these striae.

Another example of particularly well-developed growth striations in chemical vapor transport is shown in Fig.6.57. Efforts to chemically characterize the bands by electron microprobe analysis were unsuccessful.

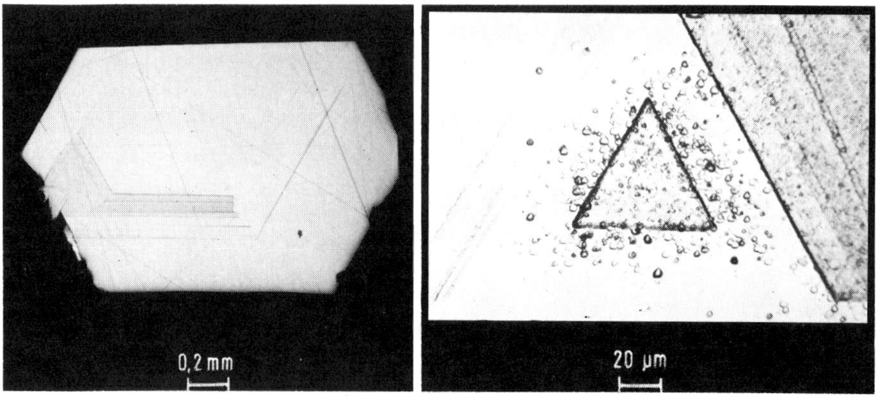

Fig.6.57. Growth striae and dislocation etch pits in $Fe_{0.7}Cd_{0.3}Cr_2S_4$ transported with $CrCl_3$. From [6.70] by permission of Siemens AG

Even in the absence of temperature fluctuations, non-steady segregation may occur when there are pressure changes which shift the melting point "significantly". As Fig.6.43 has shown, growth rate changes corresponding to $T_{melt} - T_{melting} = \Delta T$ as small as 0.01° suffice to modify k in a measurable way. Hence, rather modest pressure changes are expected to lead to micro-segregation phenomena. This has recently been demonstrated in the solidification of tin-doped InSb [6.48]. InSb expands 14% upon solidification and its melting point changes by -0.008°C/atm. As depicted in Fig. 6.58a, a molten zone was confined between two crystals that were closely fitted into a vertical quartz tube. The upper and lower crystal-melt interfaces were allowed to converge through controlled heating power reduction. Besides some convection induced striae in the upper crystal, simultaneous incorporation of dopant inhomogeneities, identical in detail in both the lower and the upper crystal, were found (independently of the melt zone heights.) These coupled inhomogeneities were ascribed to melting point changes from pressure build-up during freezing of the expanding material. Periodic discontinuities arise as the pressure mounts to a value where the friction between the ingots and container walls is overcome and one or both crystals retract. This model was confirmed in a parallel experiment with provisions for continuous pressure release during growth, as schematically indicated in Fig.6.58b. No dopant inhomogeneities were found in this mode at aspect ratios where convective fluid flow ceased (Sect. 5.8.2).

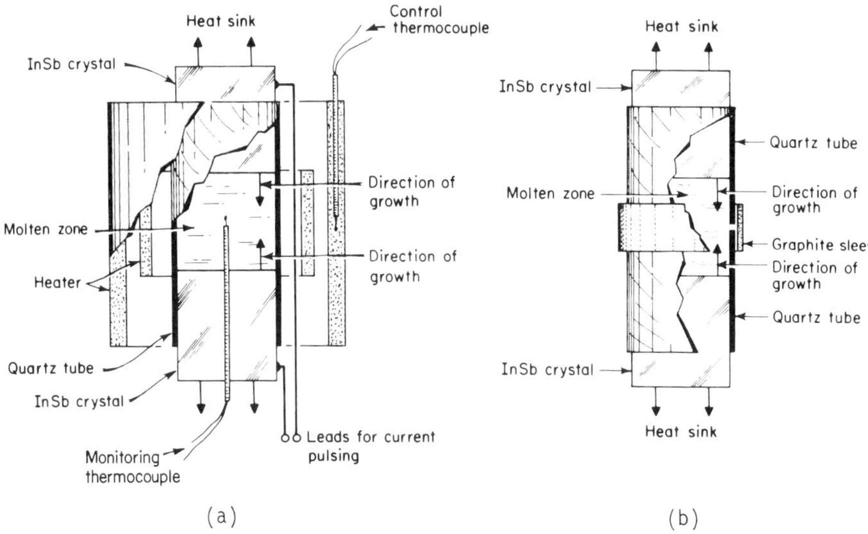

Fig.6.58. Schematic diagram of apparatus used for concurrent crystal growth under convectively stabilizing and destabilizing thermal gradients. (a) Crystal and melt contained in continuous quartz tube (b) Pressure release via graphite sleeve connecting two quartz tubes. From [6.48] by permission of Electrochemical Society

So far we have been looking at solute micro-heterogeneities only as undesirable perturbations. However, for various applications it is desirable to achieve variations in the solute concentration in a controlled, predetermined manner. One example would be the growth of p-n and more complex junctions in semiconductors. It is assumed that "accidental" fluctuations can be controlled to the extent where their effects can be ignored.

Gradual tansitions as well as abrupt changes in composition may be desired. In principle, changes of any of the parameters in the BPS relation (6.38) can be employed to change the solute concentration in the solid. Some means have been already discussed in connection with macroscopically uniform composition (Sect. 6.4.3) and in this section on the causes of undesirable striations. A detailed review of methods of "perturbing" solute concentrations in the solid is given in [6.37]. These techniques include addition of solute during growth (ΔC_∞^f), variations in stirring ($\Delta\delta$), back melting and rapid freezing (ΔV^S). To obtain a transition from p- to n-type composition (or vice versa), one has to employ two opposite-type impurities (e.g., III and V elements in a group IV matrix) whose segregation coefficients vary differently with growth rate. This so-called rate growing technique was used extensively in semiconductor preparation.

The growth rate variations that can be practically used for compositional variations are often rather limited by crystal perfection requirements. Reduction of the growth rate ($k \to k_0$) is limited by economic considerations only. However, an increase in V^s (for $k \to 1$) accomplished, for instance, by lowering of the melt temperature, is limited by the point where dislocation formation in the solid and morphological instabilities [1.1] jeopardize the desired crystallographic quality. Note also that concentration changes caused by given growth rate changes are larger, for larger interfacial segregation coefficient (see Problem 6.7).

Variations in the growth rate by external temperature control are inherently tied to long time constants. For instance, a change in overall melt temperature is impeded by the thermal inertia of the heating arrangement and the melt. Hence, it is difficult to obtain steep concentration gradients unless the growth rate is very small. Time constants given by heat capacities of typical experimental arrangements are often orders of magnitude longer than the time τ that is necessary to establish a new steady state for the solute profile in the concentration boundary layer (see below). A very elegant way to overcome this inertia is based upon the Peltier effect (inverse thermoelectric effect). In this case *the interface* between two differently composed conductors, such as a crystal and its melt, is heated or cooled, depending on the direction of a current that is imposed across it. The time lag between beginning of current flow and change in growth rate is in most systems negligible as compared to τ. It is assumed, of course, that the Peltier coefficient between melt and solid of the specific binary system is large enough to obtain a sufficient temperature change with reasonable currents. High currents, besides the possibility of undesirable electrochemical effects, lead to Joule heating that competes with the Peltier effect. The reversibility of the effect (cooling to heating and vice versa) allows for particularly rapid sequences of growth rate changes. A discussion of applications of the Peltier effect to the growth of semiconductor junctions as well as considerations of the accompanying Joule heating is given in [6.73].

The marking of interfaces and the modulation of dopant levels through current-dependent segregation (via Peltier effect, electrodiffusion, electrolysis, etc.) has become a very powerful tool in studies of growth rates and segregation not only in the melt growth of semiconductors. Fig.6.59 presents a particularly revealing example obtained in growth studies of alkali halides [6.74]. The bands shown were introduced by periodic *electrolytic* coloration (F-center formation) during the pulling of a KCl crystal from its melt and supposedly outline the momentary interface shape. Thus, this sequence illustrates the complex evolution of interfacial geometry that *can* occur during process. The reader is urged to discuss the non-uniformity in dopant distributions that results under such a drastic deviation from the one-dimensional normal freezing model; see Problem 6.8.

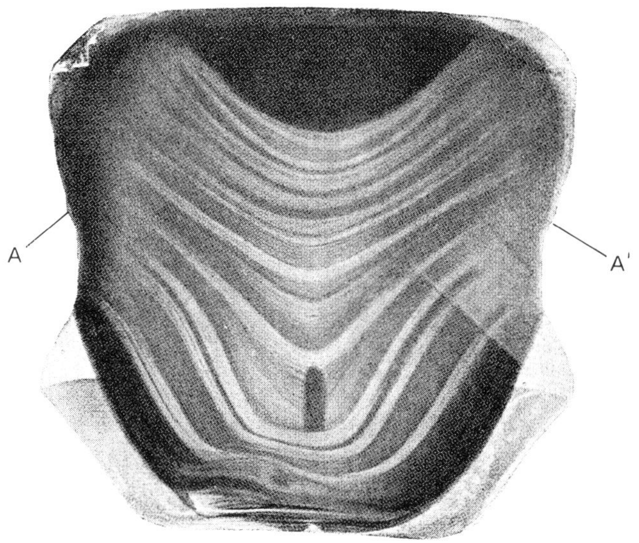

Fig.6.59. Electrolytically induced periodic banding in a melt-pulled KCl crystal. The bands mark the successive positions of the interface during growth. From [6.74] by permission of Akademie-Verlag, Berlin

More recently, Peltier cooling was also successfully applied in growth rate studies in liquid phase epitaxial processes [6.75, 76].

Electric current modified segregation can also be used in a continuous mode rather than the intermittent (marker) approach taken in most of the work referred to above. This is illustrated in the microphotograph of a lithium niobate sample shown in Fig.6.60 [6.77]. The chromium doping concentration was varied by a factor of three with remarkably low currents (50 mA/cm^2 in contrast to the 5-30 A/cm^2 typically used in Peltier modulations). Positive poling of the crystal with respect to the melt reduces the chromium doping in the crystal, whereas negative poling enhances the concentration as compared to the "no-current" segregation at the same growth rate. Periodic structures with spacings as low as 10 μm were obtained in the same work [6.77]. The potential of this capability in optical materials for light guiding structures and interference devices is very appealing.

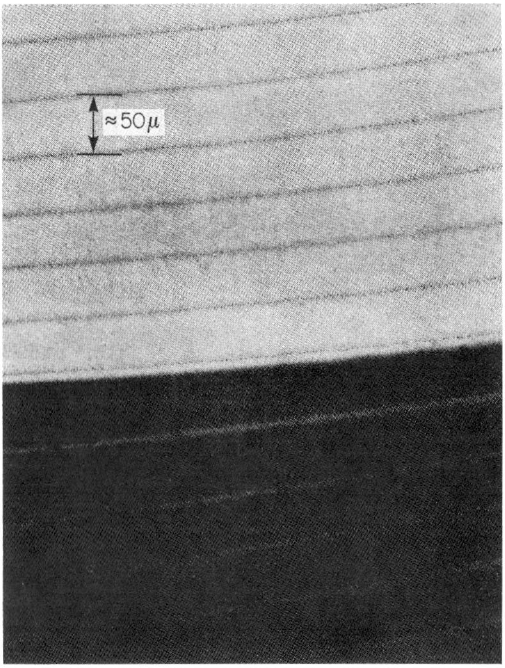

Fig.6.60. Transmission microphotograph of chromium-doped $LiNbO_3$. Doping modulation by electric current across interface. Upper (brighter) part: +50mA/cm^2 continuous except for short reduction to 5mA every 30 s. Lower part: +5mA continuous and reversed polarity markers of same frequency. From [6.77] by permission of Pergamon Press

There are, of course, other possibilities besides current induced non-steady segregation to obtain compositional time markers in crystal growth processes. If the growth rate in a system is sufficiently small, temperature programming of the whole growth arrangement can be employed. Fig.6.61 illustrates such an experiment that was used to reveal details of the flux growth history of ytterbium iron oxide. In crystal growth from vapors one has the possibility of adding periodically small concentrations of a dopant gas in

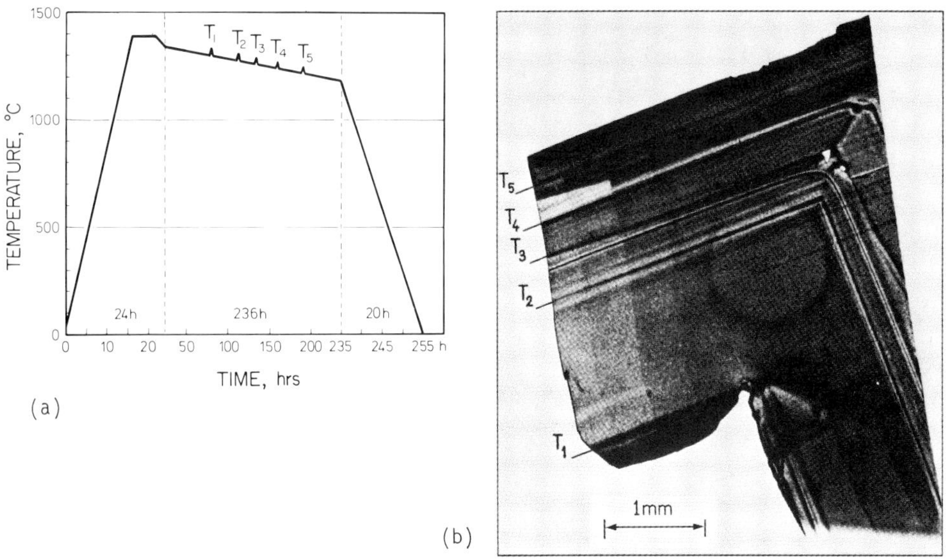

Fig.6.61. Intentionally introduced growth striations in $YbFeO_3$. (a) Temperature program showing position of induced temperature fluctuations (b) X-ray topograph of corresponding $YbFeO_3$ platelet. From [6.78] by permission of North-Holland Publ. Co.

order to obtain time markers for growth rate studies and related investigations.

Theoretical models

Table 6.6. Transient solutions to the continuity equation in segregation from melts; n.a.: not available. Adapted from [6.56]

Source of Compositional Transients	Diffusion -Only	Diffusion and Convection
1. Growth Rate Variations		
a. Initial transient	[6.81]	[6.82]
Step function	[6.81]	n.a.
b. Fluctuations (small amplitude)		
Sinusoidal	n.a.	[6.83-86]
Periodic	n.a.	n.a.
c. Fluctuations (large amplitude)		
Periodic	n.a.	[6.87]
Remelting	n.a.	n.a.
2. Boundary Layer Variations	n.a.	n.a.
3. Melt Composition Variations	n.a.	[6.27]

A compilation of references for solutions to the time-dependent continuity equation (5.120) for various transient conditions is given in Table 6.6. To this point only treatments of the impurity-in-melt case are available. Either mere diffusive mass transfer in the melt is considered, or convection is only included to the extent that a diffusion boundary layer (stagnant film) is assumed. To date no calculations that include variations in δ (or δ') have been published. Diffusion in the solid is in general neglected. We will now briefly outline some of the work listed above with emphasis on the (simplifying) assumptions made.

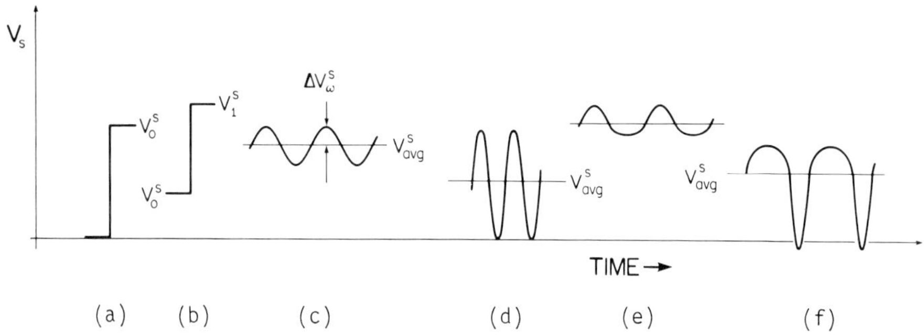

Fig.6.62. Various types of transient growth rate behavior: (a) initial step function; (b) abrupt growth rate increase; (c) sinusoidal, small amplitude; (d) sinusoidal, large amplitude with $V^S = 0$ in minimum, i.e., marginal transition state to backmelting; (e) periodic, small amplitude; (f) periodic, large amplitude with backmelting

Dopant distributions in solids after *abrupt changes in growth rate* have been calculated [6.81] for diffusion-controlled transport in the melt (with $\rho^f = \rho^S$) in the absence of convection. Results are shown in Fig.6.63a and b for growth rate increases and initial transients, respectively. From these figures one sees that a steady state in the dopant concentration in the solid after a step change in V^S from zero to V_0^S is reached approximately after a growth distance

$$x \approx \frac{3D_{AB}}{k_0 V_0^S} \quad . \tag{6.50}$$

The corresponding distance for a sudden increase from V_0^S to V_1^S is

$$x \approx \frac{10D_{AB}}{k_0 V_1^S} \frac{V_1^S - V_0^S}{V_1^S} \quad . \tag{6.51}$$

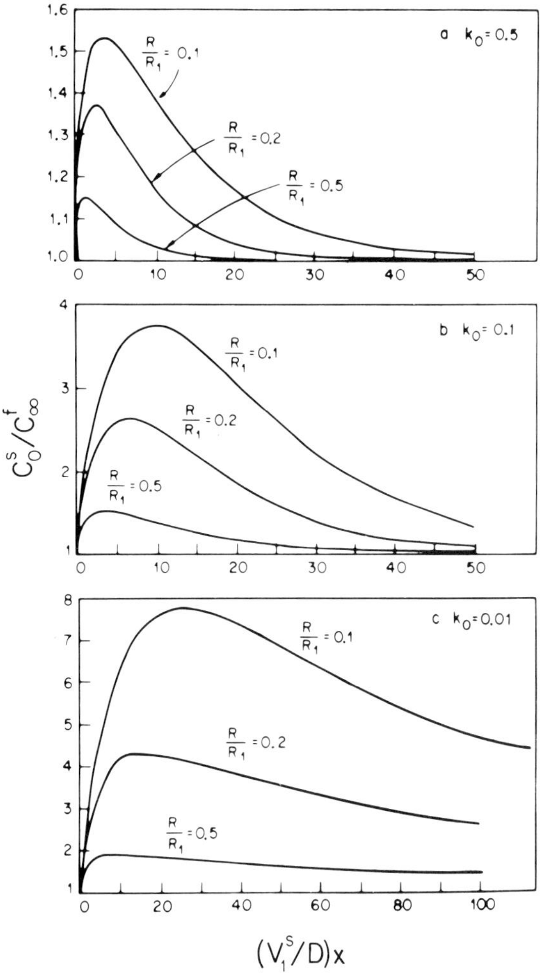

(a)

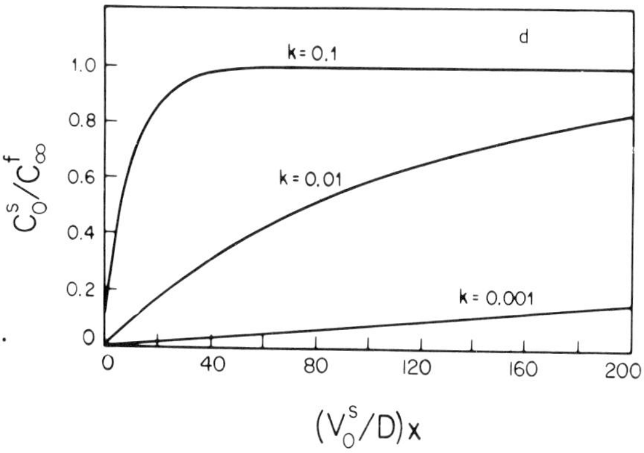

(b)

Fig.6.63. Solute distributions in a solid associated with step function changes in growth rate for various interfacial segregation coefficients. (a) After abrupt increase: $V_0^S(\equiv R) \to V_1^S(\equiv R_1)$. (b) Initial transient: $V^S = 0 \to V_0^S$. From [6.56] by permission of North-Holland Publ. Co.

As the example treated in Problem 6.9 shows, for parameters characteristic for segregation from melts one obtains thus transient distances of the order of centimeters. (See also Fig.6.15.) Note that the ordinate values in Fig. 6.63a begin at 1. Hence it is assumed that at V_0^s steady state with k = 1 had been reached (See, e.g., the diffusion-only case in curve (b) of Fig.6.15) before V^s is increased to V_1^s. During the transient a band of higher concentration than C_∞^f is produced in the solid. (For systems with $k_0 > 1$ a zone of solute-depleted solid is obtaind). After the transient, k = 1 again in both cases ($k_0 < 1$ and $k_0 > 1$) and $C_0^s = C_\infty^f$.

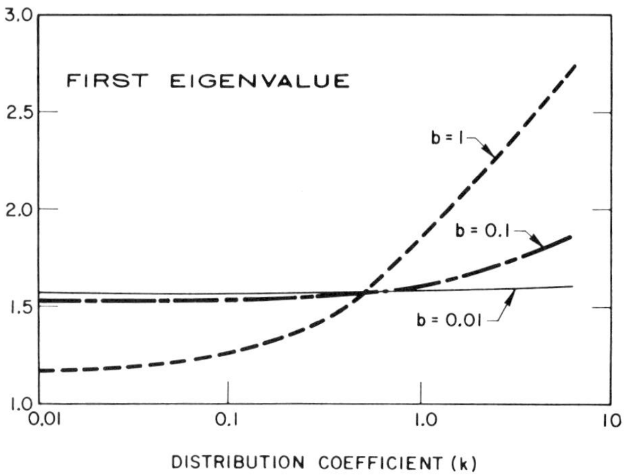

Fig.6.64. Function m_1 in (6.52) versus interfacial segregation coefficient for various values of $b = V_0^s\delta'/\mathcal{D}_{AB}$. From [6.82] by permission of American Institute of Physics

Initial transients under diffusive-convective conditions in the melt, i.e., the "incomplete mixing" case, has been treated in [6.82]. The distance in the crystal from the first solid layer formed to the point where the steady-state solute concentration is obtained to within 99% is (for $\rho^s = \rho^f$)

$$x \approx \frac{5\mathcal{D}_{AB}}{V_0^s} \frac{1}{1/4 + (m_1/\delta')^2 \cdot (\mathcal{D}_{AB}/V_0^s)^2} \tag{6.52}$$

where m_1, a function of k_0 and the dimensionless boundary layer width $b = V_0^s\delta'/\mathcal{D}_{AB}$, can be deduced from Fig.6.64. As applications of (6.52) to typical melt growth conditions show (Problem 6.9), the growth distance x is of the same order of magnitude as δ' and hence in general several orders of magnitude shorter than the distance necessary to attain steady state in the corresponding diffusion-only case.

Transient solutions for the amplitude of compositional striations resulting from *sinusoidal growth rate* oscillations were presented by several workers Small amplitude cases $\Delta V^S_\omega << V^S_{avg}$ were considered in [6.84-86] whereas [6.83, 87] included the marginal case $\Delta V^S_\omega = V^S_{avg}$; Fig.6.62 d. Back melting (Fig. 6.62f) has not been treated yet. The effect of convection is accounted for in all these papers by assuming a fixed δ'.

In the most advanced treatments in this group [6.83, 87] the growth rate fluctuations are obtained as a consequence of temperature oscillations imposed on the edge of the thermal boundary layer, rather than directly assuming an oscillatory ΔV^S_ω perturbation. This accounts for non-steady convection effects and externally imposed fluctuations equally well, since it is found that temperature changes are only very weakly damped in the convecting part of the melt. It is assumed that the growth rate is heat and solute diffusion limited. Since the freezing temperature $T_m(C)$ depends on the interfacial solute concentration (Sect. 3.4) which in turn is a function of the growth rate, heat and mass transport are coupled. Undercooling, required to drive the growth kinetics, is ignored. It is assumed that the interface responds infinitely rapidly to the thermal and solutal transport, i.e., its position is simply given by the condition $T_{interf} = T_m(C)$. This assumption of macroscopic transport-controlled growth (rather than attachment kinetics rate limitation) limits the model to metals and, under certain restrictions (non-faceted growth [1.1]), to semiconductors. The discussion is based upon a planar interface (one-dimensional problem). Mass diffusion in the solid is neglec ted. Constant but different temperature gradients are assumed in the solid and in the thermal boundary layer in the melt. Their values are related via the thermal conductivities of the two phases and the heat of solidification that is delivered at the interface proportionally to the growth rate. Simultaneous solutions to the time-dependent continuity equation (5.120) for the solute in the melt and to the time-dependent energy transport equation (5.149) for both phases are calculated to first order, i.e., under the assumption that the Fourier amplitude of the concentration changes are small as compared to the average interfacial solute concentration. The dependence of the resulting solute concentration fluctuations in the solid on amplitude and ω of the temperature fluctuations is found to be rather complex. Depending on the specific values for the various material parameters involved, resonance-like response to certain perturbation frequency ranges is predicted. Particularly strong effects are expected for solutes with $k > 1$, in accord with practical observations (see also Problem 6.7).

A particularly important finding [6.83, 87] is that the resulting spatial solute variation distribution function can strongly deviate in its shape from the originating sinusoidal (temporal) temperature variation. This is

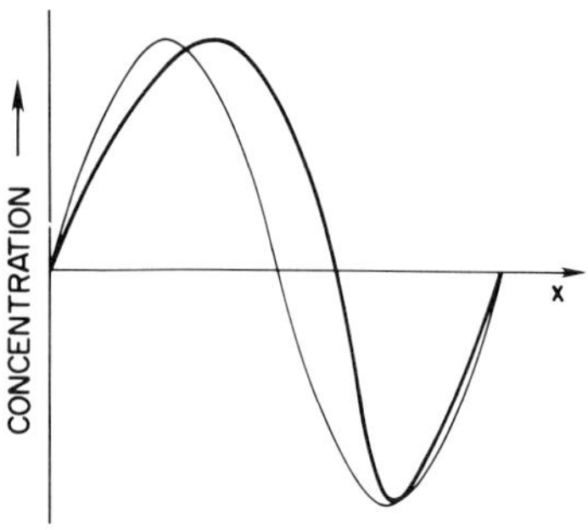

Fig.6.65. Waveform of solute distribution in the solid, resulting from sinusoidal temperature variation. Model semiconductor system with $k = 0.2$. The thin curve is a sine wave of same amplitude and frequency included for comparison. After [6.83]

illustrated in Fig.6.65 for a system resembling Ge solidification. (Similar distortations and phase shifts between growth rate variations and solute variations are observed in Fig.6.44 and 45). As a consequence the effectiveness of segregation is reduced. The average effective segregation coefficient, defined in analogy to k [see (6.2)] as

$$\bar{k} = \bar{C}^s / C^f_\infty \tag{6.53}$$

where $\bar{C}^s$ is the spatially averaged concentration in the solid is shifted towards one. For instance, for $k_0 < 1$ Fig.6.65 shows that "more crystal grows" and hence more solute is built into the solid during the faster growth period than is "deleted" during the shorter period with reduced growth rate. The net effect is a reduction in segregation. In fact, when the growth rate fluctuations are so large that the marginal state (Fig. 6.62d) is attained, $\bar{k}$ takes on half the value of k given by the BPS relation (6.39). Thus considerable errors can result when values for δ' and/or D_{AB} — or k-values are deduced from segregation data (see p. 425 and Fig.6.19) if these experimental data were obtained from samples that grew under oscillatory conditions.

For *low frequency, small amplitude* growth rate oscillations, i.e., with periods $\tau > 2\delta^2/D_{AB}$ [see (6.55)], one can obtain the concentration changes

ΔC^s associated with a certain (periodic) growth rate change ΔV^s from the steady-state BPS relation (6.39) in the form

$$\frac{\Delta C^s}{C^s_{avg}} \approx \frac{\Delta V^s}{V^s_{avg}} \frac{(1 - k_0)b \exp(-b)}{k_0 + (1 - k_0)\exp(-b)} \tag{6.54}$$

with $b = V^s_{avg}\delta'/D_{AB}$ the normalized growth rate. The derivation of (6.54) is left to the reader in Problem 6.10. Fig.6.66 displays solutions to (6.54) for three values of k_0. In order to minimize solute inhomogeneities, one will avoid experimental conditions associated with the maxima of the curves. It is, of course, tempting to work in the right-hand regime of the plot, i.e., at $\delta' V^s_{avg}/D_{AB} > 1$. In [1.1], however, we will see that the required high growth rates are prone to lead to interfacial instability problems.

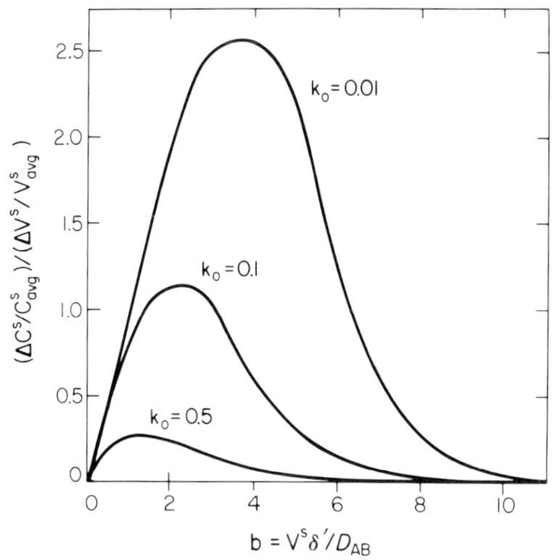

Fig.6.66. Relative compositional changes (normalized by the originating relative growth rate changes) for low frequency, low amplitude (oscillatory) growth rate fluctuations, as a function of the dimensionless growth parameter b. After [6.88]

The response of solute incorporation into a growing solid upon abrupt increase in the melt concentration has been investigated already in the BPS paper [6.27]. For a wide range of k_0 and $V^s\delta'/D_{AB}$ combinations it is found that steady-state composition is reached within a time

$$t \approx 2\delta'^2/D_{AB} \tag{6.55}$$

after the concentration has been changed at the outer edge of the diffusion boundary layer. In this analysis it is assumed that the concentration change is too small to result in significant changes in the liquidus temperature, i.e., in T_m. We see that, for instance, in a system with $D_{AB} = 10^{-4}\,cm^2s^{-1}$ and $\delta' = 3 \times 10^{-2}$cm, near steady state is reached in the solid already after 20 s.

6.5.2 Non-equilibrium segregation

It is difficult to decide whether segregation in a given situation corresponds to interfacial mass transfer equilibrium. One is tempted to assume that, in general, segregation coefficients obtained at very low growth rates represent k*. This, however, is not necessarily true. In this section we will encounter systems for which considerable anisotropies in segregation were observed even at low V^S values. Equilibrium segregation, though, cannot be anisotropic. A dependence of segregation on the crystallographic orientation of the interface *in equilibrium* would violate the First Law of Thermodynamics (Problem 6.11). Hence, we have to conclude that under typical laboratory conditions segregation proceeds quite frequently with non-equilibrium mechanisms.

A visually particularly striking manifestation of segregation anisotropy in growth from a melt is given in Fig.6.67. The depicted InSb crystal shows in the core of the longitudinal section a much higher dopant concentration than in the peripheral areas. Whereas the segregation coefficient in the core is $k_c \approx 4$, it is only $k_p \approx 0.5$ in the periphery. Other workers obtained similar results on the same system [6.90]. From the rotational striae in Fig.6.67 one can deduce the interface shape that prevailed during growth. One sees that overall the solid surface is convex. The part of the interface that corresponds to the core, however, is planar. This so-called *facet* does not follow the solidification isotherm, as schematically indicated in Fig. 6.68. The transition between core and off-core in solute concentration as well as in interface shape appears macroscopically very abrupt. Microscopic details will be seen in later figures.

At first glance one could suspect that *coring* is a fluid dynamic effect. The convection pattern indicated in Fig.6.41 could, in principle, yield a thinner boundary layer at the periphery. This would indeed result in increased segregation as compared to the not-as-well stirred center area. Note,

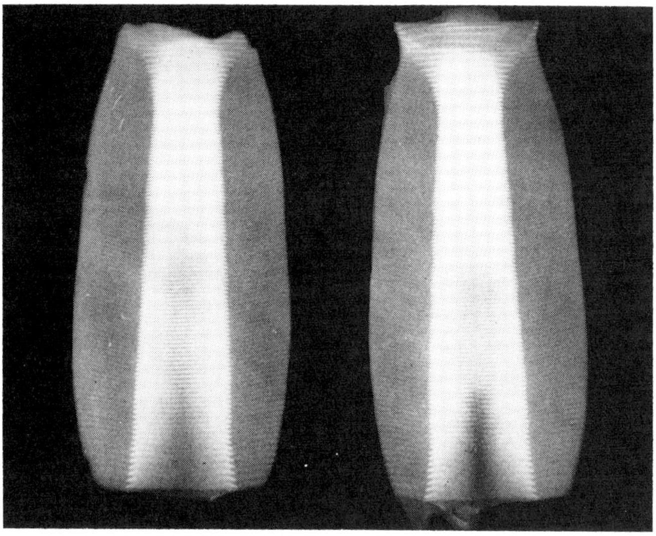

Fig.6.67. Autoradiograph of a ^{128}Te-doped InSb crystal that had been pulled from the melt in {111} direction. Diameter about 2 cm. J.B. Mullin. Royal Radar Establishment. From [6.89] by permission of North-Holland Publ. Co.

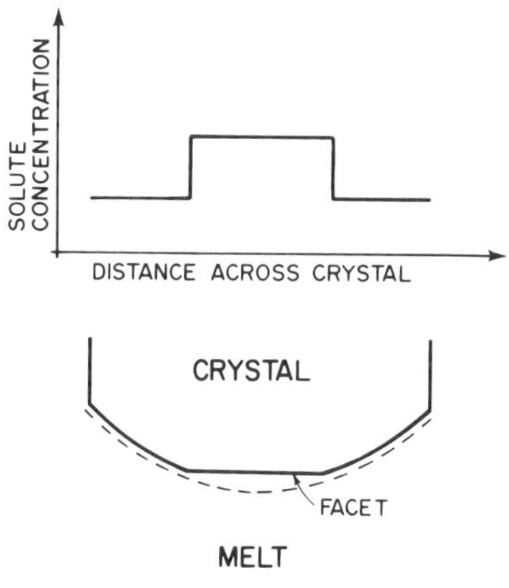

Fig.6.68. Facet effect in growth from the melt (schematic). (a) Cross-sectional solute distribution. (b) Interface shape, dashed line: melting point isotherm

however, that in the above InSb:Te system, k_c is larger than unity whereas $k_p < 1$. This reversal of k across the interface can by no means be explained dynamically with the Burton-Prim-Slichter relation (6.39). Furthermore, only very little influence of the stirring conditions on the ratio k_c/k_p was found [6.90]. All this indicates that the controlling parameters for solute incorporation must be basically different for core and off-core regions.

Figure 6.69 shows cross sections of a doped germanium crystal that, in contrast to the crystal in Fig.6.67, had been pulled with a misorientation of $\phi = 7°$ between the < 111 > direction and the pulling axis. Rotation striations on the cut and polished sections are revealed by a pulsed copper-plating technique. In the longitudinal section one sees that the "core" is not located in the crystal center anymore. As the interface shape changes gradually from convex to concave with respect to the melt, the core grows even out of the crystals. In fact, in crystals where ϕ was chosen such that a

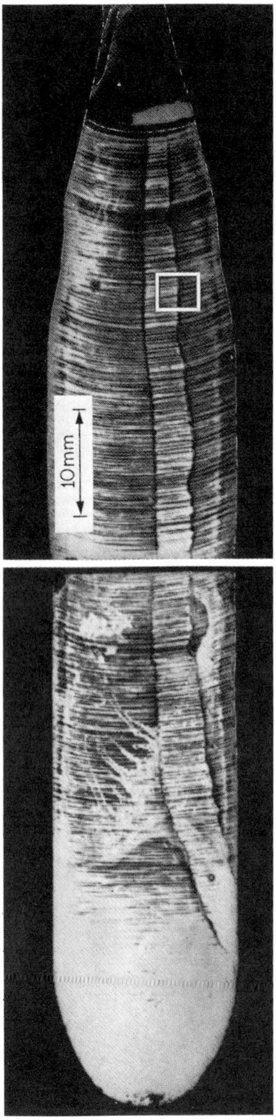

(a)

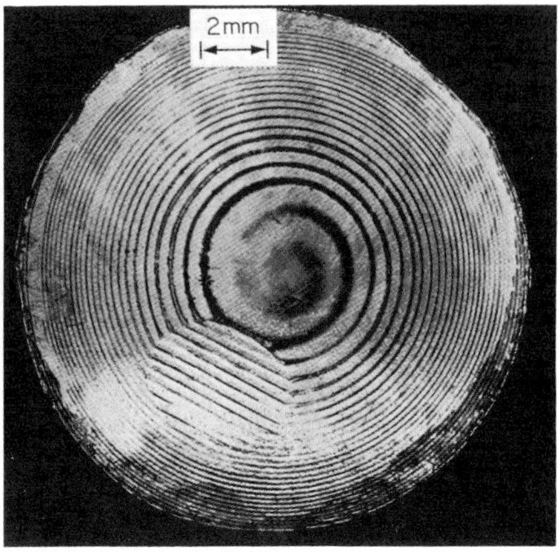

(b)

Fig.6.69. Rotational striations in a melt-grown Ge crystal with ⟨111⟩ 7° off the pulling direction. Solute striae made visible by copper plating of the polished sections. (a) longitudinal section: (b) horizontal cross section. From [6.91] by permission of Pergamon Press

{111} plane could not coincide with the curved interface at any point, coring was not observed at all. Thus, strong evidence is obtained that the observed coring effect is closely connected with the formation of flat facets on an interface. In Fig.6.69b one can even see that the rotational segregation spiral is straightened out in the core region – a beautiful illustration of the flatness of the facet from which the core grew.

Table 6.7. Anisotropy ratio k_c/k_p for segregation of various solutes in germanium. After [6.91]

Solute	k_c/k_p
P	2.5
As	1.8
Sb	1.45
Bi	1.65
Ga	0.85
In	1.4
Tl	1.2

Table 6.7 shows that there is a trend for higher solute incorporation on facets than in off-facet areas of the interface. Similar results on other systems are compiled in [Ref. 6.14, p. 121]. Note, however, the distinct exception for Ga that is more incorporated in peripheral (i.e., non-faceted) growth. On a macroscopic scale, the transition between the flat facet and the curved off-facet region as well as the difference in solute concentration appear rather sharp (Figs.6.67 and 6.69). This is not confirmed on the microscopic scale. WITT, GATOS and co-workers have performed extensive studies of on- and off-core segregation boundaries, mainly in InSb. In a sequence of papers [6.92-94], the authors were able to unravel the very complex striae patterns that occur in this transition region under various conditions of rotation and relative orientation of pulling and facet directions. Fig.6.70a depicts solute distribution fluctuations obtained in the on-core and off-core regions for various rotation rates. Fig.6.70b illustrates schematically the thermal conditions that exist at an interface with thermal assymetry in the heating arrangement. During rotation, the periodic displacement of the thermal axis about the rotation axis causes the helical solute

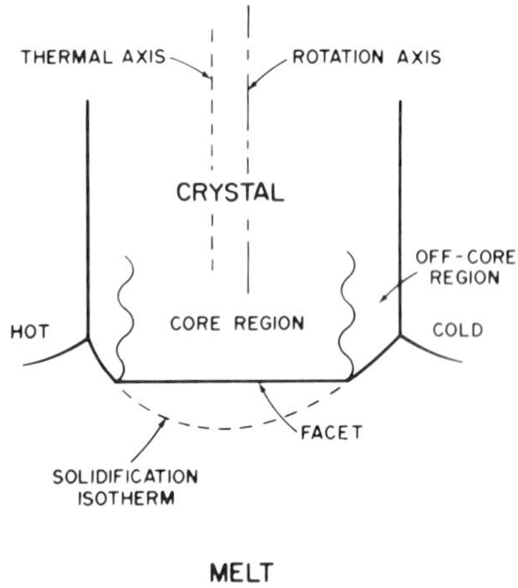

(a) (b)

Fig.6.70. (a) Rotational impurity striations in Te-doped InSb. Etched, longitudinal section. Magnification 7X. Upper, middle and lower part are pulled at 18, 4 and 45 rpm, respectively. From [6.92] by permission of The Electrochemical Society. (b) Schematic presentation of thermal conditions across faceted-non-faceted interface in the presence of thermal assymmetry. After [6.93]

banding discussed in Sect. 6.5.1 as well as a helical boundary between core and off-core region. Investigations of this transition region under higher magnification are very instructive. Employing again the Peltier effect time marking technique discussed before these workers were able to show that the *microscopic* growth rate in the core region is:

(a) very constant in spite of the rotation-induced temperature fluctuations that cause considerable growth fluctuations in the off-core region (Fig. 6.71). Correspondingly the concentration *fluctuations* are typically smaller on-core than off-core; and

(b) in most instances smaller (never greater) than in the off-core region.

These results illustrate that

(a) the impurity incorporation mechanisms on and off the facets must be different, and

(b) a higher solute concentration in the core cannot be explained merely by solute transfer effects in the melt. (We recall that according to (6.38) a lower microscopic growth rate should lead to a lower concentration for $k < 1$.)

Various mechanisms have been suggested to account for the "anomalous" segregation on facets [6.95-98]. Most models involve adsorption of the solute on the interface. The "most plausible" mechanism is based upon the different growth kinetics on faceted and non-faceted interfaces, which we will discuss in detail in [1.1]. Hence we will give here only a brief outline.

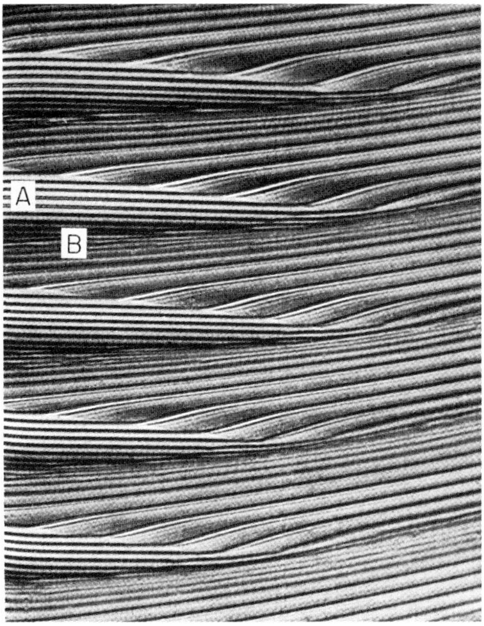

Fig.6.71. On-core (A) off-core (B) transition region in melt-pulled InSb. Magnification 380x. Rate striations introduced by Peltier cooling reveal major variations of the microscopic growth rate due to crystal rotation. From [6.94] by permission of The Electrochemical Society

Growth facets are associated with singular directions of the lattice and possess low atomic roughness. Consequently, growth onto facets proceeds via lateral layer growth rather than by statistical attachment of building blocks (normal growth) that prevails on the non-faceted, atomically rough interface parts [1.1]. This difference in growth kinetics is, for instance, reflected in the undercooling required to maintain a certain growth rate, as schematically indicated in Fig.6.68. Measurements of the undercooling at Ge-melt interfaces, revealed 1.8°, 1.2° and about 0.1° C, respectively for the faceted <111> and <110> and the non-faceted <100> direction, for a growth rate of 2.7 $\times 10^{-3}$cm s^{-1} [6.99]. During the growth of dislocation-free silicon supercoolings of about 9°C were required in <111> growth as compared to about 0.7°C for off-facet directions [6.100]. Now on facets, it is suggested, the lateral layer growth, in "sweeping" over the surface, traps more of the adsorbed solute than the random exchange mechanism prevailing on atomically rough non-faceted interfaces. Note, however, that the Ge:Ga case, as documented in Table 6.7, appears to suggest the opposite. Following the presentation of some more experimental material, we will expand the simple adsorption model and come back to this case of $k_c < k_p$.

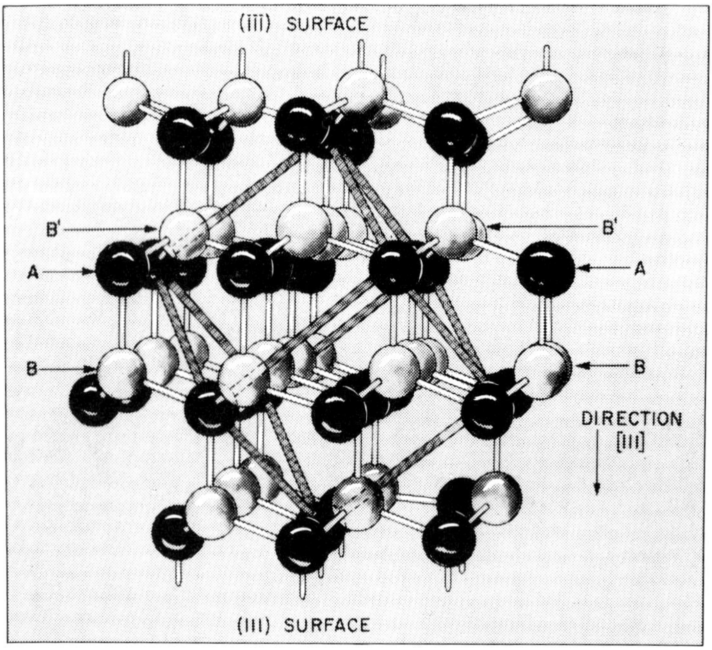

Fig.6.72. Crystal structure (zinc-blende) of the III-V intermetallic compounds; the unit cell is indicated. •, group III atoms; o, group V atoms. From [6.101] by permission of The Electrochemical Society

The adsorption models are also supported by the observation that the polarity of an interface can have an effect on (non-equilibrium) segregation. Many of the III-V intermetallic compounds and their binary solid solutions (including InSb-GaSb, GaAs-GaP, GaP-AlP, etc.), all of considerable device interest, solidify in the zinc-blende structure. As represented in Fig.6.72

there are two types of {111} surfaces in this structure that are crystallographically identical, yet chemically, i.e., in their bond character, very different. One type is terminating with group III atoms (A surfaces), and the other is terminating with group V atoms (B surfaces). Several workers observed different distribution coefficients for the two types of {111} faces. For instance, for sulfur in InSb it was found that k = 1.2 on the In {111} and k = 0.4 on the Sb {111} faces [6.102]. Also, it was observed during epitaxial growth of GaAs that the segregation coefficient for Se is greater in the As {111} direction than in the Ga {111} direction [6.103].

The bulk of the experimental data on segregation anisotropies can, at least semi-quantitatively, be accounted for by an adsorption or chemisorp-

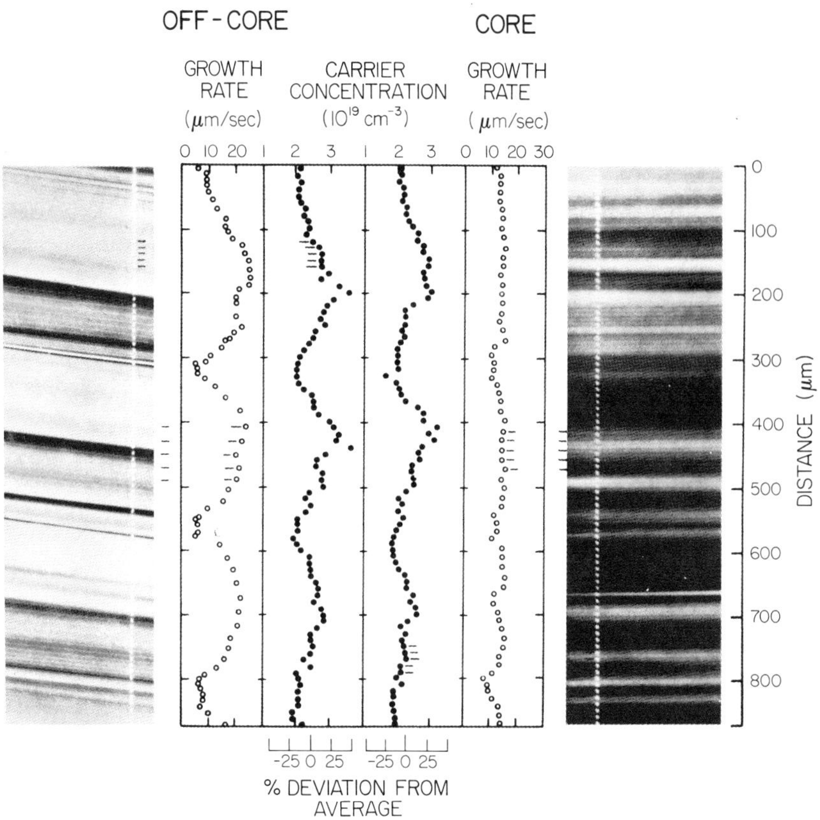

Fig.6.73. Differentially etched core area (right-hand side) and a corresponding off-core area (left-hand side) of a Ga-doped Ge crystal. Microscopic growth rates (obtained from Peltier time markings, see fine banding) and concurrent dopant concentrations (obtained from spreading resistance measurements, see impact traces on the prints of the samples). From [6.104] by permission of The Electrochemical Society

tion model (for refinements see below). Most recent high resolution experiments [6.104], however, show that the fluid dynamics in the nutrient may also play a significant role in the coring effect in a complex way which is not understood at this point. Fig.6.73, in the center part, represents microscopic growth rates and corresponding solute concentrations as obtained in pulling of a Ga-doped Ge single crystal in the core and off-core region. For experimental details see also Figs.6.44 and 45. Most surprisingly one sees that segregation inhomogeneities *can* appear in the core region with the same (rotation-induced) periodicity and equal in magnitude to those in the adjacent off-core region, *although the vertical growth rate of the core remains virtually constant*. This finding, of course, jeopardizes many of the above statements.

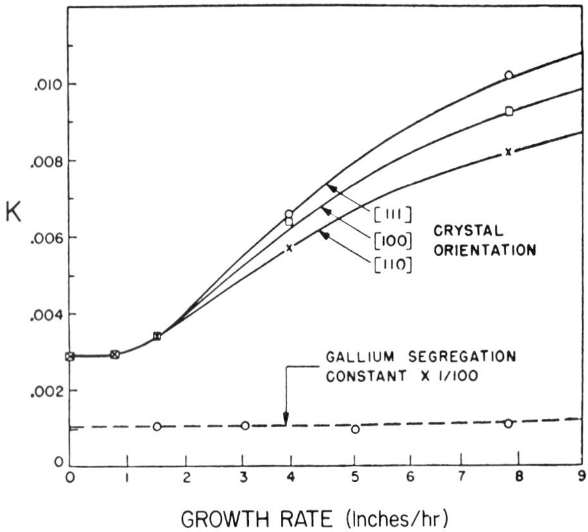

Fig.6.74. Effect of growth rate upon the segregation of Sb during melt growth of Ge crystals. From [6.105] by permission of the American Institute of Physics

Although we have argued at the beginning of this section that, for basic thermodynamic reasons, there cannot be an equilibrium segregation anisotropy, it is comforting to see some experimental verification. Fig.6.74 stems from one of the earliest investigations on this topic. HALL [6.105] has already shown in 1952 that the orientation dependence of k observed at high growth rates in the Ge:Sb system decreases and finally vanishes with decreasing growth rate.

Segregation anisotropies have also been established in other systems except semiconductors. Fig.6.75 shows lead distribution profiles obtained in CsBr crystals upon pulling from the melt in different orientations [6.106]. Alkali halides, mainly due to their poor thermal conductivity, are grown from their melts with typical rates of only a few mm/hr. Hence, in analogy to Fig.6.74 one could have naively expected that no solute incorporation anisotropy would be found in these materials. How wrong such extrapolations can be is well illustrated by Fig.6.75.

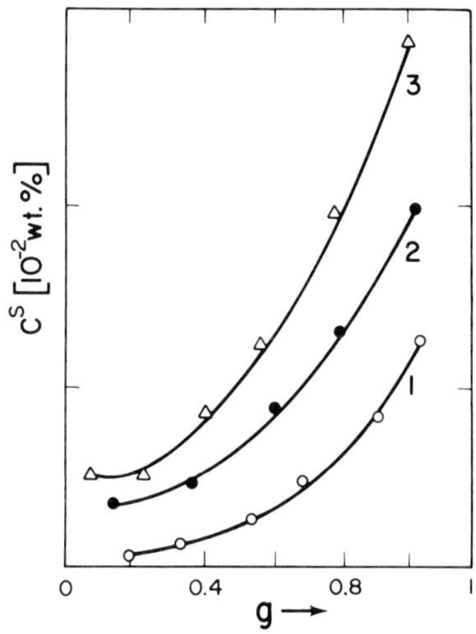

Fig.6.75. Lead concentration in melt-pulled CsBr vs. solidified fraction for various seed orientations: 1) ⟨111⟩; 2) ⟨123⟩; 3) ⟨100⟩. Growth rate 0.25 mm/hr. Lead determination by polarographic analysis. After [6.106]

In [1.1] we will see that for fundamental reasons metals crystallize from their melts with atomically rough interfaces. Hence no facet formation and consequently no anisotropy in segregation should be expected. Again, this is obviously too simple a model when we see the orientation dependence for Cd uptake into Zn depicted in Fig.6.76. This anisotropy vanishes again at very low growth rates [6.107] (not shown in Fig.6.76). Efforts to determine any significant undercooling at the interface, indicative of facet formation, failed. Measurements with very fine thermocouples that were grown into the crystals showed that the interface temperature corresponds well to the melting point of Zn over a wide range in growth rate. The authors point out that adsorption may still be involved in the observed anisotropy since the least solute is incorporated in the direction where the basal plane coincides with the interface. On the basal plane one might expect the absorbed layer concentration to be minimum since the "free bonding" is a minimum.

Most recently similar segregation anisotropies were reported for the solidification of nickel [6.108].

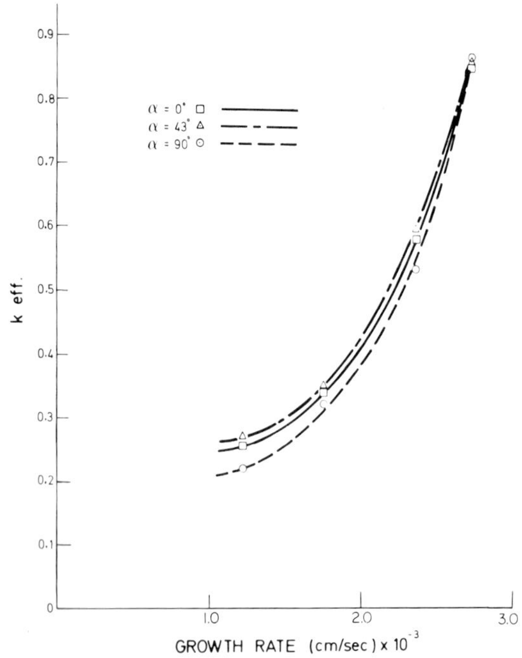

Fig.6.76. Effective segregation coefficient for Zn:2 × 10^{-4}wt% Cd as function of growth rate. From [6.107] by permission of North-Holland Publ. Co.

There are several ways to prevent or reduce the formation of solute heterogeneities due to coring in melt growth. First, one may influence the convective heat transfer conditions [1.2] in such a way that a facet covers most of the almost planar interface. Or, one can orient the seed such that a facet cannot form on any point of the interface. The necessary angle will be, of course, a function of the curvature of the interface. Also, it has been claimed [6.109], to some extent, in contrast to the findings presented in Fig.6.73, that anomalous segregation on facets disappears if there are no growth rate fluctuations present in the system. Fig.6.77 shows a photomicrograph of a selectively etched InSb single crystal. The upper part had been pulled from the melt with synchronized seed and crucible rotation. It shows the characteristic striae pattern in the on-facet off-facet transition region. In the lower part, however, where only the crucible had been rotated in order to "eliminate random thermal fluctuations in the melt", no solute inhomogeneities are found. Obviously the melt heights and temperature gradient in the melt must have been low enough in this specific run to not stimulate nonsteady convective heat transfer.

Non-equilibrium solute incorporation has also been observed in crystals grown from vapors and solutions. This is not surprising since crystallization from vapors and solutions proceeds, in general, strongly faceted. Unfortunately, just as with non-steady segregation from these phases, to this date very little well-characterized data is available.

(A)

(B)

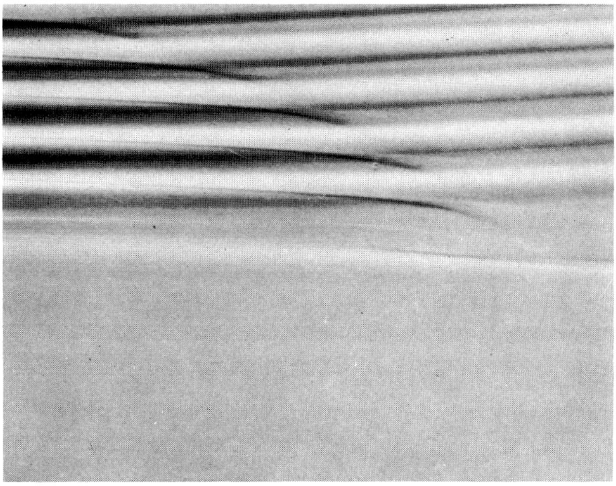

Fig.6.77. Impurity distribution in InSb grown (A) with and (B) without seed rotation. Magnification ca 350X. For details, see text. From [6.109] by permission of The Electrochemical Society

An example for orientation dependent segregation from the vapor phase is presented in [6.110]. Doped gallium arsenide was grown epitaxially by a chemical vapor transport process [1.2] onto GaAs single crystal wafers of specific orientations. Table 6.8 gives ratios of segregation coefficients found for various dopants, that had been added to the vapor phase, for different orientations of the interface. The values were deduced from electrical measurements of the carrier densities in the grown layers. Note the significant difference between the (111)A and (111)B planes that terminate with Ga and As atoms, respectively (see Fig.6.72). Note also that Zn forms an exception to the general trend that more solute is incorporated on {111} than on {100} planes.

Table 6.8. Anisotropy of segregation during vapor growth of GaAs. Ratios of k for the indicated orientations of the interface. After [6.110]

Dopant	(111)B / (100)	(111)B / (111)A	(111)B / (110)
None	11.0	15	28
Zn	0.43	0.20	0.49
Te	7.4	~20	~15
Se	6.3	--	4.4
Sn	2.3	--	--
S	1.4	--	--

An interesting observation of non-uniform impurity incorporation in growth from aqueous solutions was made recently [6.111] in a continuation of earlier work [6.79]; see also Fig.6.13a. These authors found in KCl crystals that lead was incorporated more near growth centers (microscopic hills on the interface, formed from concentric macro-growth steps, see [1.1], than in the rest of the crystal. Typical concentration profiles as obtained by polarographic analysis of thin crystal sections are plotted in Fig.6.78. The authors interpreted these results as well as their data shown in Fig.6.13 in terms of an adsorption-diffusion theory for excess impurity incorporation [6.98]. This treatment can be outlined, somewhat simplified, as follows:

The surface of a solid that is in contact with an impure fluid is partly covered with adsorbed impurity atoms. In equilibrium the coverage of available adsorption sites depends, in a given system, only on the impurity concentration in the fluid. In general, the concentration in the adsorption layer C^a will be different from the equilibrium bulk concentration C^s_e, that is determined by the thermodynamics of the dissolution of the impurity in the host solid (Sect. 6.2.1). New growth, assumed to proceed in layers of height h, which move laterally over the interface, traps a certain fraction of the adsorbed impurities. Only at "very small" growth rates will the trapped concentration in the new layer be the same as the equilibrium bulk concentration. Hence, diffusion will set in to re-establish equilibrium. Note that this is not necessarily out-diffusion of impurities through the new layer. Although not as common, the trapped concentration can as well be too low ($C^s < C^s_e$) as we have encountered in a few cases above. This diffusive readjustment will be the slower the thicker the growth layer (step)

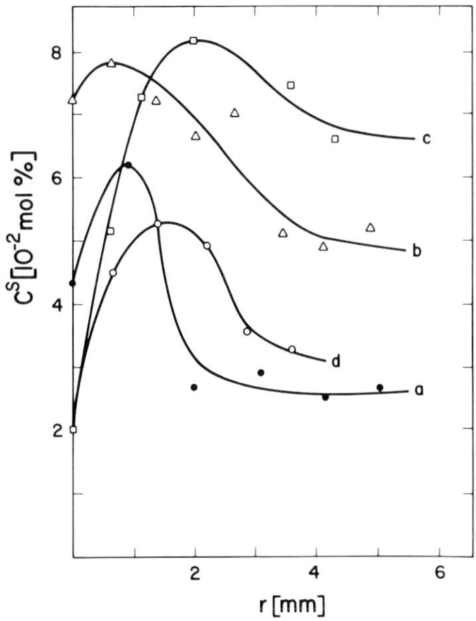

Fig.6.78. Radial Pb-distribution around growth centers in solution-grown KCl for various growth rates (in 10^{-4} μm/s): (a) 2.9; (b) 3.9; (c) and (d) 5.7. Solution doped with 10^{-3} molar parts Pb^{2+}. After [6.111]

heights h are. Also, the higher the growth rate, the more rapidly will the considered layer be covered by new ones, thus further impeding solute mass transfer. Hence, one will expect that the deviation from equilibrium (in most cases an increase over the equilibrium concentration) increases with the growth rate. Eventually, with further increase in growth rate, mass transfer limitations between fluid and interface will lead to an incomplete coverage of the possible adsorption sites. Then (in the typical case of $C^a > C^S_e$ at lower V^S) a decrease of C^S will be found with increasing V^S.

This trend of first increasing and then decreasing impurity takeup with increasing growth rate is reflected in Fig.6.12 – although somewhat obscured by a complex concentration dependence. The radial concentration distribution in growth hills can be understood along the same line. Since a detailed knowledge of the morphology (step or layer heights versus radius) is required, we postpone the discussion to later sections.

As a transition between the mere solid solutions considered so far, and macroscopic occlusions of impurities or solutes to be discussed in later sections, one should mention here the formation of "Adsorptionsmischkristalle". As suggested among others by KLEBER [6.112], incorporation of impurities into a crystal may occur by periodic impurity nucleation on adsorbed ordered impurity layers. This mechanism could lead to incorporated layers several lattice constants thick. There is some indirect experimental evidence for such a mechanism. However, attempts towards a direct observation of the claimed layers have been unsuccessful [6.113].

Non-equilibrium solid solutions are not necessarily due to non-equilibrium conditions at the interface during growth. In Sect. 6.2.1 on the thermodynamics of segregation, we have pointed out the general trend of decreasing solubility with decreasing temperature. (The opposite, retrograde solubility, is a not-too-common phenomenon). This is reflected in the gradients dX/dT of the solvus lines in the phase diagrams of Sect. 3.4. If the cooling-composition path of a system intersects such a solvus-line, compositional non-equilibrium sets in. Yet the readjustment towards equilibrium at lower temperatures is often hindered by the slow solid state diffusion. Hence, the non-equilibrium composition (excess of components) is "frozen-in". Even in systems where the phase diagram suggests "zero solubility" one can typically "quench-in" relative dopant concentrations of at least 10^{-3}. However, one must be aware that, given enough time, precipitation of the excess components will eventually occur even at room temperature.

Problems

6.1 Discuss qualitatively why the effective segregation coefficient tends towards unity, for both $k_0 < 1$ and $k_0 > 1$, as the growth rate of the solid V^s is increased.

6.2 Give a detailed derivation of (6.25). State all assumptions.

6.3 Show that (6.43) results from (6.36) for $y=0$ and $\rho^f/\rho^s = 1$.

6.4 Calculate the concentration boundary layer widths δ for the experiments in Fig.6.10b for 57, 144, 1440 RPM (for ν see Table 5.12), and compare them with the δ' values obtained in the experiments using $\mathcal{D}_{\text{Ga in Ge}} \approx 10^{-4}$ cm^2/s (Fig.5.3). Carefully discuss the origin of the differences. Calculate the actual normal fluid velocity for all three cases employing (6.40) and compare the values with V^s.

6.5 Check the ultimate distribution A in Fig.6.27 by calculating at least 3 points from the experimental parameters given in the caption.

6.6 Calculate the maximum diameter of a 2 cm long "capillary" to be used in the floating-crucible growth of Ga-doped germanium crystals for a V^S and diameter of your choice. Comment on the convective stability of the melt in the bore of calculated diameter. Suggest alternate (i.e., convectively more stable) geometries.

6.7 Show that the solute concentration changes from given, small amplitude growth rate changes increase with increasing values of k_0 — except in a certain window of the range $0 \leqslant k_0 \leqslant \infty$.

6.8 Calculate the dopant concentration at the center and periphery in the plane AA' of the crystal depicted in Fig.6.59 assuming that the initial melt concentration is C_0, the crystal forms 80% of the initial melt volume and that k_0 is concentration independent. State all assumptions necessary.

6.9 (a) Evaluate (6.50) and (6.51) for a system with $\mathcal{D}_{AB} = 10^{-4}$ cm^2/s, $V_0^S = 10^{-3}$ cm/s, $V_1^S = 2V_0^S$, and two solutes of $k_0 = 0.1$ and $k_0 = 0.01$, respectively. Comment.

(b) Calculate the initial transients x from (6.52) for systems with the above values of $\mathcal{D}_{AB}$, V_0^S, k_0 for $\delta' = 0.1$ and 0.01, respectively. Discuss the differences between x and δ'.

6.10 Derive (6.54). State all necessary assumptions.

$$\text{Hint: } \Delta C^S/C^S_{avg} = \frac{1}{k}\frac{\partial k}{\partial V^S}\Delta V^S .$$

6.11 Discuss why anisotropic *equilibrium* segregation is thermodynamically impossible.

Appendices

A. Universal Gas Constant

Values of R for various combinations of units in energy, temperature and moles.

Units of R: $\frac{[\text{Energy}]}{[\text{Temp.}] \times [\text{Mole}]}$

Energy	Temperature	Moles	R
Calories	K	gm	1.987
		lb	901.1
	°R	gm	1.1039
		lb	500.61
Kilo calories	K	gm	1.987×10^{-3}
		lb	0.9011
	°R	gm	1.1039×10^{-3}
		lb	0.50061
Joules	K	gm	8.3136
		lb	3770.2
	°R	gm	4.6187
		lb	2094.6
Ergs	K	gm	8.3136×10^{7}
		lb	3.7702×10^{10}
	°R	gm	4.6187×10^{7}
		lb	2.0946×10^{10}
BTU	K	gm	7.884×10^{-3}
		lb	3.5756
	°R	gm	4.3803×10^{-3}
		lb	1.9864
eV	K	gm	5.1892×10^{19}
		lb	2.3533×10^{22}
	°R	gm	2.8829×10^{19}
		lb	1.3074×10^{22}

Values of R for various combinations of units in pressure, volume, temperature and moles.

$$\text{Units of } R \equiv [R] \equiv \frac{[\text{Pressure}] \times [\text{Volume}]}{[\text{Temperature}] \times [\text{Mole}]}$$

Volume	Temp.	Mole	Pressure							
			kPa ≡ 10^3 Newt./m^2	Atm	psi	mm Hg ≡ torr	in Hg	in H_2O	ft H_2O	pounds/ft^2
[ft^3]	K	gm	0.293	2.90×10^{-3}	4.26×10^{-2}	2.20	8.67×10^{-2}	1.18	9.82×10^{-2}	6.14
		lb	133	1.31	19.31	999	39.3	535	44.6	2.77×10^3
	°R	gm	0.163	1.61×10^{-3}	2.37×10^{-2}	1.22	4.82×10^{-2}	0.655	5.46×10^{-2}	3.41
		lb	73.97	0.73	10.73	555	21.8	297	24.8	1.54×10^3
[cm^3]	K	gm	8.31×10^3	82.05	1.21×10^3	6.24×10^4	2.45×10^3	3.34×10^4	2.78×10^3	1.74×10^5
		lb	3.77×10^6	3.72×10^4	5.47×10^5	2.83×10^7	1.11×10^6	1.51×10^7	1.26×10^6	7.87×10^7
	°R	gm	4.62×10^3	45.6	670	3.46×10^4	1.36×10^3	1.85×10^4	1.55×10^3	9.65×10^4
		lb	2.095×10^6	2.07×10^4	3.04×10^5	1.57×10^7	6.19×10^5	8.41×10^6	7.01×10^5	4.38×10^7
[Liters]	K	gm	8.312	8.20×10^{-2}	1.21	62.4	2.45	33.4	2.78	1.74×10^3
		lb	3.77×10^3	37.2	547	2.83×10^4	1.11×10^3	1.51×10^4	1.26×10^3	7.87×10^5
	°R	gm	4.618	4.56×10^{-2}	0.670	34.6	1.36	18.5	1.55	965
		lb	2.095×10^3	20.7	304	1.57×10^4	619	8.41×10^3	701	4.38×10^5

B. Thermodynamic Functions

State Variable Pairs

	intensive		extensive
T	temperature	S	entropy
Y	mechanical variable ($-P$, $\overline{H}$, $\overline{F}$, σ)	X	mechanical variable (V, $\overline{M}$, $\overline{P}$, A)
μ_i	chemical potential	n_i	number of moles of i-th component

Thermodynamic Potentials

This text			Other common usage	
Symbol	Definition	Name	Symbol	Name
For variables				
S, X, n		*Internal Energy*		
	$U = TS + YX + \Sigma_i\mu_i n_i$		E	Total Energy
	$dU = TdS + YdX + \Sigma_i\mu_i dn_i$			Energy
S, Y, n		*Enthalpy*		Y and n constant:
	$H = U - Yx = TS + \Sigma_i\mu_i n_i$			Heat Content, Total Heat
	$dH = TdS - XdY + \Sigma_i\mu_i dn_i$			Heat Function
T, X, n		*Helmholtz Potential* or *Function*		T constant:
	$F = U - TS = YX + \Sigma_i\mu_i n_i$		A	Helmholtz Free Energy
	$dF = -SdT + YdX + \Sigma_i\mu_i dn_i$			Work Function, Work Content
T, Y, n		*Gibbs Potential* or *Function*		T and Y constant:
	$G = H - TS = U - YX - TS$		F, Z	Gibbs Free Energy
	$= F - YX$			Free Enthalpy
	$= \Sigma_i\mu_i n_i$			Thermodynamic Potential
	$dG = -SdT - XdY + \Sigma_i\mu_i dn_i$			

Chemical Potential

$$\mu_i = u_i - Yx_i - Ts_i = \left(\frac{\partial G}{\partial n_i}\right)_{T,Y,N_j}$$ Escaping Tendency

$$d\mu_i = -s_i dT - x_i dY$$ (Gibbs-Duhem Equation)

C. Derivations of some Relations Used in Sect. 5.2

Contributed by Glenn. H. Westphal

C.1 Derivation of $\sum \overline{V}_i C_i = 1$

The partial molar volumes, as any intensive state variable, depend only on the relative concentrations of the components and not on the total amount of each constituent. Hence based upon

$$dV = \sum \overline{V}_i dn_i \tag{C.1}$$

one can imagine that the total volume of a system has been established at the given P and T by simultaneously adding the components in their final ratios, thereby keeping the $\overline{V}_i$'s constant at their "final" value so that (C.1) can be readily integrated to

$$V = \sum \overline{V}_i n_i \quad \text{or} \quad \sum \overline{V}_i C_i = 1 \quad . \tag{C.2}$$

Only with ideal solutions, where $\overline{V}_i$ = const, i.e., independent of the composition of the system the descriptive designation "volume" fraction for $F = \overline{V}_i C_i$ becomes meaningful. Hence in ideal systems (C.2) is equivalent to

$$\sum F_i = 1 \quad , \tag{C.3}$$

i.e., the total volume is the sum of the subvolumes occupied by the *pure* components.

C.2 Derivation of $J_A^{\star}$ from $J_A^{\square}$

By definition

$$J_A^{\square} = C_A(v_A - v^{\square})$$

which can be written, also using (5.9) to

$$J_A^{\square} = C_A(\overline{V}_A C)(\overline{V}_B C)^{-1}[v_A - C_A\overline{V}_A v_A - C_B\overline{V}_B v_B]$$

$$= C_A\overline{V}_B C\left[\frac{v_A}{\overline{V}_B C}(1 - C_A\overline{V}_A) - X_B v_B\right]$$

where we made use of $X_B = C_B/C$. Now with $\sum C_i\overline{V}_i = 1$, i.e. (C.2), one can transform the above equation to

$$J_A^{\square} = C_A\overline{V}_B C(X_B v_A - X_B v_B)$$

which in turn, using $X_A + X_B = 1$, is

$$J_A^{\square} = C_A\overline{V}_B C[(1 - X_A)v_A - X_B v_B]$$

and with (5.8) one obtains

$$J_A^{\square} = C_A\overline{V}_B C(v_A - v^*) = V_B C J_A^* \quad .$$

Hence $J_A^* = (\overline{V}_B C)^{-1} J_A^{\square}$ which with (5.23) is also

$$J_A^* = -\mathcal{D}_{AB}(\overline{V}_B C)^{-1}\nabla C_A \quad . \tag{C.4}$$

On the other hand with (C.2)

$$\nabla C_A = (C_A\overline{V}_A + C_B\overline{V}_B)\nabla C_A \quad .$$

This, with $\sum \overline{V}_i\nabla C_i = 0$, can be transformed to

$$\nabla C_A = -C_A\overline{V}_B\nabla C_B + C_B\overline{V}_B\nabla C_A$$

and $$\nabla C_A = \overline{V}_B(C_B\nabla C_A - C_A\nabla C_B)$$

adding and subtracting $\overline{V}_B C_A\nabla C_A$ then leads to

$$\nabla C_A = \overline{V}_B(C_A\nabla C_A - C_A\nabla C_A + C_B\nabla C_A - C_A\nabla C_B)$$

which is

$$\nabla C_A = \bar{V}_B(C\nabla C_A - C_A\nabla C) \quad .$$

Applying the rule for quotient differentiation this can be rewritten to

$$\nabla C_A = \bar{V}_B C^2 \nabla(C_A/C)$$

$$= \bar{V}_B C^2 \nabla X_A \quad . \tag{C.5}$$

Finally inserting (C.5) into (C.4) one obtains

$$J_A^* = -C\mathcal{D}_{AB}\nabla X_A \quad .$$

References

1. Introduction

1.1 F. Rosenberger, *Fundamentals of Crystal Growth II: Kinetic and Morphological Concepts* (in preparation)

1.2 F. Rosenberger, *Fundamentals of Crystal Growth III: Techniques* (in preparation)

2. Thermodynamics

2.1 F.C. Andrews: *Thermodynamics: Principles and Applications* (Wiley-Interscience, New York 1971)

2.2 F.C. Andrews: *Equilibrium Statistical Mechanics*, 2nd ed. (Wiley-Interscience, New York 1975)

2.3 P.M. Morse: *Thermal Physics*, 2nd ed. (W.A. Benjamin, New York 1969)

2.4 G.N. Lewis, M. Randall: *Thermodynamics*, 2nd ed. (McGraw-Hill, New York 1961)

2.5 D.R. Gaskell: *Introduction to Metallurgical Thermodynamics* (Scripta Publishing, Washington DC; McGraw-Hill, New York 1973)

2.6 J.H. Dymond, E.B. Smith: *The Virial Coefficients of Gases* (Clarendon Press, Oxford 1969)

2.7 E.A. Mason, T.H. Spurling: *The Virial Equation of State*, The International Encyclopedia of Chemistry and Chemical Physics, Topic 10, Vol. 2 (Permagon Press, Oxford 1969)

2.8 R. Haase: *Physical Chemistry*, ed. by W. Jost, Thermodynamics, Vol. I (Academic Press, New York 1971) pp. 84-86

2.9 I. Prigogine: *Introduction to Thermodynamics of Irreversible Processes* (Wiley, New York 1967).
P. Glansdorff, I. Prigogine: *Thermodynamic Theory of Structure, Stability and Fluctuations* (Wiley, New York 1971)

2.10 K. Denbigh: *The Principles of Chemical Equilibrium*, 3rd ed (Cambridge Univ. Press, Cambridge 1971)

2.11 O. Kubaschewski, E.LL. Evans, C.B. Alcock: *Metallurgical Thermochemistry*, 4th ed. (Permagon Press, London 1967)

2.12 J.H.E. Jeffes: The Physical Chemistry of Transport Processes. J. Crystal Growth 3-4, 13 (1968)

2.13 T.B. Reed: *Free Energy of Formation of Binary Compounds* (MIT Press, Boston 1971)

2.14 F. Van Zeggeren, S.H. Storey: *The Computation of Chemical Equilibria* (Cambridge Univ. Press, Cambridge 1970)

2.15 B.I. Noläng, M.W. Richardson: The transport flux function — a new method for predicting the rate of chemical transport in closed systems. J. Crystal Growth 34, 198 and 205 (1976)

2.16 I. Barin, O. Knacke: *Thermochemical Properties of Inorganic Substances* (Springer, Berlin, Heidelberg, New York and Verlag Stahleisen, Düsseldorf 1973)

2.17 I. Barin, O. Knacke, O. Kubaschewski: *Thermochemical Properties of Inorganic Substance, Supplement* (Springer, Berlin, Heidelberg, New York 1977)

2.18 K.C. Mills: *Thermodynamic Data for Inorganic Sulphides, Selenides and Tellurides* (Butterworths, London 1974)

2.19 E. Sirtl: Graphische Methoden zur Abschätzung von Enthalpie und Entropiewerten gasförmiger Anorganischer Verbindungen. Z. Naturforsch. 21a, 2001 (1966)

2.20 D.R. Stull, H. Prophet: *JANAF Thermochemical Tables*, 2nd ed. (National Bureau of Standards, Washington 1971)

2.21 W.F. Sheehan: *Physical Chemistry*, 2nd ed. (Allyn and Bacon, Boston 1966)

3. Phase Equilibria and Phase Diagrams

3.1 P. Gordon: *Principles of Phase Diagrams* (McGraw-Hill, New York 1968). The various figures shown from this reference have been reproduced with the permission of McGraw-Hill Book Company

3.2 H.C. Yeh: "Interpretation of Phase Diagrams", in *Phase Diagrams*, Vol.1, ed. by A.M. Alper (Academic Press, New York 1970) Chap.4

3.3 F.A. Kröger: *The Chemistry of Imperfect Crystals*, 2nd ed. (North Holland, Amsterdam 1973) Chap. 2

3.4 R.E. Honig: "Vapor pressure data for the elements", in *The Characterization of High-Temperature Vapors*, ed. by J.L. Margrave (Wiley and Sons, New York 1967)

3.5 K.M. Kim, A.F. Witt, H.C. Gatos: Segregation behavior in a stationary vertical zone with converging interface: pressure induced segregation effects. J. Electrochem. Soc. 121, 448 (1974)

3.6 C.A. Knight, N.C. Knight: Superheated ice: true compression fractures and fast internal melting. Science 178, 613 (1972)

3.7 P.M. Morse: *Thermal Physics*, 2nd ed. (W.A. Benjamin, New York 1969)

3.8 W.J. Moore: *Physical Chemistry*, 3rd ed. (Prentice Hall, Englewood Cliffs, NJ 1962) p. 96

3.9 G.J. Abbaschian, S.F. Ravitz: Melting kinetics of gallium single crystals. J. Crystal Growth 28, 16 (1975)

3.10 A.B. Pippard: *Elements of Classical Thermodynamics* (Cambridge University Press, Cambridge 1966)

3.11 P.G. de Gennes: "Phase transition and turbulence: an introduction", in *Fluctuations, Instabilities, and Phase Transitions*, ed. by T. Riste (Plenum Press, New York 1975) pp. 1-18

3.12 H.M. Strong, R.H. Wentorf, Jr.: The growth of large diamond crystals, Naturwissenschaften 59, 1 (1972)

3.13 J.C. Angus, H.A. Will, W.S. Stano: Growth of diamond seed crystals by vapor deposition. J. Appl. Phys. 39, 2915 (1968)

3.14 S.P. Chauhan, J.C. Angus, N.C. Gardner: Kinetics of carbon deposition on diamond powder. J. Appl. Phys. 47, 4746 (1976)

3.15 B.V. Deryagin, D.V. Fedoseev: Epitaxial synthesis of diamond in the metastable region. Russ. Chem. Rev. 39, 783 (1970)

3.16 B.V. Deryagin, D.V. Fedoseev: The synthesis of diamond at low pressure. Scient. American 102 (November 1975)

3.17 C.D. Thurmond: Equilibrium thermochemistry of solid and liquid alloys of germanium and of silicon. J. Phys. Chem. 57, 827 (1953)

3.18 H. Seltz: J. Amer. Chem. Soc. 56, 307 (1934)

3.19 J. Steininger: Thermodynamics and calculations of the liquidus-solidus gap in homogeneous, monotonic alloy systems. J. Appl. Phys. 41, 2713 (1970)

3.20 F.J. Dunkerley, G.J. Mills: *Thermodynamics in Metallurgy* (Am. Soc. Metals, Metals Park/Ohio 1950)

3.21 G.M. Wolten, W.R. Wilcox: In *Fractional Solidification*, ed. by M. Zief, W.R. Wilcox (Marcel Dekker, New York 1967) Chap. 2

3.22 W.C. Winegard: *An Introduction to the Solidification of Metals* (Institute of Metals, London 1964)

3.23 P.G. Shewmon: *Transformation of Metals* (McGraw-Hill, New York 1969)

3.24 M. Hansen: *Constitution of Binary Alloys*, 2nd ed. (McGraw-Hill, New York 1958)

3.25 A.H. Cottrell: *An Introduction to Metallurgy* (St. Martin's Press, New York 1967)

3.26 H.W. Bakhuis Roozeboom: *Die Heterogenen Gleichgewichte*, Zweites Heft: Systeme aus zwei Komponenten (Vieweg, Braunschweig 1904) p. 125

3.27 J. Zernicke: *Chemical Phase Theory* (Uitgevers-Maatschappij Kluwer, Deventer, Antwerp, Djakarta 1955)

3.28 E.M. Levin, C.R. Robbins, H.F. McMurdie: *Phase Diagrams for Ceramists* (American Ceramic Society, Columbus 1964)

3.29 A. Reisman: *Phase Equilibria* (Academic Press, New York 1970)

3.30 R. Haase, H. Schönert: *Solid-Liquid Equilibrium*, Intern. Encyclopedia of Physical Chemistry and Chemical Physics, Topic 13, Vol. 1 (Permagon, New York 1969)

3.31 G.B. Stringfellow: Calculation of ternary and quaternary III-V phase diagrams. J. Crystal Growth 27, 21 (1974)

3.32 L.S. Palatnik, A.I. Landau: *Phase Equilibria in Multicomponent Systems* (Holt, Rinehart and Winston, New York 1964)

3.33 A.D. Pelton, W.T. Thompson: Phase Diagrams. Progr. Sol. State Chem. 10, 119 (1975).

4. Crystal Growth and Phase Diagrams

4.1 K. Nassau: "The Chemistry of Laser Crystals", in *Applied Solid State Sciences*, Vol. 2, ed. by R. Wolfe (Academic Press, New York 1971) p. 173

4.2 R.A. Laudise: *The Growth of Single Crystals* (Prentice Hall, Englewood Cliffs, N.J. 1970).
R.A. Laudise: "Techniques of Crystal Growth", in *Crystal Growth*, ed. by H.S. Peiser (Permagon Press, New York 1960) p. 3

4.3 J.W. Nielsen, R.R. Monchamp: "The Use of Phase Diagrams in Crystal Growth", in *Phase Diagrams*, Vol. 3, ed. by A.M. Alper (Academic Press, New York 1970)

4.4 L.R. Shiozawa, J.M. Jost: "Research in Improved II-VI Crystals". Document NT 1.2:AD 694 533 (1969)

4.5 R.M. Logan, D.T.J. Hurle: Calculations of point defect concentrations and nonstoichiometry in GaAs. J. Phys. Chem. Solids 32, 1739 (1971)

4.6 N.Kh. Abrikosov, V.F. Bankina, L.V. Poretskaya, L.E. Shelimova, E.V. Skudnova: *Semiconducting II-VI, IV-VI and V-VI Compounds* (Plenum Press, New York 1969)

4.7 J.R. Carruthers, G.E. Peterson, M. Grasso, P.M. Bridenbaugh: Nonstoichiometry and crystal growth of lithium niobate. J. Appl. Phys. 42, 1846 (1971)

4.8 A.A. Ballman, H. Brown, P.K. Tien, S. Riva-Sanseverino: The growth of $LiNbO_3$ thin films by liquid phase epitaxial techniques. J. Crystal Growth 29, 289 (1975)

4.9 R.L. Byer, J.F. Young, R.S. Feigelson: Growth of high quality $LiNbO_3$ crystals from the congruent melt. J. Appl. Phys. 41, 2320 (1970)

4.10 M.B. Panish: A thermodynamic evaluation of the simple solution treatment of the Ga-P, In-P and Ga-As systems. J. Crystal Growth 27, 6 (1974)

4.11a J.N. Bradley, P.D. Greene: Solids with high ionic conductivity in group I halide systems. Trans. Faraday Soc. 63, 424 (1967)

4.11b L.D. Fullmer, M.A. Hiller: Crystal growth of the solid electrolyte $RbAg_4I_5$. J. Crystal Growth 5, 395 (1969)

4.12 E.M. Levin, C.R. Robbins, H.F. McMurdie: *Phase Diagrams for Ceramists* (The American Ceramic Society, Columbus 1969) Supplement

4.13 J.C. Brice: *The Growth of Crystals from Liquids* (North-Holland Publishing Company, Amsterdam 1973)

4.14 V. Belrus, J. Kalnaj, A. Linz: Top-seeded solution growth of oxide crystals from non-stoichiometric melts. Mat. Res. Bull. 6, 899 (1971)

4.15 K. Nassau, A.M. Broyer: Application of Czochralski crystal pulling technique to high melting oxides. J. Am. Ceram. Soc. 45, 474 (1962)

4.16 S.P. Clark: Effect of pressure on the melting point of eight alkali halides. J. Chem. Phys. 31, 1526 (1959)

4.17 M. Midorikawa, Y. Ishibashi, Y. Takagi: Melt-growth of CsCl under high pressure. J. Crystal Growth 24/25, 383 (1974)

4.18 M.R. Lorenz: Preparation of CdTe crystals from near-stoichiometric and Cd-rich melt compositons under constant Cd pressure. J. Appl. Phys. 33, 3304 (1962)

4.19 F. Rosenberger: "Purification of Alkali Halides", in *Ultrapurity*, ed. by M. Zief and R. Speights (Marcel Dekker, New York 1972)

4.20 R.C. Linares: Growth of yttrium iron-garnet from molten barium borate. J. Am. Ceram. Soc. 45, 307 (1962)

4.21 H.J. Levinstein, S. Licht, W.R. Landorf, S.L. Blank: Growth of high quality garnet thin films from supercooled melts. Appl. Phys. Lett. 19, 486 (1971)

4.22 H.D. Jonker: Investigation of the phase diagram of the systems $PbO - B_2O_3 - Fe_2O_3 - Y_2O_3$. J. Crystal Growth 28, 231 (1975)

4.23 F.P. Bundy: Direct conversion of graphite to diamond in static pressure apparatus. J. Chem. Phys. 38, 631 (1963)

4.24 J.C. Joubert, J. Chenavas: "New Phases at High Pressures." in *Treatise on Solid State Chemistry*, Vol. 5, ed. by N.B. Hannay (Plenum Press, New York, London 1974)

4.25 J. Akella, S.N. Vaidya, G.C. Kennedy: Melting of sodium chloride at pressures to 65 kbar. Phys. Rev. 185, 1135 (1969)

4.26 E.M. Levin, H.F. McMurdie: *Phase Diagrams for Ceramists* (The American Ceramic Society, Columbus 1975) Supplement

4.27 L.F. Sobon, P.E. Greene: Crystal growth of some vanadium oxides. J. Am. Ceram. Soc. 49, 106 (1966)

4.28 R.C. DeVries: Epitaxial growth of CrO_2. Mat. Res. Bull. 1, 83 (1966)

4.29 W.J. Moore: *Physical Chemistry* 3rd ed. (Prentice Hall, New York 1962)

4.30 J.W. Mullin: *Crystallization*, 2nd ed. (Butterworth, London 1972)

4.31 J.M. Woodall: Isothermal solution mixing growth of thin $Ga_{1-x}Al_xAs$ layers. J. Elchem. Soc. 118, 150 (1970); and Solution grown $Ga_{1-x}Al_xAs$ superlattice structures. J. Crystal Growth 12, 32 (1972)

4.32 F.A. Kröger: *The Chemistry of Imperfect Crystals*, Vol. 1, 2nd ed. (North-Holland, Amsterdam 1973)

4.33 J.R. Arthur: Vapor pressures and phase equilibria in the Ga-As system J. Phys. Chem. Solids 28, 2257 (1967)

4.34 C.Z. Van Doorn: Method for heating alkali halides and other solids in vapor of controlled pressure. Rev. Scient. Instr. 32, 755 (1961)

4.35 V.R. Porter, R. Roy: P_{O_2}-T stability range of Magneli phases and direct evidence for Ti-interstitials in reduced TiO_2. Bull. Am. Ceram. Soc. 43, 263 (1964)

4.36 R.P. Elliott: *Constitution of Binary Alloys* (McGraw-Hill, New York 1965) First Supplement

4.37 D. DeNobel: Phase equilibria and semiconducting properties of CdTe. Philips Res. Rep. 14, 361 (1959)

4.38 J. Bloem, F.A. Kröger: The P-T-X phase diagram of the lead-sulphur system. Z. Phys. Chem. 7, 1 (1956)

4.39 A.V. Novoselova, V.P. Zlomanov, S.G. Kurbanov, O.V. Matveyev, A.M. Gas'kov: "Physico-chemical Study of the Germanium, Tin Lead Chalcogenides"; in *Progr. in Solid State Chem.*, Vol. 7, ed. by H. Reiss, J.O. McCaldin (Pergamon Press, New York 1972)

4.40 R.F. Brebrick, A.J. Strauss: Partial pressures of Hg(g) and Te_2(g) in Hg-Te system from optical densities. J. Phys. Chem. Solids 26, 989 (1965)

4.41 A.J. Strauss, R.F. Brebrick: Vapor crystal equilibrium and electrical properties of HgTe. J. Phys. Chem. Solids 31, 2293 (1970)

4.42 M. Hansen: *Constitution of Binary Alloys*, 2nd ed. (McGraw-Hill, New York 1958)

4.43 M.B. Panish: "The Use of the Phase Diagram in Investigations of the Properties of Compound Semiconductors"; in *Phase Diagrams*, Vol. 3, ed. by A.M. Alper (Academic Press, New York 1970) Chap. 2, pp. 53-86

4.44 T.B. Reed: "The Role of Oxygen Pressure in the Control and Measurement of Composition in 3d Metal Oxides"; in *The Chemistry of Extended Defects in Non-Metallic Solids*, ed. by L. Eyring, M. O'Keefe (North Holland, Amsterdam 1970) pp. 21-35

4.45 R. Roy, W.B. White: Growth of titanium oxide crystals of controlled stoichiometry and order. J. Crystal Growth 13/14, 78 (1972)

4.46 M. Muan: "The Effect of Oxygen Pressure on Phase Relations in Oxide Systems"; in *Phase Diagrams* Vol. 2, ed. by A.M. Alper (Academic Press, New York 1970) Chap. 1, pp. 1-19

4.47 A. Muan, E.F. Osborn: *Phase Equilibria Among Oxides in Steelmaking* (Addison Wesley, Reading, MA 1965)

4.48 S. Anderson, B. Collén, U. Kuylenstierna, A. Magneli: Phase analysis studies on the titanium-oxygen system. Acta Chem. Scand. 11, 1641 (1957).
S. Anderson, B. Collén, G. Kruuse, U. Kuylenstierna, A. Magneli, H. Pestmalis, S. Åsbrink: Identification of titanium oxides by x-ray patterns. Acta Chem. Scand. 11, 1653 (1957)

4.49 B.C.H. Steele, C.B. Alcock: Factors influencing the performance of solid oxide electrolytes in high-temperature thermodynamic measurements. Trans. Met. Soc. AIME 233, 1359 (1965)

4.50 E.M. Levin, C.R. Robbins, H.F. McMurdie: *Phase Diagrams for Ceramists* (The American Ceramic Society, Columbus, Ohio 1964)

4.51 B. Phillips, L.L.Y. Chang: High-temperature stability of tungsten oxide structures. Trans. Met. Soc. AIME 230, 1203 (1964)

4.52 B. Phillips, L.L.Y. Chang: Condensed-phase relations in the system Mo-O. Trans. Met. Soc. AIME 233, 1430 (1965)

4.53 W.D. Kingery, H.K. Bowen, D.R. Uhlmann: *Introduction to Ceramics*, 2nd ed. (Wiley and Sons, New York 1976) p. 402

4.54 F.A. Shunk: *Constitution of Binary Alloys*, 2nd Suppl. (McGraw-Hill, New York 1969)

5. Mass Transport and Heat Transfer

5.1 R.L. Parker:""Crystal Growth Mechanisms: Energetic, Kinetic, Transport", in *Solid State Physics*, Vol. 25, ed. by H. Ehrenreich, F. Seitz, D. Turnbull (Academic Press, New York 1970) pp. 151-299

5.2 R.B. Bird, W.W. Stewart, E.N. Lightfoot: *Transport Phenomena* (Wiley, New York 1960)

5.3 J.G. Kirkwood, R.L. Baldwin, P.J. Dunlop, L.J. Gosting, G. Kegeles: Flow equations and frames of reference for isothermal diffusion in liquids. J. Chem. Phys. 33, 1505 (1960)

5.4 C.J. Geankoplis: *Mass Transport Phenomena* (Ohio State University Bookstore, Columbus 1972)

5.5 D.D. Fitts: *Nonequilibrium Thermodynamics* (McGraw-Hill, New York 1962)

5.6 W.R. Wilcox: "A Generalized Treatment of Mass Transfer in Crystal Growth", in *Preparation and Properties of Solid State Materials*, Vol. 2, ed. by W.R. Wilcox (Marcel Dekker, New York 1976) p. 129

5.7 J.O. Hirschfelder, C.F. Curtiss, R.B. Bird: *Molecular Theory of Gases and Liquids* (Wiley, New York 1954)

5.8 I. Prigogine: *Introduction to Thermodynamics of Irreversible Processes* (Wiley, New York 1967).
P. Glansdorff, I. Prigogine: *Thermodynamic Theory of Structure, Stability and Fluctuations* (Wiley, New York 1971)

5.9 S.R. DeGroot, P. Mazur: *Nonequilibrium Thermodynamics* (North Holland, Amsterdam 1969)

5.10 G.H. Westphal, F. Rosenberger: On diffusive-advective interfacial mass transfer. J. Crystal Growth 43, 687 (1978)

5.11 J. Stefan: Über die Verdampfung aus einem kreisförmig oder elliptisch begrenzten Becken. Ann. Physik und Chemie 17, 550 (1882)

5.12 M.M. Faktor, I. Garett: *Growth of Crystals from the Vapor* (Chapmann and Hall, London 1974) p. 218

5.13 J. Bardeen, C. Herring: "Diffusion in Alloys and the Kirkendall Effect", in *Imperfections in Nearly Perfect Crystals* (Wiley, New York 1952) p. 261 ff

5.14 W.R. Wilcox: "The Role of Mass Transfer in Crystallization Processes", in *Preparation and Properties of Solid State Materials*, Vol. 1, ed. by R.A. Lefever (Marcel Dekker, New York 1971)

5.15 S. Bretsznajder: *Prediction of Transport and Other Physical Properties of Fluids* (Pergammon Press, New York 1971)

5.16 J.C. Brice: *The Growth of Crystals from the Melt* (North-Holland Publishing Company, Amsterdam 1965)

5.17 W.D. Kingery, H.K. Bowen, D.R. Uhlmann: *Introduction to Ceramics*, 2nd ed. (John Wiley and Sons, New York 1976)

5.18 K.E. Grew, T.L. Ibbs: *Thermal Diffusion in Gases* (Cambridge University Press, Cambridge 1952)

5.19 H.J.V. Tyrell: *Diffusion and Heat Flow in Liquids* (Butterworth, London 1961)

5.20 P.J. Shlichta, R.E. Knox: Growth of crystals by centrifugation. J. Crystal Growth 3/4, 808 (1968)

5.21 H. Schlichting: *Boundary-Layer Theory*, 6th ed. (McGraw-Hill, New York 1968)

5.22 I.G. Currie: *Fundamental Mechanics of Fluids* (McGraw-Hill, New York 1974)

5.23 W.G. Cochran: The flow due to a rotating disc. Proc. Cambr. Phil. Soc. 30, 365 (1934)

5.24 V.G. Levich: *Physicochemical Hydrodynamics* (Prentice-Hall, Englewood Cliffs, NJ 1962)

5.25 H.S. Lew, Y.C. Fung: Entry flow into blood vessels at arbitrary Reynolds numbers. J. Biomechanics 3, 24 (1970)

5.26 R.A. Svehla: "Estimated Viscosities and Thermal Conductivities of Gases at High Temperatures; Tech. Rpt. TR R-132, NASA (1962)

5.27 R.C. Reid, T.K. Sherwood: *The Properties of Gases and Liquids* (McGraw-Hill, New York 1958)

5.28 Y.S. Touloukian, S.C. Saxena, P. Hestermanns: *Thermophysical Properties of Matter*, Vol. 11, *Viscosity* (Plenum Press, New York 1975)

5.29 A. Klemm: "Transport Properties of Molten Salts", in *Molten Salt Chemistry*, ed. by M. Blander (Interscience, New York 1964)

5.30 R.T. Beyer, E.M. Ring: "The Viscosity of Liquid Metals", in *Liquid Metals*, ed. by S.Z. Beer (Marcel Dekker, New York 1972)

5.31 G.V. Samsonov (ed.): *Handbook of the Physicochemical Properties of the Elements* (Plenum Press, New York 1968)

5.32 D. Elwell, H.J. Scheel: *Crystal Growth from High-Temperature Solutions* (Academic Press, London 1975)

5.33 J.L. Bates, C.E. McNeilly, J.J. Rasmussen: "Ceramics in Severe Environments", in *Materials Science Research*, Vol. 5, ed. by W.W. Kriegel, H. Palmour III (Plenum Press, New York 1971) pp. 11-26

5.34 E.W. Washburn (ed.): *International Critical Tables*, Vol. 7 (McGraw-Hill, New York 1930) p. 212

5.35 R.C. Keezer, C.H. Griffiths, J.P. Vernon: Crystal growth phenomena in the selenium – tellurium system. J. Crystal Growth 3/4, 755 (1968)

5.36 K.A. Jackson, D.R. Uhlmann, J.D. Hunt: On the nature of crystal growth from the melt. J. Crystal Growth 1, 1 (1967)

5.37 B. Predel: Thermodynamische Eigenschaften, Diffusionsverhalten und Viskosität von Legierungsschmelzen. Z. Metallk. 63, 63 (1972)

5.38 L.I. Gvozdeva, A.P. Lyubimov: Variation of viscosity in systems of eutectic type. Russ. J. Phys. Chem. 43, 314 (1969)

5.39 V.P. Elyutin, V.I. Kostikov, B.S. Mitin, Yu.A. Nagibin: Viscosity of alumina. Russ. J. Phys. Chem. 43, 316

5.40 E.L. Cussler: *Multicomponent Diffusion* (Elsevier, Amsterdam 1976)

5.41 D.B. Spalding, H.L. Evans: Mass transfer through laminar boundary layers. Part 3. Similar solutions of the b-equation. Int. J. Heat Mass Transfer 2, 314 (1961).

H.L. Evans: Part 8. Further solutions to the velocity equation. Int. J. Heat Mass Transfer 5, 373 (1962)
D.B. Spalding, W.M. Pun, S.W. Chi: Further exact similarity solutions of the b-equation. Int. J. Heat Transfer 5, 79 (1962)

5.42 J.C. Brice: *The Growth of Crystals from Liquids* (North-Holland Publishing Company, Amsterdam 1973)

5.43 E.M. Sparrow, J.L. Gregg: Heat transfer from a rotating disk to fluids of any Prandtl number. J. Heat Transfer 81, 249 (1959)

5.44 E.M. Sparrow, J.L. Gregg: Mass transfer, flow and heat transfer about a rotating disk. J. Heat Transfer 82, 294 (1960)

5.45 W. Nernst: Theorie der Reaktionsgeschwindigkeit in heterogenen Systemen. Z. Phys. Chem. 47, 52 (1904).
E. Brunner: Reaktionsgeschwindigkeit in heterogenen Systemen. Z. Phys. Chem. 47, 56 (1904)

5.46 J.W. Mullin: *Crystallization* (Butterworth, London 1972) p. 203

5.47 H.S. Carslaw, J.C. Jaeger: *Conduction of Heat in Solids*, 2nd ed. (Oxford University Press, New York 1959)

5.48 J. Crank: *The Mathematics of Diffusion*, 2nd ed. (Clarendon Press, Oxford 1975)

5.49 E.R.G. Eckert, T.F. Irvine, Jr.: "Heat Transfer Reviews for 1953-1969", in *Progress in Heat and Mass Transfer*, Vol. 3 (Pergamon, Oxford 1971)

5.50 E.R.G. Eckert, T.F. Irvine, Jr.: "Heat Transfer Reviews 1970-1975" in *Progress in Heat and Mass Transfer*, Vol. 8 (Pergamon, Oxford 1977)

5.51 Y.S. Touloukian, P.E. Liley, S.C. Saxena: *Thermophysical Properties of Matter*, Vol. 3, *Thermal Conductivity – Nonmetallic Liquids and Gases* (Plenum Press, New York 1970)

5.52 N.V. Tsederberg: *Thermal Conductivity of Gases and Liquids* (MIT Press, Cambridge, Ma 1965)

5.53 Y.S. Touloukian, T. Makita: *Thermophysical Properties of Matter*, Vol. 6, *Specific Heat – Nonmetallic Liquids and Gases* (Plenum Press, New York 1970)

5.54 Y.S. Touloukian, R.W. Powell, C.Y. Ho, M.C. Nicolaou: *Thermophysical Properties of Matter*, Vol. 10, *Thermal Diffusivity* (Plenum Press, New York 1973)

5.55 R. Siegel, J.R. Howell: *Thermal Radiation Heat Transfer* (McGraw-Hill, New York 1972)

5.56 M.N. Özisik: *Radiative Transfer and Interactions With Conduction and Convection* (Wiley-Interscience, New York 1973)

5.57 E.M. Sparrow, R.D. Cess: *Radiation Heat Transfer* (Brooks/Cole Publishing Company, Belmont 1970) Revised Edition

5.58 Y.S. Touloukian, D.P. DeWitt: *Thermophysical Properties of Matter*, Vol. 7, *Thermal Radiative Properties, Metallic Elements and Alloys* (Plenum, New York 1970)

5.59 Y.S. Touloukian, D.P. DeWitt: *Thermophysical Properties of Matter*, Vol. 8, *Thermal Radiative Properties, Nonmetallic Solids* (Plenum, New York 1972)

5.60 M. Planck: *The Theory of Heat Radiation* (Dover, New York 1959) English translation of "Vorlesungen über die Theorie der Wärmestrahlung", Leipzig 1923

5.61 E.R.G. Eckert, R.M. Drake: *Analysis of Mass and Heat Transfer* (McGraw-Hill, New York 1972)

5.62 S. Goldsztaub, R. Itti, F. Mussard: Role de la diffusion dans la croissance des cristaux à partir de solutions. J. Crystal Growth 6, 130 (1970)

5.63 T.K. Sherwood, R.L. Pigford, C.R. Wilke: *Mass Transfer* (McGraw-Hill, New York 1975)

5.64 W.R. Wilcox: Validity of the stagnant film approximation for mass transfer in crystal growth and dissolution. Mat. Res. Bull. 4, 265 (1969)

5.65 T.B. Reed: Transparent furnace for vapor crystal growth. Solid State Res. Lincoln Lab. MIT 1, 21 (1969)

5.66 R. Viskanta, E.E. Anderson: "Heat Transfer in Semitransparent Solids", in *Advances in Heat Transfer*, Vol. 11, ed. by T.F. Irvine, Jr., J.P. Hartnett (Academic Press, New York 1975)

5.67 L.I. Rubinstein: *The Stefan Problem*, Transl. Mathem. Monographs, Vol. 27 (American Mathematical Society, Providence 1971)

5.68 S.G. Bankoff: Heat conduction or diffusion with change of phase. Adv. Chem. Eng. 5, 75 (1964)

5.69 J.C. Muehlbauer, J.E. Sunderland: Heat conduction with freezing or melting. Appl. Mech. Rev. 18, 951 (1965)

5.70 American Institute of Physics Handbook, 3rd., ed. by D.E. Gray (McGraw-Hill, New York 1972)

5.71 M.G. Velarde: "Hydrodynamic Instabilities in Isotropic Fluids", in *Fluid Dynamics, Les Houches, 1973*, ed. by R. Balian, J.L. Peube (Gordon and Breach, New York 1977)

5.72 H. Bénard: Les tourbillons cellulaires dans une nappe liquide transportant de la chaleur par convection en regime permanent. Ann. Chim. Phys. 23, 62 (1901)

5.73 E.L. Koschmieder: On convection on a uniformly heated plane. Beitr. Phys. Atmosph. 39, 1 (1966)

5.74 Lord Rayleigh: On the convection currents in a horizontal layer of fluid when the higher temperature is on the under side. Phil. Mag. 32, 529 (1916)

5.75 M.J. Block: Surface tension as a cause of Bénard cells and surface deformation in a liquid film. Nature 178, 650 (1956)

5.76 J.R.A. Pearson: On convection cells induced by surface tension. J. Fluid Mech. 4, 489 (1958)

5.77 E.L. Koschmieder: "Bénard Convection" in *Advances in Chemical Physics*, Vol. 26, ed. by I. Prigogine, S.A. Rice (John Wiley and Sons, New York 1973) p. 177

5.78 C.O. Hoard, C.R. Robertson, A. Acrivos: Experiments on the cellular structure in Bénard convection. Intern. J. Heat Mass Transfer 13, 839 (1970)

5.79 R. Krishnamurti: Finite amplitude convection with changing mean temperature. J. Fluid Mech. 33, 445 and 457 (1968)

5.80 T.C. Bannister, P.G. Grodzka, L.W. Spradley, W.V. Bourgeois, R.O. Hedden, B.R. Facemire: Apollo 17 heat flow and convection experiments, final data analysis results. NASA Technical Memorandum X-64772 (1973)

5.81 J.R. Carruthers: "Thermal Convection Instabilities Relevant to Crystal Growth from Liquids" in *Preparation and Properties of Solid State Materials*, Vol. 3, ed. by W.R. Wilcox, R.A. Lefever (Marcel Dekker, New York 1977)

5.82 J.R. Carruthers: "Origins of convective temperature oscillations in crystal growth melts." J. Crystal Growth 32, 13 (1976)

5.83 H.v. Tippelskirch: Über Konvektionszellen, insbesondere in flüssigem Schwefel. Beitr. Physik Atmosph. 29, 37 (1956)

5.84 S. Chandrasekar: *Hydrodynamic and Hydromagnetic Stability* (Clarendon Press, Oxford 1961)

5.85 J. Boussinesq: *Théorie analytique de chaleur*, Vol. 2 (Gauthier-Villars, Paris 1903)

5.86 J. Mihaljan: A rigorous exposition of the Boussinesq approximation applicable to a thin layer of fluid. Astrophys. J. 136, 1126 (1962)

5.87 A. Oberbeck: Über die Wärmeleitung der Flüssigkeiten bei der Berücksichtigung der Strömungen infolge von Temperaturdifferenzen. Ann. Phys. Chem. 7, 271 (1879)

5.88 A. Pellew, R.V. Southwell: On maintained convective motion in a fluid heated from below. Proc. Roy. Soc. 176, 312 (1940)

5.89 A.S. Eddington: *Internal Constitution of the Stars* (Cambridge University Press, London 1926) p. 201

5.90 E. Palm: Nonlinear thermal convection. Ann. Rev. Fluid Mech. 7, 39 (1975)

5.91 J.T. Stuart: Nonlinear stability theory. Ann. Rev. Fluid Mech. 3, 347 (1971)

5.92 B. Gebhardt: "Natural Convection Flows and Stability" in *Advances in Heat Transfer*, Vol. 9, ed. by T.F. Irvine, Jr., J.P. Hartnett (Academic Press, New York 1973) p. 273

5.93 D.D. Joseph: On the stability of Boussinesq equations. Arch. Ration. Mech. Anal. 20, 59 (1965)

5.94 D.T.J. Hurle, E. Jakeman, E.R. Pike: On the solution of the Bénard Problem with Boundaries of Finite Conductivity, Proc. Roy. Soc. 296A, 469 (1967)

5.95 E.A. Spiegel, G. Veronis: On the Boussinesq approximation for a compressible fluid. Astrophys. J. 131, 442 (1960)

5.96 R.J. Schmidt, S.W. Milverton: On the instability of a fluid when heated from below. Proc. Roy. Soc. 152A, 586 (1935)

5.97 E. Jakeman: Convective instability in fluids of high thermal diffusivity. Phys. Fluids 11, 10 (1968)

5.98 E.L. Koschmieder, S.G. Pallas: Heat transfer through a shallow horizontal convecting fluid layer. Int. J. Heat Mass Transfer, 17, 991 (1974)

5.99 C. Christophorides, S.H. Davis: Thermal instability with radiative transfer. Phys. Fluids 13, 322 (1970)

5.100 D.L. Turcotte, A.T. Hsui, K.E. Torrance, G. Schubert: Influence of viscous dissipation on Bénard convection. J. Fluid Mech. 64, 369 (1974)

5.101 J.D. Verhoeven: Convection effects in the capillary reservoir technique for measuring liquid metal diffusion coefficients. Trans. Met. Soc. AIME 242, 1937 (1968)

5.102 S. Ostrach: "Laminar Flows with Body Forces" in *Theory of Laminar Flows*, ed. by F.K. Moore, Vol. 4 of High Speed Aerodynamics and Jet Propulsion (Princeton University Press, Princeton 1964)

5.103 E.H. Cheng, M.N. Özisik: Radiation with free convection in an absorbing, emitting and scattering medium. Int. J. Heat Mass Transfer 15, 1243 (1972)

5.104 R. Krishnamurti: Some further studies on the transition to turbulent convection. J. Fluid Mech. 60, 285 (1973)

5.105 F.H. Busse, J.A. Whitehead: Instabilities of convection rolls in a high Prandtl number fluid. J. Fluid Mech. 47, 305 (1971)

5.106 R.M. Clever, F.H. Busse: Transition to time dependent convection. J. Fluid Mech. 65, 625 (1974)

5.107 L.N. Howard: "Convection at high Rayleigh number" in *Proc. 11th Intern. Congress Applied Mechanics, München 1964*, ed. by H. Görtler (Springer, Heidelberg) p. 1109

5.108 J.W. Elder: The unstable thermal interface. J. Fluid Mech. 32, 69 (1968)

5.109 F.H. Busse: The oscillatory instability of convection rolls in a low Prandtl number fluid. J. Fluid Mech. 52, 97 (1972)

5.110 R. Krishnamurti: On the transition to turbulent convection. Part 1. The transition from two- to three-dimensional flow. J. Fluid Mech. 42, 295 (1970)

5.111 R. Krishnamurti: On the transition to turbulent convection. Part 2. The transition to time-dependent flow. J. Fluid Mech. 42, 309 (1970)

5.112 R.V. Birikh, G.Z. Gershuni, E.M. Zhukhovitskii, R.M. Rudakov: Hydrodynamic and thermal instability of a steady convective flow. J. Appl. Math. Mech. (PPM) 32, 246 (1968)

5.113 G.Z. Gershuni, E.M. Zhukhovitskii: Stability of plane-parallel convective motion with respect to spatial perturbations. J. Appl. Math. Mech. (PPM) 33, 830 (1969)

5.114 S.A. Korpela: A study on the effect of Prandtl number on the stability of the conduction regime of natural convection in an inclined slot. Int. J. Heat Mass Transfer 17, 215 (1974)

5.115 J.E. Weber: On thermal convection between non-uniformly heated plates. Int. J. Heat and Mass Transfer 16, 961 (1973)

5.116 S.P. Bhattacharyya, S. Nadoor: Stability of thermal convection between non-uniformly heated plates. Appl. Sci. Res. 32, 1412 (1976)

5.117 R.V. Birikh, G.Z. Gershuni, E.M. Zhukhovitskii, R.N. Rudakov: Stability of the steady convective motion of a fluid with a longitudinal temperature gradient. J. Appl. Math. Mech. (PPM) 33, 937 (1969)

5.118 C.K. Crawford: High efficiency high-temperature radiation heat shields. J. Vac. Sci. Techn. 9, 23 (1972)

5.119 F. Gambale, A. Gliozzi: Formation of dynamic patterns in a fluid layer. J. Phys. Chem. 76, 783 (1972)

5.120 E.D. Burger, L.M. Blair, J.A. Quinn: Intermittent convection: confirmation of a model for mass transfer into stratified fluid layers. Chem. Eng. Sci. 29, 1545 (1974)

5.121 D.T.J. Hurle, E. Jakeman: Significance of Soret effect in the Rayleigh-Jeffreys' problem. Phys. Fluids 12, 2704 (1969)

5.122 S.H. Smith, D. Elwell: Growth of nickel ferrite crystals from barium borate by a pulling method. J. Crystal Growth 3,4 471 (1968)

5.123 B. Baranowski, A.L. Kawczynski: Hydrodynamic stability in liquid electrochemical systems with concentration polarization. Roczn. Chem. 44, 2447 (1970)

5.124 J.S. Turner: Double-diffusive phenomena. Ann. Rev. Fluid Mech. 6, 37 (1974)

5.125 J.S. Turner: Laboratory experiments on double-diffusive instabilities. Adv. Chem. Phys. 32, 135 (1975)

5.126 J.S. Turner: *Buoyancy Effects in Fluids* (Cambridge University Press, Cambridge 1973)

5.127 P.G. Baines, A.E. Gill: On thermohaline convection with linear gradients. J. Fluid Mech. 37, 289 (1969)

5.128 R.S. Schechter, M.G. Velarde, J.K. Platten: The two-component Bénard problem. Adv. Chem. Phys. 26, 265 (1974)

5.129 W.R. Lindberg, R.D. Haberstroh: On simultaneous transport of heat and mass in natural convection. AIChE J. 18, 243 (1972)

5.130 H.E. Huppert, D.R. Moore: Nonlinear double-diffusive convection. J. Fluid Mech. 78, 821 (1976)

5.131 C.F. Chen: Double-diffusive convection in an inclined slot. J. Fluid Mech. 72, 721 (1975)

5.132 C.F. Chen, R.D. Samford: Stability of time-dependent double diffusive convection in an inclined slot. J. Fluid Mech. 83, 83 (1977)

5.133 T.G.L. Shirtcliffe: Thermosolutal convection: observation of an overstable mode. Nature 213, 489 (1967)

5.134 D.T.J. Hurle, E. Jakeman: Soret-driven thermosolutal convection. J. Fluid Mech. 47, 667 (1971)

5.135 R.S. Schechter, M.G. VeLarde, J.K. Platten: The two-component Bénard problem. Adv. Chem. Phys. 26, 265 (1974)

5.136 J.K. Platten, G. Chavepeyer: Nonlinear two-dimensional Bénard convection with Soret effect: free boundaries. Int. J. Heat Mass Transfer 20, 113 (1977)

5.137 D.R. Caldwell: Experimental studies on the onset of thermohaline convection. J. Fluid Mech. 64, 347 (1974)

5.138 D.R. Caldwell: Thermosolutal convection in a solution with large negative Soret coefficient. J. Fluid Mech. 74, 129 (1976)

5.139 J. Bdzil, H.L. Frisch: Chemical instabilities. VI. Hydrodynamic instabilities of the dissociating fluid $A_2 \rightleftarrows 2A$. Phys. Fluids 14, 2048 (1971)

5.140 E. Fitzer: Dynamische Instabilitäten bei heterogenen Gas/Feststoff-Reaktionen. Chem. Ing. Techn. 41, 331 (1969)

5.141 W. Fritz: Oscillations during hot wall pyrolysis. High Temp. High Press. 2, 291 (1970)

5.142 R.N. Noyes, R.J. Field: Oscillatory chemical reactions. Ann. Rev. Phys. Chem. 25, 95 (1974)

5.143 D.A. Nield: Surface tension and buoyancy effects in cellular convection. J. Fluid Mech. 19, 341 (1964)

5.144 K.A. Smith: On convection instability induced by surface-tension gradients. J. Fluid Mech. 24, 401 (1966)

5.145 L.E. Scriven, C.V. Sternling: On cellular convection driven by surface-tension gradients: effect of mean surface tension and viscosity. J. Fluid Mech. 19, 321 (1964)

5.146 T.C. Bannister, P.G. Grodzka, L.W. Spradley, S.V. Bourgeois, R.O. Hedden, B.R. Facemire: NASA Report No. TMX-64772 (1973)

5.147 R.V. Birikh: Thermocapillary convection in a horizontal layer of liquid. J. Appl. Mech. Techn. Phys. 7, 43 (1966)

5.148 S. Ostrach: Motion induced by capillarity. *Proc. Int. Conf. Physicochemical Hydrodynamics (Levich Conference) 1977*. (Advance Publications, St. Peter Port/Guernsey, UK 1978)

5.149 A. Passerone: "The Basic Principle of Wetting Processes", in *Science and Technology of Surface Coating*, ed. by B.N. Chapman, J.C. Anderson (Academic Press, London 1974) p. 194

5.150 *Moving Boundary Problems*, ed. by D.G. Wilson, A.D. Solomon, P.T. Boggs (Academic Press, New York 1978)

5.151 G.Z. Gershuni, E.M. Zhukhovitskii: *Convective stability of Incompressible Fluids* (Keter Publ., Jerusalem 1976)

5.152 C. Normand, Y. Pomeau, M.G. Velarde: Convective instability: A physicist's approach. Rev. Mod. Phys. 49, 581 (1977)

5.153 C.L. Strong: Experiments with salt fountains and related instabilities in water. Scientific-American 224(6) 124 (1971)

5.154 J. Walker: The salt fountain and other curiosities based on the different density of fluids. Scientific American 237(4) 142 (1977)

5.155 R.C. Jones, W.H. Furry: The separation of isotopes by thermal diffusion. Rev. Mod. Phys. 18, 151 (1946)

6. Segregation

6.1 C.P. Thurmond, J.D. Struthers: Equilibrium thermochemistry of solid and liquid alloys of germanium and silicon. J. Phys. Chem. 57, 831 (1953)

6.2 K. Lehovec: Thermodynamics of binary semiconductor-metal alloys. J. Phys. Chem. Solids 23, 695 (1962)

6.3 K. Weiser: Theoretical calculation of distribution coefficients of impurities in germanium and silicon, heats of solid solution. J. Phys. Chem. Solids 7, 118 (1958)

6.4 R.N. Hall: Variation of the distribution coefficient and solid solubility with temperature. J. Phys. Chem. Solids 3, 63 (1957)

6.5 J.C. Brice: *The Growth of Crystals from the Melt* (North-Holland, Amsterdam 1965)

6.6 M. Ikeya, N. Itoh, T. Suita: Distribution coefficients of various impurities in alkali halides. Japan. J. Appl. Phys. 7, 837 (1968)

6.7 U. Gross: Zonenschmelzen von Alkalihalogeniden. Thesis, University of Stuttgart (1970)

6.8 T.B. Douglas: Calculated heats of dilute solid solution among the alkali halides other than cesium salts. J. Chem. Phys. 45, 4571 (1966)

6.9 M. Blander: Some calculations for a one-dimensional salt mixture. J. Chem. Phys. 34, 697 (1961)

6.10 A.P. Ratner: On the theory of the distribution of electrolytes between a solid crystalline and a liquid phase. J. Chem. Phys. 1, 789 (1933)

6.11 F. Vaslow, G.E. Boyd: Thermodynamics of coprecipitation: dilute solid solutions of AgBr in AgCl. J. Am. Chem. Soc. 74, 4691 (1952)

6.12 F. Rosenberger, H.G. Riveros: Segregation in alkali halide crystallization. J. Chem. Phys. 60, 668 (1974)

6.13 F.A. Kröger: *The Chemistry of Imperfect Crystals*, Vol. 1 (North-Holland, Amsterdam 1973) pp. 5-6

6.14 J.C. Brice: *The Growth of Crystals from Liquids* (North-Holland, Amsterdam 1973)

6.15 F. Rosenberger: "Purification of Alkali Halides", in *Ultrapurity*, ed. by M. Zief, R. Speights (Marcel Dekker, New York 1972)

6.16 W.L. McIntire: Trace element partition coefficients. Geochim. Cosmochim. Acta 27, 1209 (1963)

6.17 H.H. Schock: Bestimmung sehr kleiner Verteilungs-Koeffizienten von Cs, Na und Ba zwischen Lösung und KCl-Einkristallen mittels radio-aktiver Isotope. Contr. Mineral. and Petrol. 13, 161 (1966)

6.18 J.C. Brice: "Phase Relationships in Semiconductors", in *Atomic Diffusion in Semiconductors*, ed. by D. Shaw (Plenum Press, London 1973)

6.19 F.A. Kröger: *The Chemistry of Imperfect Crystals*, Vol. 2 (North-Holland, Amsterdam 1973)

6.20 C. Wagner: Physical chemistry of ionic crystals involving small concentrations of foreign substances. J. Phys. Chem. 57, 738 (1953)

6.21 W.D. Edwards: The interaction between oxygen and boron in liquid germanium. J. Appl. Phys. 39, 1784 (1968)

6.22 K. Nassau, G.M. Loiacono: Calcium tungstate-III, trivalent rare earth substitution. J. Phys. Chem. Solids 24, 1503 (1963)

6.23 R.A. Laudise: *The Growth of Single Crystals* (Prentice Hall, Englewood Cliffs, NJ 1970)

6.24 J.C. Brice: Some thermodynamic aspects of the growth of strained crystals. J. Crystal Growth 28, 249 (1975)

6.25 W.A. Wilcox: "The Role of Mass Transfer in Crystallization Processes" in *Preparation and Properties of Solid State Materials*, ed. by R.A. Lefever (Marcel Dekker, New York 1971)

6.26 W.R. Wilcox: Crystallization flow. J. Crystal Growth 12, 93 (1972)

6.27 J.A. Burton, R.C. Prim, W.P. Slichter: The distribution of solute in crystals grown from the melt. Part I. Theoretical. J. Chem. Phys. 21, 1987 (1953)

6.28 W.R. Wilcox: "Mass Transfer in Fractional Solidification", in *Fractional Solidification*, ed. by M. Zief, W.R. Wilcox (Marcel Dekker, New York 1967)

6.29 J.A. Burton, E.D. Kolb, W.P. Slichter, J.D. Struthers: The distribution of solute in crystals grown from the melt. Part II. Experimental. J. Chem. Phys. 21, 1991 (1953)

6.30 W.G. Cochran: The flow due to a rotating disk. Proc. Cambr. Phil. Soc. 30, 365 (1934)

6.31 J.C. Brice, P.A.C. Whiffin: Solute striae in pulled crystals of zinc tungstate. Brit. J. Appl. Phys. 18, 581 (1967)

6.32 M.C. Flemings: *Solidification Processing* (McGraw-Hill, New York 1974)

6.33 H. Kelting, H. Witt: Über KCl Kristalle mit Zusätzen von Erdalkalichloriden. Z. Physik 126, 697 (1949)

6.34 G.A. Andreev, B.P. Aleksandrov: Flotation study of the distribution of singly charged impurities in NaCl. Sov. Phys. Solid State 7, 135 (1965)

6.35 R.H. McFee: Foreign ion rejection in the growth of sodium chloride crystals from the melt. J. Chem. Phys. 15, 856 (1947)

6.36 M. Krumnacker, W. Lange: Investigation of the concentration ratios at the solid-liquid interface. Kristall and Technik 4, 207 (1969)

6.37 W.G. Pfann: *Zone Melting*, 2nd ed. (Wiley, New York 1966)

6.38 M. Zief, W.R. Wilcox (ed.): *Fractional Solidification* (Marcel Dekker, New York 1967)

6.39 W.G. Pfann: Principles of zone melting. Trans. Am. Inst. Mining Met. Engrs. 194, 747 (1952)

6.40 A.F. Witt, H.C. Gatos, M. Lichtensteiger, M.C. Lavine, C.J. Herman: Crystal growth and steady-state segregation under zero gravity: InSb. J. Electrochem. Soc. 122, 267 (1975)

6.41 F.V. Dean, J.R. Kerr, A. Hellawell: Factors affecting the solute distribution during the normal freezing of lead-antimony alloys. J. Inst. Metals 90, 234 (1962)

6.42 C.E. Shoemaker, R.L. Smith: "Survey of Inorganic Materials", in *Fractional Solidification*, ed. by M. Zief, W.R. Wilcox (Marcel Dekker, New York 1967)

6.43 H. Schildknecht: *Zone Melting* (Verlag Chemie-Academic Press, New York 1966)

6.44 J.S. Shah: "Zone Melting and Applied Techniques", in *Crystal Growth*, ed. by B.R. Pamplin (Pergamon Press, Oxford 1975)

6.45 D. Fischer: A study on zone refining: solid-phase impurity diffusion and the influence of separating the impure end. J. Appl. Phys. 44, 1977 (1973)

6.46 V.Ya. Khaimov-Malk'ov: Distribution of volatile impurities in various methods of crystallization from melts: KI-TlI system. J. Crystal Growth 35, 302 (1976)

6.47 W.R. Wilcox: "Heat Transfer in Fractional Solidification", in [6.38]

6.48 K.M. Kim, A.F. Witt, H.C. Gatos: Segregation behavior in a stationary vertical zone with converging interfaces: pressure induced segregation effects. J. Electrochem. Soc. 121, 448 (1974)

6.49 F. Rosenberger: Preparation of alkali halide single crystals of highest purity by zone refining. Mat. Res. Bull. 1, 123 (1966)

6.50 F. Rosenberger: "Preparation of ultrapure alkali halide single crystals", in *Crystal Growth*, ed. by H.S. Peiser (Pergamon Press, Oxford 1966) p. 141

6.51 K.-Th. Wilke: *Methoden der Kristallzüchtung* (VEB Deutscher Verlag der Wissenschaften, Berlin 1963)

6.52 A.Z. Knittel: Vapour growth of crystals with a steady state source. J. Crystal Growth 26, 33 (1974)

6.53 S.V. Airapetyants, G.I. Shmelev: Method for growing uniform monocrystals of alloyed semiconductor materials, solid solutions, and intermetallic compounds of a given composition determined by the composition of the melt. Sov. Phys. Solid State 2, 689 (1960)

6.54 E.N. Da C. Andrade, R. Roscoe: Glide in metal single crystals. Proc. Phys. Soc. (London) 49, 152 (1973)

6.55 L.G. Van Uitert, W.A. Bonner, W.H. Grodkiewicz, L. Pitroski, G.J. Zydzik: Garnets for bubble domain devices. Mat. Res. Bull. 5, 825 (1970)

6.56 J.R. Carruthers and A.F. Witt: "Transient Segregation Effects in Czochralski Growth", in *Crystal Growth and Characterization*, Proceedings of the ISSCG2 Springschool, Japan, 1974, ed. by R. Ueda, J.B. Mullin (North-Holland, Amsterdam 1975)

6.57 K.M. Kim, A.F. Witt, H.C. Gatos: Crystal growth from the melt under destabilizing thermal gradients. J. Electrochem. Soc. 119, 1218 (1972)

6.58 A.F. Witt, M. Lichtensteiger, H.C. Gatos: Experimental approach to the quantitative determination of dopant segregation during crystal growth on a microscale: Ga doped Ge. J. Electrochem. Soc. 120, 1119 (1973)

6.59 J.T. Yue, F.W. Volt: Influence of gravity-free solidification on solute microsegregation. J. Crystal Growth 29, 329 (1975)

6.60 J.T. Yue, private communication (1976)

6.61 H. Beleites, F. Fröhlich: Autoradiographic investigation on the incorporation of Ca ions in KCl crystals during Kyropoulos growth. Kristall und Technik 12, 1329 (1972)

6.62 J.-Y. Boniort, C. Brehm, G. Desplanches, J.-Y. Barraud, P. Margotin: Crystal growth of strontium barium niobate $Ba_xSr_{1-x}Nb_2O_6$. J. Crystal Growth 30, 357 (1975)

6.63 R.L. Barns: A survey of precision lattice parameter measurements as a tool for the characterization of single-crystal materials. Mat. Res. Bull. 2, 273 (1967)

6.64 A.B. Chase, W.R. Wilcox: Temperature fluctuations and striations in flux crystal growth. J. Am. Ceram. Soc. 50, 332 (1967)

6.65 B.M. Wanklyn: "Practical Aspects of Flux Growth by Spontaneous Nucleation", in *Crystal Growth*, ed. by B.R. Pamplin (Pergamon Press, Oxford 1975)

6.66 A.A. Chernov, V.E. Khadzhi: Trapping of colloidal inclusions in the growth of quartz crystals. J. Crystal Growth 3,4, 641 (1968)

6.67 B.J. Curtis, J.P. Dismukes: Effects of natural and forced convection in vapor phase growth systems. J. Crystal Growth 17, 128 (1972)

6.68 J.P. Dismuskes, B.J. Curtis: in *Semiconductor Silicon*, ed. by H.R. Huff, R.R. Burgess (The Electrochemical Society, Princeton, N J 1973) p. 258

6.69 J.C. Marinace: Epitaxial vapor growth of Ge single crystals in a closed-cycle process, IBM J. Res. Develop. 4, 248 (1960)

6.70 A. Meyer: Gastransport und Charakterisierung von Einkristallen aus $CdCr_2S_4$, $FeCr_2S_4$ und $Cd_{1-x}Fe_xCr_2S_4$. Paper at DGKK Meeting, Freiburg (1972)

6.71 E. Fitzer: Dynamische Instabilitäten bei heterogenen Gas/Feststoff-Reaktionen. Chemie-Ing.-Tech. 41, 331 (1969)

6.72 W. Fritz: Oscillations during hot wall pyrolysis. High Temp. - High Press. 2, 291 (1970)

6.73 W.G. Pfann, K.E. Benson, J.H. Wernick: Some aspects of Peltier heating at liquid-solid interfaces in germanium. J. Electronics 2, 597 (1957)

6.74 H. Bethge, F. Fröhlich: Über die bei der Kristallzüchtung nach dem Nacken-Kyropoulos-Verfahren auftretenden Wachstumsformen und die Herstellung von Alkalihalogenid-Bikristallen. phys. stat. sol. 3, 55 (1963)

6.75 M. Kumagawa, A.F. Witt, M. Lichtensteiger, F.C. Gatos: Current-controlled growth and dopant modulation in liquid phase epitaxy. J. Electrochem. Soc. 120, 583 (1973)

6.76 D.J. Lawrence, L.F. Eastman: Electric current controlled growth and doping modulation in GaAs liquid phase epitaxy. J. Crystal Growth 30, 267 (1975)

6.77 A. Räuber: Doping modulation by electric currents in lithium niobate during crystal growth. Mat. Res. Bull. 11, 497 (1976)

6.78 J.P.M. Damen, J.M. Robertson: Induced non-periodic growth striations in flux-grown magnetic oxide single crystals. J. Crystal Growth 16, 50 (1972)

6.79 L. Malicskó, L. Jeszensky: Über den Einbau von Zusätzen in KCl Einkristalle bei der Züchtung aus Blei- und Zinnhaltigen Wässrigen Lösungen. J. Crystal Growth 7, 13 (1970)

6.80 G.A. Andreev: Distribution of Impurities in Crystallization of NaCl, KCl and KBr from Aqueous Solutions. Sov. Phys. Cryst. 12, 82 (1967)

6.81 V.G. Smith, W.A. Tiller, J.W. Rutter: A mathematical analysis of solute redistribution during solidification. Can. J. Physics 33, 723 (1955)

6.82 W.R. Wilcox: Incomplete liquid mixing in crystal growth from the melt. J. Appl. Phys. 35, 636 (1964)

6.83 D.T.J. Hurle, E. Jakeman, E.R. Pike: Striated solute distributions produced by temperature oscillations during crystal growth from the melt. J. Crystal Growth 3,4, 633 (1968)

6.84 J.R. Carruthers: Solute incorporation during cyclic solidification of silicon. Can. Met. Quart. 5, 55 (1966)

6.85 W.P. Slichter, J.A. Burton: "The Distribution of Solute Elements: Transient Conditions", in *Transistor Technology*, ed. by H.E. Bridgers, J.H. Scaff, J.N. Shive (Van Nostrand, Princeton NJ 1958) Chap. 6

6.86 K.J. Berg, F. Fröhlich, M. Schmuntzsch: Berechnung der Verteilungsfunktion für einen periodisch veränderlichen Verteilungskoeffizienten beim Kristallwachstum nach dem Kyropoulos-Verfahren. Kristall und Technik 9, 1349 (1974)

6.87 D.T.J. Hurle, E. Jakeman: Effects of fluctuations on the measurement of distribution coefficients by directional solidification. J. Crystal Growth 5, 227 (1969)

6.88 J.R. Carruthers: "Crystal Growth from the Melt", in *Treatise on Solid State Chemistry*, ed. by N.B. Hannay, Vol. 5 (Plenum Press, New York 1975) Chap. 7

6.89 D.T.J. Hurle: "Melt Growth", in *Crystal Growth*, ed. by P. Hartmann (North-Holland, Amsterdam 1973)

6.90 J.B. Mullin, K.F. Hulme: Orientation-dependent distribution coefficients in melt-grown InSb crystals. J. Phys. Chem. Solids 17, 1 (1960)

6.91 J.A.M. Dikhoff: Gross-sectional resistivity variations in germanium single crystals. Solid State Electron. 1, 202 (1960)

6.92 A.F. Witt, H.C. Gatos: Impurity striations in InSb as revealed by interference contrast microscopy. J. Electrochem. Soc. 113, 808 (1966)

6.93 K. Morizane, A.F. Witt, H.C. Gatos: Growth characteristics and impurity incorporation during facet growth. J. Electrochem. Soc. 115, 747 (1968)

6.94 R. Singh, A.F. Witt, H.C. Gatos: Application of the Peltier effect for the determination of crystal growth rates. J. Electrochem. Soc. 115, 112 (1968)

6.95 R.N. Hall: Segregation of impurities during the growth of germanium and silicon crystals. J. Phys. Chem. 57, 836 (1953)

6.96 A. Trainor, B.E. Bartlett: A possible mechanism of crystal growth from the melt and its application to the problem of anomalous segregation at crystal facets. Solid State Electron. 2, 106 (1961)

6.97 P.J. Holmes: A competitive adsorption model of steady state growth of a crystal from a lightly-doped melt. J. Phys. Chem. Solids 24, 1239 (1963)

6.98 A.A. Chernov: "Excess Impurity Trapping During Crystal Growth", in *Growth of Crystals* (Rost Kristallov), ed. by A.V. Shubnikov, N.N. Sheftal, Vol. 3 (Consultants Bureau, New York 1962) p. 35

6.99 J.C. Brice, P.A.C. Whiffin: The temperature distribution in pulled germanium crystals during growth. Solid State Electron. 7, 183 (1964)

6.100 T. Abe. The growth of Si single crystals from the melt and impurity incorporation mechanisms. J. Crystal Growth 24/25, 463 (1974)

6.101 H.C. Gatos, M.C. Lavine: Characteristics of the {111} surfaces of the III-V intermetallic compounds. J. Electrochem. Soc. 107, 427 (1960)

6.102 E.V. Skudnova, M.S. Mirgalovskaya: Partition coefficient of sulfur in indium antimonide. Inorg. Materials 1, 165 (1965)

6.103 H. Beneking, W. Vits: Proc. 2nd Int. Symp. on Gallium Arsenide, Inst. Phys. Soc. Conf. Ser. No. 7, 96-100

6.104 A.F. Witt, M. Lichtensteiger, H.C. Gatos: Application of interface demarcation to the study of facet growth and segregation: germanium. J. Electrochem. Soc. 121, 787 (1974)

6.105 R.N. Hall: p-n junctions produced by growth rate variation. Phys. Rev. 88, 139 (1952)

6.106 G.F. Dobrzhanskii, O.L. Kreinin, L.E. Nikolaeva, K.M. Rozin, M.P. Shaskol'skaya: Anisotropy of impurity introduction into CsBr single crystals. Sov. Phys.-Crystallogr. 16, 581 (1962)

6.107 J.A. Spittle, M.D. Hunt, R.W. Smith: Orientation dependence of the partition coefficient in zinc-base alloy single crystals. J. Crystal Growth 3,4, 647 (1968)

6.108 A.A. Kralina, V.A. Sazonova: The influence of orientation of growth directions on the impurity distribution in nickel single crystals and on their substructure. Sov. Phys. Cryst. 22, #4 (1977)

6.109 A.F. Witt, H.C. Gatos: Homogeneous impurity incorporation during crystal growth from the melt. J. Electrochem. Soc. 116, 511 (1969)

6.110 F.V. Williams: The effect of orientation on the electrical properties of epitaxial gallium arsenide. J. Electrochem. Soc. 111, 886 (1964)

6.111 L. Malicskó, L. Jeszensky: Investigations on inhomogeneous impurity distribution caused by growth centers. J. Crystal Growth 15, 243 (1972)

6.112 W. Kleber: Über den Einlagerungsmechanismus bei Adsorptionsmischkristallen, Z. Phys. Chemie 212, 222 (1959)

6.113 U. Steinike: Zur Bildung von Adsorptionsmischkristallen. Kristall und Technik, 6, 7 (1971)

6.114 M.M. Lukina, L.A. Chernyaev: Intake of iron in hydrothermal zincite crystals. Sov. Phys. Cryst. 13, 979 (1969)

Subject Index